SECOND EDITION

Energy and Society

AN INTRODUCTION

—| SECOND EDITION |—

Energy and Society

AN INTRODUCTION

Harold H. Schobert

CRC Press
Taylor & Francis Group
Boca Raton London New York

CRC Press is an imprint of the
Taylor & Francis Group, an **informa** business

CRC Press
Taylor & Francis Group
6000 Broken Sound Parkway NW, Suite 300
Boca Raton, FL 33487-2742

© 2014 by Taylor & Francis Group, LLC
CRC Press is an imprint of Taylor & Francis Group, an Informa business

No claim to original U.S. Government works

Printed on acid-free paper
Version Date: 20130816

International Standard Book Number-13: 978-1-4398-2645-4 (Paperback)

Library of Congress Cataloging-in-Publication Data

Schobert, Harold H., 1943-
  Energy and society : an introduction / Harold H. Schobert. -- Second edition.
      pages cm
  Includes bibliographical references and index.
  ISBN 978-1-4398-2645-4 (pbk. : alk. paper) 1. Power resources. 2. Force and energy.
  3. Energy consumption--Social aspects. 4. Power resources--Environmental aspects.
  I. Title.

TJ163.2.S34528 2014
333.79--dc23   2013033042

Visit the Taylor & Francis Web site at
http://www.taylorandfrancis.com

and the CRC Press Web site at
http://www.crcpress.com

Contents

Preface

Our use of energy and our dependence on energy supplies pervade every activity that we undertake and every aspect of our society. Agriculture, transportation, manufacturing, and domestic activities all require, in various ways, electricity and many kinds of fuels. Energy is frequently in the news, with such issues as energy costs, especially for petroleum products; availability of supplies, particularly in countries that rely heavily on imports; and its availability in those parts of the world that still lack an adequate energy infrastructure.

New energy sources and new ways of using energy have had transformative effects on society. Very likely, daily life for most people into the mid-eighteenth century was not very much different from that of their ancestors centuries earlier. Then came the invention of the steam engine and the discovery of ways to produce and transmit electricity. The late nineteenth and early twentieth centuries saw the rise of the internal combustion engine and its applications in automobiles and airplanes. In mid-century, nuclear energy became available. Now, in the early twenty-first century, we see increasing interest in nonpolluting energy sources that are unlikely to be depleted: solar energy, wind energy, and energy from plant products.

Particularly since the latter decades of the twentieth century, there has been an increasing awareness that our use of energy is inextricably linked with impacts on the environment. The most massive of these impacts is the likely human effects on the global climate. Perhaps the most spectacular have been the tragic accidents at nuclear reactors and the continual debate about the storage of nuclear waste. However, the important lesson is that every energy source, no matter what it is, has some technological advantages and some disadvantages, some negative impacts on the environment and sometimes advantageous effects, and some economic drawbacks and some incentives. As a society, our challenge is to navigate our way through this sequence of positive and negative issues, choosing the optimum (or least bad) solution for our locality. Despite the claims of advocates of many energy sources, there is no one-size-fits-all answer.

This is the second edition of a book that grew out of a course on energy and society, which I taught for many years at Penn State University. The course was intended for students not majoring in science or engineering to fulfill partially their requirements for science courses. Experience with the first edition has shown that this book can also be read by persons wanting to learn more about energy and its impacts, reading on their own without using the book as a textbook. I have assumed that most readers, whether students or interested laypersons, will have minimal background in science and mathematics, not beyond high school algebra and science. Readers without even that background can still enjoy and learn from almost all the material in this book, picking up the needed science "on the fly."

As the energy and society course evolved over many years of revisions, it seemed to me (and, I think, to most of the students) to present the material in a roughly chronological fashion, beginning with the era in which almost everything relied on human or animal muscles and ending with the early twenty-first century. Because of this approach, the book as a whole can also serve as a history of technology, albeit with special focus on energy technologies.

However, I have tried also to make the chapters, or short sequences of chapters, accessible so that they can be read on their own or form the basis for a course more narrowly focused than a broad-brush history of energy technology and its impacts on society and on the environment.

I have benefitted greatly from the help of the many people listed in the Acknowledgments, from student feedback on course evaluations, and from informal discussions with colleagues at Penn State, North-West, and elsewhere. Any errors or shortcomings are entirely my responsibility.

Acknowledgments

It is a great pleasure to acknowledge the tremendous contributions of my wife, Nita, who helped locate most of the illustrations that are new for this edition and secured permissions for their use. She assisted in many other ways, with word processing, correspondence, and many related chores. I could not have done it without her. This book derives from lectures presented as a course on energy and society that was offered for many years at Penn State University. I have certainly benefitted from exchanges of ideas and material with my colleagues at Penn State who also taught this course and from useful discussions with my colleagues at North-West University. Some of the newly added text was originally typed by Carol Brantner from my handwritten lecture notes. Many people helped by making illustrations available or by granting permission for their use; they are mentioned in the captions of the appropriate illustrations. Any errors of fact or omission are solely mine.

Author

Harold Schobert is professor emeritus of fuel science at Penn State University (University Park, Pennsylvania) and extraordinary professor of natural sciences in the Coal Research Section at North-West University (Potchefstroom, South Africa). He has nearly 40 years' experience in energy education and research, primarily in coal science and technology. Professor Schobert is the author of 10 other books and about 140 papers in the peer-reviewed literature. His work has been recognized by his election as a fellow of the American Chemical Society; the Henry H. Storch Award for lifetime excellence in fuel chemistry; awards from Penn State's College of Earth and Mineral Sciences for excellence in teaching and in research; and an award from the Golden Key Honor Society as Penn State's outstanding faculty member. Professor Schobert has also served on numerous energy-related committees or advisory boards at the state, national, and international levels.

1 Introduction

ENERGY AND US

Consider the beginning of a typical day. As we get out of bed, we're probably concerned with whether we feel warm enough, with turning on lights so we can see, and with cooking something for breakfast.

These activities all require energy. Many of us live in homes with electric heating; if not, heat is provided by burning a fuel such as natural gas or heating oil. Most people rely on electricity for lighting, but without electricity, the flames of wax candles or kerosene lamps provide illumination. Electric stoves or electric microwave ovens are popular for cooking; so are stoves that use fuels such as wood, coal, or natural gas. Cooking breakfast means that we have to have something to cook. In many parts of the world, few people subsist entirely on what they can raise or catch themselves. Having food available first requires planting, cultivating, and harvesting on farms. Many farmers use a variety of agricultural machinery that operates with gasoline or diesel fuel, and others rely on using animals that require their own food sources. Harvested food has to be transported to facilities where it is processed, prepared, and packaged. Various sources of energy might be used in the processing of foods, including electricity, natural gas, or heating oil. Then the packaged food is transported to warehouses and stores for sale to us, the consumers. Transportation from farm to processing facility and then to stores might require gasoline or diesel fuel.

As we dress and begin to go about our daily routine, we depend on a huge variety of manufactured articles. Very, very few of us weave our own cloth, turn logs into boards to make wooden articles, or make any of the other items we use throughout the day. Manufacturing begins with the production of raw materials, such as producing metals from their ores. Many kinds of energy might be used in such operations, including electricity, heating oil, or coal. Then the raw materials must be fabricated into useful articles. Fabrication might involve cutting, casting, machining, or weaving (as but a few examples). Many of these operations use electrically powered machinery, and coal, heating oil, or natural gas as heat sources. Then the manufactured articles must be transported to stores.

On most days, we must get out of the house and around for work, shopping, or socializing. If we walk or bicycle, we use energy from our own muscles for transportation. Carts or carriages pulled by animals use the energy of their muscles instead of ours. Cars or light trucks have engines that use gasoline or diesel fuel. Electric automobiles or hybrid electric/gasoline cars are coming onto the market too. If we take the bus, we are probably relying on diesel fuel or natural gas. Trains may have diesel locomotives or operate using electricity. Small airplanes may use gasoline as fuel; others use jet fuel, a refined form of kerosene.

FIGURE 1.1   This photograph of the skyline of Philadelphia illustrates the enormous use of energy in modern society—electricity in the many lights in the buildings and gasoline in the cars on the road on the left side of the picture. (Reprinted from first edition, and provided by Philadelphia Printing, Willow Grove, Pennsylvania.)

The many uses of energy in the home for warmth, cooking, and lighting, together with farming, manufacturing, and transportation consume prodigious quantities of energy (Figure 1.1).

Many people in the industrialized world are able to surround themselves with electrical appliances and gadgets: television set, microwave oven, music system, personal computer, electric razor, hair dryer, refrigerator, lamps, coffee maker, electric clock, electric pencil sharpener, electric tooth brush, power tools, and radio—and, for some of these items, often more than one of each. In the kitchen, for example, one can expect to find a stove, a refrigerator, a dishwasher, and a microwave oven. What else? A coffee maker, espresso machine, electric can opener, pasta maker, bread maker, slow cooker, electric carving knife, toaster or toaster oven (or both), and a blender or food processor. No one worries about whether or not there will be enough electricity to operate these gadgets. If there's a problem, it's how to find counter space to use all this stuff or cabinet space to store it.

The idea of even asking whether there would be "enough" electricity to operate a gadget we're buying when we get it home might sound silly. We simply assume that we can purchase and plug in a limitless number of electrically operated items. Usually, if we think of any limit at all, it's the number of electric outlets available for plugging items into. We can even solve that problem if we remember to buy some power strips (that provide five or more electrical outlets from an original single outlet). Though it's very unwise, and maybe even illegal in some places, some of us even connect two or more power strips together.

Our assumption about the eternal availability of essentially unlimited quantities of electricity is tested when there is a power failure. When the electricity supply suddenly fails, we might have a momentary bit of panic until we assure ourselves that we're unhurt, but then many of us react to a power failure with a feeling of annoyance or anger. We were watching that television show, or cooking that meal, or reading that book, feeling that we could do those sorts of things as much as we wanted, any time we wanted, and now, suddenly, no electricity.

It's important to recognize that not everyone in the world gets to enjoy the lifestyle that's just been described. There are places in the world where, if electricity is available at all, it is only "on" for a certain period each day. Currently, something like a quarter of the world's population—1.5 billion people—lack ready access to electricity. Worst hit is sub-Saharan Africa, where in some places, such as Burundi and Rwanda, more than 90% of the population lack access to electricity entirely.

In the United States, we usually have the same cavalier attitude toward gasoline. We expect to be able to drive around without ever once worrying about whether we will be able to buy gasoline whenever and wherever we need it, and as much as we want. (Probably only a few of those who recall the gasoline shortages during the oil embargoes of the 1970s may challenge this attitude.) Even at two o'clock in the morning on Christmas Day there will be someplace to buy gasoline. Gasoline seems to be as easily and widely available as water. And gasoline shortages produce the same anger as electricity outages.

As with electricity, assumptions we make about the availability of gasoline are not valid everywhere; they apply only to the so-called developed or industrialized nations. The cheapest gasoline in the world is in the Persian Gulf region, especially in those countries having significant amounts of oil: Saudi Arabia, Qatar, Bahrain, and Kuwait as examples. In many industrialized nations with strong economies, such as Japan and Western Europe, gasoline costs two to three times as much as in the United States.*

But for those people fortunate enough to live in reasonably prosperous, industrialized nations, energy is readily and always available at the flick of a switch, stopping at a filling station, or plugging an appliance into a wall socket. The key idea that should come from thinking about how we get through the day is this:

Energy is ubiquitous in our lives and is so common that we seldom even think about it.

---

* But understand that a comparison of gasoline prices can be tricky. The cost of gasoline reflects not only the price of petroleum, an internationally traded commodity, and the cost of refining petroleum into gasoline, which is approximately the same everywhere, but also the cost of transporting the gasoline from the refinery to the consumers, and the whole package of local, state or regional, and national taxes. These last two figures can vary quite widely from place to place. Nevertheless, the United States enjoys remarkably low gasoline prices relative to most other modern industrialized nations.

FIGURE 1.2   William Stanley Jevons (1835–1882), the British economist who studied the utility of energy in society. (From http://en.wikisource.org/wiki/File:PSM_V11_D660_ William_Stanley_Jevons.jpg#filelinks.)

The great English economist William Stanley Jevons* (Figure 1.2) expressed the idea this way:

> *Many commodities which are most useful to us are esteemed and desired but little. We cannot live without water, and yet in ordinary circumstances we set no value on it. Why is this? Simply because we usually have so much of it that its final degree of utility is reduced nearly to zero. We enjoy every day the almost infinite utility of water, but then we do not need to consume more than we have. Let the supply run short by drought, and we begin to feel the higher degrees of utility, of which we think but little at other times.*

—**Jevons**[1]

Another way of illustrating our dependence on energy is to consider it from the other perspective: How we would live if the supply of electricity, petroleum products, and natural gas suddenly was not available any more?

---

* William Stanley Jevons (1835–1882) was one of the great economists of the nineteenth century. Probably his most enduring work is his book *The Coal Question*, which focuses on the rate of consumption of energy resources and the efficiency with which we use them, questions that we will come back to in detail in later chapters. Jevons also devised a curious forerunner to the modern computer, the "logic piano," which was intended to be able to solve any problem in logic. Jevons's career was cut short by his death in a drowning accident in 1882.

- What would we eat? Probably the best we could do would be foods raised by ourselves or foraged in the woods. Many people living in large cities would probably starve.
- How would we get around? Most of us would be confined to an area that would be accessible by walking. We could use horses, if we had them. We could ride bicycles, until they broke or wore out. Perhaps those who lived near the coast or near a navigable river, could travel by boat—one that we paddled or rowed, or that used the energy of the wind to sail, or a steamboat that used wood as fuel.
- How would we stay warm? We could use firewood, if we had access to it and for as long as the wood lasted. A few clever persons might rig up solar energy collectors or figure out how to use windmills or waterwheels to operate electrical generators.
- What would we use in our daily lives? When our clothes, tools, and utensils broke or wore out, we would have no replacements, except for things that could be made of wood (assuming we hadn't burned it to stay warm), or wool or cotton cloth.
- How would we regulate our days? Most of us would rise at dawn and go to bed at dark, because there would be little artificial light other than fires.

In essence, the vast majority of people—especially city dwellers—would quickly freeze and starve in the dark. Of the survivors, most would be reduced to a fairly brutish existence not unlike that experienced by the poor during medieval times. A very few, those who were competent at subsistence farming and at manufacturing or repairing of small tools and machinery (e.g., the Amish), might "make it."

ENERGY AND NATIONS

Energy consumption is an absolutely necessary component of the industrial society. Before the Industrial Revolution of the eighteenth century, most energy use relied on three sources: human and animal muscles (Chapter 3), the chemical energy in firewood (Chapter 6), and the energy of wind and water available in nature (Chapters 8 and 9). Since then, there have been three major historical transitions in the way we use energy. The first was the steam engine (Chapter 10), for which the principal fuel was coal. The second was the development of electricity (Chapters 12 through 16). Electricity provided a double benefit: It was the first form of energy in which the place where it is produced could be separated by many kilometers—indeed, many hundreds of kilometers—from the place where it is used. It is also the only energy source that is easily converted into light, heat, or mechanical work wherever it is used. The third major development was the internal combustion engine, which introduced an enormous mobility into society (Chapters 19 through 22), but also a significant dependence on petroleum products. Petroleum (Chapters 18 through 22 and 28) gradually took over the position of being the dominant global energy source by the end of the twentieth century.

At any particular time, there is a tendency for a correlation between income growth and growth in energy consumption among the countries of the world (e.g., Figure 1.3).

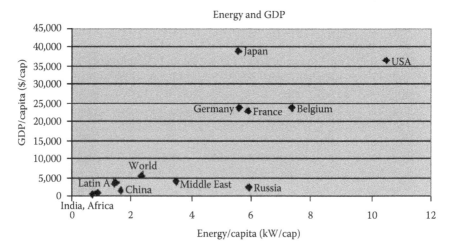

FIGURE 1.3   This graph illustrates the increase in economic activity, as measured by gross domestic product, with per capita energy use for various countries. Generally, the greater the energy use, the higher the GDP. (From http://secondlawoflife.wordpress.com/2007/05/17/energy-consumption-and-gdp/.)

This relationship holds for many countries if income and energy use are examined over a period of years. The correlations are not perfect. Different countries may have different ways of reporting economic statistics. Some factors, such as fuel supplies obtained by bartering, may not even be counted in official statistics. Over a period of years, a country may change the way it collects and uses statistics. Even though these uncertainties exist, until about 1975, there seemed to be fairly good correlation between the economic well-being of a country, as measured by its total output of goods and services (i.e., its gross domestic product, GDP), and its energy consumption. The countries with the largest GDP tended to use the most energy, while undeveloped economies used less energy.

The ratio of energy consumption to GDP is defined as the energy intensity of the economy. Regardless of whether we consider the present energy situation, the history of energy use, or predictions of energy use in the future, we find that population growth, economic development, and technological progress are the major considerations affecting economic development. As a country develops over a period of time, growth of GDP occurs through an increase in population—with its consequent demand for housing, transportation, consumer goods, and services—and results in an increase in energy consumption too. During that time of development, the energy intensity of that country's economy is nearly constant. The growth of GDP virtually parallels that of energy consumption.

The last quarter of the twentieth century witnessed a remarkable change in the relationship between GDP and energy use, especially in developed countries. This can be looked at from two different perspectives. The rate of energy consumption per dollar of GDP generated has been falling. A period of great change occurred between 1975 and 1985, when energy used per dollar of GDP dropped by one-fifth while GDP grew by 30%, based on data for the United States. Around the turn of the century, energy use per dollar of GDP was continuing to drop at about 2% per year. This parameter measures, essentially, the amount of energy required to generate a dollar of GDP. A decline means that, collectively, we are getting more efficient at generating goods and services from the amount of energy we use.

The alternative way of considering the same relationship between GDP and energy use is to look at GDP per unit of energy use. This parameter measures the number of dollars of GDP that can be produced for a given amount of energy consumed. This value should rise as we do a better job of producing goods and services from a given amount of energy (essentially, it is "upside down" from the parameter discussed in the previous paragraph). Indeed, it is doing exactly that—slowly, thanks to the world economic crisis that hit in 2008, but increasing nonetheless.

Several factors have contributed to this change, which has been experienced generally by most of the developed nations. First, from the end of the Second World War until the early 1970s, the cost of energy (referenced to one specific time, so as to adjust for inflation) dropped. During that period, many utilities encouraged consumers, even via cash incentives, to use as much electricity or natural gas as possible. The oil price shock in the early 1970s caused industries to develop ways of operating their various processes in ways that would reduce energy consumption (there had been no incentive to do this in an era when the cost of energy was dropping). A second major factor that "decouples" the use of energy from growth in GDP is the transition of the developed countries to the so-called postindustrial society. This change represents a transition from an industry-based economy to a service-based one, requiring much less energy. In addition, those industries that do remain tend to be ones, like making computers, that add considerable value to the raw materials or components but consume relatively little energy; the heavy, energy-intensive industries such as steel or cement move out.

We must recognize that these comments do not apply to all the countries or all the people of the world. There are enormous differences in the levels of economic development, standards of living, and access to energy around the world. The richest one-fifth of the world's population consumes about four-fifths of the world's goods and services, and uses about half the world's energy. The poorest one-fifth of the world's people consume 1% of goods and services, and get by on about 5% of the world's energy. Per capita GDP differences among countries reflect differences in economic structure—agricultural, industrial, or postindustrial—and in government. Human development indicators (HDI) take into account not only per capita GDP but also such factors as longevity, quality of education, crime rates, and maternal and infant mortality. Currently, Australia and Norway top the HDI list—advanced modern economies with stable democratic governments. Zimbabwe, a heavily agricultural nation presently ruled by a brutal and corrupt government, is dead last.

WHERE WE'RE GOING

Hopefully, this book will help dispel concerns that gaining an understanding of "energy" is either mysterious or difficult, and there is no need to feel overwhelmed or intimidated by politicians, salesmen, hucksters, or demagogues. What if, for example, an electric utility were to propose to construct a nuclear power plant in your town? Likely, you would be bombarded with an entire range of arguments, pro and con, ranging from the assurance that nothing can ever possibly go wrong to the assertion that if this reactor is built, you and your families will be exposed to so much radiation that you'll glow in the dark. How can we determine where the truth lies? (Perhaps it's already obvious to most readers that in this specific example the truth lies somewhere between these extremes—but where, exactly?) In fact, rather than being at their mercy, *you* can become very frightening to the politicians, salesmen, hucksters, and demagogues of this world, because by far the most truly dangerous person is someone equipped to think for herself or himself.

> *Freedom of thought is the only guarantee against infection of people by mass myths, which, in the hands of treacherous hypocrites and demagogues, can be transformed into bloody dictatorships.*

—**Sakharov**[2]

Because the world and the way we use energy are changing at such an incredible pace, it is vital to be able to continue to learn throughout our lives. Fifty years ago, digital watches and calculators didn't exist. Forty years ago, these items cost multiple hundreds of dollars each. Now, either can be purchased from displays at the checkout stands of discount stores for less than ten dollars. Fifty years ago, vinyl records dominated the music industry. Since then, we have seen a progression from vinyl records to tape cassettes to compact disks to music downloaded from the "cloud." Adding compounds of lead to improve gasoline performance was an industry standard for close to fifty years. Leaded gasoline vanished in about a decade.

It's hoped that this book will help the reader feel confident to continue reading elsewhere, continue learning, and continue probing. When we visit another country, we can better appreciate its culture and customs if we can understand something of the language. Unfortunately, it seems to many people that science itself has now become another country, or another culture.* The technology of energy and the issues surrounding energy use are all around us. Many energy-related terms are in common use: acid rain, greenhouse effect, cold fusion, the China syndrome, meltdown, and semiconductors are a few examples. This book helps the interested reader to understand what these terms mean. This is especially so because, as we will soon see (Chapter 2), a scientist sometimes uses a word in a slightly different way or with a more restricted meaning than when the same word is used in everyday conversation.

The second point is that there are limits as to what can be accomplished with energy. We can't create energy out of nothing. The best we can hope to do is to utilize the amount of energy available to us and convert it from one form to another. Ideally, we would hope to be able to convert energy completely from one form to another, with no waste or losses. As we'll see, we'll never really be able to

---

* The concept of a society divided into two cultures, of which one is science, has been argued eloquently and most famously by the British writer and physicist C.P. Snow (1905–1980) in his appropriately titled book, *Two Cultures and the Scientific Revolution*, first published in 1959.

attain the ideal of a device that is 100% efficient in converting energy from one form into another. Common electric motors come pretty close, changing about 90% of the electric energy we supply to them to doing the work we want done. An electric power plant that burns coal is much worse; only about 35% of the coal's energy comes out of the plant as usable electricity. Steam locomotives, which at their zenith represented mighty examples of brawn and brute force, are terrible, with less than 10% of the energy of the coal fuel becoming available to drive these behemoths.

The third point that will be made is that there are limits to what scientists and engineers can accomplish, and what they can tell us. We will see many examples in the coming chapters of the close connection between energy use and its impacts on the environment. Possible strategies for reducing environmental effects include, as examples, removing a potential pollutant from a fuel before it is burned or capturing it after the fuel is burned but before it can escape to the environment. Scientists can develop methods for removing or capturing potentially harmful materials before they can impact the environment. Engineers can design plants for carrying out these methods on a large scale. Economists can calculate the increase in costs that will result from building and operating these plants. But none of these people can tell you, or calculate for you, let alone dictate to you, what your optimum trade-off would be between increased electricity bills or fuel prices on one hand and accepting greater environmental degradation on the other hand. It is up to the individual citizen, or groups of citizens working together, to make the choice between, say, increased electricity costs vs. forest destruction from loosely regulated pollution from electric generating plants. Some people might be willing to accept substantially higher electric bills in exchange for protecting the environment. Others might adopt the "if you've seen one tree, you've seen 'em all" attitude. The choice is not something that comes from scientific or engineering calculations. The choice is something that each person must make individually by weighing many factors.

As another example, petroleum is used both as a source of fuels (such as gasoline, diesel fuel, and home heating oil) and as a source of many of the synthetic materials ubiquitous in daily life. Likely, petroleum availability will decline sometime in the middle of this century. If that happens, should we continue to burn scarce petroleum, or should we save it to make plastics? This question has no right answer—though zealots on one side or the other will claim there is. Each person will have to think through this issue carefully and then express an opinion by voting, or by attempting to persuade elected representatives to vote in some particular way.

REFERENCES

1. Jevons, W.S. *The Theory of Political Economy*. Penguin Classics: New York, 1970; Chapter III. Also cited in Heilbroner, R. *Teachings from the Worldly Philosophers*. Norton: New York, 1996; Chapter V.
2. Sakharov, A. *Progress, Coexistence, and Intellectual Freedom*. W.W. Norton: New York, 1970.

FURTHER READINGS

Nakićenović, N.; Grübler, A.; McDonald, A. *Global Energy Perspectives*. Cambridge University: Cambridge, U.K., 1998. This book examines the balance, and trade-offs, among fossil fuels, alternative fuels, efficiency increases, and energy conservation in meeting increasing consumer demands for clean, convenient energy. Chapter 3 discusses issues of energy intensity, and gross domestic product.

Smil, V. *Energy at the Crossroads*. MIT Press: Cambridge, MA, 2003. Chapter 2, on "energy linkages," discusses such issues as energy intensity, relationships of energy use with GDP, and how energy use impacts the quality of life.

Smil, V. *Energy in Nature and Society*. MIT Press: Cambridge, MA, 2008. Chapter 12 provides a useful discussion of energy and the economy, and relationships between energy use and quality of life.

2 Energy, Work, and Power

Many people in today's world can reasonably expect being able to visit another country. This might be done for a vacation or a short business trip, or possibly a longer stay, such as an academic term abroad or a sabbatical leave from work. In many of the major cities and resorts, it's possible to get along modestly well without even being able to speak the dominant language of the host country. But to get around the countryside, to understand something of the culture and the literature of the country, it's helpful to know some of the language that one is hearing or trying to read. In the same way, understanding something of the culture and literature of science is greatly helped by knowing some of the language used in science. This notion may seem peculiar at first, because in any country scientists speak the same language as all the other citizens, or at least they appear to do so. Where the difficulty arises, or where inadvertent confusion may be caused, is that sometimes scientists will use a common word—*energy* being a fine example—but with a meaning or connotation somewhat different from the way the word is used in everyday speech. That's especially the case in today's world, because scientists use a variety of jargon and acronyms for efficient communication with colleagues in the same scientific discipline. In fact, scientists trained in one discipline often have difficulty reading professional papers concerning a different discipline unless they first learn the "lingo."

Many words used casually in everyday conversation have more restricted or specialized definitions when used in a scientific context. This is very important to keep in mind when reading scientific material or when having discussions with scientists. For effective communication, it's crucial to make sure that everyone is using words in the same way.

> *How hard one must work in order to acquire his language—words by which to express himself!... With the knowledge of the name comes a distincter recognition and knowledge of the thing.... My knowledge was cramped and confined before, and grew rusty because not used—for it could not be used. My knowledge now becomes communicable and grows by communication. I can now learn what others know about the same thing.*
>
> **—Thoreau**[1]

Though this understanding of the meaning of terms is crucial for reading and learning about science and technology in our modern world, its importance of meaning in this regard has been recognized at least since the time of Aristotle (Figure 2.1),*

---

* Aristotle (384–322 BCE) was a scientist, philosopher, and one of the first persons (indeed one of the few in all of history) of whom it might reasonably be said that he had a command of the entire range of human knowledge. Some would argue that Aristotle, more than any other thinker, is responsible for providing the form and original content of Western civilization. His writings covered medicine, law, art, biology, metaphysics, rhetoric, natural history, and many other subjects. Various anthologies of his writings are available, including *The Basic Works of Aristotle* (Modern Library, 2001) and *The Complete Works of Aristotle* (Oxford University Press, 1984).

FIGURE 2.1 Aristotle (384–322 BCE), the Greek philosopher who argued the need for using correct terminology to convey scientific information. (From http://en.wikipedia.org/wiki/File:Aristotle_Altemps_Inv8575.jpg#filehistory.)

who argued, some 2400 years ago, that scientific knowledge cannot be communicated if the terms that are being used are not used in their correct senses.

WORK, ENERGY, AND POWER IN THE SCIENTIFIC CONTEXT

Three terms will be of special importance throughout this book: energy, work, and power. All three are quite commonly used in informal everyday conversation and writing, often to mean different things. Consider these examples:

- I should wash my clothes today, but I just don't have the energy.
- If you eat this candy, it will give you some quick energy.
- That band's performance was just packed with energy.
- Hard work never killed anybody.
- That test was a lot of work!
- I have to go to work today at 4 o'clock.
- That golfer can really hit a ball with a lot of power.
- The Ferrari is a very powerful car.
- Oh no! The power just went off!

These, and hundreds of similar examples, can easily be gleaned from the popular press and news media. The examples here should be clear and easy to understand, yet a careful look will show that the same word is not being used in quite the same way in each case. Surely the "power" with which a golfer hits a ball is not the "power" that operates the television set and lamps. The definitions of these three key terms that we will use throughout the remainder of this book are more restricted than the ways in which we use them in everyday discourse.

We begin with the scientific definition of work (i.e., the meaning scientists use to define the word):

> WORK is causing an object to move into, or out of, some position, especially when it moves against a resistance.

In places where it's important to emphasize that a common word is being used in its more restricted definition, we'll write it in small capitals, as WORK. Eventually, it will be possible to infer from the context whether we're using a word in its restricted sense or in its more common, informal usage.

Some examples of WORK include

- Carrying boxes of belongings or furniture upstairs, against the resistance of gravity
- Plowing soil, creating furrows by breaking apart the natural resistance of the soil
- Driving down the road, against the resistances of friction caused by tires moving against the road surface and the mechanical parts of the car against each other
- Converting sheets of metal or plastic into useful objects, against the resistance of the natural stiffness or strength of the material
- Sending electricity from the generating station to our homes, against the electrical resistance of the wires

In fact, just about everything we try to accomplish in our daily lives—manufacturing items, transporting ourselves or our goods, using electricity—represents doing WORK. The first real concept of WORK as we now use the term derives from a German philosopher, Gottfried Leibniz (Figure 2.2).* A general understanding of WORK dates from the early decades of the nineteenth century. To accomplish the things we do, and obtain the things we use, requires doing WORK. But, to do this WORK, we need ENERGY.

> ENERGY is the capacity for doing work.

---

* Gottfried Wilhelm Leibniz (1646–1716), born in Leipzig, Germany, was, like Aristotle, a near-universal genius, who contributed to such fields as mathematics, logic, geology, and linguistics. He is perhaps best remembered today for his invention of the calculus, done apparently independently of the almost simultaneous development of the subject by Isaac Newton. A controversy raged for centuries (and probably still does in some quarters) on the issue of which of these two great men had priority of invention and whose system of calculus was better.

FIGURE 2.2 Gottfried Wilhelm Leibniz (1646–1716), the German philosopher who developed the first concept of WORK as we now use the term. (Figure courtesy of Edward Zalta, Stanford University.)

ENERGY, therefore, is the key to accomplishing, or having the capacity for doing, all of the activities that form the basis of our modern industrial society. The word "energy" is only about two centuries old. Thomas Young (Figure 2.3) proposed it in the early nineteenth century.*

> *"Then idiots talk," said Eugene, leaning back, folding his arms, smoking with his eyes shut, and speaking slightly through his nose, "of Energy. If there is a word in the dictionary under any letter from A to Z that I abominate, it is energy. It is such a conventional superstition, such parrot gabble!..."*

> —**Dickens**[2]

In several of the following chapters, we'll also see that it's helpful to think of some concepts related to energy by using analogies related to money. It's

---

* Following the tradition of Aristotle and Leibniz, the English physician and scientist Thomas Young (1773–1829) made significant contributions to many fields. Young could read by the time he was two years old and is said to have read the Bible completely through twice by age four. He produced one of the first translations of the Rosetta Stone, which provided important information on how to understand and decipher Egyptian hieroglyphics. Young's contributions to the theory of color are used today in both color television and color photography. His name survives particularly in the concept of Young's modulus, which describes the elasticity of a material, i.e., its ability to resist changing its length when it is subjected to being compressed or stretched. Young's modulus is a central concept in structural engineering.

FIGURE 2.3 Thomas Young (1773–1829), an English scientist who made contributions to numerous fields and developed the concept of ENERGY. (From http://commons.wikimedia.org/wiki/File:SS-young.jpg#file.)

obvious—sometimes painfully so—that you can't spend unless you have *money*. Put in a different way, *money* is the capacity or ability to spend. *Money* and *spending* have the same relationship as ENERGY and doing WORK. Why do we do WORK? It's to get something done, to accomplish something, and to make something happen. We need ENERGY to do this. Why do we *spend*? It's to acquire things, to obtain some service, and to get what we need (or want) for our daily lives. We need *money* to do this.

Let's continue the money–energy analogy a bit further. What happens when we buy something? We transfer money from ourselves, or from our bank accounts, to the seller. The act of *spending* involves a transfer of *money*. In exactly the same way, the act of doing WORK involves a transfer of ENERGY. As we explore different applications of ENERGY in the coming chapters, we will also occasionally stop to examine what kinds of ENERGY are being transferred and how the transfer takes place. There's another important aspect of what happens when we buy something. The total amount of money does not change. True, the amount of *your* money drops, and the amount of money that the seller has goes up. However, the total amount of money in circulation in the country doesn't change. In other words, *spending* involves a transfer of *money* but not a change in the total money supply. Similarly,

FIGURE 2.4 This Jaguar is an example of an automobile with a great deal of POWER. (From http://www.jaguar.com/us/en/r1/.)

doing WORK involves a transfer of ENERGY, but doesn't change the total amount of ENERGY available in the world.*

The concept of POWER addresses the question of how quickly we can do WORK. A powerful car (Figure 2.4), for example, is one that can do a great deal of WORK and, usually, consumes a great deal of ENERGY, in a short time.

POWER is the rate of doing work.

We can express this relation mathematically as

$$Power = \frac{Work}{Time}$$

When we say that a particular item (e.g., a car) is more powerful, that is, has more POWER, than another, what we mean is that either it can do more WORK in a given amount of time, or it takes less time to do the same amount of WORK.

To recap, WORK is required for accomplishing virtually every aspect of our society: agriculture, manufacturing, transportation, and electricity production and distribution. ENERGY gives us the ability to do that WORK, and POWER tells us how quickly (or slowly) we can get it done. Anything that we attempt to accomplish in our daily lives—doing WORK with our own bodies, transporting ourselves or the goods that we consume, manufacturing items, or distributing electricity—involves doing WORK. Everything that we want to do, or want to have done for us, in our daily lives involves doing WORK. But we need ENERGY to do that WORK. The reason that ENERGY is important lies in the fact that it is what we need to do WORK. POWER tells us how rapidly, or how slowly, we can accomplish a given amount of WORK (or, from the other perspective, how much WORK we can do in a given amount of time).

---

* It's important not to push this analogy too far. It is true, indeed one of the most fundamental laws of science, that the amount of ENERGY in the universe is constant. We can neither create new ENERGY nor destroy the ENERGY that now exists. All we can hope to do is move ENERGY into different forms. However, governments do have the ability to increase or to decrease the amount of money in circulation as part of their financial policies. Whether doing so is a good idea or not is a fierce debate among economists and politicians. Following the debate would get us way too far afield from the present topics, but interested readers will find no shortage of magazine articles, books, "op-ed" pieces, blogs, websites, and just plain rants on the topic.

To sum up, doing WORK lies at the heart of every imaginable activity that provides the food, the goods, and the services we enjoy; ENERGY is what is absolutely necessary to be able to do that WORK; and POWER tells us how fast we can do it.

HOW MUCH WORK GETS DONE

What does the doing of WORK actually involve? We've described the concept qualitatively as moving an object against some resistance, gravity being a familiar example. The concept of WORK is usually defined quantitatively as the application of a force of some sort through a distance. That is,

$$\text{Work} = \text{Force} \times \text{Distance}$$

This relationship implies two things. First, if there isn't a force being exerted, or if there isn't a distance being traversed, then WORK isn't being done. Imagine helping a friend move into a new dwelling. You're handed a heavy box, or heavy piece of luggage, with the instruction, "Here, hold this," while your friend fumbles around to find the door key. You may be holding that item till it feels like your arms are ready to pop out of their sockets, but if you're not moving it, you're not doing WORK. All of us have had the experience of being faced with a very difficult examination question and having to think as hard as we can for some time to come up with an answer. Sitting still in a classroom seat, exerting no force, nor moving anything through a distance, is not WORK. Of course it would be fully justified in both instances to exclaim, "That was a lot of work!" in our everyday connotation of the word. But, in our restricted definition of the term, it really wasn't any WORK at all.*

Another, possibly more subtle, implication also lies in the equation relating WORK to the product of force and distance. A large force applied through a small distance can amount to the same amount of WORK as a small force applied through a large distance. Disregarding the units involved (so long as they are the same for each force and each distance), clearly

$$50 \times 2 = 2 \times 50$$

In other words, applying a force of fifty units over a distance of two units represents the same amount of WORK as applying a force of two units over a distance of fifty units.

To see why this is important, imagine being given the task of shifting some object that weighs 200 kg. It might be a big rock, a large piece of furniture, or a safe. Very, very few of us are capable of exerting the force needed to lift and move a 200 kg object. Our early ancestors solved this problem, probably doing it intuitively without even knowing of the underlying physics. The secret of moving a 200 kg object a short distance is to exert a force equivalent to moving, say, 20 kg over ten times the distance. How do we do this? With three simple devices that date from antiquity, or possibly prehistory: the lever, the inclined plane, and the pulley-and-rope. These simple tools,

---

* This is true when we consider the human body as an entity, and in this example, the body is not doing WORK. At a different organizational level, that is, in the individual organs of the body, WORK is being done.

which we still use, derive from the concept that we can do an enormous amount of WORK simply by applying a small force through a relatively large distance. Almost certainly, the unknown individuals who figured out how to use sturdy sticks to help move heavy rocks or logs had no scientific concept of WORK, but likely an intuitive understanding quickly developed. This was of enormous importance to those early ancestors when the WORK to be done far exceeded the ability of relatively puny human muscles.

Many of us don't even think of activities like climbing steps or lifting small objects as WORK. Many others, though, still do a large amount of WORK every day, especially manual laborers who work on farms, on construction jobs, or in factories. In our modern industrialized society, the proportion of people who do a great deal of WORK is steadily decreasing. However, for the overwhelming portion of human history, and still today in many parts of the world, the only way that humankind had of accomplishing WORK was to rely on the muscles of the human body.

BRIEF HISTORY OF THE HUMAN USE OF ENERGY

The chapters that follow develop the concept of energy in a historical context, progressing from the kinds of ENERGY available to the earliest humans up through the most modern developments, and finally looking a bit into the future. To set the stage for that discussion, we will consider briefly a "time line" of human development and energy use.

Hominids, the family of primates of which we are a member, are about 5,000,000 years old. Modern humans (*Homo sapiens*), people that would be essentially similar to us in appearance, thought patterns, behavior, and genetics, appeared in Africa about 200,000 years ago. Even if we restrict our attention only to so-called modern humans, we still have a long time period to consider.

Archeological evidence suggests that hominids have been making use of fire for 400,000 years (slightly less than one-tenth the time that hominids have been in existence). Some evidence suggests that hominids may have been using fire for much longer, up to 1,700,000 years, but the further the date is pushed back, the less consensus about its validity. The earliest uses of fire required that the fire be started naturally, such as by lightning, and then kept going. If the fire went out, there was no way of restarting it, a fact that may be responsible for religious rituals and folk rites concerning guarding the fire to ensure that it didn't go out. Recent evidence from the Wonderwerk Cave in South Africa suggests that our early ancestors may have been able to control fire a million years ago.

Some few animals may have been domesticated for probably as long as there have been hominids. Domestic animals may originally have been kept as pets; archeological evidence suggests that domestication of many animals that we use for WORK, such as horses, occurred about five thousand years ago. Elephants and camels were domesticated some three to four thousand years ago. For the first 197,000 years of human existence, we relied on our own muscles to do WORK. We then figured out how to have animals help us, but they too rely on the action of their own muscles.

Waterwheels used to do WORK, such as grinding grain, first came into use about 3000 BCE. In other words, it is only in the past three thousand years—representing just 1.5% of human existence—that we have had any source of ENERGY other than human or animal muscles. The first practical windmill in Western nations that can be dated with certainty was put into operation in 1185. Windmills had been developed in the Middle East several centuries earlier, possibly developed independently in China even before that. Steam engines for doing practical WORK were developed about three hundred years ago. Crude engines, for pumping water from coal mines, were built in 1698; an improved version was developed in 1712. The first clear demonstration of the sustained generation of electricity happened in 1831. The first oil well in the United States was drilled in 1859. The first sustained nuclear chain reaction was achieved in 1942.

To put these dates into perspective, we can imagine what it would be like if the whole ENERGY history of *Homo sapiens* were condensed into a single year (Figure 2.5). Figure 2.5 shows that we humans have had to rely on our own muscles in order to do WORK for almost the entire span of human history—all but nine "days" of the human "year." All but about four and a half "days" of the human year relied either on human or animal muscles. Another way of interpreting this compression of human use of ENERGY into a single "year" considers that for 97% of the entire time that humans have existed, we have relied on our own muscles for doing WORK. In the next chapter, we will examine the source of the ENERGY that gives our bodies the ability to do WORK.

FIGURE 2.5 If all of human history were compressed into a single year, the use of ENERGY sources by humans would not have begun until December 20.

REFERENCES

1. Thoreau, H.D. *Journals*. Volume XI, 1858, p. 137. Also conveniently found in Walls, L.D. (Ed.) *Material Faith: Henry David Thoreau on Science*. Houghton Mifflin: Boston, MA, 1999, pp. 82–83.
2. Dickens, C. *Our Mutual Friend*. Penguin Classics: New York, 1997, pp. 29–30. Many other editions of this book are also readily available.

FURTHER READINGS

Cutnell, J.D.; Johnson, K.W. *Physics*. John Wiley: New York, 2007. This is a university-level introductory textbook on physics, requiring more mathematical sophistication than other books listed here. Chapter 6 provides a solid introduction to energy, work, and power.

Lindsay, J. *Blast-Power and Ballistics*. Barnes and Noble: New York, 2009. This book provides a discussion of the development of concepts of energy, force, and related topics in the ancient world, including the work of Aristotle. An interesting contribution to the history of science. Chapters 1 and 2 are particularly relevant.

Pielou, E.C. *The Energy of Nature*. University of Chicago: Chicago, IL, 2001. The first two chapters of this book—a book on energy written from the point of view of a naturalist— relate to this present chapter. Written at a level for readers of, say, *New Scientist* or *Scientific American* articles.

Purrington, R.D. *Physics in the Nineteenth Century*. Rutgers: New Brunswick, NJ, 1997. As the title implies, this book provides a history of the development of physics in the 19th century. It will be quite interesting to anyone curious about the topic, but assumes considerably more scientific background from the reader than do the other books listed here. Chapter 5 covers topics relating to energy and work.

Smil, V. *Energies*. MIT Press: Cambridge, MA, 1999. Chapter 4 discusses the importance of horses and cattle in pre-industrial societies.

3 Human Energy

If the 200,000-year history of *Homo sapiens* could be compressed into a single year (Figure 2.5), for almost all of that year—all the way to about December 22—we relied on our own muscles to do WORK. Either alone, or organized into groups, humans can do remarkable amounts of WORK, in farming, construction, and manufacturing. Charles Darwin has described the incredible feats of workmen manually removing ore from deep mines:

> *Captain Head has described the wonderful load which the "Apires," truly beasts of burden, carry up from the deepest mines. ... The load was considered under weight when found to be 197 pounds. The apire had carried this up eighty perpendicular yards—part of the way by a steep passage, but the greater part up notched poles, placed in a zigzag line up the shaft. According to the general regulation, the apire is not allowed to halt for breath, except the mine is six hundred feet deep. The average load is ... rather more than 200 pounds, and ... one of 300 pounds by way of a trial has been brought up from the deepest mine! At this time the apires were bringing up the usual load twelve times in the day; that is 2400 pounds from 80 yards deep; and they were employed in the intervals in breaking and picking ore.*
>
> *These men, excepting from accidents, are healthy, and appear cheerful. Their bodies are not very muscular. They rarely eat meat once a week, and never oftener, and then only the hard dry charqui. Although with a knowledge that the labour was voluntary, it was nevertheless quite revolting to see the state in which they reached the mouth of the mine; their bodies bent forward, leaning with their arms on the steps, their legs bowed, their muscles quivering, the perspiration streaming from their faces over their breasts, their nostrils distended, the corners of their mouth forcible drawn back, and the expulsion of their breath most laborious. Each time they draw their breath, they utter an articulate cry of "ay-ay," which ends in a sound rising deep in the chest, but shrill like the note of a fife. After staggering to the pile of ore, they emptied the "carpacho;" in two or three seconds recovering their breath, they wiped the sweat from their brows, and apparently quite fresh descended the mine again at a quick pace. This appears to me a wonderful instance of the amount of labour which habit, for it can be nothing else, will enable a man to endure.*

—Darwin[1]

The Egyptian pyramids provide perhaps the greatest testament to the scale at which human WORK can be accomplished (Figure 3.1). But WORK can only be done if we have to have the capacity or ability to do it—in other words, we need ENERGY. Where does that ENERGY come from?

Likely even prehistoric peoples appreciated that there is some correlation between the intake of food and human ENERGY. Sometimes crops failed and food

FIGURE 3.1 The construction of the pyramids in Egypt represents the enormous scale on which human WORK can be accomplished. (From http://www.ask-aladdin.com/Pyramids-of-Egypt/index.html.)

became scarce. During wars, food supplies could be cut off, or prisoners be allowed to starve. People might choose to undergo extended fasting, as for religious purposes. In all of these situations, the same observation can be made: human ENERGY decreases. The long-term consequence is that a person who doesn't eat can reach a point at which the person is no longer able to fend for, or even care for, himself. Death becomes inevitable. World history contains many dreadful tales of human suffering and death during famine or deliberate starvation: the Irish potato famine of the 1840s, Stalin's "liquidation of the kulaks" in the Soviet Union in the 1930s, the starvation of concentration camp inmates during the Holocaust of World War II, extensive famine in China as a result of Mao Zedong's "great leap forward" in the 1950s, and, in recent years, the inability of North Korea to provide adequate nourishment for its citizens.

Early chemists appreciated the need for accurate information on the compositions of the substances they worked with, began to learn how to perform accurate chemical analyses, and began to understand the physiology of the human body.* They made the following observations:

---

* An important example is the German chemist Justus von Liebig (1803–1873), who made numerous contributions to agricultural chemistry and to the beginnings of the field of biochemistry. He also did much to establish the teaching of organic chemistry, including the use of laboratory experiments as part of the educational experience, that is, chemistry lab. Liebig invented the useful material we now recognize as bullion and is also credited—or blamed, depending on one's taste—with the invention of Marmite.

- The major components of foods are carbon, hydrogen, and oxygen.
- Many foods contain small amounts of nitrogen and sulfur, and various other chemical elements in still smaller quantities.
- We inhale air, which is essentially a mixture of nitrogen and oxygen. When we exhale, we find that the amount of oxygen in the air is depleted, and our breath contains measurable amounts of carbon dioxide and water vapor.

At about the same time, other scientists were studying what happens when fuels such as wood or coal are burned. They too made three key observations:

- The major components of fuels are carbon, hydrogen, and oxygen.
- Some fuels contain small amounts of nitrogen and sulfur, and various other elements in still smaller quantities.
- A fuel needs air to burn. The air surrounding the burning fuel becomes depleted in oxygen and contains carbon dioxide and water vapor.

Clearly, there are some significant similarities between foods and fuels!

We recognize intuitively that ENERGY of some sort is liberated when a fuel burns, because we can feel the heat. (Establishing the relationship that heat is a form of ENERGY is a very crucial point. For now, we gloss over this relationship, but it will come back in later chapters.) From the comparison of fuels and food, and the observation that consuming a fuel liberates ENERGY, we can write two simple relationships:

$$\text{Fuel} + O_2 \rightarrow CO_2 + H_2O + \text{ENERGY}$$

and

$$\text{Food} + O_2 \rightarrow CO_2 + H_2O$$

The similarities between foods and fuels, in terms of their composition and the processes that seem to occur when each is consumed, lead to the reasonable assumption that consuming food also liberates ENERGY. In other words, we may amend our statement for food by writing

$$\text{Food} + O_2 \rightarrow CO_2 + H_2O + \text{ENERGY}$$

That is, there is an analogy between burning a fuel to get ENERGY from a fire and converting food to ENERGY in our bodies. The analogy is not exact; we will examine why later in this chapter. We can observe from personal experience that different foods seem to provide different amounts of ENERGY. For example, eating candy or other sugar-rich foods often provides a quick burst of ENERGY.

Science is always on the firmest ground when accurate measurements can be made, when the results can be expressed numerically, and when quantitative relationships can be established among the measurements being made. So, we question how much ENERGY can be released from food, and how various kinds of foods compare as sources of ENERGY. The experimental apparatus that allows us to determine the amount of ENERGY in foods—or fuels—is a *calorimeter* (Figures 3.2 and 3.3).

FIGURE3.2 AmoderncalorimeterforthedeterminationoftheENERGYcontentoffoodsorfuels. (From http://www.ika.com/-Lab-Eq/Calorimeters-Oxygen-Bomb-calorimeter-cph-330/.)

An accurately weighed sample is placed in the sample container with air or pure oxygen. The container holding the sample is surrounded by an accurately weighed amount of water. The sample is then ignited, and the heat that is produced as it burns passes into the water. Thanks to the thick insulation, heat cannot escape from the water, so the temperature of the water rises. The amount of heat that causes an increase in temperature in a known quantity of water is known very precisely (it's called the specific heat of water*). Having measured the weight of water before start-ing the experiment and having observed the temperature rise during the experiment, we can then figure out the amount of heat (ENERGY) that was produced from our sample of food or fuel.

The energy content of various foods is expressed in two systems of units. One uses the calorie as the measurement of ENERGY. One calorie is the amount of heat that will raise the temperature of one gram of water by one degree Celcius.† Unfortunately,

---

* Specific heat, also known as specific heat capacity, is a characteristic physical property of all substances. It is expressed numerically in units of the amount of energy required to raise the tempera-ture of a known mass of substance by one degree. For example, specific heat can be expressed as joules per kilogram per degree Celsius, or British thermal units per pound per degree Fahrenheit (or in other units as well). Values of specific heats can be used to determine the amount of heat needed to produce a desired temperature increase in some given amount of material. Water has one of the highest known specific heats, higher than virtually all other common substances. Many kinds of steel have specific heats that are only about one-tenth that of water. It takes only one-tenth the amount of heat to achieve a given temperature increase in steel as in water, or, from the other perspective, the same amount of heat will cause a temperature increase ten times greater in steel than in water.

† Strictly speaking, this definition of calorie holds for a temperature of 15°C. However, the variation with temperature is so slight that it's a reasonable approximation to consider that this definition holds for liquid water at any temperature.

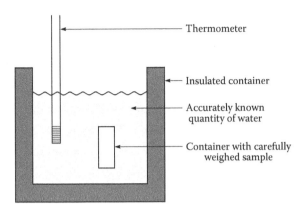

FIGURE 3.3   This sketch illustrates the essential components of a calorimeter.

nutritionists have defined their own "calorie." It is called the *nutritional calorie*, or sometimes the "large calorie." It should be—but usually isn't—written with an upper case C: Calorie. One nutritional calorie (Calorie) is the amount of heat that will raise the temperature of 1000 g of water 1°C. Thus we have the relationships that

$$1 \text{ nutritional calorie } (\text{Calorie}) = 1000 \text{ calories} = 1 \text{ kilocalories}$$

The unit of nutritional calories, or Calories, nowadays survives mainly in the United States. Almost all other nations use the unit of kilojoules (kJ) to express the ENERGY content of food. One kilojoule represents ENERGY equivalent to raising the temperature of 239 grams of water by 1°C. One nutritional calorie is exactly 4.184 kJ, but for *rough* conversion, it's adequate to remember that the kilojoule is four times bigger than the Calorie. Those who check the ENERGY content of foods—when trying to lose weight, for example—can easily convert back and forth between the units by using four as the conversion factor.

The amount of ENERGY available to us depends not only on what kind of food we eat, but also on how much we eat. Sugar is a prime example of a high-energy food. Yet we will certainly derive more energy from eating an apple than, for example, allowing a single tiny grain of sugar to dissolve on the tongue. A comparison of the amounts of ENERGY available from different foods (or from different fuels) needs to be made on the basis of equal quantities, in order for any comparison to be fair and sensible. Unfortunately, one can find all sorts of zany units, such as Calories per ounce (mixing metric and British units) or Calories per cup (mixing the metric system with arbitrary cooking units). The important thing to remember is that whether using kilojoules or Calories, what unit the "per" refers to—cups, kilograms, pints, or whatever—must always be the same for the comparison to be fair.

The dietary ENERGY allowance varies with age and gender, the type of WORK he or she is doing, and climate. It takes about 4000 kJ per day simply to sustain life. An adult in a temperate climate doing a typical range of daily activities (but not heavy labor) would require about 7000 kJ per day. (This is about equivalent to a 75-watt light bulb being on continuously.) A worker who spends his or her day in heavy manual labor will require about 16,000 kJ per day.

ENERGY IN FOODS AND ENERGY IN FUELS

A calorimeter can be used to measure the energy released from foods. It can also be used to measure the energy released from fuels. A given amount of food can be considered to be the source of a certain amount of joules or calories. So can a given amount of fuel. This discussion neglects the essential question of what the real difference is between the energy released from a food or from a fuel. Fortunately, this question has a simple answer: there isn't any difference.

There is no difference in the ENERGY released from a substance whether it is eaten as a food or burned as a fuel.

As two examples, suppose we could be so extravagant that we could burn sugar in a home furnace or an automobile engine. (This also supposes that we could figure out the engineering details for doing that!) We would find that exactly the same amount of energy is released when sugar is burned as a fuel as when the same amount is consumed as food. Granted, the specific chemical pathways by which sugar is converted to carbon dioxide and water in the human body or is converted to carbon dioxide and water in a furnace are quite different (as illustrated by the fortunate fact that we don't set ourselves on fire by eating), but the energy output for a given amount of sugar is exactly the same. Next, consider the hypothetical case that the human body was able to metabolize coal (or, put crudely, that we could eat coal). If that were actually true, the energy that would be released to our bodies by eating some amount of coal would be exactly the same as we would get by burning that amount of coal in a furnace. As an example, a standard slice of cheese pizza contains 600 kilojoules. The energy in 24 slices of pizza is equivalent to that in 500 grams of coal. The mythical average person, who needs about 10,000 kilojoules per day, could obtain this ENERGY—in principle—by eating about 350 grams of good-quality coal.

The way in which we came to understand the relationship between consuming food for energy and burning a fuel derives from a collaboration between two of the greatest of French scientists: Antoine Lavoisier* and Pierre-Simon Laplace.† Lavoisier had already observed that carbon dioxide is exhaled by animals, certainly including humans. Lavoisier and Laplace put a guinea pig in a calorimeter of which the outer jacket was packed with ice. They observed that, after ten hours in the calorimeter, the

---

* Antoine Lavoisier (1743–1794) is regarded by many historians of science as being the person most responsible for transforming chemistry into a truly scientific discipline, from a hodgepodge of scattered experimental observations. Among his numerous contributions were establishing that sulfur is a chemical element, establishing that mass is conserved in chemical processes, giving names to hydrogen and oxygen, and writing the first-ever introductory textbook on chemistry. During the Reign of Terror after the French Revolution, Lavoisier was executed by guillotine, presumably for some of the positions he had held under the monarchy. Responding to many protests about the possible execution of this brilliant scientist, the judge in his case supposedly said, "The Republic needs neither scientists nor chemists; the course of justice cannot be delayed." The harassment, imprisonment, or execution of scientists on political grounds continues in various parts of the world 220 years later.

† Pierre-Simon Laplace (1747–1827) is best known nowadays for his contributions to mathematics and astronomy. He speculated on the possible existence of the expansion of the universe more than a century before it was actually proven and postulated the existence of what have now come to be called black holes.

natural heat of the guinea pig had melted 370 grams of ice. Then, they experimented with burning pieces of charcoal (which is nearly pure carbon) to see how much ice could be melted by various amounts of charcoal. This information let them determine how much charcoal it would take to melt 370 grams of ice. These two sets of experiments could be connected by measuring the amount of carbon dioxide exhaled by the guinea pig during its confinement in the "ice calorimeter" and the amount of carbon dioxide made by burning enough charcoal to melt the same quantity of ice. The two results agreed within the limits of error of their equipment. The warmth generated by a guinea pig digesting its food and exhaling carbon dioxide, and the heat generated by burning an equivalent amount of charcoal to carbon dioxide were the same.

We consume food: the food goes through that strange set of vessels and organs within us, and is brought into various parts of the system, into the digestive parts especially; and alternately the portion which is so changed is carried through our lungs by one set of vessels, while the air that we inhale and exhale is drawn into and thrown out of the lungs by another set of vessels, so that the air and the food come close together, separated only by an exceedingly thin surface: the air can thus act upon the blood by this process, producing precisely the same results in kind as we have seen in the case of the candle. The candle combines with parts of the air, forming carbonic acid,* and evolves heat; so in the lungs there is this curious, wonderful change taking place. The air entering, combines with the carbon (not carbon in a free state, but, as in this case, placed ready for action at the moment), and makes carbonic acid, and is so thrown out into the atmosphere, and thus this singular result takes place; we may thus look upon the food as fuel... the oxygen and hydrogen are in exactly the proportions which form water, so that a sugar may be said to be compounded of 72 parts of carbon and 99 parts of water; and it is the carbon in the sugar that combines with the oxygen carried in by the air in the process of respiration, so making us like candles; producing these actions, warmth, and far more wonderful results besides, for the sustenance of the system...

*   ... A candle will burn some four, five, six, or seven hours. What then must be the daily amount of carbon going up into the air in the way of carbonic acid? What a quantity of carbon must go from each of us in respiration!... A man in twenty-four hours converts as much as seven ounces of carbon into carbonic acid; a milch cow will convert seventy ounces, and a horse seventy-nine ounces, solely by that act of respiration. That is, the horse in twenty-four hours burns seventy-nine ounces of charcoal, or carbon, in his organs of respiration to supply his natural warmth in that time. All the warm-blooded animals get their warmth in this way, by the conversion of carbon, not in a free state, but in a state of combustion. And what an extraordinary notion this gives us of the alterations going on in our atmosphere. As much as 5,000,000 pounds, or 548 tons, of carbonic acid is formed by respiration in London alone in twenty-four hours. And where does all this go? Up into the air.... As charcoal burns it becomes a vapour and passes off into the atmosphere, which is the great vehicle, the great carrier for conveying it away to other places. Then what becomes of it? Wonderful is it to find that the change produced by respiration, which seems so injurious to us (for we cannot breathe air twice over), is the very life and support of plants and vegetables that grow upon the surface of the earth.*

—Faraday[2]

---

* The "carbonic acid" that Faraday is speaking of here is carbon dioxide. Today a solution of carbon dioxide in water is sometimes referred to as carbonic acid. As we will see in later chapters, carbon dioxide is possibly the most important chemical compound associated with energy use today.

WHY CAN'T WE EAT COAL AND DRINK OIL?

The amount of energy released when, say, sugar is eaten is the same as it would be if we burn it in the fireplace at home. The amount of energy released when we burn coal in the fireplace is the same as would be released if coal were used as food. So, why can't we eat coal? Or why can't we drink oil?

He gave out an oily-alcoholic odour; it was said of him that he drank petroleum.

—**Mann**[3]

The answer derives from the way that we digest food. There's an old saying, "you are what you eat." This may be true in a certain sense, in that the kinds and amounts of food we eat affect our weight, stamina, and health. But on the other hand, we are not made of pieces of beef or broccoli or biscuits, and milk does not course through our veins.

You may be an undigested bit of beef, a blot of mustard, a crumb of cheese, a fragment of an underdone potato. There's more of gravy than of grave about you, whatever you are!

—**Dickens**[4]

The components of foods include proteins, carbohydrates, and fats. Though they comprise different kinds of chemical families, they all undergo one particular chemical reaction after the food is eaten. In each case, the first step in the digestion process involves these food components reacting with water. The breaking apart of molecules by reaction with water is called hydrolysis. Hydrolysis of proteins produces amino acids; hydrolysis of carbohydrates, sugars; and hydrolysis of fats, fatty acids. Many subsequent reactions then convert these amino acids, sugars, and fatty acids into the new chemical components of living cells or tissues. Some of these simple compounds, especially sugars, may be used immediately by our bodies as a source of ENERGY. Others, such as fats, are stored as a reserve supply of ENERGY for future needs.

The chemical processes of digestion are facilitated by special compounds inside our bodies, called enzymes.*,† One family of enzymes is responsible for the hydrolysis reactions of fats, proteins, and carbohydrates. Enzymes interact only with very specific chemical structures. The way a key fits into and opens a lock is a

---

* In the various biochemical processes occurring in living organisms, enzymes function as catalysts. A catalyst is a substance that increases the rate of a chemical process without itself being consumed or permanently altered by that process. Enzymes are superb catalysts and can sometimes increase the rates of reactions a billion-fold. They function by changing the sequence of steps leading from the reacting substance, a carbohydrate, for example, to the product, a sugar. In the presence of the enzyme, one or more of the individual steps in the reaction changed to a new one that requires much less energy than would be needed if no enzyme were present.

† Unfortunately, not all humans possess the complete set of enzymes needed for digestion of the many things we normally eat. As an example, digestion of the sugar that occurs in milk, which is called lactose or milk sugar, is facilitated by the enzyme called lactase. Many people—especially of African, Asian, or Jewish ancestry—do not have lactase and are unable to digest milk or products made from milk. This situation, called lactose intolerance, can lead to such problems as diarrhea or stomach cramps. A solution that helps some people is to take a lactase supplement as pills or as a food additive.

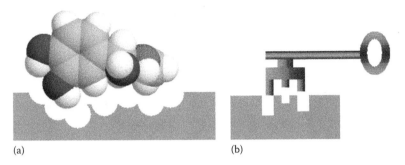

(a)   (b)

FIGURE 3.4 An enzyme has a very specific molecular configuration to accommodate the substance with which it is reacting (a) just like a lock has a specific configuration that will accommodate only a particular key (b). (From http://pandasthumb.org/archives/2007/11/i-see-red-not-q.html.)

common analogy for the way that enzymes facilitate chemical reactions. To unlock a door, a very specific pattern of notches or grooves on the key must fit exactly into the corresponding parts of the mechanism of the lock (Figure 3.4). Otherwise, the lock can't be opened. In the same way, in a reaction facilitated by an enzyme, the reacting molecule—a protein, for example—fits exactly into the geometric structure of the enzyme molecule. If it does not, the molecule won't react. Because such reactions depend so very highly on specific aspects of molecular geometry, many enzymes can facilitate hydrolysis of only one particular chemical linkage in the food molecule. Thus a host of different enzymes might be required to digest a single protein molecule.

Starch is a common carbohydrate (Figure 3.5). A molecule of starch consists of many subunits of simple sugar molecules connected to each other via chemical bonds to oxygen atoms. The linkage sugar–oxygen–sugar has a very specific

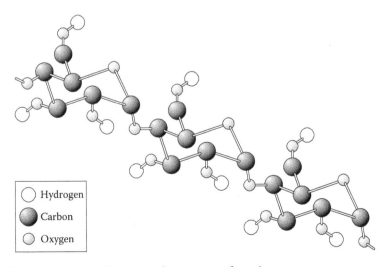

Hydrogen

Carbon

Oxygen

FIGURE 3.5 A portion of the molecular structure of starch.

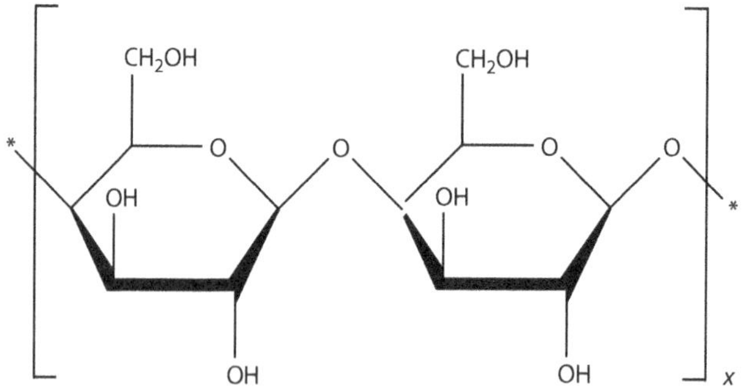

FIGURE 3.6 A portion of the molecular structure of cellulose. The key distinction between cellulose and starch (Figure 3.5) is the geometrical arrangement of the oxygen atoms linking the rings of carbon atoms.

geometrical arrangement. Without worrying about details, let's call this as linkage type *a*. Cellulose is another carbohydrate (Figure 3.6). It too consists of many subunits of sugar molecules connected via oxygen linkages. However, the sugar–oxygen–sugar linkage in cellulose has a different geometrical arrangement than the one in starch. Let's call this as *b* linkage, again without bothering about the specific details. Our bodies have enzymes that assist in the hydrolysis of the *a* linkage, but we lack the kind that will hydrolyze *b* linkages. Thus we can eat and digest starches, but not substances in which cellulose is a major constituent, such as wood or grass.

In the same way, even though coal and petroleum have elemental compositions similar to some foods and even though foods liberate the same energy when eaten or burned, we can't eat coal because we don't have the enzymes that will allow us to digest coal. For the same reason, we can't drink oil.

WHY DON'T WE CATCH ON FIRE WHEN WE EAT?

The ENERGY available from a particular substance—whether it be sugar or coal—is the same whether we eat that substance (assuming that we can digest it!) or burn it. We all know that spectacular fires can occur in the burning of fuels. Some of us have direct experiential evidence, from outdoor grills or kitchen stoves, that foods can also produce remarkable flames if accidentally ignited. We might then ask why, when the ENERGY is liberated from food in our bodies, we don't just catch on fire ourselves.

The chemical pathways of burning fuel and digesting food lead to the very same products, the important ones being carbon dioxide and water vapor. The details of these two processes, examined at the molecular level, are quite different. Burning usually involves a reaction with oxygen molecules from the surrounding air.* Digestion of a food proceeds through a sequence of biochemical reactions, facilitated by enzymes. Different chemical pathways, even though they ultimately lead

---

* In some cases, the fuel being burned will be partially decomposed by the heat of the fire, and the actual combustion reaction itself will involve the reaction of the decomposition products with oxygen. This is the case for chemically complex fuels such as wood or coal. With chemically simpler fuels, such as natural gas, molecules of the fuel itself will react with oxygen.

to the same products, proceed at different rates. Digestion is much slower than burning. A bit of fat accidentally ignited on your stove top burns quickly; that same amount of fat taken into your body may require hours to digest. The total ENERGY released may be the same, the products may be the same, but the rates of these processes are very different.

The slow release of ENERGY from digestion helps keep our normal body temperature at 37°C. The rapid rate of combustion might lead to a fireball with a temperature of hundreds or thousands of degrees. That is, there is a critical contribution from the rate at which ENERGY is produced. The difference in rates of burning and digestion has an important implication for human activity. The relatively slow release of ENERGY from digestion means, in effect, that the amount of WORK we can do in a given time—per second, per minute, per hour—is limited. The amount of WORK done in a given time is POWER. Since the WORK we can do in a given time is limited, and therefore so is our POWER. Put another way, we humans are not very POWERful.

Imagine having the job of clearing a field of rocks. Let's assume that it's a bit of a strange field, so that all the rocks are the same size, say 10 kilograms. Suppose that, to get rid of the rocks, you need to pick them up and carry them a hundred meters (the minimum length of a soccer field for international play). Allowing time for trudging back and forth, by the end of a day, you'll have shifted a hundred of these rocks. A hundred 10 kilograms rocks constitutes a tonne of rock. Your POWER output for the day is a tonne per eight hours (which is 480 minutes). Few of us would need to be on this job very long before getting the idea that there must be a *much* easier way to move these rocks than picking them up one by one and carrying them a hundred meters. Why not load them all in the back of a truck and drive the truck a hundred meters (Figure 3.7)?

Suppose it takes the truck a minute to go the hundred meters. This is equivalent to an average speed of only six kilometers per hour. In comparison to what we could do with our bodies, the truck takes only 1/480th of the time; from the other perspective, the truck is 480 times as powerful as a human (Figure 3.8).

A human is not as powerful as a truck. From the definition of POWER (Chapter 2), a human therefore does not use ENERGY as fast as a truck. In a human, the ENERGY source, food, is digested (converted to ENERGY) rather slowly, and we of course do not catch on fire when we eat. In a truck, the ENERGY source, gasoline or diesel fuel, is consumed relatively quickly and does burn. Fast consumption of ENERGY leads to greater POWER.

Of course, this example is contrived. What kind of field has rocks of identical size? None. Wouldn't it make more sense to carry several rocks at a time? Certainly. Why carry every single rock exactly a hundred meters? No good reason. The value of this contrived example is that it illustrates a process of arriving at an estimated solution to a problem, or a process of reasoning through a situation, by making a simple set of assumptions and using numbers that are easy to manipulate, even mentally. This approach to solving problems is sometimes referred to as a "back-of-the-envelope" calculation. The term implies that we can work out a solution without needing a calculator or computer, simply by jotting a few numbers on the back side of an old envelope.

Suppose that instead of a hundred 10-kg rocks, there was only one 1-tonne rock to be moved. Nobody is going to pick that up and move it. Even if you had a truck of

FIGURE 3.7 This large truck is a lot faster and easier way to do WORK than relying on our own muscles. This mining scene is near Kemerovo, Siberia. (From http://commons. wikimedia.org/wiki/File:Dump_truck_in_Russia_gets_load.jpg.)

FIGURE 3.8 The comparative sizes of the person and loader in this photograph also indicate their relative amounts of POWER. This photograph is from a mine in New South Wales, Australia. (Permission courtesy of Coal Age magazine, E&MJ/Coal Age-Mining Media.)

the right size, you would still be confronted by the 1-tonne rock. How could you get a rock that big into the back of a truck? One way might be to construct a long, gently sloping ramp,* and slowly roll the rock up the ramp into the truck. The ENERGY required to move the rock is exactly the same whether we lift it in one mighty, gorilla-like yank, or whether we slowly roll it, centimeter by centimeter, up a gentle slope. However, the POWER needed is much, much less in the second case—because it takes a greater amount of time to use that ENERGY.

The pyramids of ancient Egypt represent the most remarkable example of engineering construction almost entirely due to human ENERGY (Figure 3.1). No one knows for sure how the pyramids were actually constructed. Though some ludicrous suggestions have been made that the pyramids were constructed by ancient astronauts from outer space, or other "New Age" nonsense, it seems clear that the pyramids were built with no other ENERGY source than human muscles. The Great Pyramid of Khufu (a ruler also known as Cheops) took about 23 years to build. The construction involved about 1250 workmen working for the entire 23 years. The stone blocks also had to be quarried and hauled on sledges—pulled by men—to the construction site. Some of the blocks weigh fifty tonnes. Overall, the combined effort in quarrying and transporting the stone, and then building the pyramid may have involved 10,000 men. The pyramids remain an enduring testimonial to the ability of the human body to do WORK.

Slavery represents the negative aspect of the problem of using humans to do WORK. Slaves have been used throughout most of human history to provide a variety of basic services, some amenities, and even luxuries. Ancient Rome was one of the societies most dependent on slaves, including for such fundamental activities as mining and maintaining the water supply. In the relatively recent history of the United States, slaves were used as the foundation of the agrarian society of the southern states. Sadly, the institution of slavery, though poorly and incompletely documented, is believed to exist even now in parts of the Arabian Peninsula, southeastern Asia, and Africa.

We needn't do a back-of-the-envelope calculation to realize that we humans are not as powerful as a large truck (Figure 3.8). Long, long before there were trucks or other sorts of machinery, early humans learned to harness (literally) many kinds of animals to use the ENERGY of their muscles rather than our own. We are not as powerful as the various kinds of animals that were pressed into service for doing WORK. For heavy, slow jobs, the primary animal of choice was the ox. Depending on the specific job, an ox can exert about ten times as much force as a human. Horses were preferred for lighter work, but in situations in which speed was important. The horse can exert about the same amount of force as an ox—at least for a short time—but can do the work much faster. In terms of POWER, the ox may be some seven to eight times as powerful as a human, but the horse is some ten to twelve times as powerful.

We have made use of plows for at least six thousand years. Probably the earliest plows were pulled by humans, but eventually systems of harnesses were developed

---

* Chapter 2 mentioned three key inventions of ancient humans to help accomplish WORK—the lever, pulley, and inclined plane. The ramp used in this hypothetical situation to help load the rock into the truck is a classic example of an inclined plane.

FIGURE 3.9 The use of human muscles for doing WORK is illustrated in this picture of workers from ancient times. (From http://commons.wikimedia.org/wiki/File:Maler_der_ Grabkammer_des_Rechmir%C3%AA_002.jpg.)

to allow animals to pull the plow. Another crucial contribution of animal ENERGY to early agriculture was the use of oxen, donkeys, or camels hitched to waterwheels to provide irrigation water for fields. At one time or another, in various parts of the world, other large animals such as elephants, llamas, and buffaloes have also been used to do WORK for us.

Until about 1700, human and animal muscles remained the dominant way of doing WORK in society (Figure 3.9). A major drawback of this ENERGY source is that the ENERGY produced, or the POWER output, must be applied within a very short distance of its source (i.e., of the human or animal doing the WORK). A society of this kind is sometimes referred to as a "low-energy society." Low-energy societies are often very stable societies, but also tend to be resistant to change.

WHERE DOES THE ENERGY IN FOOD COME FROM?

We use food as a source of ENERGY. So far, we've neglected the issue of where the ENERGY in food comes from in the first place. Many of us eat a varied diet from both plant and animal sources. Animals that we consume as food obtained their nourishment by eating plants. We also eat plants or their products—vegetables, grains, nuts, fruits, or berries. Whether we have a mixed diet, a vegetarian or vegan diet, or an all-meat diet doesn't really matter. One way or another, all of our foods derive ultimately from plants. So, the question of where the ENERGY in food comes from can be narrowed to asking where the ENERGY in plants comes from. It comes from the sun.

The crucial process in the growth and development of plants is *photosynthesis*. Plants absorb carbon dioxide from the atmosphere and, by using ENERGY from sunlight, combine it with water to form simple sugars. The overall process can be represented by the equation

$$6CO_2 + 6H_2O \rightarrow C_6H_{12}O_6 + 6O_2$$

where $C_6H_{12}O_6$ is the molecular formula of a simple sugar molecule.[*][†] This is likely the most important chemical equation in the world. The growing plant uses the sugar molecules to form all of the many other chemical substances needed for the growth and life processes of the plant. The prefix "photo-" signals us that light is somehow important in the photosynthesis process. Indeed, photosynthesis cannot take place in the absence of sunlight.

Energy from the sun drives the photosynthesis process. Photosynthesis produces the chemical energy source that plants need for their life processes. We eat plants (or we eat the animals that ate the plants) to obtain the ENERGY for our bodies to function. In other words,

Human ENERGY ultimately derives from solar energy.

Unfortunately, we humans are not very efficient solar energy converters. The atmosphere absorbs about 20% of the solar energy reaching the Earth (another 30% is reflected back to space immediately, before ever getting into the atmosphere). Then, plants convert only about 0.2% of that ENERGY into the substance of the living plants. We consume only a tiny fraction of the plant or animal population on Earth. Overall, our human ENERGY represents about two millionths of the solar energy received on Earth.[5]

Since human ENERGY derives from foods, changes in diet can have a potentially significant impact on society. For example, the late Middle Ages—particularly the twelfth century—were characterized by a tremendous flourishing in many areas of human activity, including architecture, politics, mechanical invention, and literature. Some historians have related this flourishing to the increased consumption of beans (and similar vegetables of the legume family) about that time. The same change is believed to have affected the civilization of the Anasazi people[‡] around the year 600.

---

[*] The ability of plants to remove carbon dioxide from the atmosphere in this way is a vital component in the movement of carbon around and through Earth's atmosphere, water systems, and living organisms. We will revisit this important role later, when we discuss the buildup of carbon dioxide in the atmosphere and the serious implications that buildup has for the global climate.

[†] Though photosynthesis can be represented in a simple way by this equation, the detailed chemical pathway by which carbon dioxide is converted to sugar is extremely complicated. Unraveling the complex chemistry of photosynthesis resulted in the American chemist Melvin Calvin (1911–1997) receiving the 1961 Nobel Prize in Chemistry for his work in unraveling the complex pathways of photosynthesis.

[‡] The Anasazi civilization flourished in what is now the southwestern part of the United States for about 1200 years, from 100 A.D. to 1300. Their culture survived until the coming of the Spanish occupancy around 1600. Their name derives from a Navajo word meaning "ancient ones." Their modern descendants are the Hopi and Zuni. An excellent essay on this remarkable culture, "Searching for Ancestors," can be found in the collection *Crossing Open Ground* by Barry Lopez (Vintage, 1989).

Beans provided an excellent supplement to a diet that was based largely on cereals, because beans are good sources of vitamins and the essential nutrients* that are not available in high concentrations in the cereals. The improved nutrition led directly to improvements in health and longevity.

REFERENCES

1. Darwin, C. *The Voyage of the Beagle*. Modern Library: New York, 2001, p. 304.
2. Faraday, M. *A Course of Six Lectures on the Chemical History of a Candle*. Chatto and Windus: London, U.K., 1861. Modern paperback editions of this book are available.
3. Mann, T. *Buddenbrooks*. Vintage: New York, 1952.
4. Dickens, C. *A Christmas Carol*. McElderry: New York, 1995. There are many other editions of these four excellent books available, including relatively inexpensive paperbacks.

FURTHER READINGS

Caret, R.L.; Denniston, K.J.; Topping, J.J. *Principles and Applications of Inorganic, Organic, and Biological Chemistry*. Brown: Dubuque, IA, 1997. This is an introductory textbook in college-level chemistry mainly—but not exclusively—intended for students majoring in health sciences. Many of the chapters of this book, notably 22 and 23, provide a much more detailed view of the chemistry of respiration, metabolism, and human energy production.

Dikötter, F. *Mao's Great Famine*. Walker: New York, 2010; Keneally, T. *Three Famines*. PublicAffairs: New York, 2011; Reid, A. *Leningrad*. Walker: New York, 2011. These three books document the horrific effects of famine on human energy, health, and society. Any one of them should help the reader realize the good fortune of living in any country where food is relatively abundant.

Faraday, M. *The Chemical History of a Candle*. Collier: New York, 1962. This is one of the classic examples of scientific explanation for persons who are not professional scientists—one of the great books of science. It is seldom out of print, and available in editions other than the specific one cited here. Lecture VI discusses the analogy between the burning of a candle and the 'burning' of food in the body.

Gebelein, C.G. *Chemistry and Our World*. Brown: Dubuque, IA, 1997. This excellent introductory chemistry textbook has two useful chapters on the chemistry of food and of digestion processes. Chapter 15 provides a discussion of the chemistry of food.

Mann, C.C. *1493*. Alfred A. Knopf: New York, 2011. This fine work of history has extensive discussions on slavery and why, for example, slaves were preferred to indentured servants in 17th century America. There is also a discussion of silver and mercury mining, centuries before Darwin's observations, and of the extraordinarily brutal working conditions.

Morton, O. *Eating the Sun*. HarperCollins: New York, 2008. This excellent book provides a discussion of photosynthesis, the origin of plant life on Earth, and how plants affect both human energy and other energy sources on which we depend.

Smil, V. *Energy in Nature and Society*. MIT Press: Cambridge, MA, 2008. Chapter 5 of this book discusses human energy, with an enormous wealth of detail. A fine resource for persons wishing to dig deeper into the topic.

---

* Beans are a good source of B vitamins, which have various roles in metabolism. Deficiency of B vitamins can lead to eye problems, dermatitis, beriberi, and pellagra. "Essential" nutrients are ones that must be included in the diet because they cannot be made from other chemicals in the body. Beans provide the essential nutrient lysine, which is one of the building blocks of proteins.

4 The Energy Balance

The last chapter introduced the concepts that the source of ENERGY for the human body is food, and that foods consumed for ENERGY have many points in common with the burning of fuels. Now we will build on these ideas to develop two new concepts. The first is that we can keep track of ENERGY in the same way as we keep track of money and follow changes in ENERGY similarly to balancing a checkbook or bank account. The second is that, when ENERGY is used, some of it is wasted rather than being converted to useful WORK.

We'll start with an analogy. Suppose that you get a hundred dollars (as a paycheck, a gift, or some other source), and suppose that you buy an item that costs fifty dollars. Now, what happened to the other fifty dollars? Of course it didn't disappear or "evaporate"; this is money that you have saved, whether by keeping it on your person or by actually putting it into your bank account. You have stored that remaining fifty dollars someplace. If we express this relationship mathematically, we could write

$$\text{Money } in - \text{Money } used = \text{Money } stored$$

But now let's take another example. Suppose that again you get a hundred dollars from some source, and you want to buy some item that costs $125. This transaction is not possible unless you are able to come up with the "missing" $25. In this example, the "money stored," as would be calculated from the simple equation shown earlier, is actually a negative number, −$25. What does the negative number mean? The "missing" $25 has to be provided by removing it (as implied by the minus sign) from the money you have saved or "stored" earlier; in other words, the total amount of money you have in savings goes down.

This is a very straightforward approach to keeping track of personal finances. Many of us develop an intuitive understanding of the relationships among the amount of money we receive, the amount that we use, and what we have saved as soon as we are old enough to be entrusted with money of our own. So why is it worth considering such apparently simple examples? Because the situation with human ENERGY is exactly analogous:

$$\text{Food energy } in - \text{Food energy } used = \text{Food energy } stored$$

Before we examine the possible solutions to this new equation, we might ask how food ENERGY is stored in our bodies. This storage occurs in several ways. Short-term ENERGY storage (analogous to money we might keep in a checking account for almost immediate use) is accomplished by storing carbohydrates in the liver. Our long-term reserves of ENERGY, akin to investments of money, are established because our bodies store up a reserve supply of ENERGY as fatty tissue or fat.

There are three possible general solutions to this equation. First, consider the case where the "food energy in" is exactly equal to the "food energy used." Regardless

FIGURE 4.1 The eventual effect of storage of energy in the body as fat, when "energy in" far exceeds "energy out."

of the numerical values of these terms, subtracting two identical numbers always leads to a difference of zero. In other words, in this case, "food energy stored" is zero. We have neither increased nor decreased the amount of fat stored; in practical terms, and therefore, we would neither gain nor lose weight. Second, suppose that the amount of food energy taken in is greater than the amount used. That means we are increasing the amount of ENERGY stored in our "food energy bank" in the same way (but unfortunately not in the same form!) as we store money in our financial banks. We see this storage manifested as a gain in body weight. In extreme cases, this continuous storage of fat leads to serious obesity (Figure 4.1) that, in turn, can have serious health effects. The third solution is the case in which the ENERGY that we use for doing WORK is greater than the amount we take in. Now the "food energy stored" term becomes a negative number. The only way that we can account for doing this much WORK with less ENERGY taken in is by "withdrawing" ENERGY from our "energy bank." This is analogous to the example in which in order to purchase a $125 item with a hundred-dollar paycheck, we must withdraw $25 from the money stored in our financial (savings) bank. In fact, the analogy goes further. There are people in the world who are in such dire poverty that this sort of financial transaction is not possible, because they simply do not have the additional $25 saved someplace. Tragically, people who are in the last stages of starvation (Figure 4.2) may no longer have extra food energy stored in their bodies; for them, it would be impossible to do more WORK than they can "pay for" with food energy taken in that day. When the

FIGURE 4.2   These survivors of the Dachau concentration camp are persons suffering from starvation. (Figure courtesy of Francis Rbert Artz Merle Spiegel, United States Holocaust Memorial Museum.)

WORK to be done means getting up from bed to obtain food and even that seemingly little bit of WORK is impossible, then death is inevitable.*

The most practical application of the food energy balance is that it allows us to regulate our body weight. The ENERGY contents of various food items can be measured in a calorimeter (Chapter 3). Tables of this information are readily available; extensive charts or tables can be found on various websites. Table 4.1 provides examples of the kind of information that can be found.

Doing various kinds of WORK requires ENERGY. Information on this is also widely available, especially on the web, with examples shown in Table 4.2.

Anyone can use information of this kind to regulate his or her body weight. Tables 4.1 and 4.2 provide the numerical data for solving the food energy balance equation. The outcome tells whether that person has added to ENERGY in storage, withdrawn ENERGY from the body's energy bank, or stayed about the same.† We can keep track of the human ENERGY that we use, and the human ENERGY that we take in as food, in exactly the same way that we keep track of our personal finances.

---

* As horrific as this situation is, it can get even worse. As an example, during the siege of Leningrad (now Petersburg) in the former Soviet Union during World War II, food was severely rationed. When there was no one left in a family who had enough ENERGY to go out, stand in a queue, try to obtain food, and get it back home, then the entire family was doomed.

† Unfortunately, the situation is not quite so simple. Because fat is produced as the body's *long-term* energy storage medium, initially the body is likely to use up some of its short-term energy stored in the liver. Many days or weeks of careful exercise and watching one's food intake might be needed to effect a noticeable withdrawal from the long-term storage and a corresponding weight loss. Also, weight can be affected by the body's water balance—perspiring a great deal or drinking large quantities of liquids can produce temporary weight fluctuations.

TABLE 4.1
Energy Contents of Some Common Foods

Item	Weight grams	kilojoules	Nutritional Calories
Apple	100	150	35
Bacon, fried	40	1975	475
Beef sirloin	120	1180	285
Bread, brown	25	1015	240
Cheese soufflé	100	1030	250
Cod, fried in batter	100	835	200
Dates, dried	35	1055	250
Doughnut	40	1465	350
Egg, fried	60	960	230
Frankfurter	100	1135	275
Grapefruit	200	45	10
Macaroni and cheese	180	725	175
Meat pie	180	960	230
Olives	20	340	80
Peanut butter	20	2580	625
Popcorn, plain, no salt	15	1605	380
Rice, boiled	160	1560	370
Salami	90	2030	490
Scone	30	1500	370
Tuna, canned in oil	120	1200	290

Note: Keeping track of the amount of each food eaten during the course of a day gives us a numerical value for "food energy in."

TABLE 4.2
Energy "Cost" of Various Forms of Work or Sport per Hour of Activity

Body Weight, kg (lb)	55 (120)		90 (200)		125 (280)	
Energy Used	kJ	cal	kJ	cal	kJ	cal
Chopping wood	1155	276	1925	460	2694	644
Cycling (16 km/h)	1355	324	2259	540	3163	756
Farm work (heavy)	1607	384	2678	640	3749	896
House cleaning	803	192	1339	320	1874	448
Jogging	2109	504	3514	840	4920	1176
Sawing wood	1657	396	2761	660	3866	924
Swimming	1757	420	2929	700	4100	980
Walking	1205	288	2008	480	2812	672

But there is really nothing special about food ENERGY; burning sugar as a fuel would liberate the same amount of ENERGY as we obtain when we eat sugar. In fact, the importance of understanding the balance of food ENERGY is that it is perfectly and generally applicable to any kind of ENERGY. This is an important and critical concept.

Before we explore this point in more detail, it will help to digress to introduce the concept of SYSTEM.

> A SYSTEM is whatever portion of the universe is under our specific observation or study.

For example, when we are concerned about balancing human energy, the SYSTEM is our body. In a calorimeter, the SYSTEM is the sample and its container, the water, the thermometer, and the insulated box. In exactly the same way as we can speak of balancing our checkbooks or bank accounts, or seeing whether we've gained or lost weight, we can speak of the *energy balance* for a SYSTEM. Let's use the symbol E to represent any type of energy. Then we can speak of the energy balance as being

For any SYSTEM, $E_{in} - E_{used} = E_{stored}$

If more ENERGY is put into a SYSTEM than is taken (or comes) out, then some has to be stored inside the SYSTEM. On the other hand, if we take more ENERGY out of a SYSTEM than we put into it, then some of the ENERGY that had previously been stored in the SYSTEM must have been used. What if there is no ENERGY previously stored in the SYSTEM? This transaction is impossible—and analogous to a person dying of starvation. This leads to a general, and extremely important, statement:

> If no ENERGY is stored or "hidden" in a SYSTEM, a process that produces more ENERGY than was originally put in is impossible.

Many of us have experienced being outdoors on a cold winter day and find it comfortable to do a bit of exercising, because doing so helps us to feel warmer. We also know that in the summertime, we can become uncomfortably hot when doing a lot of WORK outdoors. (Perspiring—Figure 4.3—is one of our body's mechanisms for trying to get rid of this excess heat. In extreme cases, our body can become so over-heated that it "shuts itself off," i.e., we can pass out from heat stroke.) These same effects occur with inanimate objects. After driving a car for a while, the hood will feel perceptibly warm. An electric motor running steadily will also feel slightly warm.

But what does your body's getting hot have to do with helping you do WORK? What does the hood of a car getting warm have to do with the WORK of moving the car down the road? What does the casing of an electric motor getting warm have to do with helping it turn a fan (or whatever WORK it's doing)? The answer to every one of these questions is exactly the same: absolutely nothing.

FIGURE 4.3 Perspiration is one of the body's ways of dealing with waste heat generated as ENERGY is used to do WORK, as shown in this photograph of a marathon runner. (Photograph courtesy of Reece Forrester, Shakopee, Minnesota.)

Consider the cases of doing some exercise to stay warm outdoors in the winter or of getting "overheated" when working outdoors in the summer. Where does this heat come from? Not from the sun, since these phenomena can certainly occur on cloudy days or at night. The only source of this heat is the ENERGY of our bodies, which we know comes from food. So far we've emphasized that food ENERGY is converted to WORK. But this common observation of becoming warm or overheated actually suggests a process that we could write as

$$\text{ENERGY} \rightarrow \text{WORK} + \text{Waste} \left(\text{as heat}\right)$$

This relationship is one of the most profound observations we can make about the world and how it works, so profound and so thoroughly pervasive in everything we do that rarely do we even think about it. One way that we can express this relationship is by the following:

Whenever ENERGY is converted into WORK, a portion is wasted (usually as heat).

A simple example is illustrated in Figure 4.4; here most of the ENERGY put into turning the drill is converted to waste heat—enough to ignite a fire.

FIGURE 4.4   This young man is converting the WORK supplied by his hands and arms into ENERGY in the form of heat. With lots of WORK and patience, it is possible to start a fire in this way. (Courtesy of Steve Sanford.)

An alternative way of saying the same thing is this:

ENERGY cannot be converted completely to WORK. Some ENERGY is inevitably wasted.

The ENERGY of our bodies is obtained from the food we eat. As we've seen in the last chapter, the energy in food ultimately derives from solar energy, through the vital process of photosynthesis. When we use our human ENERGY, a portion can be converted into WORK, and the remainder appears as waste heat. Where does that energy that is apparently wasted go? A portion of the heat can be absorbed by nearby objects, but eventually it becomes radiated away from Earth into outer space. There's no way for us to get it back or to make any further use of it. Essentially the waste heat that we generate becomes a part of an enormous reservoir of heat energy in the universe—but the crucial point is that such energy is no longer available to us.*

---

* The fact that the energy is no longer available to us is not because we lack the appropriate kind of spaceship to fly somewhere and bring it back. Rather, the reason is much more fundamental. The kind of energy associated with heat can only be made to move (in other words, used) if there is a difference in temperature between two objects. The waste heat radiated away from the Earth becomes part of an unimaginably vast reservoir of heat energy in the universe, which happens to be at uniform temperature. There is no way it can ever be used again.

Our planet is home to more than six billion people, nearly two billion cars and light trucks, and thousands or millions of factories, airplanes, railway locomotives, heavy trucks, ships, and electric generating plants. All of these things, including us, generate heat every day. If a mechanism did not exist to get rid of excess heat, Earth would have become uninhabitable long ago. But now we might finally be on the brink of doing just that, because we are changing the composition of the atmosphere in such a way as to trap more of this heat on Earth, rather than allowing it to radiate away into space. This problem is popularly known as the greenhouse effect. We will examine it in detail later in this book and will find that an understanding of the concept of the energy balance is one of the keys to understanding the greenhouse effect.

Recall another aspect of our analogy of energy and money: ENERGY is the capacity to do WORK, and money is the capacity to spend. Spending involves the transfer of money, and doing WORK involves the transfer of ENERGY. The ENERGY that can easily be used to do WORK is ordered, while heat, which is not so easy to use, is disordered. What does this mean? Imagine lifting a heavy box of books from the floor to a table. As you do WORK on the box, the box and its contents are moving in a single direction (up). This is an example of ordered energy. If we had a microscope so powerful as to let us watch the motions of isolated atoms and used it to examine a hot object, we would see that the individual atoms are also all moving—but not in any concerted way all in the same direction; they would be jumping all around in all sorts of directions. For this reason, we speak of heat as being disordered energy. We all have abundant information from daily experience that it's easy to disorder something. It's easy to make a mess. We also know that it's hard to put order into something; given a messy room, or house, or lawn, or car, it takes effort to convert the disorder in the mess back into order. In other words, if we've created a mess, it's hard to undo what we have just created. It is relatively easy to convert ordered energy into disordered energy (e.g., waste heat). It is difficult to convert disordered energy— heat—into ordered forms of energy.

FURTHER READINGS

Atkins, P. *Four Laws*. Oxford University Press: Oxford, U.K., 2007. A superb introductory book on the fundamental science of heat and energy. Chapters 2 and 4 are relevant to the discussion here.

Brown, L.; Holme, T. *Chemistry for Engineering Students*. Brooks/Cole: Belmont, CA, 2011. Chapter 9 of this introductory text discusses energy, heat, work, and the unavoidable formation of waste energy.

Krogh, D. *Biology: A Guide to the Natural World*. Pearson Prentice Hall: Upper Saddle River, NJ, 2005. Unit 2 of this introductory text discusses the importance of energy in living organisms, including us, and how energy is derived from food.

Smil, V. *Energies*. MIT Press: Cambridge, U.K., 1999. Chapter 3, on "People and Food," discusses, among other topics, human energy and how it is used in various kinds of activities such as walking, running, and doing various sorts of work.

Smil, V. *Energy in Nature and Society*. MIT Press: Cambridge, U.K., 2008. Chapter 5 discusses energy use by humans, backed up by a vast array of facts and data.

5 Fire

The heat produced in a fire was the first ENERGY source used by humans, other than the muscles of human or animal bodies. A fire involves the chemical process of combustion, the reaction of some substance—the fuel—with oxygen from the air. As this reaction takes place, heat and usually light are produced, along with the chemical products of the reaction.

Fire was either the first or the second chemical reaction to be exploited by humans.* The development of an understanding of how to use and control fire ranks with the invention of stone tools as one of the crucial steps forward in the development of early humans. Chapter 3 gives a brief historical overview of the relationship of humans and our hominid ancestors with fire. No one knows exactly how or where our ancestors first discovered fire and developed an appreciation of the benefits of the ENERGY produced in a fire. Very likely this happened independently at many places around the world at different times. Some anthropologists suggest that the knowledge of fire may have been discovered, lost or forgotten, and then rediscovered. For the past several hundred thousand years, every culture has used fire, although likely not all knew how to make a fire as needed. Remarkably, the pygmies in the tropical rain forest on the Andaman Islands (off the south coast of Burma) have, until recent years, carefully tended fires ignited by lightning or other natural sources, because they knew no methods for making a fire when they needed it.

Most of the peoples of the world have a myth or legend about how they first acquired fire. Since these early cultures were not likely in contact with one another, the worldwide prevalence of fire myths provides evidence supporting the idea that fire was discovered independently in many places around the world. Examples are the Prometheus legend of the ancient Greeks and the similar legend of the ancient Vedics about Pramantha. In the Greek version, Prometheus was the person who stole fire from the gods for the benefit of mankind (Figure 5.1). Prometheus did this because he took pity on the sorry state that humans were in without fire.

The importance of fire to primitive humans is reflected in many ways. It is used in the ceremonies of many religions. All of us, sooner or later, discover for ourselves that a fire can cause a painful burn. Thus, in some religions, fire is a symbol of divine punishment. If not fire itself, then a symbol of fire, such as candles, might be used in ceremonies. Various religious and ceremonial rituals involve the concepts of the "sacred flame" or the "eternal flame." For a long period of human history, it was of utmost importance that a fire, once started, be kept going, an importance that is also reflected in many cultures. According to legend, when the Greeks destroyed Troy, the hero Aeneas (supposedly the son of Aphrodite and the greatest of the Trojan

---

* Besides fire, the other chemical process exploited by early humans must have been fermentation.

FIGURE 5.1 Prometheus bringing fire to humankind, an 1817 painting by the German artist Heinrich Füger (1751–1818). (From http://ru.wikipedia.org/wiki/%D0%A4%D0%B0%D0%B 9%D0%BB:Heinrich_fueger_1817_prometheus_brings_fire_to_mankind.jpg.)

heroes after Hector) carried the sacred fire from Troy to Italy, where it became preserved in Rome. Ancient Romans worshipped the goddess Vesta, an adoption of the earlier Greek goddess, Hestia. Worship of Vesta led to the cult of the Vestal Virgins, women consecrated to the service of Vesta and charged with keeping the flame perpetually burning. In the Jewish religion, the lighting of one candle from another in the eight-day celebration of Hanukkah also indicates the importance of perpetual fire. Vestiges of these concepts survive in our modern times, as, for example, in the "eternal flame" around graves (such as for the tombs of unknown soldiers, and in the Olympic torch). The importance and religious significance of an eternal flame may derive from the considerable difficulties that primitive humans must have experienced in igniting fires at will.

EARLY USES OF FIRE

Though nowadays we think of fire as something that destroys, that burns things up, in many primitive cultures, fire was regarded not as a destroying agent, but as a transforming agent—something that causes important changes. Most primitive cultures worldwide used fire for the same purposes: to keep warm; to drive off animals, especially predators; to clear woods and forests for agriculture; to cook; to dry and harden wood and clay; and to heat and split stones to make them easier to move when clearing land or digging wells or mineshafts.

Humankind's first contact with fire was likely with those ignited by natural phenomena, such as lightning or volcanoes. In prehistoric Japan, for example, the Ainu people worshiped a fire goddess who, they thought, lived in a volcano. The steps in learning how to use fire could have come from observing natural phenomena. Reducing the amount of fuel available can control fire, or it can be invigorated and spread by blowing on it (as the wind does). Fire can be put out by water, as the rain does. Perhaps even the technique of lighting fire by rubbing pieces of wood together (Figure 4.4) may have been learned by watching tree branches being rubbed together by wind.

The first use of fire might have been for safety, such as to frighten wild animals away from caves or other shelters. Probably the second use of fire was to provide warmth. It may be that the gathering of humans, or our prehuman ancestors, around the fire to keep warm and safe would have increased bonding among the group. This bonding would be essential to get the group working cooperatively to kill larger, stronger animals for food, to organize migrations, or to make plans for planting crops. Such bonding around the fire could also have contributed to the development of language.

In temperate and cold climates, the heat from the fire was vital. We humans are much more susceptible to cold than are most animals. (Consider animals that survive outdoors through harsh winter conditions; few of us could do that.) It is likely that simply having a warm, dry shelter improved the health of early humans and our hominid ancestors. There appears to be a collective subconscious urge in the human psyche to get warm, or stay warm, when we are ill. Usually when we are sick (unless with a very high fever), many of us feel better—at least emotionally or psychologically if not physically—wrapping in blankets or putting on warm clothes. Small wonder, then, that in many places, the name for the most common ailment that afflicts us is a *cold*.

Many early religions reflect the importance of cooking and warmth to human well-being. Prehistoric Japanese actually worshipped cooking ovens. The Greek goddess Hestia and her Roman counterpart Vesta were the patronesses of fires in homes and in temples. Even nowadays, we sometimes hear the expression "hearth and home" describing the comforts and pleasures of home and family life; in this expression, it's the hearth—the floor of the fireplace or cooking area—that gets precedence.

Fire also provides light. Archeologists have found primitive lamps, small bowls made of soft stone in which animal fat could be burned. Some of the remarkable wall paintings in caves dating from the last Ice Age were done in parts of the caves that daylight could not reach, suggesting that lamps must have been used to help make these paintings.

The next major use of fire probably by humans is cooking. In legend attributed to the Australian Aborigines, bodies of kangaroos killed in a tribal hunt had been put in a pile. Lightning ignited the surrounding grass. Some brave soul then discovered that cooked kangaroo tasted a whole lot better than raw kangaroo. Cooking may even have appeared to be magical to primitive peoples. Though nowadays some nutritionists recommend eating much of our food, especially fruits and vegetables, raw, many ancient cultures have stories, admonitions, or proscriptions about the virtues of cooked food relative to raw food. Cooking can destroy harmful germs or other parasites. Cooking provides a greater range of tastes. Cooking makes more substances potentially available to us as foods, because some plants are inedible or poisonous if used raw. In addition, cooking provides ways of preserving food. The process of smoking, especially of meat, is a preservative technique. Examples are the jerky prepared by Native Americans and bakkwa produced in China. This smoking technique may have been one of the first benefits of early humans' use of fire in food preservation. Cooked food can also be preserved if sealed away from air. Even today people do this, in the process of home canning. Primitive techniques involved storing cooked food in the bladders of animals that had been killed or in hollow sections of bamboo that were then sealed with tallow.

The ability to preserve food represented a major step forward for humankind. Storing preserved food eliminates the cycle of "feast or famine." Before there were ways of storing food, people had to eat as much food as they could while it was available (and before it spoiled) and then go hungry in those times when none was available. Preserving food frees people from the constant foraging and hunting and provides time for other activities, such as making tools, utensils, and other artifacts.

Directly related to cooking itself is the use of fire for hardening wood. Humans do not have the strong canine teeth and jaw muscles of carnivorous animals, so we are not as good at chewing up raw or partially cooked carcasses of food. Sharpened and fire-hardened sticks help to tear up raw or partially cooked meat. Probably it did not take much for someone then to recognize that the same sharpened and fire-hardened sticks could be used to tear up carcasses on the hoof—that is, as hunting weapons.

Two other discoveries, both probably accidental, arose from the use of fire. Someone (again, most likely more than one person at different times in different parts of the world) noticed that parts of the soil around the fire were converted into fairly hard, watertight material. This process takes place with clays, which are plentiful in the soils of most parts of the world. Primitive ceramics produced whenever a fire happened to be built on a patch of ground containing abundant clay could be used for making cooking vessels, containers for storage of foods, or even containers for transporting preserved foods when the group moved from one area to another. Prehistoric clay vessels from Japan, found on the island of Honshu, are the oldest known pottery, produced about 13,000 years ago.

Somewhere, sometime, ores of metals such as tin, lead, or copper fell into a fire. Pieces of metal were found in the ashes. These observations led, however gradually, to the deliberate production of metals and then of alloys.* Metals could be used as tools for manufacturing things, as farming implements, and, of course, as weapons. The discovery of metals and their use marks the transition in human history from the Stone Age to the Bronze Age. Metals are today so common in our everyday lives that we lose sight of the fact that this discovery was one of the greatest of technical leaps forward in human history.

---

* An alloy is a mixture of two or more metals, in many cases a "solid solution" of one metal in the other. Alloys are usually produced deliberately so that some desired property of the alloy is superior to that of its constituent metals. In ancient times, the most important alloy was bronze, which is produced from copper and tin. Bronze is harder than pure copper, yet melts at a lower temperature, making it easier to cast. Bronze is more resistant to corrosion than iron. Nowadays the most important family of alloys is the steel alloys.

FIGURE 5.2 Fire figured prominently in the work of the alchemists, forerunners of today's chemists, as shown in this painting by David Teniers the younger (1610–1690). (From http://commons.wikimedia.org/wiki/File:L%27alchimiste_-_David_Teniers_the_Younger.png.)

It became appreciated that fire was not only something for domestic applications of cooking, lighting, keeping animals at bay, and warmth. Fire also was a medium whereby raw materials—clay minerals in the soil and ores of metals—could be changed into useful new products (Figure 5.2).

Where rock was an obstacle in clearing land for farming, for digging wells, or in early efforts at mining, it could be broken apart by a technique called fire-setting. A fire was started on, or around, the rock to heat it. Pouring cold water on it could then splinter the hot rock. The rapid cooling of the rock caused a contraction that set up very severe internal forces that would shatter the rock. In medieval Europe, miners would leave fires burning over a weekend to get rock hot enough to shatter by pouring cold water on it when the workweek began.

THE PROCESS OF BURNING

Let's consider some common observations that can easily be made about fire. Fires require air to burn. (This is useful to us both in kindling a fire, because we need to make sure abundant air is available, and in trying to stop a fire, because we can put a fire out by smothering it.) Something is consumed—burned up—in the process. In some fires, the burning material appears to be totally consumed, as in the case of burning gasoline, and in others, some small amount of an incombustible residue, ash, remains behind. Certainly we notice that heat is given off, as is light. Sometimes sound is also given off, as the fire "crackles" as it burns.

The "something" that is consumed in the fire is the *fuel*. A fuel is defined as being a material that is consumed to produce energy and, most commonly, a material that is burned to produce heat. Since these two statements must be equivalent, that is, a fuel produces energy and most commonly produces heat, embedded in the definition is an extremely important fact that we will revisit later:

Heat is a form of ENERGY.

The evidence of our senses is that a fire destroys or consumes something. This is especially the case when the fuel is a solid or liquid, such as wood, coal, or petroleum products. If the fire is allowed to burn until all the fuel has been consumed, we see quite clearly that the fuel is gone and nothing, except perhaps for some incombustible ash, remains. Various ideas were developed over the years to explain the action of a fire. Beginning with a few careful experimental observations in the seventeenth century, and culminating with the studies of the French scientist Antoine Lavoisier* (Figure 5.3) in the late eighteenth century, scientists established that, in fact, nothing is truly destroyed in a fire.

Very precise measurements showed that the weight of the fuel plus the weight of the air involved in the fire equaled exactly the weight of the products of combustion (including the leftover air). This relationship holds to the highest degree of precision we can possibly measure. Further, Lavoisier showed that fires don't require air to burn—strictly speaking, they require the oxygen that's in the air. When a solid or liquid fuel is burned, matter appears to be destroyed in the fire because the principal products of combustion, carbon dioxide and water vapor, are invisible gases. Nothing is really destroyed in a fire, in the sense that the weight of the fuel burned and the weight of the oxygen used to consume the fuel are equal to the weight of all of the chemical products of combustion. But something seems to be produced: ENERGY. We can feel it as heat, see it when the flame is luminous, and sometimes hear it, as in the crackling of firewood.

To consider in more detail this observation that ENERGY is given off from a fire, we must first select a SYSTEM for study. We can take one of the simplest fuels, natural gas. The main ingredient of natural gas is methane, CH_4. The other component of the SYSTEM will be oxygen. We can imagine a container that holds

---

* Antoine Laurent Lavoisier (1743–1794) is considered by many historians of science to be the father of modern chemistry. His detailed experimental work represented a turning point between the scattered, empirical testing of earlier workers and a modern science based on theories and detailed understanding of chemical processes. In the mid-1780s Lavoisier demonstrated clearly that there is no change of mass in a system undergoing chemical reaction. In 1789 he published a definitive work on the distinction between elements and compounds. Although other chemists had prepared oxygen before Lavoisier, he recognized that oxygen is an element, and that the process of combustion is that of chemically combining a fuel with oxygen. Lavoisier even coined the name "oxygen," on the mistaken belief that oxygen is a component of all acids (the name comes from the Greek words for "acid" and "to form"). Because Lavoisier had held some government positions during the reign of Louis XVI, he was sent to the guillotine during the French Revolution. His international reputation as one of the most illustrious scientists of the eighteenth century was not enough to save him. (See also note on page 26, Chapter 3.)

FIGURE 5.3 Antoine Lavoisier (1743–1794), the brilliant French scientist who is generally regarded as the founder of modern chemistry. (Reprinted from first edition, courtesy of Visual Image Presentations, Bladensburg, Maryland.)

a mixture of methane and oxygen in the exact proportions needed to combust the methane completely to carbon dioxide and water.* This SYSTEM of methane and oxygen can be kept for years with nothing happening inside it. However, if we choose to supply a small amount of ENERGY to the SYSTEM, as, for example, by inserting a lighted match, or from an electric spark,† we would find that the methane–oxygen mixture burns with great rapidity and provides a large quantity of energy.

---

* When there is exactly enough oxygen present to completely burn the fuel to carbon dioxide and water, we speak of this condition as being one of stoichiometric combustion. The term "stoichiometric" derives from a Greek word meaning "element." It is not necessary for stoichiometric conditions to exist in order for a fuel to burn. Situations can occur in which a system has more oxygen than is needed for complete burning, or has less oxygen. The latter case is undesirable, since, instead of producing carbon dioxide, the carbon in the fuel might be converted to soot or, worse, the potentially lethal carbon monoxide. We will return to this concept when we discuss the burning of gasoline in automobile engines.

† Many processes that appear spontaneous will proceed readily, and often rapidly, provided that we give a "nudge" to get the process started. A physical example is provided by a brick or block standing on its small end. A block standing in this position can topple over, and once it begins falling, it will continue to do so unless we somehow intervene (by catching it, for example). However, the block will stand in the upright position essentially forever unless it is somehow provided with a small nudge to get it started. In the example of combustion discussed in this chapter, the match or electric spark provides the nudge. We call this "nudge" the activation energy.

The important parts of the SYSTEM are the fuel (in this case, methane) and oxygen. Recall the energy balance:

$$E_{in} - E_{out} = E_{stored}$$

In a fire, the term E_{out} usually will be quite large. On the other hand, the term E_{in} is usually very small, indeed trivial. If we ignite a methane fire with a match, the E_{in} from the match is much, much smaller than E_{out} from the burning methane. We've also said that a process that produces more ENERGY (i.e., E_{out}) than was put into the SYSTEM is impossible, unless ENERGY was somehow stored or "hidden" in the SYSTEM to begin with. Certainly a fire is not impossible. Then since E_{out} is much, much greater than E_{in}, we need to enquire what this E_{stored} term is.

The chemical equation for the combustion of methane* is

$$CH_4 + 2O_2 \rightarrow CO_2 + 2H_2O \left(+heat!\right)$$

$$Methane + oxygen \rightarrow carbon\ dioxide + water \left(+heat!\right)$$

This equation tells us that one molecule of methane reacts with two molecules of oxygen to produce a molecule of carbon dioxide and two molecules of water. This equation can be written in a slightly different, but completely equivalent, form, showing the way the atoms are connected to each other (Figure 5.4). Regardless of which of these two equations we look at, we should be able to see that it satisfies

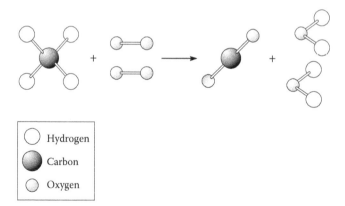

Hydrogen

Carbon

Oxygen

FIGURE 5.4 The molecular structural changes accompanying the complete combustion of methane in oxygen, producing carbon dioxide and water.

---

* The equation for the combustion of methane appears to represent an extremely simple chemical process. But the combustion of even a small five-atom molecule like methane is a devilishly complicated business that may involve over a hundred intermediate steps on the route from methane and oxygen to carbon dioxide and water.

the requirement that no matter is actually destroyed. On the left-hand side, there are one carbon atom, four hydrogen atoms, and four oxygen atoms. On the right-hand side, there are still one carbon atom, four hydrogen atoms, and four oxygen atoms. What has changed? Only the way in which the atoms are linked (i.e., chemically bonded) together.

The collection of a carbon atom, four hydrogen atoms, and four oxygen atoms represents the components of our methane–oxygen SYSTEM. In the combustion process, this system has undergone a change. Originally, the four hydrogen atoms are bonded to the carbon atom, and the oxygen atoms are bonded to each other. At this point, we introduce a new bit of jargon: *state*. A STATE is the condition in which a SYSTEM happens to exist.

We arbitrarily call the assemblage of atoms in which hydrogen is bonded to carbon in methane, and the oxygen atoms are bonded to themselves, STATE 1 of the SYSTEM. At the end, the carbon atom is now bonded to two oxygen atoms, and each of the remaining two oxygen atoms is bonded to two hydrogen atoms. Let's call this STATE 2. Therefore, in very general terms, the burning of methane in oxygen to make carbon dioxide and water represents a change of this SYSTEM from STATE 1 to STATE 2.

When this change happens, heat is given off. Since heat is a form of ENERGY, we could speak more broadly to say that energy is given off or is released from the SYSTEM when it changes from STATE 1 to STATE 2. There is one other important consideration. We know from repeated experience that the conversion of methane and oxygen into carbon dioxide and water will proceed spontaneously once we have supplied a small amount of ENERGY—all we need to do is strike a match. We also know from experience that the reverse change, the conversion of a mixture of carbon dioxide and water into methane and oxygen, does not happen spontaneously. Sticking a lighted match into a bottle of methane and oxygen will ignite a fire or even an explosion. Sticking a lighted match into a bottle of carbon dioxide and water vapor will put the match out. So, going in one direction—from methane and oxygen to carbon dioxide and water—is relatively easy. Going the other way—carbon dioxide and water to methane and oxygen—is not impossible, but we can expect it to be very difficult.

Let's represent this with a diagram (Figure 5.5). This diagram applies to the particular case of methane combustion. We can change this diagram to make it much more general (Figure 5.6). This generalized diagram is a pictorial representation of an extremely important principle:

> When a SYSTEM undergoes a spontaneous change from one STATE to another, ENERGY is released.

We'll see later how we can build on this diagram and this principle to extend the understanding of various ways in which ENERGY is obtained and used. For now, recall that we need ENERGY in order to be able to do WORK, and WORK provides all of the things that we rely on in our daily lives: transportation, manufacturing goods, and

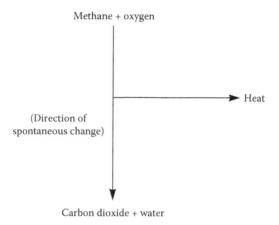

Methane + oxygen

(Direction of
spontaneous change)

→ Heat

Carbon dioxide + water

FIGURE 5.5 An energy diagram for the combustion of methane. The direction of spontaneous change is from methane and oxygen to carbon dioxide and water; heat is given off.

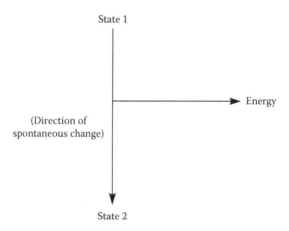

State 1

(Direction of
spontaneous change)

→ Energy

State 2

FIGURE 5.6 A generic energy diagram for any SYSTEM.

transmitting electricity, for example. *Getting the* ENERGY *needed to do* WORK *means that we need to figure out how to take advantage of spontaneous changes in* SYSTEMS.

There remains one other issue—where does the ENERGY of a fire come from? On the basis of the energy balance equation, ENERGY must be hidden or stored in the SYSTEM (methane + oxygen) before the fire is started. We say this because the value of E_{out} in the energy balance (the energy released by the fire) is much greater than the value of E_{in} (provided by the match). As we saw in Chapter 4, a situation in which E_{out} is greater than E_{in} is impossible unless some ENERGY was stored in the SYSTEM. Most certainly, a methane fire is not impossible. Where can the stored ENERGY be? There is only one place it can be. It must be stored in the only thing that changes as the methane burns: the chemical bonds between atoms. Since ENERGY comes out of a burning SYSTEM, then, continuing our example of methane, it must be that more

ENERGY is stored in four C–H and two O–O bonds than is stored in four H–O and two C–O bonds. Generalizing from this specific example,

The ENERGY released during combustion comes from ENERGY stored in the chemical bonds of the fuel and the oxygen.

FURTHER READINGS

Faraday, M. *The Chemical History of a Candle.* Dover Publications: Mineola, NY, 2002. An inexpensive paperback reprint of lectures originally given in the 1860s, this book still remains an excellent introduction to combustion, particularly for those with little scientific background.

Gribbin, J. *History of Western Science.* The Folio Society: London, U.K., 2006. Chapter 7 discusses, among other topics, the contributions of Lavoisier to establishing the modern science of chemistry.

Ihde, A. *The Development of Modern Chemistry.* Dover Publications: Mineola, NY, 2012. A reprint of an excellent work of history first published in 1964. Many of the topics of this chapter, and the scientists who contributed to their understanding, are discussed.

Lavoisier, A. *Elements of Chemistry.* Dover Publications: Mineola, NY, 1984. A reprint—albeit translated into English—of the first-ever chemistry textbook, and a discussion of the role of oxygen by the great scientist himself.

6 Firewood

Historically, the first fuel that humankind relied on in large amounts was wood. There are several reasons for this. In many places of the world, wood is abundant. It flourishes on the Earth's surface (Figure 6.1), so we do not need to drill or dig into the Earth to find it. Wood is easy to handle and store, unlike, for example, natural gas, which requires pipes, valves, and storage tanks. Wood—unless it's wet—is easy to ignite, unlike some kinds of coal. Wood is easy to burn in open fires. Anyone can figure out how to get a wood fire started (given enough matches and enough patience), but other fuels require fairly sophisticated appliances—gas stoves or oil furnaces, as examples. Wood was the primary fuel that sustained humanity through most of our history. It was the dominant fuel well into the mid-nineteenth century, when it was supplanted by coal.

HOW DOES WOOD BURN?

The last chapter introduced an example of combustion, the burning of methane. Chemically, methane is the simplest fuel in common use; a molecule of methane contains only five atoms and only two elements. Wood is a much more complex material than methane, so it will be no surprise that the combustion of wood is, in turn, a much more complex process.

Almost all wood contains some amount of water in the form of free water molecules. As much as 50% of the weight of a piece of freshly cut, "green" wood can be water.* Since water has no value as a fuel, its presence reduces the amount of heat released when compared to burning the same weight of dried wood. When a piece of green wood is put on the fire, these molecules of water boil off. Then the thoroughly dried wood begins to break down chemically as the molecules of the components of the woody material itself are heated to higher and higher temperatures. As heat causes these molecules to break down, two kinds of products form. One appears as vapors of various compounds; collectively, these vapors are referred to as *volatiles*. As the volatiles make their way out of the solid wood, they cause the crackling and hissing noises that sometimes accompany a wood fire. The other product is a solid material very rich in carbon, called *char*. At this point, no burning has yet occurred; all that has happened is that the components of wood have been decomposed by heat into volatiles and char. Ignition of the volatiles produces the flame that can be seen dancing over the fire. Ignition of the solid, carbon-rich char produces the glowing embers or "coals." Figure 6.2 shows the burning of volatiles and char.

Consider what happens when a gas burner (on a gas stove, for example) is lighted. Touching a lighted match to a source of gaseous fuel causes instantaneous ignition. A flame appears at once. Lighting a candle is not quite so easy. It's necessary to hold

---

* In this context, the word *green* has nothing to do with color. It denotes the fact that the material is not dried.

FIGURE 6.1   This flourishing pine forest in Sweden illustrates how abundantly wood occurs in many parts of the world. (From http://commons.wikimedia.org/wiki/File:Pine_forest_in_Sweden.jpg.)

a match to the wick for several moments before the candle is lit. The difference lies in the fuel. A gaseous fuel consists of small molecules that mix easily with air and ignite readily. The solid wax in a candle contains molecules that can contain more than twenty carbon atoms. These molecules have to be broken down by heat from the match and form a vapor. Because it is harder to do this than it is to ignite a small, gaseous molecule, there is an evident time delay in lighting a candle. The large, complex molecules in wood are even bigger and harder to break apart than are the wax molecules in a candle, causing wood to be even harder to ignite.

Heat released from burning wood comes from the same source of "stored energy" as discussed for the burning of methane (Chapter 5). Many of the chemical bonds in wood are carbon–hydrogen bonds, as in methane, or carbon–carbon bonds. The interaction of the volatiles from wood with oxygen follows essentially the same processes as with methane. Formation of new carbon–oxygen and hydrogen–oxygen bonds releases ENERGY. The ENERGY supplied to the wood by the match, or by other pieces of burning wood, begins to break down the carbon–hydrogen and carbon–carbon bonds to allow the new carbon–oxygen and hydrogen–oxygen bonds to form.

Ignition of the volatiles produces a luminous flame sometimes accompanied by abundant soot. Once the fire is started, it continues because some of the heat of the fire breaks down more molecules of wood components, producing more volatiles. But this process depends on the pile of burning wood being well constructed, so that

FIGURE 6.2   A wood fire, showing flames from the burning of volatiles and glowing embers from burning of the char. Even in the modern era of clean and efficient household appliances, many people find something aesthetically or emotionally satisfying about a wood fire. (From http://commons.wikimedia.org/wiki/File:Firewood_with_flame_ash_and_red_embers.jpg.)

some of the heat can be utilized in this way. If too much heat is lost, the unburned wood cannot break down chemically, and the fire will go out. Over centuries, many useful rules of thumb have evolved for building and maintaining good wood fires. As examples, three logs in good physical contact with each other burn better than if the same three logs were spaced apart. This occurs because each of the logs in contact can share its heat with the others. It's generally good to allow the ashes to stay underneath the fire. Ashes are very good at trapping heat (notice how long the ashes in a charcoal grill stay hot) and can radiate this heat back to the fire.

Woods differ greatly in the amount of heat produced per unit weight of wood burned (Figure 6.3). There are also considerable differences in moisture content and the yield of ash. Two premium woods for combustion are hickory and oak. A cubic meter of hickory or oak provides as much heat as a quarter-tonne (250 kg) of coal or 210 liters of fuel oil. Softwoods, such as pine and poplar, are not nearly so good, taking up to twice as much wood to provide the same heat as hickory or oak. Of course, other factors need consideration in selecting wood for a fire—ease of igniting, amount of smoke produced, and how quickly the wood is consumed—are all important.

WHERE DOES THE ENERGY IN WOOD COME FROM?

Wood comes from trees, members of the plant kingdom. The crucial step in the life processes of plants is photosynthesis. Photosynthesis converts carbon dioxide and water into sugar molecules, with oxygen as a second product. Chapter 5 discussed

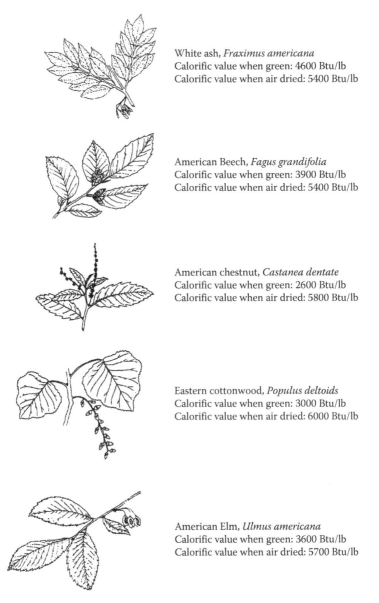

White ash, *Fraximus americana*
Calorific value when green: 4600 Btu/lb
Calorific value when air dried: 5400 Btu/lb

American Beech, *Fagus grandifolia*
Calorific value when green: 3900 Btu/lb
Calorific value when air dried: 5400 Btu/lb

American chestnut, *Castanea dentate*
Calorific value when green: 2600 Btu/lb
Calorific value when air dried: 5800 Btu/lb

Eastern cottonwood, *Populus deltoids*
Calorific value when green: 3000 Btu/lb
Calorific value when air dried: 6000 Btu/lb

American Elm, *Ulmus americana*
Calorific value when green: 3600 Btu/lb
Calorific value when air dried: 5700 Btu/lb

FIGURE 6.3 Not all kinds of wood provide the same amount of heat when burned. This chart illustrates the differences among some kinds of wood to convert Btu/lb to MJ/kg, multiply by 0.0023. (Reprinted from first edition.)

how making carbon dioxide and water from methane and oxygen was easy, but that converting carbon dioxide and water to methane and oxygen would be very difficult. In photosynthesis, carbon dioxide and water are converted to a product much more complex, chemically, than methane. We should expect this to be a difficult process. The process of photosynthesis needs the light of the sun—solar ENERGY—to drive the conversion of carbon dioxide and water to simple sugars. Many individual reaction steps have to occur to build up a sugar molecule from carbon dioxide. Plants use sugar molecules as sources of ENERGY to synthesize many other molecules required for the life processes of the plant, including the chemical components of wood. Since the growth and health of trees depend on photosynthesis, which requires sunlight, wood ultimately derives from the ENERGY of the sun.

Wood ENERGY is a form of stored solar ENERGY.

PROBLEMS ASSOCIATED WITH WOOD COMBUSTION

In comparison with coal and many fuel oils, wood contains virtually no sulfur, so produces no harmful sulfur oxides on combustion.* However, wood fires often produce smoke. The potential pollution caused by smoke from these fires needs to be considered. Many people in the industrialized world consider the smoke from wood-burning fireplaces or outdoor bonfires with some nostalgia. That indeed may be the case for a single, isolated, small wood fire. Adding smoke from thousands of wood-burning households to a city atmosphere that is already seriously polluted with motor vehicle exhaust fumes, power plant emissions, and factories would greatly multiply the environmental problems.

In addition to being an atmospheric pollution problem, excessive smoke production can cause a fire hazard by leaving soot and creosote deposits in a flue. Wood creosote, a liquid of high boiling temperature (above 200°C), forms when components of wood are broken apart chemically by the heat of the fire, rather than being burned. Wood creosote contains a variety of chemical compounds. It has some valuable uses, for example, as an antiseptic. In many chimneys, particularly in home fireplaces, vapors of creosote leave the region of the fire and condense to form a tar-like liquid on the chimney walls. This sticky condensed creosote can trap and accumulate particles of soot, a product of the incomplete burning of the wood. Soot is nearly pure carbon. The mixture of creosote and soot is a superb fuel. If the upper part of the chimney becomes hot enough, and if enough oxygen is available, the creosote–soot mixture can ignite inside the chimney. This makes the chimney even hotter, and in serious cases, the hot chimney can in turn ignite floors or walls of the house, resulting in a disastrous house fire.

A wood fire generally leaves only a small amount of ash for disposal. If one wants to be very self-reliant, it's possible to use the wood ash for making soap and as a fertilizer. Ash contains a number of compounds, one of which is potassium carbonate. This compound dissolves in water. By saving the ashes from wood fires, and then stirring them with hot water, it's possible to make a concentrated solution of potassium carbonate (along with a mess). In a separate step, this potassium carbonate

---

* The formation of sulfur and nitrogen oxides and their effects on the environment are considered later in this book, particularly in the chapters on electricity generation from coal and on acid rain.

solution can be reacted with waste fats or oils from the kitchen, such as used cooking oil or the grease produced when cooking bacon. The principal product of this reaction is a form of soap, which works very well for household applications. Potassium compounds are essential for plant growth. Using the ash from a wood fire as a source of potassium provides a crude, but useful, form of fertilizer.

Wood smoke contains a variety of large organic molecules that are known or suspected to cause cancer.* Smoke particles may increase susceptibility to respiratory illness. Atmospheric emissions from many small wood-burning stoves and fireplaces, localized in individual homes, are much more difficult to trap than emissions from large, centralized electric plants. Because of the difficulty of controlling emissions from many small units vis-à-vis a single, large, centralized unit, if a large fraction of residential heating were to be converted back to wood burning (ignoring for the moment the issue of where the wood would come from), pollution due to particles in smoke and to potential carcinogens would be serious. This problem, combined with the question of wood supply, could be a major detriment to a return to large-scale use of wood.

ENERGY CRISES MADE OF WOOD

Wood is often considered to be a "renewable" energy source, renewable because, in principle, more wood can be grown to replace the amount harvested for immediate use. However, in the case of wood, unlike crops that can be regrown annually, this principle only works with very careful management of the wood supply. It is very easy for an increasing population, with increasing energy demands, to outstrip the ability of the wood supply to keep up with demand. Even in the time of ancient Rome, fuel wood got to be in scarce supply, and in the Middle Ages, people were already mining coal to alleviate their dependence on wood. The issue of the ownership of the wood is so crucial that it has left a mark on our language yet today. In medieval England, serfs originally were limited to whatever firewood they could pick up from the ground, such as fallen trees or dead limbs that had come down. The land, the trees, and the wood all belonged to the local lord. Eventually, the scarcity of wood resulted in this rule being relaxed. The feudal lords realized that dead serfs (who had died as a result of not having fuel for warmth and cooking) were of no use to them.† Shepherds were allowed to knock or pull down dead branches using their crooked staffs. Farm workers could do likewise, using a billhook, an implement that's used for rough pruning or for clearing brush. In other words, these peasants were allowed to have all the wood they could collect, *by hook or by crook*. Today the

---

* These chemicals include benzo(a)pyrene, dibenz(a,h)anthracene, benzo(b)fluoranthene, benzo(j)fluoranthene, dibenzo(a,l)pyrene, benz(a)anthracene, chrysene, benzo(e)pyrene, and indeno(1,2,3-cd)pyrene. They are all known or suspected carcinogens. Of course, their formation by pyrolysis is not limited to wood. Many kinds of plant material that pyrolyze as part of the burning process produce these cancer-causing substances. An excellent source is tobacco.

† Actually, there is an interesting counterargument to the concept that dead serfs are useless to their master. Nikolai Gogol eloquently expresses this argument in his book *Dead Souls*. (Numerous editions of this book, including inexpensive paperbacks, are available.)

expression has a more general meaning, of accomplishing something in any way, or by any means, necessary.

Good King Wenceslas looked out
On the feast of Stephen
When the snow lay round about
Deep and crisp and even
Brightly shone the moon that night
Though the frost was cruel
When a poor man came in sight
Gath'ring winter fuel

—**Neale**[1]

This traditional carol is based on the life of Saint Wenceslaus, who lived in the tenth century. Much later, Leo Tolstoy described the dreadful plight of Russian peasants:

... the Russian landowners last year [1892], their efforts to combat the famine, which they had caused, and by which they profited, selling not only bread at the highest price, but even potato haulm at five rubles the dessiatine [about 2.8 acres] for fuel to the freezing peasants.

—**Tolstoy**[2]

In the seventeenth century, Britain was still almost entirely dependent upon wood. A timber shortage occurred as early as 1580. Furnaces for smelting iron ore in the Forest of Dean were being dismantled around the time of the Restoration, 1660, for lack of wood. Because of increasing shortage of wood in Britain, the first colonists of America were astonished when they saw the apparently limitless forests of New England. (They also recognized an opportunity for exporting iron back to the home country.)

Here is good living for those that love good Fires. Though it bee here somewhat cold in the winter, yet here we have plenty of Fire to warme us, and that a great deale cheaper than they sel Billets and Faggots in London. Nay, all Europe is not able to afford so great Fires as New-England.

—**Higginson**[3]

Likewise, many Mediterranean countries were so dependent on wood that the hillsides were stripped of their trees to provide firewood and lumber (Figure 6.4), causing a range of environmental problems of the kinds that we will discuss later in this chapter.

Although it seemed as if the forests of America were limitless, it didn't take long for reality to overcome perception. By 1637, only seven years after Francis Higginson's claim of "good living for those that love fires," poor people in Boston were already suffering from a scarcity of firewood.

In their Country Plantations, the wood grows at every Man's Door so fast that after it has been cut down, it will in seven Years time grow up again from seed to substantial Fire Wood.

—**Trefil**[4]

FIGURE 6.4   The overuse of forested land for firewood leads to the environmental problem of deforestation, shown here. Compare this figure with Figure 6.1. This picture illustrates deforestation in Nepal. (Figure courtesy of Ian Swarbirck, www.imagesfromthewild.ch.)

It did not take long—less than a century—for the young, expanding cities to get trapped in America's first energy crisis. As the population grew, so too did the need for ENERGY (wood). But as more and more of the wood was cut, fresh wood had to be transported for longer distances to reach the consumers in the cities. Eventually, a point came at which supply could no longer adequately match demand. In the middle of the eighteenth century, the price of wood in New York City quadrupled in a few years. This was an energy crisis, though the term had not yet come into use in the eighteenth century. In an energy crisis, as true now as it was then, the crisis does not result from running out a particular ENERGY source. Early America possessed vast quantities of wood. The existence of a supply was not a problem. Rather, the problem lay in the fact that the wood could not be harvested and shipped fast enough to meet the demands in the growing cities. Supplies couldn't reach consumers.

> An energy crisis does not result from running out of an ENERGY supply. It results from real or contrived difficulties in distribution and matching supply with demand.

The situation became so serious in some places that even the gathering of driftwood from beaches became subject to legislation. One New England town passed legislation in 1765 "to choose four persons to take care of the town's beach, in order to prevent other towns' people from carrying off sand, rockweed, and clams, and trash (driftwood), or other rubbish of said beach."[5] Throughout the colonial period and well into the nineteenth century, this problem affected the great cities of the eastern seaboard of the United States. Philadelphia, for example, suffered firewood shortages during the war of 1812. From 1635 to 1850, Concord, Massachusetts, used wood equivalent to having burned every tree in the town three times over. In three centuries (up to 1930), Americans burned approximately 45 billion cubic meters of wood.

People responded to this energy crisis in ways remarkably like the better-known energy crisis of the 1970s–early 1980s, which was caused by embargoes on the sale of oil on world markets. When any energy source becomes in short supply, one of the first, and indeed most effective, responses is to find ways of conserving the resource, or using it more efficiently. In the 1970s, the responses included switching away from the large "gas guzzlers" to smaller cars with much greater fuel efficiency and to take greater care with energy consumption in the home. In eighteenth-century America, Benjamin Franklin, with his invention of the so-called Franklin stove (Figure 6.5), contributed one of the "efficiency responses" in the 1740s. One of the improvements made by Franklin was an opening near floor level that allowed cold air from the room to enter an "air box" around the stove, be warmed there, and then be returned into the room via a second, higher opening. A second response to an energy crisis is to find an alternative fuel, to replace the one in short supply. The 1970s and early 1980s saw considerable interest in many countries in switching away from expensive, imported petroleum to less expensive, domestic coal. In the 1740s and after, the thrust was away from expensive, scarce wood to use instead—coal.

It is difficult nowadays to appreciate the immense amount of labor involved in running an entire household using wood as the only source of energy. Before there could even be a fire, there was lots of effort in preparation. This preparatory work involved cutting down trees, hauling them home, chopping the wood into fireplace- or stove-sized pieces, stacking it, and finally bringing it into the house and putting

FIGURE 6.5 The Franklin stove was a major innovation in the use of firewood in the home. (Reprinted from first edition, courtesy of Good Times Stove Company, Goshen, Massachusetts.)

it on the fire. Everybody in the family had a job. Men cut the trees and chopped and split the wood, children gathered kindling wood and carried wood to keep woodbox full in the house, and women transferred the wood to the fire as more was needed. Starting at the fireplace, the wood had to be arranged properly on the hearth. Usually this began with a foundation of large logs, including an exceptionally large log—the backlog*—at the rear of the fireplace. Additional wood could be piled onto this foundation of logs as it was needed. The fire was the source of energy for heating the home, cooking food, and boiling water. Tending to the fire was a constant job in the household. It could not have been pleasant to work right over the hot fire and to try to manipulate hot cooking utensils in the fire. The fireplaces in some medieval dwellings were so huge that the cook worked *in* the fireplace! A wood fire is almost certain to produce some soot and ashes, and these are almost certain to get on the floors and furniture. Cleaning was a permanent chore.

WOOD IN THE INDUSTRIALIZED WORLD TODAY

For most industrialized nations, wood was almost totally abandoned as a major energy source from the time that coal, and then oil and gas, became available. The main exception was in fireplaces, which many people seem to find to be more of a decoration than a component of the home heating system. Without a doubt, electric heat, or oil or gas fuels, is much more convenient for the homeowner than is wood. However, with proper care for the design and uses of wood burners or fireplaces, wood can provide supplemental heating and can offer cost savings in areas where it is less expensive than gas or oil.

With careful design, a wood stove can supply a wide variety of household needs. The traditional Russian stove provides an example. These stoves were made of brick or other masonry and often set up in the middle of the peasant's hut. The stove was used for cooking, heating, baking, drying foods for preservation during the winter, and even converted into a steam bath. In winter, some people sat or slept on their stoves.

Wood heating in the home can be achieved in either open or enclosed heaters. Fireplaces are examples of open heaters. Enclosed heaters are the closed stoves, usually made of metal. They are much superior in terms of providing heat. For one thing, it's easier to control the combustion process in an enclosed heater, such as by opening and closing air vents. In fact, enclosed heaters are sometimes referred to as "complete combustion" burners. Also, enclosed heaters do a better job of radiating heat into the room. However, few, if any, of the enclosed heaters can provide the aesthetic pleasure that comes from watching an open fire in a fireplace.

A well-built metal stove can provide about three times as much heat to a room as can a fireplace, using the same amount of wood in each. This is the importance of the invention of the Franklin stove. It helped to alleviate the first energy crisis in America because it could generate much more heat from a given amount of wood

---

* The original meaning of the word *backlog* was simply a very large log rolled into position at the back of a fireplace. There, it could provide a large reserve supply of fuel that would take a long time to consume. Our modern usages of the word, as an accumulation of work not yet done, or as a reserve supply of something, derive from this original meaning.

(or, from the other, more important, perspective, could generate the same required amount of heat using less wood). By using less wood to meet the needs of a given amount of heat in a home, the Franklin stove, in effect, meant that dwindling wood supplies could last much longer. For all the enjoyment we might get from having a fireplace in the home and watching the fire, fireplaces are actually not very good means of providing heat. On cold winter evenings, it's even possible that a fireplace in operation can *cool* the rest of the house, by pulling cold air in from outside.

When efficiency of wood use and providing useful heat are the main concerns, then an airtight stove (a complete combustion burner) is unquestionably the best choice. The best models provide numerous features to obtain the most efficient use of wood. These features include the control of air intake by thermostats; a so-called secondary intake system, which brings unburned gases back through the stove for a second attempt at combustion; and water coils to generate hot water using the hot gases from the fire. Such stoves represent the best approach to using wood effectively. But they have no way of providing the aesthetic pleasure that many people derive from watching a wood fire in a fireplace. There is something about an open, wood-burning hearth that strikes a powerful chord in many people. A fireplace is one of the few improvements that one can make to a home that increases its market value by more than the cost of the added fireplace itself (Figure 6.6).

Aside from the aesthetic appeal or pleasure, and the value it may add to the home, there is otherwise little good to be said about a fireplace. Assuming a tightly sealed house, many of the products of burning the wood remain in the room and must be breathed by the people in it. At the same time, the fire is using up oxygen from the room. An open fire is also a potential safety hazard.

FIGURE 6.6 A modern Mors⌀ high-efficiency fireplace from Denmark. (Used with permission of Helle Hyldig S⌀rensen, Mors⌀ Jernst⌀beri A/S, Nyk⌀bing, Denmark.)

For the individual householder relying on wood, the ideal source of firewood is a managed woodlot. In favorable cases, the simple acts of pruning and picking up broken limbs can yield about nine cubic meters of wood per hectare every year. Harvesting trees for firewood, as well as for timber, requires careful management, generally referred to as "sustained-yield." Sustained-yield management allows nature each year to grow back the amount of wood that has been removed. Of course, trees cannot immediately be replaced on a one-for-one basis. That is, if we were to cut down a mature tree for firewood now, planting a seed or even a replacement sapling will not result in the replacement of that mature tree in only one year. Because of this, at least several acres of sustained-yield forestland are needed to supply firewood for a typical home. For example, a house of rather modest size, say 90–110 square meters in a region that has cold but not subzero winters, would need three to four hectares to produce enough wood for one winter's heating. In turn then, this means that planning to have all of the population of a large industrialized nation rely completely on firewood for home heating is hopelessly unrealistic—there just isn't enough land to grow sustained-yield forests for all of us.

A managed woodlot of hardwood trees will provide roughly nine to ten cubic meters per hectare. A rough rule is that the "average" house will need about eighteen cubic meters of wood per winter. Of course, there is no such thing as an "average" house, and the actual need varies with the size of the house, the quality of its insulation, the type of wood being burned, and the temperature that the occupants expect to maintain. However, with this rough rule as a guideline, each household needs two hectares of *renewable* woodlot. Not all of us live in houses. Apartment buildings would require much more available wood. A thirty-story apartment building might require the equivalent of 2.5 square kilometers of woodlot. Since so many of us now live in cities, and in apartment buildings, the idea of a large-scale return to wood as an energy source in most industrialized nations would mean that much land now used for agriculture or forestry would have to be converted to plantations, or managed forests, of fast-growing trees used exclusively for firewood. (We should take care to remember that all of this wood would have to be cut, dried, and transported into the cities. Each of these steps would itself have a significant requirement for ENERGY.) Speaking of wood as a renewable resource does not mean that trees would be cut and forests *as we know them today* would be regenerated. Forests might exist, but they would be managed to provide maximum yields of firewood. In many places, the specific kinds of trees in the forests would be different. This change would in turn result in a significant change in the ecosystem, that is, the organisms of a natural community—the forest—with their environment.

Despite the several drawbacks or limitations to using wood as a fuel, wood has enjoyed a resurgence in popularity. Many people perceive that something "natural" or "organic" about wood that cannot be said of other fuels. In some rural areas, the individual householder can obtain quite adequate supplies of wood, often at low cost. As we've seen, the best of the wood-burning stoves are much more efficient than fireplaces. When all factors, pro and con, are considered though, it still seems very unlikely that wood can displace the fossil fuels or electricity as a major source of domestic energy in most industrialized nations.

WOOD IN THE DEVELOPING WORLD

Even today, wood is the dominant fuel in many parts of the developing world. Developing countries use about 75% of all the wood consumed for fuel. In terms of consumption of wood by households, the leading country is India, followed by Indonesia, Nigeria, Ethiopia, and the Democratic Republic of the Congo. In 2010, African nations consumed nearly a half-billion cubic meters of wood, with demand expected to grow another ten percent by 2030. Countries such as Burundi and Rwanda already have a deficit of wood for fuel use. At least a third of the world still relies on wood (Figure 6.7). The heavy reliance of these countries on wood has several important consequences.

Broadly, developing countries have two kinds of energy needs. One is the energy that will help the country develop industrial production, transportation, and mechanized agriculture. The other is energy needed for simple survival of the populace, energy for such domestic uses as cooking, and heating, for example. We can balance the energy needs against the possible sources of energy that could meet these needs. Energy sources can be put into three categories. *Commercial* energy sources include oil, natural gas, coal, and electricity. *Traditional* energy sources, sometimes also called noncommercial sources, include wood; charcoal, which is usually made from wood; agricultural crop wastes (an example being bagasse, the straw-like material left over after sugarcane has been processed to extract its sugary juices); and animal wastes. In addition, some countries may be developing *new* energy sources, solar energy being one example.

Commercial energy sources are the ones that can meet the energy demands for development. However, there is a "catch." The nation must have sufficient financial capital resources, either internally or supplied by foreign aid, to develop the commercial energy sources. These are expensive. Drilling an oil well can cost about two thousand dollars per meter drilled. An electric power plant of modest size can easily cost hundreds of millions of dollars. The situation in Indonesia is an example. Indonesia is richly endowed with energy—coal, petroleum, and natural gas. It has enough energy supplies not only to meet its own needs for many years to come, but also to be a significant exporter, indeed one of the top coal-exporting nations of the world. Yet domestically, wood is a very important energy source, in part because of the lack of capital to invest in the extensive development of the coal, gas, and petroleum.

Usually, the traditional energy sources are the ones used to meet needs of energy for domestic applications or even outright survival. Often, this is because such energy sources are not bought and sold as commodities, but rather can be gathered freely from the local environment. In many parts of the world, wood falls into this category. Wood is the classic example of a traditional energy source. It is the energy source that supplies most people's needs for cooking, heating, and domestic comfort. In Kenya, for example, fuel wood consumption represents at least one kilogram of wood per person per day. Even a small family will consume over a tonne per year.

Many developing countries are faced with problems of burgeoning populations. As more and more people need more and more wood, several factors come into play regarding wood availability. The first is the distance people have to travel to

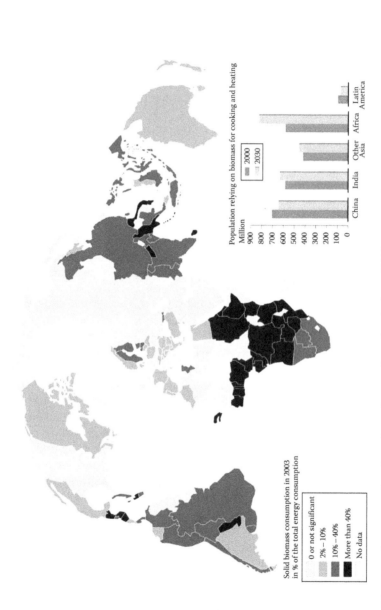

FIGURE 6.7 This map and accompanying chart shows the use of wood and other solid biomass fuel in developing countries. (From http://www.grida.no/graphicslib/detail/solid-biomass-consumption-including-woodfuel_9764.)

collect wood. The second is having access to the wood—who owns it or owns the land on which the wood can be found. A third consideration that may also be important is whether any public- (government) or private-sector constraints exist concerning use of the wood.

In Africa, population growth and the increasing amount of urbanization drive a developing energy crisis. That is, not only are there more people, but more of them are moving to cities. As we have seen in considering America's first energy crisis, one possible response is switching to a different kind of fuel. Unfortunately, this is not an attractive option in some African countries because of the comparatively high price of petroleum products such as kerosene (paraffin).

Firewood availability is also a gender issue. In most Third World nations, collecting wood is considered to be "women's work" (Figure 6.8). Again, as wood becomes increasingly scarce, more and more time has to be devoted to gathering it. In Kenya, women and children can spend six hours per day gathering firewood. In some cases, the bundles of wood to be carried back home may weigh about fifty kilograms. This time is taken away from family activities, such as child rearing. It also takes away from assisting with farming, from planting, through watering and weeding, to harvesting. The loss of time also severely impacts and restricts women's ability to improve their lot, such as by working outside the home or by obtaining an education.

As wood is used up in the vicinity of a village and becomes scarce, negative aspects of the situation become apparent. People have to go farther distances, and therefore spend much more time, to get wood. The time spent doing this is taken away from other essential tasks. People may be forced to switch to poorer-quality fuels, such as agricultural wastes. The fuel quality will affect the performance of stoves or burners used for heating and cooking. People may instead wind up buying fuels, such as kerosene, instead of collecting and using free wood. Their limited financial resources may be used up, at least partially, to buy fuel rather than to purchase other necessities such as food or clothing.

FIGURE 6.8   Women in Kyambogo, Uganda, hauling firewood. (Photograph courtesy of Brian D. Martin, www.treeswaterpeople.org.)

There are environmental problems associated with the use of firewood. Often, all the available trees might be cut down over wide areas, for example, in circles of 15–20 kilometer radius around a village. This extreme devastation is called *deforestation*. Deforestation has a number of consequences. It can result in the loss of habitat for a wide variety of plants and animals, essentially, the destruction of an ecosystem. In fact, even a small stand of trees, which one might not think of as a "forest," can still have significant benefit to the local ecology, that is, the interrelationships of organisms with one another and with the environment in which they live. Some kinds of tough conifers can be planted on land that was ruined by erosion resulting from the clearing of previous hardwood forest. The soil can be restored by these hardy trees to a point at which other forest species can again thrive. Besides controlling erosion and rejuvenating soil, trees can have an important influence on the microclimate of land by providing shade, helping to moderate temperatures and humidity, and serving as windbreaks. When used as a windbreak, trees can also help to reduce heating needs for a home by as much as forty percent. Destruction of the trees means that their root systems, which held the soil in place, are gone too. In rainy periods, the fertile topsoil can be washed away. This loss of soil can result in mudslides and, if it's washed into rivers or lakes, can cause pollution of these bodies of water. Alternatively, destruction of the ecosystem can result in the spread of deserts (a process called desertification). Deforestation also contributes to the problem of global warming, to be discussed in detail in Chapter 29.

The Norwegian philosopher and ecologist Arne Naess* has documented the tragic consequences of deforestation in Nepal:

> *Deforestation above 9,000 feet began in earnest. Without trees as cover, erosion started on a grand scale. Today, practically nothing is left. Nepal became an "export country", exporting hundreds of millions of tons of soil to India. The rivers brought much of the soil to the giant Indian dams, built largely through misguided efforts at "development," and became filled with silt.*
>
> *With no cultural restraints and little Western scientific knowledge, the destruction of Nepal gradually reached gigantic proportions. Immigration from the countryside to the city of Kathmandu accelerated. The young were not willing to walk all day for fuel. They would rather live without dignity in Kathmandu.*

—Drengson and Devall[6]

The environmental problems of deforestation and desertification should not be minimized. However, cutting trees for firewood is not the principal cause of deforestation in most places. The major source of the problem is the cutting of forests to open up land for agriculture, either for growing crops or for grazing animals. Even commercial logging is not so severe as the cutting of forests for agriculture. In fact, in many Third World countries, firewood does not come from cutting forests; rather, it often

---

* Arne Naess (1912–2009) is regarded as the founder of the "deep ecology" movement. Naess had the distinction of being the youngest person ever to attain the rank of full professor at the University of Oslo, and, for a time, was not only the youngest, but the only, professor of philosophy in Norway. He was also an expert mountaineer.

comes from the practice of "agroforestry."* In some cases, the wood burned for fuel is the leftover scraps of wood remaining from other applications.

Numerous human factors impact deforestation. Overpopulation is an obvious and severe one. The national priorities in many impoverished developing nations might favor agriculture (especially for raising crops or animals that can be exported for hard currency). In some regions, especially Africa, national boundaries imposed quite artificially by long-departed colonial powers make it difficult for nomadic peoples to migrate across borders.† And we should never downplay the impact of utter managerial incompetence. In Brazil, for example, probably the dominant cause of deforestation in the Amazon basin are schemes of resettling of groups that have already failed once in agriculture because soil conditions were not adequate. Production of charcoal may also cause deforestation in some places. For example, in Brazil, trees are cut to produce charcoal for use in the steel industry. The exploitation of the resource by humans is made all the worse by the fact that many forest ecosystems are very fragile. Once the ecosystem has been damaged, it has difficulty recovering.

To sum up,

The poor people of the Third World have an ENERGY crisis—the lack of availability of firewood—that threatens their survival and their environment.

REFERENCES

1. Neale, J.M. *Carols for Christmas-Tide*. 1853. Available also in a vast number of other song books or web sites http://www.carols.org.uk/good_king_wenceslas.htm.
2. Tolstoy, L. *The Kingdom of God Is Within You*. Barnes and Noble Books: New York, 2005, p. 308.
3. Higginson, F. In: Hazen, M.H.; Hazen, R.M. *Keepers of the Flame*. Princeton University: Princeton, NJ, 1992, p. 3.
4. Trefil, J.A. *Scientist in the City*. Doubleday: New York, 1994, pp. 103–107.
5. Hazen, M.H.; Hazen, R.M. *Keepers of the Flame*. Princeton University: Princeton, NJ, 1992, p. 159.
6. Drengson, A.; Devall, B. *The Ecology of Wisdom: Writings by Arne Naess*. Counterpoint: Berkeley, CA, 2008, p. 282.

---

* Agroforestry is a land management practice in which trees are combined with crops, animals, or both, in a comprehensive approach to land use. The land yields a whole range of products rather than a single product (such as, say, wood, or grain, or meat). Ideally, agroforestry would rely on trees that are well suited to the region and that have multiple uses—providing shade for animals or crops, yielding fruits or nuts, and eventually being harvested for wood. Experience in some countries, like Kenya, indicates that, given help and support, farmers will actually work to increase the amount of wood growing on their farms—not, as sometimes imagined, ruthlessly chop it all down.

† Contrary to what is often asserted about nomads or the nomadic way of life, a true nomad does not wander aimlessly about, but tends to follow a relatively fixed pattern of seasonal migration. A splendid explanation is provided by Bruce Chatwin's article, "Nomads," in his collection of essays *What am I Doing Here* (Penguin: New York, 1989).

FURTHER READINGS

Bryce, R. *Power Hungry*. Public Affairs: New York, 2010. Several chapters relate to wood energy, including Chapter 4 on the transitions from wood to coal to oil as the dominant energy source, and Chapter 20 on the highly unlikely prospect of replacing coal as an energy source with wood.

Franklin, B. *Autobiography*. One of the enduring classics that is always in print and available in numerous editions. Franklin's discussion of stoves and fireplaces is particularly relevant.

Mann, C. *1493*. Alfred A. Knopf: New York, 2011. Chapter 5 discusses aspects of the sad history of deforestation in China, over a period of many centuries, culminating in the disastrous policies of Mao Zedong's regime. An excellent book.

Sloane, E. *A Reverence for Wood*. Dover Publications: Mineola, NY, 2004. An excellently illustrated book on the importance of wood in early America, and the many uses to which it was put.

Smil, V. *Energies*. MIT Press: Cambridge, MA, 1999. Chapter 4 provides some information on the importance of wood, along with charcoal and straw, to pre-industrial societies.

7 Combustion for Home Comfort

Most of us use fuels in our daily lives in two ways: for transportation and for domestic comfort—heating and cooking. Many kinds of fuels are used for domestic heating. These include liquefied petroleum gas (LPG), fuel oil, and kerosene, all of which are made from petroleum, natural gas, coal, and wood. Some people live in "all electric" homes with electric heating, but as we'll see in later chapters, most of the electricity that we use is generated in power plants that burn coal, petroleum products, or natural gas.

CENTRAL HEATING IN HISTORY

Central heating uses one central furnace to provide heat, which then is distributed through the house by heated air, hot water, or steam. Central heating offers many conveniences compared to the alternative of having a separate fireplace or stove in every room. The idea goes back at least 2000 years, to the Han Dynasty (202 BCE–AD 220) of ancient China. In northern China, homes were built with a raised floor, for sleeping, over the furnace or oven. Pottery models that show how this system worked have been found in archeological excavations of tombs from the Han Dynasty.

At about the same time (around 100 BCE), a concept of central hot air heating was developed in Rome (Figure 7.1). The person credited with the invention is Caius Sergius Orata, who was in the business of fish and oyster farming. To keep his livestock warm and healthy, Orata used a furnace to produce hot air, which passed underneath the bottoms of the fish or oyster tanks. The original concept, intended simply to help the fish and oyster business, was quickly extended to keep tanks of water comfortably warm for humans—that is, to provide the heating for public baths. The next step in development extended the concept further, to heat an entire house rather than to heat only the bath. As the Roman Empire expanded, particularly into the north and west of Europe, where the climate makes heating for comfort very desirable, so too did the expansion of hot air central heating.

The largest application of central heating in ancient times, in terms of size, was the use of central heating for public baths. In Constantinople (modern-day Istanbul), a naturally occurring petroleum-like material* was burned in open

---

* This material may have been a natural bitumen, which occurs in several places in Turkey. Bitumens are very viscous (i.e., thick) substances more dense than petroleum and boiling at higher temperatures than most petroleum products. Exactly how bitumens form is still a matter of debate; they may represent oil that seeped to the Earth's surface and was then chemically altered by contact with air or bacteria.

FIGURE 7.1   This image shows the remains of a central heating system, a hypocaustum, from the era of Roman occupation near modern-day Kempten, Bavaria, Germany. (From http://fr.wikipedia.org/wiki/Fichier:Kempten_Hypocaustum_2.jpg.)

rooms underneath the baths. The heat warmed both the floors and the water baths themselves. This system worked too well, in a sense, because in some baths, the bathers had to wear sandals with thick soles to avoid getting burns on the feet. The largest of Constantinople's public baths was used by more than 2000 people a day. Operation of all of the public baths must have involved a prodigious consumption of fuel.

In the United States, central heating began to be incorporated in homes toward the end of the nineteenth century. Many of the earliest central heating systems were of similar design to boilers used to generate steam in factories. The improvement in comfort, relative to earlier heating systems, was tremendous. Before the advent of central heating, those rooms with a fireplace or stove might seem to be blazingly hot, yet other rooms in the house would be miserably cold. A central heating system allowed heat sources, such as radiators, to be located at many points throughout the house, resulting in a much more even heating. The central furnace was not so wasteful of fuel as a fireplace, which often sent much of the heat straight up the chimney. By the early twentieth century, a centrally heated American home, even a small, modest dwelling owned by people of limited financial means, was far more comfortable in the wintertime than the grandest of European manors and palaces with their scattered fireplaces.

The ancient Romans were masters of central heating. Few other early cultures developed central heating, possibly because many were located in regions of moderate climates where the modest amount of heat from an open fire was adequate. Central heating largely disappeared with the decline and eventual demise of the western part of the Roman Empire, around AD 300–450. Remarkably, it took about 1500 years before central heating was revived in the West.

STOVES

During the fifteenth century, in the regions that are now part of southern Germany, a stove was developed to burn coal or wood. It looked roughly like an iron box that someone had bolted together. A major advantage of a closed stove, relative to the more common fireplace, was that it would put out much more useful heat than the fireplace or produce the same amount of heat with less fuel. On the other hand, a closed stove does not provide the same emotional comfort or satisfaction as from an open fire.

The first so-called Franklin stove was developed in the 1740s (Figure 7.2). The Franklin stove did not achieve immediate popularity. Its most serious drawback was that it could not be used for cooking, because early designs called for it to be installed in the fireplace, replacing the open fire and taking up enough space to make cooking on or around it difficult. With the relatively primitive metallurgy of the time, the plate iron of which the stove was made tended to crack. The market for the Franklin stove was limited to the rich, who could afford to own both a fireplace for cooking and a separate stove for heating. An iron stove is heavy, which meant that it would be expensive to transport, especially so in the days before railroads, when freight transport was limited to horse-drawn or ox-drawn wagons. The transportation problem effectively limited a given market to a very small one that could be served by small manufacturers selling only to customers in the immediate locality.

Solutions to these two problems came in the middle of the nineteenth century. First, a rapidly expanding system of transportation made iron and coal available over a wide area, but at still relatively low prices. The new transport came in the form of canals (especially in Britain and the United States) and, by mid-century, railroads.

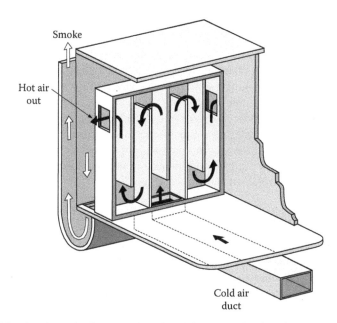

FIGURE 7.2   A schematic diagram of the Franklin stove, showing the cold air intake, air circulation, and the exit for the heated air. A picture of a Franklin stove is shown in Figure 6.5.

FIGURE 7.3 An ornate wood stove, the 1891 Radiant Home model, built in Erie, Pennsylvania. (Used with permission of Doug at the Barnstable Stove Shop, West Barnstable, Massachusetts.)

Canals and railroads made the stoves themselves readily available to consumers over a wide geographical area and helped in a second way by lowering the price of the fuel (coal) that would be used in them. Around the same time, it was learned how to work iron to produce parts that wouldn't crack. As an added benefit, the improved manufacturing allowed making a stove that could be used for both heating and cooking (Figure 7.3). The mid-nineteenth century was also a time of the great westward emigration in the United States. The same thing was happening elsewhere in the world. Russians were pushing east into Siberia. Boers moved north in what is now South Africa. These new iron stoves were ideal for the pioneers traveling by wagon or ox cart, because they could be taken apart, transported in the wagon, and then reasonably easily reassembled and got working.

COMBUSTION: THE STORY SO FAR

Several lines of evidence suggest a role for air in combustion. A supply of air facilitates the ignition and expansion of a fire, for example, by blowing on it or by fanning it. A fire can be put out by smothering it to exclude air. Oxygen is the component of air that reacts with the fuel in the combustion process. Three possible combustion processes can take place; which one dominates depends on the amount of air (oxygen) available. Here we continue the example of methane combustion first introduced in Chapter 5. With abundant oxygen, complete combustion occurs:

$$CH_4 + 2O_2 \rightarrow CO_2 + 2H_2O$$

Methane + oxygen → carbon dioxide + water

(recall Figure 5.4). Notice that the carbon in the fuel has become carbon dioxide, CO_2. Of the three reactions we discuss here, this one provides the greatest amount of heat per unit of methane burned.

If not enough oxygen is available for complete combustion, we obtain incomplete combustion (Figure 7.4):

$$CH_4 + 1.5O_2 \rightarrow CO + 2H_2O$$

Methane + oxygen → carbon monoxide + water

Notice that in this case, the carbon-containing product is carbon monoxide, CO. *Carbon monoxide is a deadly poison.* Its formation in combustion processes that are not supplied with an adequate amount of air can lead to accidental death. This can happen with stoves or furnaces that are not properly adjusted to allow sufficient air for the amount of fuel present, or sometimes may happen when an amateur car mechanic is working on a car with the engine running in a closed garage, not obtaining adequate ventilation. This incomplete combustion reaction yields less heat per unit of methane burned than does complete combustion.

In very limited amounts of oxygen, we obtain oxygen-starved combustion (Figure 7.5):

$$CH_4 + O_2 \rightarrow C + 2H_2O$$

Methane + oxygen → carbon (soot) + water

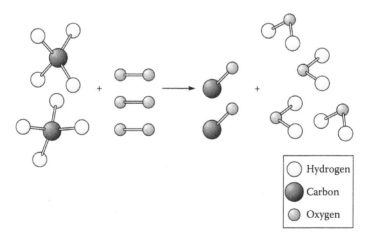

◯	Hydrogen
●	Carbon
◯	Oxygen

FIGURE 7.4 The molecular structural changes accompanying the incomplete combustion of methane to carbon monoxide and water.

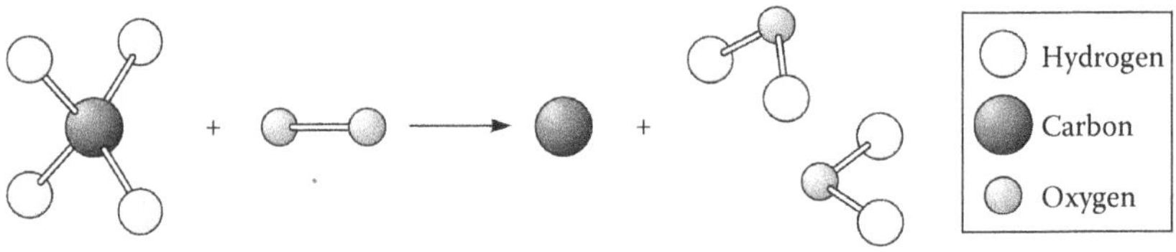

FIGURE 7.5 The molecular structural changes accompanying the oxygen-starved combustion of methane to soot (carbon) and water.

Carbon usually forms as soot. As a minimum, soot formation is aesthetically undesirable. In addition, soot can contribute to air pollution or, if inhaled, to human respiratory tract problems. This reaction produces the least amount of heat per unit of methane burned.

THE FUELS

WOOD

After serving as the main home heating fuel for centuries, wood was largely abandoned when coal, oil, and gas became available for domestic use in industrialized nations. Fireplaces have continued to be a decorative feature in homes, but their role in heating has largely been neglected. Still today the open wood-burning fireplace seems to provide powerful emotional or psychological comfort, as well as some nostalgic link to a simpler (and, as often presumed, better) time. Wood fires in the home, in fireplaces or so-called "wood-burners," can be a useful supplement to central heating methods. Properly managed, wood combustion as an auxiliary source of home heat can cut the consumption of oil or natural gas. It's questionable whether wood would ever experience a resurgence that would make it again a dominant choice for central heating in the home. Wood-fueled units simply can't provide the convenience of oil and gas heaters, which require no significant amount of time or work on the part of the householder. And, it's doubtful whether any populous area could possibly grow enough wood to meet demand.

COAL

Cast iron stoves for cooking and heating with coal were developed in the 1830s and became fairly common by the 1850s. Like the use of wood, the domestic use of coal also required an immense amount of labor. Coal had to be cleaned (to remove rocks, for example), the pieces sorted by size, and hauled from the coal bin to the stove or fireplace. The fire had to be kept going by periodic additions of more coal. Ashes had to be raked out and taken out of the house.

With both coal and wood, there was also the problem of cleaning the chimney to remove accumulations of soot and creosote that could catch on fire there. A fire in the chimney could ignite the rest of the house. Various methods were used for cleaning the chimney, such as a brush on a long handle, or a chain lowered down the chimney and rattled to knock soot and creosote deposits loose, or, in what must

FIGURE 7.6 A Swedish chimney sweep, from the Stockholm area, photographed in the 1880s. (From http://commons.wikipedia.org/wiki/File:Sotar1%C3%A4rlingen_Viktor_Norin_Stockholm_kabinettsbild_1880-tal_-_Nordiska_-_Museet_-_NMA.0041502.jpg#file.)

have been a memorable day in the household, stuffing a live goose up the chimney. In human terms, though, the worst method was the use of "climbing boys" by professional chimney sweeps (Figure 7.6).

In the days long before occupational health and safety laws, and before child labor laws, young boys would be put to work either to climb upward through the chimney or to be lowered downward on a rope. They would scrape away the accumulated soot and creosote. This job must have been about one step removed from actually being in hell. The first disease ever traced to occupational exposure was discovered in chimney sweeps and their climbing boys: cancer of the scrotum. Some of the chemicals that occur in the creosote or tar that accumulates in the chimney are carcinogenic. Steady exposure to such chemicals on a daily basis leads to this dreadful form of cancer. Worse, sometimes boys might be put into the chimney *while it was still hot*. This horrific testimony was recorded in England in hearings of the Parliamentary Committee on Climbing Boys (1817):

> *On Monday morning, 29 March 1813, a chimney sweeper of the name of Griggs attended to sweep a small chimney in the brewhouse of Messrs. Calvert and Co. in Upper Thames Street; he was accompanied by one of his boys, a lad of about eight years of age, of the name of Thomas Pitt. The fire had been lighted as early as 2 o'clock the same morning, and was burning on arrival of Griggs and his little boy at eight. The fireplace was small, and an iron pipe projected from the grate some little way into the flue. This the master was acquainted with (having swept the chimneys in the brewhouse for some years), and therefore had a tile or two broken from the roof, in order that the boy might descend the chimney. He had no sooner extinguished the fire than he suffered the lad to go down; and the consequence, as might be expected, was his almost immediate death, in a state, no doubt, of inexpressible agony. The flue was of the narrowest description, and must have retained heat sufficient to have prevented the child's return to the top, even supposing he had not approached the pipe belonging to the grate, which must have been nearly red hot; this however was not clearly ascertained on the inquest, though the appearance of the body would induce an opinion that he had been unavoidably pressed against the pipe. Soon after his descent, the master, who remained on the top, was apprehensive that something had happened, and therefore desired him to come up; the answer of the boy was, "I cannot come up, master, I must die here." An alarm was given in the brewhouse immediately that he had stuck in the chimney, and a bricklayer who was at work near the spot attended, and after knocking down part of the brickwork of the chimney, just above the fireplace, made a hole sufficiently large to draw him through. A surgeon attended, but all attempts to restore life were ineffectual. On inspecting the body, various burns appeared; the fleshy part of the legs and a great part of the feet more particularly were injured; those parts too by which climbing boys most effectually ascend or descend chimneys, viz., the elbows and knees, seemed burnt to the bone; from which it must be evident that the unhappy sufferer made some attempts to return as soon as the horrors of his situation became apparent.*

—**Carey**[1]

Coal required a great deal of domestic labor, well over an hour each day to sift the ashes, lay the fire, add coal to it, carry coal and empty the ashes, and put "blacking" on the stove. (Blacking was a paste applied to the stove to give it a uniform, glossy black polish.) Using coal introduced a split in the way labor is applied to energy supply. With wood, all of the necessary labor could be—and quite frequently was—supplied by the household, even beginning with cutting the trees if need be. As coal became important, things began to change. Some of the work relating to the use of coal had to be supplied not by the householder, but by specialists—specifically, coal miners and those who are employed in distributing and selling coal.

Nowadays coal is rarely used as the primary source of heat in homes, at least in most industrialized countries. Several perceived problems with coal diminish its popularity and usefulness for domestic heating. It produces the least amount of heat, per unit weight, relative to natural gas or petroleum products (though it is higher than wood). Solids, like coal, cannot be conveniently metered or controlled in domestic furnaces or stoves. Compared to oil or gas, coal produces a substantial amount of ash residue on combustion, which is undesirable because it must be collected and hauled away, and a fine dust of ash particles can get through the house. Handling lumps of coal produces some coal dust, from lumps knocking or grinding against each other, and this too can be a nuisance inside the house. Some coals burn with a smoky flame, again adding to pollution in, and outside, the home. In contrast to oil and natural gas, it takes some effort to kindle a coal fire, and once it is going, it burns best if kept at a steady heat output. That is, it is not easy to "turn down" or "turn up" a coal fire, and coal lacks the quick on/off response of gas or oil. An abundant amount of storage space, usually in the basement, must be set aside for a coal bin. One very positive aspect of coal is that, at least in some places, it can be by far the least expensive fuel in terms of its cost per unit heat produced.

PETROLEUM PRODUCTS

Oil products have numerous advantages relative to coal or wood. These include the facts that oil furnaces offer essentially instantaneous on/off control, eliminating the laborious task of "building" a coal or wood fire and waiting some time for it to give off abundant heat; no dust or dirt in handling the fuel, and no ashes to be collected and disposed of. The rise of oil heating in homes was helped greatly by the invention of thermostats in the 1920s. Thermostats allowed totally automated home heating, controlled to a chosen temperature. The cleanliness of oil relative to coal and the ease of controlling an oil furnace helped oil displace coal for home heating, especially shortly after the Second World War, as oil pipelines spread through the United States and heating oil became readily available. Even in the heart of coal territory, such as the anthracite-mining region of northeastern Pennsylvania, it became a status symbol in the 1950s to have one's old coal furnace torn out and replaced by an oil burner.

Several fuels derived from petroleum are in use for domestic heating. Fuel oil is the most popular of these. Kerosene is a useful fuel, especially for auxiliary space heaters, such as might be used to heat a garage or basement area.

> *...if a stove was not burning satisfactorily you poured a gallon or two of kerosene on the flames. Women were burned bald, and children were consumed like little phoenixes (although they never rose again from their ashes) and even fathers of households were Called to Their Reward blazing like heretics in the fires of the Inquisition, and shrieking to be extinguished. But with the cost of kerosene nowadays only the very rich would be able to afford this form of spectacular demise.*

> —Davies[2]

Fuel oil shares many of the desirable features of natural gas. It has a high calorific value (though not as high as gas). It is a fluid and is relatively easy to monitor, control, and automate. It generally provides a quick on/off response. On the other hand, fuel oil has some slight disadvantages. Many fuel oils contain some small amount of sulfur, which contributes to air pollution. Fuel oils may also contain small amounts of incombustible material that accumulates in the furnace as ash, requiring a periodic cleaning and maintenance of the furnace. Fuel oil does require some space to be set aside for its storage, and some arrangements to be made with the local fuel company for delivery.

NATURAL GAS

Gas heaters were introduced early in the twentieth century. Natural gas is now the most popular fuel for domestic heating in the United States. Gas shares many of the advantages of oil as a source of domestic heat: it can be automated and controlled by thermostats, there is no dust or ashes, and there is an instant on/off capability. A furnace working properly produces no soot. Gas also has two additional advantages that oil does not have. First, oil burners are complex. The oil has to be "atomized" into droplets and sprayed into the air in the furnace. Gas mixes freely with air, so generally a gas furnace is simpler and more trouble free than an oil furnace. Second, oil has to be stored at the home, in tanks in the basement or outdoors near the house. Gas, though, is supplied straight from the distribution system, so there is no storage problem and no wasted space in the home to store fuel (such as fuel oil tanks or a coal bin). Also, there's no need for the householder to have to make arrangements for delivery of the fuel. Finally, natural gas has the highest calorific value, per unit weight, of any common fuel.

If cost were the only consideration, gas could not compete with coal. In terms of cost per useful amount of energy, coal is the cheapest fuel. Rather than compete on the basis of price, the gas industry stressed many of the advantages of gas that we have listed earlier. In terms of price, coal is the easy choice. In terms of convenience and cleanliness, gas is the easy winner. Gas remained the preferred fuel even after price increases in the 1970s. Now, with the discovery and development of huge amounts of so-called shale gas in many countries, the price of gas may drop to a point where it can compete not only in terms of convenience and cleanliness but also on cost.

LPG is a useful gaseous fuel, particularly in areas not served by natural gas pipelines. Propane often serves as a clean domestic fuel in such areas. Its use usually involves one or more cylindrical tanks directly adjacent to a house or sitting in the backyard. The propane tanks themselves can be refilled by delivery from a tank truck specially set up to handle liquefied gas, but otherwise not much different from the delivery of home fuel oil. Except for the need for periodic resupply, using propane is otherwise about as convenient as natural gas.

The switch from coal to oil and gas as preferred fuels for home heating also meant an essentially complete shift in the application of labor, from the householder to specialized labor. With oil and gas, essentially all of the labor is supplied by specialists, the oil well drillers, the people who work in refineries, those who work with distribution and supply systems, as examples. For wood, all of the energy supply of the household could be obtained by those in the home, starting with the cutting of the trees. Though few people would actually try to mine or collect their own coal, it can nevertheless be done. For example, some very impoverished people managed to heat their homes by walking along railroad tracks and collecting coal that had fallen off railroad cars. *Nobody* is likely to drill his or her own oil well and set up a backyard oil refinery. In the course of the transition from wood to coal to oil or gas, individual homeowners have changed from energy producers to energy consumers.

A century ago, heating, cooking, and lighting of homes were hard physical labor for the homeowner or servants. With the advent of oil and natural gas, the homeowner no longer had to chop wood, break coal, bring wood or coal into the home, and collect and haul ashes. Now it was only a matter of hitting a switch or twisting a thermostat. But because of the shift in the application of labor, the homeowner could no longer provide energy for himself or herself. The homeowner now had to rely entirely on workers in the oil and gas industries. The hard physical work of the nineteenth century also meant that a home was essentially "energy autonomous" because, if need be, the whole energy supply could be provided by the homeowner. The physical labor was traded for the great convenience of gas and oil. The price was surrendering autonomy and relying on specialized labor—and, of course, the monthly utility bills.

THE SMOKE PROBLEM

One of the most persistent problems in home heating is what to do about the smoke that is often produced, especially from open fires (Figure 7.7). Anyone who has sat too close to a campfire or to a smoky fireplace in the home may have experienced sore eyes; damaged clothing; food tasting of smoke; clothes or hair smelling of smoke; smoke and ashes on clothes, in the hair, or around the house; even damaged furniture. Benjamin Franklin complained about "skins almost smoked to bacon."

FIGURE 7.7 This outdoor fire, at a "burning of the winter" celebration in Zurich, Switzerland, is producing plenty of smoke. The aroma of the smoke will be evident in the clothing and hair of the crowd. (From http://commons.wikimedia.org/wiki/File:%27B%C3% B6%C3%B6gg%27_-_2012_Sechsel%C3%A4uten_2012-04-16_18-11-42.JPG.)

Persons who already have respiratory problems can find them made all the worse by exposure to smoke.

The problem has been with us since the Stone Age. The first solution was a fairly simple and direct one—simply to cut a hole in the roof to let the smoke escape. This of course has some rather obvious disadvantages, especially when it's raining. A much better solution was the invention of the chimney. A chimney is mentioned in the writings of Theophrastus (a pupil of Aristotle) in Greece in the early fourth century BCE. However, chimneys only became common some 1800 years later. By the twelfth century, building construction had advanced to a point where large rooms in which the feudal aristocracy held court were now being installed on the second story of buildings. To accommodate this change in design, it was also necessary to move the central fireplace or hearth from its customary position in the center of the room so that, instead, it was up against an outside wall. A central hearth would have been quite inconvenient in a two-story house except perhaps in a very lofty main room or hall that provided sufficient space for the smoke to escape. Even worse, on an upper level, a central hearth would have been positively dangerous without a solid flooring of noncombustible material such as stone.

At the same time, the fireplace or hearth was also covered with a canopy or hood in an effort to collect the smoke. Because matches did not exist at that time (matches were not invented until the mid-nineteenth century), it was easier to keep the fire lit continuously, rather than to have to light a new fire each time one was wanted. Keeping the fire lit was not safe in a central hearth at night without some means of protecting the inhabitants from the fire. The protection was provided by a so-called fire cover placed on the fire. Many of these covers were made of pottery, a circular lid with holes for ventilation. In many of the large houses, a bell would be rung to signal the inhabitants that the fire was about to be covered and further signaled that the time had come to cease household activities for the night. After the transition from central hearths to fireplaces mounted against an external wall, the ringing of the fire-cover bell itself, rather than the covering of the fire, was the signal that activities were over for the night. In French, the fire cover, and the associated bell, was known as the *couvre-feu*. The anglicized version is *curfew*.

Smoke remained a chronic problem for homemakers for many more centuries (Figure 7.8). In the eighteenth century, six hundred years after chimneys began to become commonplace in the large buildings of northern and western Europe, people were still struggling with smoke in the home. A significant part of the overall problem was recognized by Benjamin Franklin. It was simply the poor design of many of the fireplaces and chimneys. Franklin addressed this problem by inventing a particularly effective type of stove, as we will discuss in the next section. Benjamin Thompson (Count Rumford) redesigned the common fireplace to prevent downdrafts (currents of air) from blowing smoke into the room (Figure 7.9). (As we will see in later chapters, both of these Benjamins have had several important roles in the development of energy.)

Another inventor of the time, Charles Peale, developed the so-called smoke eater. Smoke contains a great deal of soot, formed by essentially carbonizing the fuel rather than burning it completely. Under the right conditions, soot can be burned,

FIGURE 7.8 When most households burned wood, or coal, air pollution due to smoke was a serious problem in cities. (From http://commons.wikimedia.org/wiki/File:English_ School,_19th_Century,_Snow_Hill,_Holburn,_London.jpg.)

FIGURE 7.9 The fireplace designed by Benjamin Thompson (Count Rumford) was intended to reduce significantly the smoke produced from wood fires. (From http://en.wikipedia.org/ wiki/File:RumfordFireplaceAlc1.jpg.)

since it is mainly carbon. Peale's smoke eater forced smoke back through the fire to remove the soot by burning it.

Improving the way the fireplace was designed and built addressed only part of the problem. Numerous other factors also had a role. For example, wet or green wood can be counted on to produce smoke, an observation that many a camper has verified. An excessive buildup of soot in the chimney could cause smoke (or, worse, could contribute to a chimney fire). Even the prevailing weather was important, because sometimes winds could affect the way a chimney would "pull" smoke. The situation was helped somewhat as cooking and heating stoves began to replace fireplaces as the principal source of heat. Stoves had the advantages that the fire was enclosed and that smoke could be controlled by regulating the amount of air allowed into the stove via dampers. Use of stoves instead of fireplaces was not a complete solution. Smoke production and emission were especially hard to regulate when several stoves throughout the house were connected into a single chimney. For example, sometimes a gust of wind can produce enough air pressure at the chimney to prevent smoke from escaping; some of the smoke can be forced back down the chimney into the fireplace. When multiple stoves are connected into a common chimney, the problem is worsened, because smoke can blow from a stove in one room partway up the chimney and then be forced out into a different room.

ABOUT THE HUMBLE MATCH

Until about two centuries ago, the actual creation of fire was still a very difficult job. Fires were started by using flint and steel, rotating "fire sticks," or simply by resorting to "borrowing" fire from a neighbor. The ordinary match was developed, in a reasonably practical form, in 1836. It was so remarkable an invention, in terms of giving people the ability to create fire as needed, that some surveys of nineteenth-century technology listed the match as one of the three greatest inventions of the century (the others being the steam boat and the cotton gin).

Matches were first invented in 1827. To be useful, matches need to be able to be ignited very easily. At first, a very inflammable form of phosphorus* was the chief ingredient of matches. However, this material can be very hazardous, leading to severe problems with the gastrointestinal tract and liver, circulatory problems, and even death. Its use was outlawed by many countries during the nineteenth century. Matches are of two types: the "strike-anywhere" match, which can be ignited on any handy surface, and the "safety" match, which requires a special surface on the matchbook or box for use.

---

* Phosphorus, like some other elements, can exist in very different forms. Often the different forms, called allotropes, display very different physical properties and chemical reactivities. The best-known allotropes occur with the element carbon: diamond and graphite. The phosphorus used in the original matches was the allotrope commonly called white phosphorus. It is extremely poisonous, as little as about one hundred milligrams being fatal. If handled with unprotected fingers, it can produce deep slow-to-heal skin lesions. White phosphorus will ignite spontaneously in air at temperatures as low as 30°C–35°C. Suffice it to say that a material this nasty and dangerous to handle has found a variety of military uses and uses in rat poison.

Strike-anywhere matches typically contain a material that is easily ignited by friction, a material to provide that friction, an oxidizer to help the burning take place, and glue, to hold the other ingredients in place. Often phosphorus sulfide is the material to be ignited, finely ground glass helps provide the friction, and potassium chlorate serves as the oxidizer. When the match is rubbed on any surface, the phosphorus sulfide is ignited and the potassium chlorate helps to sustain a vigorous burning. Although the combustion of the phosphorus sulfide takes place in a very short time, it usually generates enough heat to ignite the wood of the matchstick. In some matches, the stick will have been dipped in paraffin to help it burn more easily.

Safety matches contain the ground glass for friction and a material to assist ignition (usually a different, somewhat less dangerous form of phosphorus*) on the dark-colored strip found on the side of the box or at the bottom of the matchbook. The tip of the match contains the oxidizer, potassium chlorate, and a substance to initiate burning, such as antimony sulfide. In this case, the most combustible substance in the system, the phosphorus, is separated from the oxidizer. When the match is struck, friction produces enough heat to ignite a bit of the phosphorus on the side of the box. It in turn quickly ignites the antimony sulfide–potassium chlorate mixture on the match itself. The burning of a bit of the phosphorus on the igniting strip each time a match is struck is why, after such a strip has been used many times, it becomes difficult to light other matches on it. Safety matches get their name from being relatively safer than the strike-anywhere matches, because of the separation of the phosphorus and the oxidizer. Either kind can of course lead to property destruction, injury, or loss of life if misused.

REFERENCES

1. Caray, J. (Ed.) *Eyewitness to History.* Harvard University Press: Cambridge, MA, 1987, pp. 280–281.
2. Davies, R. *The Papers of Samuel Marchbanks.* Viking: New York, 1986, p. 369.

FURTHER READINGS

Bloomfield, L. *How Things Work.* John Wiley & Sons: New York, 2010. Chapter 7 includes a useful discussion of how wood stoves operate, with related concepts of heat, temperature, and combustion reactions.

Norton, T. *Smoking Ears and Screaming Teeth.* Pegasus Books: New York, 2010. The chapter entitled "Suffer" touches on the potentially lethal effects of carbon monoxide. It is not a good idea to try any of the experiments in this book at home.

Tenner, E. *Why Things Bite Back.* Knopf: New York, 1996. This book examines the "revenge of unintended consequences" in technology—how a technological advance sometimes leads to consequences no one had ever foreseen. Chapter 4 includes a discussion of the unintended indoor air pollution arising from stoves and furnaces.

Wood, M. *The English Mediaeval House.* Studio Editions: London, U.K., 1994. The evolution of English houses up to about 1540. Chapters 17–19 particularly relate to arrangements for cooking and heating.

---

* The friction strip used to ignite safety matches contains a different allotrope of phosphorus, called red phosphorus. It is not nearly so reactive nor dangerous to handle as white phosphorus—unless it contains traces of white phosphorus as an impurity, it is relatively safe to handle and work with.

8 Waterwheels

Likely, the second major ENERGY source to be developed and exploited for doing WORK was the ENERGY that is available in a flowing body of water.

KINETIC ENERGY

Start with a simple observation that all of us make, often in childhood. When wading in a stream of water, we can feel the water pushing against our legs. In a very rapid stream, the "push" of the water might be great enough that we have to be careful not to be knocked down by the rushing water. Wading on the coast, we can feel the push or pull of the tide on us. Sometimes we can get swept off our feet by a big wave coming in. Certainly if we were to stand under a waterfall (Figure 8.1), we would feel the rush of water over our bodies.

Now consider a simple wheel with paddles on it (Figure 8.2). If this wheel were partially immersed into a flowing stream, water will push against it, just as it pushes against us (Figure 8.3). By pushing against only one paddle (in this illustration), the water will cause the wheel to turn. Then, connecting the shaft of the wheel to some piece of machinery, we can do useful WORK, such as the grinding of grain. This simple device illustrates the concept that is the basis of the waterwheel.

But this simple waterwheel would not operate in every situation. If we were to dip it into a swimming pool or pond, the wheel wouldn't turn. Similarly, standing or wading in a swimming pool or a pond, we don't think about being knocked down by the water. What's the difference? In a swimming pool, pond, or other body of water that has no flow, the water isn't moving. Only moving water can push against the waterwheel or push against us. To make a waterwheel move means doing WORK. Only moving water has the capacity for doing WORK. Water that is standing still, as in a pool, does not. This means that moving water must possess some kind of ENERGY that allows us to do this WORK.

The ENERGY associated with moving bodies is called kinetic energy.

Of far more importance for society is to find a way to transfer the WORK done by the water on the waterwheel into some kind of useful output. That is, we would like to capture this WORK and use it to supplement or replace WORK that would otherwise have to be done by direct human labor. Many kinds of tasks can benefit from this WORK: sawing wood, crushing rocks, or grinding grain provide examples.

Kinetic energy can be used to do WORK.

FIGURE 8.1   This person is experiencing the kinetic energy in a waterfall, in this case near Banos, Ecuador.

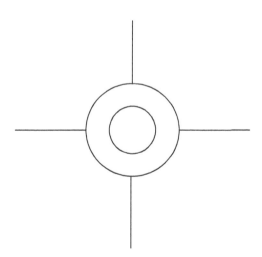

FIGURE 8.2   Attaching a set of paddles to a wheel provides a simple way of making a primitive waterwheel.

POTENTIAL ENERGY

To use water to do WORK requires that the water be in motion. Waterfalls provide some of the best sources of water in motion. The biggest and most spectacular even become tourist attractions (Figure 8.4).

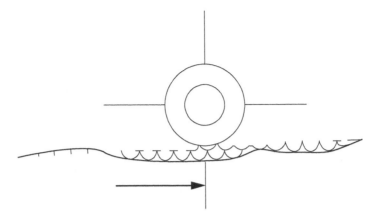

FIGURE 8.3   Partially immersing a simple waterwheel (like Figure 8.2) into a stream allows the kinetic energy of the moving water to turn the wheel.

FIGURE 8.4   Victoria Falls, on the Zambezi River in southern Africa, is one of the great waterfalls of the world. (From http://www.sxc.hu/photo/471565.)

An ambitious way of using a waterwheel would take advantage of a waterfall. The kinetic energy of falling water turns the wheel and thus—with appropriate machinery—does WORK.

Consider a volume of water just at point **A** of a waterfall, as indicated in Figure 8.5. Can that bit of water do WORK? Not in the waterwheel (at least not yet), because it does not have kinetic energy in the direction downward along the waterfall.* We do know, though, that the water at **A** could *potentially* do WORK. All we need to do is let that bit of water acquire kinetic energy, by falling off the cliff over the waterfall.

---

* Strictly speaking, the water at point **A** does indeed possess some amount of kinetic energy, because it is moving with the stream that is about to go over the falls. We deliberately neglect that bit of kinetic energy in the discussion because it is of no use to the waterwheel.

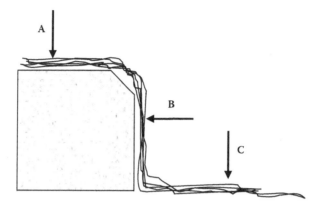

FIGURE 8.5 This cartoon sketch illustrates the relationships between potential and kinetic energies in a waterfall. At point **A**, the water has high potential energy. At point **B**, some of the potential energy has been transformed to kinetic energy, where it is available to do WORK. At point **C**, the water has lower potential energy than at point **A**, but it also retains some kinetic energy, because it is flowing away from the waterfall.

In other words, at **A**, the water has no kinetic energy, but *potentially* it could acquire some. To describe this situation, we say that the water at **A** has *potential energy.*

What makes the water fall? It is pulled, toward the center of the Earth, by the force of gravity. To be complete, we would say that the water at **A** has *gravitational potential energy*, because its potential is due to its position in Earth's gravitational field. Gravitational potential energy is the energy associated with an object by virtue of its position in a gravitational field.

Now let's add three more points. First, potential energy can be converted into kinetic energy. Second, point **B** (Figure 8.5) is the point at which the water is hitting the blades of the waterwheel. The kinetic energy of the water is used here to impart kinetic energy to the wheel; in turn, the motion of the wheel can be used to do WORK. Put colloquially, point **B** is "where the action is." Third, consider what has happened to our volume of water at point **C**. We might be tempted to conclude that it no longer has potential energy. However, that's not quite true, because it could—in principle, at least—encounter a further waterfall somewhere down the stream. The water still has some gravitational potential energy at point **C** because it retains the potential of falling still further. In fact, it will always have some gravitational potential energy unless it happens to be at the center of the Earth.

To recap, water at point **A** had high gravitational potential energy. In passing over the waterfall, some of that potential energy was converted to kinetic energy. The kinetic energy of the falling water was able to do WORK, turning the waterwheel. Then, the water was at a new lower gravitational potential energy.

SPONTANEOUS CHANGE AND THE ENERGY DIAGRAM

We know from many, many personal observations that things always fall *down*. That includes us, when we stumble or trip. Anything that is able to move will spontaneously fall from a position of high gravitational potential energy to a new position of low gravitational potential energy. We also know from abundant observations that things never spontaneously fall *up*. (Sometimes we see an effect on television or in the movies where a film is run backward and people or things do "fall up." Usually this is done for its humor. We find it funny because it is so contrary to our usual expected experience.)

> *For every natural event, there is an allowed direction and a forbidden direction. Apples fall; they don't jump up to their branches. Sugar lumps dissolve in your coffee; they don't form by themselves in a cup of sweetened coffee.*
>
> **—de Duve**[1]

Objects move spontaneously from high gravitational potential to low gravitational potential. We've also seen that we can extract kinetic energy—in this case, by using a waterwheel—when that change from high to low potential energy occurs.

Figure 8.5 depicts a SYSTEM consisting of water, a waterfall, and a waterwheel. The essentials of this SYSTEM can be represented by a simple diagram (Figure 8.6). This diagram resembles, but is not identical to, a diagram introduced earlier (Figure 5.5) for another SYSTEM—a carbon atom, four hydrogen atoms, and four oxygen atoms—from which we could also extract ENERGY.

In the case of the chemical changes involved in combustion, the SYSTEM was in a state of high "chemical" potential energy at the beginning, analogous to a SYSTEM in a state of high gravitational potential energy for a waterwheel in a waterfall. In fact, these two diagrams are so similar that they can be combined into a "generic energy diagram" (Figure 8.7) that applies to both SYSTEMS.

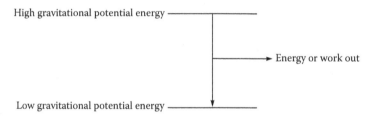

FIGURE 8.6 The energy diagram for a waterwheel. Compare Figures 5.5 and 5.6.

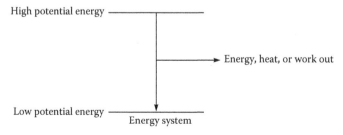

FIGURE 8.7 The generic energy diagram, expanded from the version shown in Figure 5.6.

Figure 8.7 is a crucial diagram that will explain the behavior of a large number of systems that will be encountered throughout the rest of this book. The central concept that goes along with this diagram is this:

When a SYSTEM undergoes a spontaneous change from high potential to low potential, we can extract ENERGY (or WORK) from the SYSTEM.

WATERWHEELS IN THE ANCIENT WORLD

The waterwheel allows us to extract ENERGY from a SYSTEM—moving water—undergoing a spontaneous change from high potential to low potential energy (in this specific case, gravitational potential energy). By connecting some sort of device to the shaft or axle of the waterwheel, its kinetic energy can be used to do WORK. In fact, the waterwheel was the first practical device that freed humans from reliance on their own muscles or on draft animals for WORK. We don't know where or when waterwheels were first used, or who developed them. Like the other inventions or discoveries from prehistory or antiquity, there could well have been multiple independent discoveries at various places around the world and at various times.

Around 300 BCE, there may have been some use of waterwheels connected, via a chain drive, to a set of buckets that could scoop up water and deliver it for household use or for irrigation. This is the simplest, and almost certainly the earliest, application of waterwheels. The wheel itself was a simple paddle wheel that would have been driven by the current of the stream. This type of device is called a noria (Figure 8.8). The noria may have originated in the Mediterranean region. Some were of quite large size, about thirty meters in diameter.

Greece laid claim to the first mill, supposedly owned by the son of the first ruler of Sparta. If such a mill actually existed, it would go back to the seventh century BCE. By about 100 BCE, waterwheels were in use in several scattered regions of the world—northern Turkey (probably the first), Greece, and India—for grinding grain and extracting oil from olives. These wheels were so-called horizontal wheels mounted so that the shaft stuck up vertically (Figure 8.9).

The horizontal waterwheel is also sometimes called the Norse wheel. For grinding grain or other materials, the shaft of this wheel was used to turn a large, heavy, flat stone, called the millstone.* A second millstone was mounted atop the one that was turned by the waterwheel. Grain or other materials fed between the two heavy turning stones were ground into much finer sizes, such as flour. Even today, it is possible to purchase, in the baking supplies section of a supermarket, flour that is advertised as being "stone ground." Horizontal wheels had the practical problem of the need to build some sort of channel so that the water hits only one side of the wheel, to get it to turn.

The horizontal wheel was most likely the first appliance for the "mechanization" of the home. Until the horizontal or Norse wheel was developed, grain had to be

---

* The millstones were very heavy, and usually large, which made them rather difficult to move into place or to handle. The word *millstone* has evolved into everyday language to refer to anything or any situation that represents a heavy and oppressive burden.

FIGURE 8.8 A noria waterwheel in Hama, a city on the Orontes River in modern-day Syria. (Photograph courtesy of Alex Chan, Auckland, New Zealand.)

ground (to make flour) in hand-operated mills. The women of the household had to do this job, and in large households, this could keep the women occupied all day. The water-driven mill for grinding grain changed this. It may seem absurd from our vantage point two millennia later, but the introduction of this labor-saving device, and the freedom from the day-long drudgery of grinding grain by hand, caused some moralists to be concerned about what the women would do with their new-found free time. The first literary reference occurs in Greek, about 30 BCE. The mill was praised in the following lines:

> *Stop grinding, ye women who toil at the mill,*
> *Sleep on, though the crowing cocks announce the break of day.*
> *Demeter has commanded the water nymphs to do the work of your hands.*
> *Jumping onto the wheel, they turn the axle*
> *Which drives the gears and the heavy millstones.*

—Antipater of Thessalonica[2]

The most basic design of a waterwheel, shown as a simple concept in Figure 8.3, is the undershot wheel. The prefix *under* indicates that the flow of water is under the shaft or axle on which the wheel rotates. The earliest practical design was indeed just a simple paddle wheel that was immersed in a stream. The undershot wheel was in use in places such as northern Denmark and China (where it was used for milling rice) by about AD 30. At about the same time, waterwheels were connected

FIGURE 8.9 A horizontal waterwheel. This particular waterwheel has been moved to an outdoor cultural center south of Santa Fe, New Mexico. It was originally in Las Truchas. (From http://edgardurbin.com/SantaFe2009/SFFrame.htm.)

to bellows that were used to supply a high volume of air to iron-making furnaces. This first happened in the region of modern-day Nanjing, in east-central China, at about this same time. In ancient China, the waterwheels were also used about this time for operating trip hammers* that would be used to crush ore or pound iron into shape. Another twelve hundred years passed before this technology was adopted in medieval Europe, where it may have been invented independently of the much earlier Chinese development.

The first clear description of any type of waterwheel occurs in a book by the Roman engineer, author, and architect Marcus Vitruvius Pollio—the first author to chronicle an energy-related device whom we can actually identify in the historical record (Figure 8.10).

Vitruvius describes the design and use of this innovation in his book *De architectura*, which was first published about 27 BCE and is still available.† Vitruvius claimed

---

* A trip hammer is a type of hammer—usually a very large and heavy one that could not be handled by a person—used in such applications as pounding cloth or leather, crushing ores, or hammering metals into desired shapes. The hammer is raised by a mechanical device like a waterwheel or steam engine (Chapter 10). When the hammer has been raised to a certain point, a "trip" (a generic term for a device that releases a mechanism) allows the hammer to fall, striking a blow. As the waterwheel continues to operate, the hammer is then raised up again, "tripped" again, and the cycle repeats to provide a continuous series of hammer blows.

† Vitruvius's book, *The Ten Books on Architecture*, is available as a two-volume bilingual edition from the Loeb Classical Library. Other editions are available, e.g., from Dover Publications. This is the seminal book on architecture, still in print some 2000 years after it was written.

FIGURE 8.10 Vitruvius, the great Roman architect and engineer (about 75–15 BCE), who was probably the first writer in antiquity to describe the use of gear systems. (From http:// mk.wikipedia.org/wiki/%D0%9F%D0%BE%D0%B4%D0%B0%D1%82%D0%BE%D1% 82%D0%B5%D0%BA%D0%B0:Vitruvius.gif.)

that the machines he described were in common use and attributes their invention to Greek engineers.

Vitruvius tells of a vertical wheel that is immersed into a stream, creating an undershot wheel similar to that shown in Figure 8.3. With this type of wheel to turn the millstone, the shaft of the wheel must be connected to the shaft of the stone via a system of gears (Figure 8.11). When we do this, we are no longer obliged to drive the millstone at the same rate as the waterwheel is turning. By a proper selection of the relative sizes of the gears, we can drive the stone faster (which is usually what's desired) or slower than the wheel.

> *Then at dawn we came down to a temperate valley,*
> *Wet, below the snow line, smelling of vegetation,*
> *With a running stream and a water mill beating the darkness,*
> *And three trees on the low sky.*

> **—Eliot**[3]

The development of gears and their use to change the speed of rotation were the first major achievement in the design of continuously operated machinery. It occurred roughly 2100 years ago. That seems like a very long time ago from our perspective, but from a different point of view, it means that we humans took about 198,000 years to develop machinery. Even though the undershot wheel and its gear system shown in Figure 8.11 may seem very primitive to us nowadays, it's a very recent development in the long span of human history.

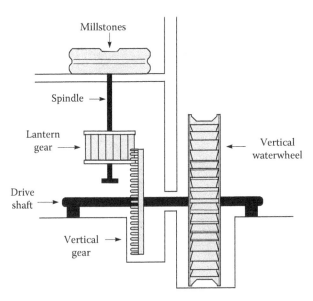

FIGURE 8.11 A diagram of a water-operated gristmill. The gears make it possible to change the direction and speed of motion of the various parts as ENERGY is being transformed into WORK.

The undershot waterwheel also has some practical problems. The WORK done by the undershot wheel derives almost entirely from the kinetic ENERGY of the water, which in turn is dependent on the water velocity. The WORK and POWER of the wheel are directly affected by seasonal changes in the rate of flow of the stream or river. In hot weather, or times of drought, the stream level may drop so low that there is not enough water to turn the wheel. In times of flood, there may be so much water that the wheel can't turn, because it is totally submerged. Despite these potential drawbacks, the undershot wheel was used in many places around the world for centuries. Its longevity very likely was due to the ease, simplicity, and relatively low cost of its construction.

In an overshot wheel, the water flows down over the wheel. Figure 8.12 illustrates this and compares overshot and undershot waterwheels. The overshot wheel is superior to the undershot in cases where the stream flow is scanty. In fact, the overshot wheel can work in any stream, provided that we build a dam upstream of the mill. Doing so creates a millpond. This lets us keep a constant flow of water the year round. In a dry season, we have a reserve in the millpond; in a wet season, we can let the excess water past the dam via a sluice and sluice gates. The earliest historical record of the overshot wheel is a mural in the Roman catacombs, from the third century AD.

The overshot wheel is also vertical on a horizontal axle. In this case, its name implies that water flows over the wheel, rather than underneath it. The blades of the wheel serve to divide it into bucket-like compartments. Water pours into these compartments or buckets from above. In an overshot wheel, only a small part of the WORK is a result of the impact of the water hitting the blades. Most of the WORK derives from the weight of the water as it descends in the "buckets" as the wheel turns.

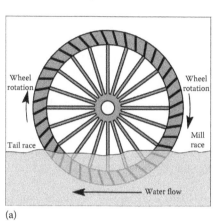 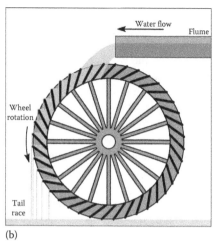

(a)   (b)

FIGURE 8.12   (a) An undershot waterwheel, compared with (b) an overshot waterwheel. The prefixes under- and over- tell us the position of the water flow relative to the axle of the wheel. (http://en.wikipedia.org/wiki/File:Undershot_water_wgeel_schematic.svg, and http://enwikipedia.org/wiki/File:Overshot_water_wheel_schematic.svg.)

The so-called floating mill illustrates the cliché that "necessity is the mother of invention." In 537, the Ostrogoths, a Germanic people moving from the region of the Black Sea, invaded Italy and laid siege to Rome. Most of Rome's watermills were supplied from the extensive system of aqueducts used to bring water into the city. The besieging Goths thought that it would be possible to starve Rome into surrender by cutting off the water supply from the aqueducts. Without water to operate the mills, the Romans could not grind grain to make flour, so could not bake bread. The Byzantine general Belisarius, assigned the task of defending Rome during the siege, had paddle wheels mounted in boats moored in the Tiber River. The Tiber flows through Rome, and the Goths had no way of halting its flow. By connecting these wheels with grinding mills, the supply of flour was assured. This design slowly spread over the next thousand years, first apparently to Venice and Baghdad. Three floating mills were built under bridges in the Seine River in Paris. By the end of the Middle Ages, many floating mills were in use, mainly for grinding grain or pumping water. Floating mills built in the Rhine River at Cologne in the fifteenth century were maintained in use for at least three centuries. Victor Hugo's classic novel *Notre Dame of Paris* mentions in several places pedestrians being splashed by water from the bishop's mills set in the Seine.

WATERWHEELS IN THE MEDIEVAL WORLD

During the Middle Ages, the use of waterwheels burgeoned. William the Conqueror in 1086 ordered a survey of England, compiled in *The Domesday Book*. It showed 5624 watermills operating in England, one on average for about every fifty families, or one mill per 200–400 people (from which we can infer that the individual mills must likely have been small and of low POWER). Some of these were still in use seven hundred years later, and a few were modified and were still working into the twentieth century! Will our descendants in the year 2700 be using any device built in this century?

The importance of water is illustrated, in an odd way perhaps, by the fate of technology when the western half of the Roman Empire collapsed around 400. Most of what the Romans achieved, such as their spectacular aqueducts and baths, slowly fell into disuse. In marked contrast, waterwheels were used more widely, as indicated by *The Domesday Book*, and their design continued to be improved.

While the Norman conquerors of England were compiling the Domesday survey, the Chinese were developing large water-operated clocks. The largest was operated by a waterwheel about 3.3 meters in diameter. The wheel revolved at exactly one hundred revolutions each day. The mechanism must have been very precise. Before 1300, the Chinese developed the "great water-driven spinning machine," for use in making cloth from hemp or ramie.*

At about the same time, water-powered mills became very important in the Islamic world. By 900, they were supplying about ten tonnes of ground grain every day to a city still in the news today—Baghdad. Around 990, a water-operated mill was introduced in the region of modern-day Iran and Iraq, for processing sugar cane. Such cane-crushing mills are known to have existed at Basra, a city frequently mentioned during the Iran–Iraq and the Persian Gulf wars. Other applications of waterwheels developed in this region during the Islamic period were similar to those being developed in Europe, as, for example, fulling cloth† and preparing pulp for paper.

Waterwheels and water-operated mills were an excellent source of wealth for their owners. Typically, the miller was paid for his efforts by being allowed to keep—and then sell or trade—some fraction of the grain being ground. As a result, and as seems to happen at many times in many societies, the wealthy often became the target of jokes, ill feeling, or invective from those who were less well-off. Medieval millers were no exception, as eloquently documented in "The Miller's Tale" in Geoffrey Chaucer's immortal *Canterbury Tales*:

> *The miller was a burly fellow—brawn*
> *And muscle, big of bones as well as strong,*
> *As was well seen—he always won the ram*
> *At wrestling matches up and down the land.*
> *He was barrel-chested, rugged and thickset,*
> *And would heave off its hinges any door*
> *Or break it, running at it with his head.*
> *His beard was red as any fox or sow,*
> *And wide at that, as though it were a spade.*
> *And on his nose, right on its tip, he had*
> *A wart, upon which stood a tuft of hairs*

---

* Ramie is a tropical Asian plant (*Boehmeria nivea*). Fibers from its stems can be woven to make fabric, cords, or rope.

† Fulling is an operation in the manufacture of cloth, especially woolens. In the early days of clothing manufacture, woven woolen cloth was often quite porous. To make the cloth feel more substantial or "fuller," it was pounded with a special type of clay still known as "fuller's earth" to fill the tiny pores. With modern weaving machinery, fulling is no longer important, but fuller's earth is still used, e.g., as a component of talcum powder.

Red as the bristles are in a sow's ears.
Black were his nostrils; black and squat and wide.
He bore a sword and buckler by his side.
His big mouth was as big as a furnace.
A loudmouth and a teller of blue stories
(Most of them vicious or scurrilous),
Well versed in stealing corn or trebling dues,
He had a golden thumb—by God he had!

—Chaucer[4]

The kinetic energy of water was the crucial ENERGY source for the society of medieval Europe, and likely in China at the same time. Water-operated machinery was used for making flour by grinding grain; cloth, by fulling; leather, by crushing oak bark used for tanning; beer, by crushing the malt and hops; paper, by shredding and pulping rags; and sawn lumber. By the end of the Middle Ages, waterwheels were used to raise buckets of ore from mines, in crushing minerals and ores, operating the bellows on iron-making furnaces and hammering the iron to shape, operating grindstones for shaping and sharpening tools, and making wire by drawing metal through dies. Perhaps the ultimate application was the use for a watermill for polishing gems (in Paris, in the 1530s).

Water was as vital an ENERGY resource to medieval Europe as oil is to us today.

The use of water in medieval society stands in great distinction with its use in the ancient world. Medieval societies wholeheartedly adopted water as a source of ENERGY for all sorts of mechanical WORK. In the ancient world, little, if any, mechanical WORK was done by water except for the processing of agricultural products (grinding grain and pressing olives). Two factors may have contributed to this difference. First, ancient Greece and Rome relied on slaves for much of the hard labor. (A two-meter diameter wheel could grind 1500 kilograms in ten hours this same WORK required forty slaves.) The supply of slaves was renewed by captures made during war. The second issue is geography. Many parts of southern Italy and southern Greece do not have reliable supplies of fast-moving water that will run the year around. Therefore, water as an energy source, to replace the human muscles of slaves, was not an attractive option in the ancient world.

We owe two other things to the watermills of medieval Europe. First, building a mill with an overshot wheel required a very substantial initial investment, for the mill itself, the wheel, the dam, and sluices. It became difficult for a single individual to have the capital necessary to build a mill. By about 1100, it was common for two to five people to own a mill collectively, taking equal shares in the profits. In twelfth-century France, a company was established in which there were "shares" in the mill. The value of each share fluctuated with business conditions. These shares could be bought, sold, traded, or bequeathed. This practice gave rise to the modern stock corporation and the stock exchange. Beginning in the early thirteenth century, it appears that no millers—the people who actually operated the mill and ground the grain—were themselves among the shared owners (or as we would say nowa-days, the stockholders) in the mill. Thus we see the beginnings of a major division in society: that between "capital," the people who invest the money and hold the ownership, and "labor," those who do the work.

Second, it was quite common for more than one mill to be built along a given stream. This arrangement worked fine provided that nobody later interfered with the water flow. For example, if the upstream mill raised the height of its dam, more water would be impounded in the "millpond" and correspondingly less would be available for the other mills downstream. On June 8, 1278, a lawsuit was filed by a "down-stream" mill against the "upstream" mill, which had illegally raised the height of its dam. Suits, counter-suits, and appeals followed, until the case was finally settled when one of the two companies went bankrupt, and there was only one surviving descendant of the original litigants—in 1408, 130 *years* later!

WATERWHEELS IN THE EARLY MODERN WORLD

Throughout the eighteenth and much of the nineteenth centuries, waterwheels pro-vided WORK or generated POWER about as well as the primitive steam engines of the time. In situations where several factories could be built near each other at a site of reliable water flow, it was cheaper to rely on water than for each of the factories to have its own steam engine. At least into the 1830s, the kinetic energy of water was still the major source of ENERGY for industry in Britain, as was true in the United States into the middle of the nineteenth century.

The pioneer mechanized factory produced yarn for making stockings, set up at Derby (England) in 1702. Its machines were driven by a waterwheel about four meters in diameter. This was the prototype of the textile factory whose widespread adoption throughout northern England was a foundation of the Industrial Revolution. It was worked by a single undershot waterwheel, which, by means of gearing, drove the factory machinery, which was in turn tended by some three hundred factory workers. For its time, there was no precedent for this scale of mechanization. Although not a commercial success itself, the factory became the model for the use of water to drive spinning machines and the standard technology in the cotton industry.

The waterwheel was the ENERGY source that allowed for mechanization of the textile industry, one of the three major technological developments that constituted the Industrial Revolution in Britain. (The others were the steam engine—to be dis-cussed in Chapter 10—and the emergence of technologies based on iron and coal.) The waterwheel allowed the textile industry to concentrate in large spinning mills located along swiftly moving streams. Those early factories were, by all accounts, dreadful places. Child labor was prevalent. The workweek consisted of six twelve- or thirteen-hours days; even in one establishment near Manchester that was considered a "model mill," the standard workweek for children was 74 hours. The lack of safety guards on machinery caused mutilations such as broken or lost limbs. The damp environment was a breeding ground for tuberculosis.

And was Jerusalem builded here Among these dark Satanic Mills?

—**Blake**[5]

As factories clustered together around good water sources, the people who worked in them clustered around the factories. This meant a shift of population, with a migration of people into cities that grew up around the factories. In the United States, water power was still the major energy source for factories, textile mills, and metal forges into the 1850s. Industrialization of America began in New England because of the water resources that could be used for water-operated machinery (Figure 8.13). Thousands of such installations were in operation in New England during the early decades of the nineteenth century.

If we consider the United States as it was in early 1850, with about thirty states mainly on the eastern seaboard, rivers that were suitable for watermill installations were located primarily in New England and, to a lesser extent, in New York and Pennsylvania. The biggest, "The Niagara of Waterwheels," stood 18 meters high, weighted about 320 tonnes, and could generate 375 kilowatts. Because of the concentration of watermills, most of the factories were in the north and northeast. And because the population concentrated in cities around manufacturing areas, most of the people were in the north and northeast as well. Consequently, when the American Civil War broke out in 1861, the northern states had an overwhelming advantage in men and material. The southern states had several military advantages, such as being able to fight on their home terrain, having internal lines of communication and transport, and, in many cases, superior generals. However, once the Civil War became a war of attrition (especially after the Battle of Gettysburg), it was inevitable that the north would win, simply on the basis of men and material. This has obviously had very profound implications for American, indeed world, history and particularly for the status of African Americans.

From our vantage point, early in the twenty-first century, we would find that the massive waterwheels, engineering marvels for their time, provided remarkably little POWER. A waterwheel might produce about 0.4 kilowatts, that is, enough to light four 100 watts lamps. By far, the largest and most elaborate installation of the early modern era was at Marly, in France. It was built in the 1680s, requiring the labor of nearly two thousand workers for eight years. Its purpose was to supply water to the fountains of the French court at Versailles. The pumping station itself had 235 pumps. The pumps were operated by thirteen undershot waterwheels, each about fourteen meters in diameter, which pumped water from the river at a rate of about four million liters per day. The POWER output of this installation was about 93 kW. Only about 10% of the kinetic energy of the water was converted to WORK. In the early 1800s, this gigantic complex of machinery was replaced by a single 110 kW steam engine. Another measure of the relatively low output of POWER and WORK—relative to the size of the wheel—is the estimate that an overshot wheel two meters in diameter with a water flow of eleven liters per second would generate about 2 kW, equivalent to a small 250 cc motorcycle engine.

Because of its greater efficiency at converting kinetic energy to WORK, the overshot wheel has been the most widely used design since the eighteenth century. However, the cost of the dam, the necessary millraces, and other equipment for handling the water flow had to be justified by the return on the amount of capital invested. Despite its technical disadvantages relative to the overshot wheel, the undershot wheel sometimes proved to be a better choice. It is cheaper to build and install than the overshot

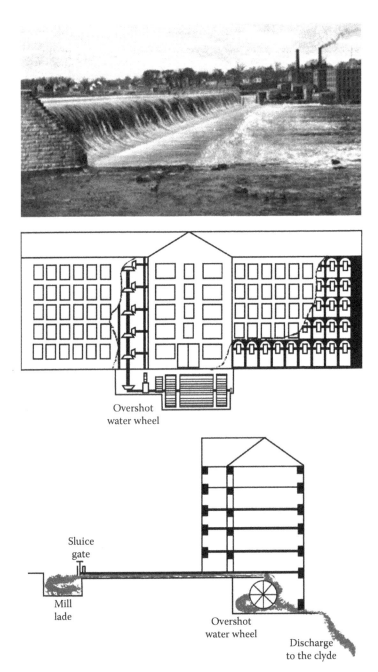

FIGURE 8.13 Mills like these used an exceptionally complex series of gears, pulleys, and belts to operate machines throughout the plant from the motion of the waterwheel.

wheel, especially if it's necessary to build a dam to provide water to the overshot wheel. If not much capital is available for investment, the undershot wheel is very attractive. Further, given a fast-moving stream with a reliable flow that is not subject to drastic fluctuations, the POWER of the undershot wheel can be very high, despite its low efficiency. To manage this, the wheel needs to be large. At the end of the nineteenth century, the largest waterwheel anywhere in the world was a 22 meter-diameter behemoth, an overshot wheel on the Isle of Man, which produced only twenty kilowatts.

WATER "POWER" AS A FORM OF SOLAR ENERGY

Where does the gravitational potential energy of the water come from in the first place? Imagine a simple SYSTEM of a river that flows over a waterfall and, from there, into the ocean. Given enough time, sooner or later, all the water in the river will have to wind up in the ocean—*unless* the water in the upper portion of the river can somehow be replenished. As water in the ocean is heated by the sun, some of it evaporates into the atmosphere as water vapor. Since the air over the lake or ocean is also heated, the moisture-laden warm air rises into the atmosphere. Some of this air will be carried over land. At some point, the air will cool, causing the water vapor to condense and to fall to the ground as rain or other precipitation. The rain will accumulate eventually in streams, which run into larger streams, which run into a river, which, in this simple SYSTEM, runs into the ocean. The cycle can begin all over again, and will run for eons, unless perturbed by some major geological change.

The cycle described by this SYSTEM provides a model of the movement of liquid water and water vapor on Earth. There is water of high gravitational potential energy in the region of the river upstream of the waterfall because it was deposited there by precipitation of condensed water vapor from the atmosphere. Water vapor was in the atmosphere because water evaporates, especially if warmed. The warmth causing water to evaporate comes from the sun. The model SYSTEM discussed here could be expanded to account for transport of water throughout the worldwide ocean, all of the world's streams, rivers, ponds, lakes, and even glaciers. The model describing the transport and fate of water throughout the world is known as the *hydrologic cycle* (Figure 8.14). The ENERGY that allows the hydrologic cycle to function is from the sun—solar energy.

In a sense, the ENERGY in water is an indirect form of solar energy. Instead of using solar energy directly to operate machinery, a dream that has still not been fully attained, the energy engineers who exploit water as a means of doing WORK are using solar energy in a roundabout way, relying on the sun to drive the hydrologic cycle to ensure a steady supply of water with the necessary potential energy.

We can consider water POWER to be an indirect form of solar energy.

WATER "POWER" AND THE ENVIRONMENT

The kinetic energy of water is one of the cleanest forms of ENERGY available to us. The water itself is almost pollution free (assuming, of course, that it's not already

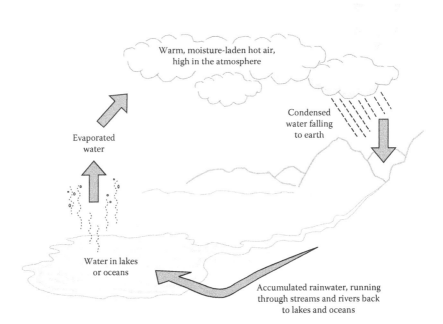

FIGURE 8.14   The Earth's hydrologic cycle is operated by solar energy evaporating water, which later condenses and returns to Earth.

contaminated). There are no by-products from extracting WORK from the water. There are no emissions to the atmosphere, no contamination of the water, nor any radioactivity. As a fluid, water is relatively easy to control. Although most water-wheels are of very low POWER, the best designs provide a conversion of 80%–90% of the kinetic ENERGY in water to useful WORK. This is excellent performance in comparison to many other energy systems. Efficiency of converting ENERGY to WORK matters greatly when one has to purchase fuel; an advantage of waterwheels is that running water is free. Another advantage is that many times even very small streams can be utilized, with an appropriate-sized wheel, to produce WORK. In fact, water-wheels will operate in streams even that have large fluctuations in the flow rate during the course of a season or year.

Although water as a source of ENERGY has numerous advantages, being cheaper than purchased fuels, renewable, and nonpolluting, it is not without its disadvantages as well. Certainly the wheel's operation is halted by freezing, getting "drowned" in times of flooding, or loss of water entirely in a sustained drought. Most notably, many installations, or prospective sites for installing a waterwheel, require the construction of a dam upstream to provide adequate amounts of water at a controlled flow. While the waterwheel installation itself is very clean, the damming of a river can bring with it severe environmental issues. We will examine this point in more detail in Chapter 15.

REFERENCES

1. de Duve, C. *Vital Dust*. Basic Books: New York, 1995, p. 35.
2. Antipater of Thessalonica, in Strandh, Sigvard. *The History of the Machine*, Dorset: New York, 1979.
3. Eliot, T.S. Journey of the Magi. In: Michie, J. (ed.), *The Folio Golden Treasury*. Folio Society: London, U.K., 1997.
4. Chaucer, G. *The Canterbury Tales*. Oxford University: Oxford, U.K., 1998, p. 15. Many other editions of Chaucer are readily available.
5. Blake, W., 'Jerusalem' in Heaney, S.; Hughes, T. *The Rattle Bag*. Faber: London, U.K., 1982, p. 221. Blake's poetry is available in many anthologies and collected editions of Blake's works.

FURTHER READINGS

Fagan, B. *Elixir*. Bloomsbury Press: New York, 2011. Subtitled "A history of water and humankind," this book traces the impacts of water on human society over a period of about 5000 years. Many aspects of waterwheel technology are covered.

Hill, D. *A History of Engineering in Classical and Medieval Times*. Barnes and Noble: New York, 1997. Chapters 3, 8, and 9 provide information relating to the use of water as an energy source. The book covers the period from 600 BCE to about 1450.

Smil, V. *Energies*. MIT Press: Cambridge, U.K., 1999. Chapter 4 provides, among other things, a short discussion on waterwheels, with some very nice illustrations.

Smil, V. *Energy in Nature and Society*. MIT Press: Cambridge, U.K., 2008. A discussion of waterwheels and windmills (which we will discuss in the next chapter) occurs in Chapter 7, with some illustrations.

Usher, A. P. *A History of Mechanical Inventions*. Dover Publications: Mineola, NY, 2011. One of a series of useful, inexpensive paperbacks from this publisher, dealing with the history of inventions and of engineering. Chapter VII discusses waterwheel technology through the year 1500.

9 Wind Energy

Waterwheels were a primary energy source for most of medieval Europe. Water was as important to the industrial economy of that time as oil is to us today. However, waterwheels have significant limitations. They can only be located near good streams or rivers. They become useless if the water freezes in winter or dries up in summer. In times of war, an army attacking a city or castle could starve it by stopping the flow of water to the local mills.

Another force in nature also possesses kinetic energy, without being subject to the problems associated with water as an ENERGY source—wind. Wind is a mass of air in motion. All of us occasionally experience the force of the wind in various ways—scattering papers that we've dropped, blowing a hat off, turning an umbrella inside out, or even making difficulty in walking into a strong wind. These experiences are no different than some of the ways in which we can personally experience the force of moving water. And, just as moving water can be harnessed to do practical WORK for us in many ways, we can also harness the wind.

WHERE DOES WIND COME FROM?

Wind is a mass of air in motion relative to the surface of the Earth. Like any other body or object in motion, the wind possesses kinetic energy. Using a windmill, we can capture some of that kinetic energy and transform it to useful WORK as, for example, in pumping water. As wind flows, it loses some of its kinetic energy as a result of friction with the Earth's surface and—though not so readily apparent—of friction within the wind itself. If the energy losses due to these kinds of friction were not restored, then eventually so much kinetic energy would be lost that the wind would "run down" and stop blowing. Of course, this does not happen. We all know that wind blows every year, and, in some parts of the world, such as the Great Plains of the United States and Canada, or in Patagonia, it seems to blow all the time, day in and day out. From our discussion of the energy balance (Chapter 4), we know that we cannot extract an unlimited amount of ENERGY from a SYSTEM, because at some point all of the stored energy would become depleted and any further removal of ENERGY would be impossible. Somehow the ENERGY that is lost from the wind, or that we extract from it to do WORK, must be replenished.

The ultimate ENERGY source that causes wind, and that replenishes the ENERGY lost from wind, is the sun. Wind is caused by the unequal heating of various parts of Earth's atmosphere. ENERGY from the sun is absorbed by the land surfaces and by water, which is dominated by the ocean. Different substances—soil, the green canopies of forests, seawater as examples—absorb the sun's radiation to different extents. Furthermore, not all locations on Earth's surface are heated to the same

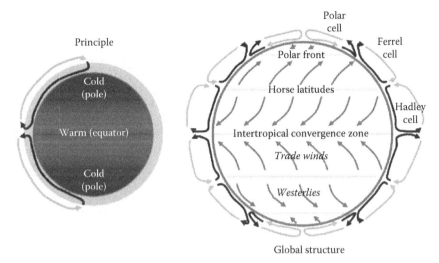

FIGURE 9.1   The winds on Earth derive from unequal heating of different parts of the atmosphere by solar energy. (From Rodrigue, J.-P. et al., *The Geography of Transport Systems*, 2nd edn., Routledge, London, U.K., 352pp., 2009, http://people.hofstra.edu/Jean-paul_Rodrigue/.)

extent by the sun. This is easy to confirm by checking a daily temperature forecast for the major cities of the world. The heated land features and water heat the air above them. Because of these two factors, different absorption of solar radiation by different substances, and the different amount of solar radiation supplied to various places on Earth, the air is not heated to the same extent everywhere in the world. Air that is above the hotter locations expands and, as it does so, rises. As this warm air rises into the atmosphere, air from cooler locations—those that have not received so much heating by the sun—flows toward the area where the warm air is rising, establishing wind currents. Thus,

Winds are a manifestation of solar ENERGY.

This is illustrated in the diagram in Figure 9.1.

WINDMILLS IN THE ISLAMIC WORLD

Certainly the ancient Greeks and Romans knew of the ENERGY of the wind, because they had sailing ships. Cloth sails were known to the Sumerians at least by 3500 BCE (Figure 9.2). The sailing ship represents the first time in history that a source of ENERGY available in nature was successfully harnessed to do a type of WORK. (The use of fire as an ENERGY source predates harnessing the wind, but fire was used to supply heat for warmth and cooking, not for the doing of mechanical WORK.) Egyptians also had various kinds of sailing ships. As far as we know, the classical Greeks and Romans never developed the sail (from ships) into any rotary device—a

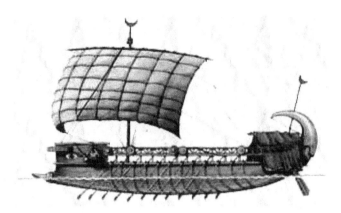

FIGURE 9.2 A very early sailing ship, used by the ancient Phoenicians around 700 BCE, used in the Mediterranean Sea. (From http://phoenicia.org/ships.html; http://phoenicia.org/imgs/ships10.jpg.)

windmill—for doing mechanical WORK. Hero of Alexandria (Egypt) describes, around AD 50, a wind organ that was supplied air by a piston moved by a wheel equipped with sail-like blades to collect air. Hero's description is about the only mention of windmills in the ancient world. It seems that nobody ever thought that the little toy described by Hero could somehow be scaled up so that the WORK and POWER of wind could be used to relieve humans or draft animals from their chores.

The first practical development of windmills apparently came in the Islamic countries of the Middle East during the seventh century. The first extensively written description of a windmill is by Abul-Hasan Al-Hasudi, one of the greatest of scientists and historians of the Muslim world, in the tenth century. The Banû Mûsà,* writing in Baghdad about 850, mentioned a small "windwheel" that was used to operate a fountain for decorative purposes. Other works of medieval Islamic authors also mention windmills. Al-Istakhrî described windmills in the region that is nowadays the western part of Afghanistan. Windmills are described in Syria by the great geographer al-Dimashqî, writing in the latter part of the thirteenth century. Al-Mas'ûdî, writing around the end of the tenth century, tells the story of a Persian inventor who claimed to Caliph Úmar I that he was able to build a windmill. The caliph (a secular and religious leader of a Muslim community or country) called his bluff and made him substantiate his claim by actually doing it. Úmar I is believed to have ruled in the years 634–644, so this story had been around for some three hundred years before al-Mas'ûdî recorded it. According to legend, a Caliph "Omar" (possibly the

---

* The Banû Mûsà brothers, Muhammad, Ahmad, and al-Hasan, were the sons of noted astronomer, Mûsà Shâkir. When the father died, the three brothers were entrusted to the care of Caliph al-Mámûn; they became trusted advisors to this caliph and his successors in Baghdad. They made many contributions to science, especially mathematics and astronomy, and to engineering in waterwheel and windmill technology. Some historians consider that their competence in the technology of devices that operate by differences in wind pressure or water pressure has only been matched in the past century. One of their most important books was the *Book of Ingenious Devices*, published in Baghdad around 850, which contained descriptions of about a hundred mechanical devices.

FIGURE 9.3   Museum model of an Islamic vertical-type windmill. (From http://museum. kaust.edu.sa/explore-4-technology.html.)

aforementioned Úmar I) put heavy tax on windmills, making the inventor so outraged that he assassinated the caliph.

The stimulus for inventing the windmill in the Islamic world was the need to grind corn in areas lacking steadily flowing streams of water. Fortunately, an abundant supply of wind energy occurs in many parts of the Islamic region—sometimes at speeds up to 160 km/h.[1] The Islamic vertical mill (Figure 9.3) looked somewhat like our modern revolving doors. Probably the vertical mill was invented in what is now modern-day Iran in the early seventh century. Some aspects of the design were likely copied, or derived, from sailing ships. One reason for thinking this is that the first sails of windmills were made of cloth, like the sails on ships.

The vertical windmill spread throughout Islam, and eventually to India and then to China. In addition to its most important use of grinding grain, it was also used to process sugar cane in medieval Egypt. In China, the windmills were used for pumping water and for hauling canal boats.

Part of the success of the Islamic mill design comes from the fact that, in regions where these designs were used, the wind generally blows from the same direction most of the time. Wind also blows from the same direction, but for just part of the year, in the monsoon regions of southern Asia. There, iron-making furnaces were developed that could take advantage of the high-velocity winds of the monsoon, channeling them into the furnace to help achieve high temperatures.

THE MEDIEVAL POST MILL

The design that many people may mentally associate with windmills apparently was an independent invention in the West, and not a derivative of the Islamic vertical mill.

The idea that it would be possible to use the ENERGY of wind (i.e., flowing air), rather than of flowing water, may have come into Europe from the Islamic world. However, the actual design and development likely derive from water-operated mills described by Vitruvius. That inference is drawn from the fact that many of the mechanical details of European windmills and the Vitruvius mill are fairly similar.

The idea must have originated at least in the early twelfth century. A deed, dated around 1180, recorded a gift of land lying near a windmill to an abbey in Normandy (part of modern-day France). A windmill was built and was operating in Yorkshire, England, in 1185, based on records of rent payments. Other historical records from the end of the century also mention windmills.

Around the end of the twelfth century, Pope Celestine III ruled that windmills should pay tithes to the church. In 1191, a British knight of Bury St. Edmunds refused to pay from his mill. This did not sit well with the local abbot. It represented a challenge to, and disrespect for, the authority of the church. More seriously, the abbot apparently was the owner of the only other mills (operated by animals) in the vicinity, and the knight's refusal also represented a business problem for the abbot, since wind is free, but animals must be fed and sheltered. The knight's argument was based on the proposition that anyone has access to the wind. The abbot replied, "I thank you as I should thank you if you had cut off both my feet. By God's face, I shall never eat bread till that building be thrown down. You are an old man, and you should know that neither the King nor his Justiciar can change or set up anything within the liberties of this town without the assent of the Abbot and the Convent."[2] The knight ordered his servants to tear down the mill.

During the thirteenth century, windmills became common features on the plains of northern and western Europe. As an example, about 120 windmills were built in the vicinity of Ypres (a small city in what is now the southwestern part of Belgium). Illustrations dating from about 1270 show post mills, indicating that, by then, they must have been fairly common. Figure 9.4 provides an example.

Wind differs in one crucial aspect from water: water flows in a single direction, as determined by the bed of the stream or river, but in many parts of the world, the direction of wind can shift. In some places, including ancient Persia, the prevailing wind direction is relatively constant. But in many parts of the world, including much of medieval Europe, the wind direction can change daily, and even sometimes during a single day. A method had to be found to allow the windmill to take advantage of the energy of the wind, regardless of its direction. Medieval engineers solved this problem by mounting the entire windmill on a large, upright post, so that the whole mechanism is free to turn with the prevailing wind direction.

These devices are called post mills. In post mills, the entire structure rotates so that the vanes or sails will face the wind. A few working post mills, which seem to have stepped right out of the illustration of a medieval manuscript, still exist in Britain and Europe. They are not economically useful any longer, but are remarkable testaments to the ingenuity of medieval engineers and to the longevity of their construction practices. Indeed, the post mill is a remarkable engineering challenge.

FIGURE 9.4   A medieval post mill. (From http://www.machine-history.com/node/569.)

Imagine architects and engineers nowadays proposing to balance an entire structure, complete with its machinery for capturing ENERGY and converting it into useful
WORK, on a single post—not only balancing it on the post, but also allowing it to
rotate around on that post. Almost certainly, we would think them to be incompetent, or possibly completely out of their senses. And yet—based on the surviving
examples in Britain and Europe—these devices were not only conceived and built
by medieval engineers, they actually worked for about seven hundred years! We
can pose the same question as for early waterwheels (Chapter 8): will any factory or
industrial product built now still be functioning in the year 2700?

The post mill design spread throughout *northern* Europe. There, it had the same
kinds of applications as did waterwheels. The advantages of windmills made them
especially important for the cold, windy plains of northern Europe. The design was
so successful that it was "exported" to the Middle East by the crusaders (1189–1192).
Post mills were probably three to five times as powerful as the Islamic mills. Just as
did waterwheels, windmills greatly improved living standards in the medieval world
by freeing women from the very time-consuming job of grinding grain by hand. The
importance of windmills to the industrial economy of northern Europe was so great
that in 1341, the bishop of Utrecht tried to obtain a monopoly on the wind blowing
in his diocese.

The windmill has several advantages relative to the overshot waterwheel. No
capital investment is needed for a dam, sluices, and other water-control equipment.
The windmill can operate in arid areas, or in areas where there are no fast-moving
streams. It will continue to function even in the freezing winter weather of northern
Europe. The fact that no one can "shut off" the wind means that windmills could be
used in besieged castles for grinding grain, which is why some medieval castles had
windmills built on their walls.

Windmills were not nearly so important in the warm mountainous regions
of southern Europe, with its fast-moving streams that seldom, if ever, froze.

Several centuries after the windmills were common in northern Europe, they were still rare in the south, as exemplified by Don Quixote's astonishment at seeing windmills (in Cervantes's novel of about 1605) in Spain. The late appearance of windmills in Spain, which had been an Islamic country for many centuries, suggests that their design migrated into Spain from northern Europe, since windmills with vertical sails were not known to have been used in medieval Islam.

At that moment they caught sight of some thirty or forty windmills, which stand on that plain, and as soon as Don Quixote saw them he said to his squire: "Fortune is guiding our affairs better than we could have wished. Look over there, friend Sancho Panza, where more than thirty monstrous giants appear. I intend to do battle with them and to take all their lives. With their spoils we will begin to get rich, for this is a fair war, and it is a great service to God to wipe such a wicked brood from the face of the earth."
"What giants?" asked Sancho Panza.

—de Cervantes[3]

THE TURRET MILL

Despite steady engineering improvements in post mills—such as the invention of the brake, and a way of adjusting the clearance of the millstones—post mills became too unwieldy as they got bigger. This results from the fact that the entire mill had to be turned, and the bigger they got, the more unmanageable the job of turning became. They were also likely to be wrecked in severe windstorms. Continued application of the post mill, especially its evolution to larger and larger sizes, became rather impractical because of these difficulties. As has happened many times in the history of science and technology, the solution to the problem came from an entirely unexpected source, a person working on a different problem in a seemingly unrelated area.

While the post mill continued to be developed, some visionary individuals were considering an entirely different application of wind energy: the possibility of human flight. From time immemorial, our human and hominid ancestors saw birds flying by the simple act (as it appeared) of flapping their wings. It seemed reasonable to expect that some device ought to allow humans to fly too, by using muscles of their bodies. The most notable such attempt may be embodied in the legend of Daedalus and Icarus—how Daedalus fashioned "wings" for himself and his son Icarus by attaching feathers to their arms with wax to escape from Crete, only to have Icarus fly too close to the sun and plunge into the sea because the softened wax allowed his feathers to fall off.*

By the fifteenth century, it was obvious, at least to one scientist and inventor living in Milan and, later, Florence, that there were two parts to the problem of flight. One part was simply the propulsion necessary to get off the ground in the first place.

---

* The story of Daedalus and Icarus dates from about 630 BCE. Remarkably, the first successful human-powered flight did not occur until 1961, in a "man-powered aircraft" designed at Southampton University in England. The flight covered about 650 meters.

FIGURE 9.5   A turret windmill for grinding grain. This mill is in Victoria, Texas. (From http://commons.wikimedia.org/wiki/FileVictoria_grist_windmill_2008.jpg.)

The second part was the "lift" provided by the air flowing past the wings. He managed to find a fairly complete solution to the problem of lift, but never solved the problem of generating enough propulsion energy from human muscles to actually take off. Since many of his notebooks have survived to the present day, we know that eventually his sketches turned from the issue of how to solve the combined propulsion-and-lift problem to new designs for windmills. He is credited with the idea of having only the top part of the mill turn with the wind. This design is called a turret mill or tower mill (Figure 9.5). In the turret mill, the main portion of the building of the mill is a nonmoving solid structure that could be made of brick or stone instead of wood. On top of this, solid portion of the building is a revolving turret that holds the sails. With this improvement, mills immediately could be made much larger and more efficient. With this new design, builders were no longer limited to using the relatively lightweight material, wood, for the mill. They now could be made larger and built of more permanent materials, such as stone or brick.

The inventor? A person who, possibly, was the greatest genius who ever lived: Leonardo da Vinci (Figure 9.6).

A left-handed, vegetarian, homosexual bastard, Leonardo da Vinci (1452–1519) contravened most of the accepted norms of his day. Reared by his peasant grandparents in a remote Tuscan village, he had minimal schooling. He was apprenticed as a painter because his illegitimacy debarred him from respectable professions. (Painting in fifteenth-century Tuscany was regarded not as "creative art" but as a lowly trade,

FIGURE 9.6 Leonardo da Vinci (1452–1519), possibly the greatest genius in history. (From http://en.wikipedia.org/wiki/File:LEONARDO.JPG#file.)

fit for the sons of peasants and artisans.) Lacking literary culture he was scorned in the highbrow Florence of the Medicis. This turned him towards science and observation. "Anyone who invokes authors in discussion is not using his intelligence but his memory," he contended.

He was insatiable for newness, both in art and science. His first known drawing was also the first true landscape drawing in western art. He was the first painter to omit haloes from the heads of figures from scripture and show them in ordinary domestic settings, and he was the first to paint portraits that showed the hands as well as the faces of the sitters. His Leda (which does not survive) was the first modern painting inspired by pagan myth. His notebooks, of which over 5,000 pages survive, are all written backwards in mirror writing, and are dense with intricate drawings. They record his observations on geology, optics, acoustics, music, botany, mathematics, anatomy, engineering and hydraulics, together with plans for many inventions, including a bicycle, a tank, a machine gun, a folding bed, a diving suit, a parachute, contact lenses, a water-powered alarm clock, and plastics (made of eggs, glue and vegetable dyes).

It is true that Leonardo was not strictly a scientist, not always as original as he seems. His war-machines had already been designed by a German engineer, Konrad Keyser; his "automobile" by an Italian, Martini. Though he came close to formulating some scientific laws, his insights were sporadic and untested by experiment. He thought of looking at the moon through a telescope a century before Galileo, but he did not construct one. He knew no algebra, and made mistakes in simple arithmetic. His man-powered flying machine, designed to flap its wings like a bird, could never have flown. Apart from anything else it must have weighed about 650 lbs (as against 72 lbs for Daedalus 88, the man-powered aircraft which flew 74 miles over the Aegean in 1988).

Despite these reservations his notebooks give an astonishing preview of the new world science was to open.

—Carey[4]

The turret mill was invented by Leonardo. He designed it around 1500–1502, but it is not certain whether his design was ever actually built. The turret mill did not come into general use for about another two hundred years. Its practical development was left to the people we often stereotypically associate with windmills—the Dutch. The turret mill originated independently in northwestern Europe in the early fifteenth century; written descriptions appear in France around 1420. In Holland, the most important use of the windmill was for pumping water, to reclaim low-lying wetlands. Other uses included raising materials out of mines, crushing the ores, stretching fibers to make hemp, and operating sawmills. This last application, sawing wood, was especially important for the Dutch navy and its large fleet of commercial merchant ships. Between 1500 and 1650, the amount of land available for farming in the Netherlands increased by about forty percent, thanks to the drainage provided by the windmills. In these early Dutch windmills, the sails supplied WORK to move large wooden wheels (designed much like the waterwheels used for irrigation discussed in the previous chapter) that removed water from fields in buckets or scoops. This windmill-driven "scoop wheel" may have been invented as early as the fourteenth century. The use of the windmill for pumping water was common in England by the end of the seventeenth century.

The Dutch windmills were one of the first—possibly *the* first—mechanical devices to be the subject of a detailed engineering analysis. This work was done by Simon Stevin.* He designed new mills that featured much larger scooping wheels, which revolved more slowly, and geared transmission systems that used more efficient designs of gears. He figured out the design of the gears—that is, how big the wheeled portion should be and how many teeth each gear should have. He worked out the minimum wind pressure required on a given surface area of each sail or vane that would be needed to lift the water to whatever height would be required for dumping it. On the basis of these calculations, Stevin could also calculate how much water would be raised with each complete revolution of the vanes.

WIND ENERGY IN THE EARLY MODERN AGE

Windmills and waterwheels share the limitation that the WORK or POWER has to be used essentially on the spot where it is produced. In fact, it took nearly to the end of the nineteenth century to figure out how to do WORK at, or transmit POWER over, long

---

* Simon Stevin (1548–1620) contributed much to the development of Dutch science and mathematics. Like many prolific inventors and engineers, not all of his ideas were practical. Perhaps the most conspicuous of these was the "land yacht," essentially a wind-powered automobile. In 1600, Stevin, along with almost thirty passengers, achieved a successful test run on a smooth, level stretch of beach sands. Nowadays the mechanical engineering students at the Technical University of Eindhoven (Netherlands) have a Simon Stevin club, said to own an entire fleet of land yachts.

distances, as we will see later. As late as 1850, most of the corn ground in England and Wales, the most populous parts of Europe in those days, was still ground in hundreds of little wind- or watermills scattered over the countryside.

As a rule, water is a much better source of ENERGY than wind. An important reason for the relatively limited applications of wind is that they can be badly affected by fluctuations in the ENERGY supplied by the wind. This would not be true for, say, operating a factory in cases where a steady reliable output of mechanical WORK was needed all day long. Even for milling grain, water (if available) was preferred during medieval times. For example, records for the county of Hampshire in England show waterwheels outnumbering windmills by a ratio of nearly 25 to 1 during that time. Why then build a windmill? One obvious reason would be a lack of a water source nearby. Another would be a steady reliable source of wind in the specific locality chosen for the mill. A third factor might be ease of access. That is, if a farmer had to transport his grain a long distance to get to a watermill and then haul the flour back, it might just be easier to put a windmill on his own property.

Even though the steam engine (Chapter 10) eventually became the dominant source of WORK and POWER in industry, improvements were still being made to windmills and watermills well into the nineteenth century. By the nineteenth century, for example, the state-of-the-art windmill design included vanes mounted on springs, a fantail (described later), and a governor to help regulate speed. It's possible that even if the steam engine had never been invented, the nineteenth century could have achieved significant industrial growth solely on the basis of wind and water.

The fantail was invented in 1745 in England by Edmund Lee.* The fantail is mounted at a right angle to the main blades of the windmill (Figure 9.7). As the wind shifts direction, it strikes the vanes of the fantail and causes the turret to turn so that the blades are again moved into the wind. In essence, the fantail is a separate, little windmill mounted on the rear of the turret. If the wind shifts away from the main set of vanes, the fantail will likely catch the wind instead. Then, through a sequence of gears, the wind-operated fantail will turn the turret of the main mill so that it will stay facing the wind. Any slight shift in wind direction will be compensated by the fantail's immediate correcting motion of the turret. This invention, from nineteenth-century Britain, is one of the first examples of using "feedback" to establish a self-regulating machine.

Windmills and watermills share another limitation. In a windmill, for example, the wind moving past the blades of the mill still has some kinetic energy. If it didn't, the air would have to be perfectly still (i.e., have zero kinetic energy) just downwind of the mill. The windmill functions by having the kinetic energy of the wind do WORK on the sails or blades. Since the wind moving past the mill still possesses some kinetic energy, the kinetic energy was not converted completely to WORK. The same argument applies to waterwheels. Here too the water leaving the wheel still has some kinetic energy (because it is moving). Thus we have not converted the kinetic

---

* Edmund Lee was a blacksmith working at a place called Brock Mill, near Wigan, a city in the northwest of England. He received a patent for the fantail in 1745. The design was perfected by trials on various mills in the area around the cities of Leeds and Hull.

FIGURE 9.7 Windmill showing a fantail. This particular windmill is in Sønderho, Denmark. (From http://commons.wikimedia.org/wiki/FileDK_Fanoe_Windmill01.JPG.)

energy of the water *completely* to work. (Again, if we had done so, the water would have to stop completely.) This pair of observations leads to a conclusion of enormous importance, which we will explore in more detail later:

It appears that it is impossible to convert ENERGY completely (i.e., 100%) into WORK.

We've already seen something very much like this (in Chapter 4):

ENERGY cannot be converted completely to work (i.e., with 100% efficiency). Some ENERGY is inevitably wasted.

There seems to be a trend here!

In the eighteenth and nineteenth centuries, as many as a hundred thousand windmills may have been operating in Europe. Despite this enormous number of operating windmills, and despite their advantages, it eventually proved that they could not compete with the steam engine. By the middle of the nineteenth

FIGURE 9.8 A working windmill in the twentieth century. Many such mills have continued working worldwide, often for pumping water for livestock or irrigation of crops. This example is at Toowoomba, Queensland, Australia, circa 1925. (From http://commons.wikimedia.org/wiki/File:StateLibQld_1_255391_Inspecting_a_windmill_at_Toowoomba,_ca._1925.jpg.)

century, milling grain and doing the many other mechanical jobs of the windmill had largely been taken over by the steam engine. The steam engine didn't have to depend on weather conditions and the prevailing wind. The steam engine—as a locomotive—could also provide cheap transportation of the grain and flour, largely wiping out the issue of distance. Over the last part of the nineteenth century, and into the twentieth, windmills steadily faded as useful devices. The "last stand" of the traditional windmill was in remote farms in rural areas, where windmills were used to pump underground water to the surface for livestock (Figure 9.8). Even into the early twentieth century, windmill-operated pumps were used on oil wells in the region around Baku, Azerbaijan.

REFERENCES

1. Calculated from Andrews, J. and Jelley, N. *Energy Science*. Oxford University Press: Oxford, U.K., 2007, p. 101.
2. DeBlieu, J. *Wind*. Houghton Mifflin: Boston, MA, 1998, pp. 214–216.
3. de Cervantes, M. *Don Quixote*. Penguin Books: London, U.K., 1950, p. 68.
4. Carey, J. *The Faber Book of Science*. Faber and Faber: London, U.K., 1995.

FURTHER READINGS

Brooks, L. *Windmills*. MetroBooks: New York, 1999. An extensively illustrated "coffee table" overview of windmills. Chapter 2 treats their history.

de Cervantes, M. *Don Quixote*. Penguin Books: London, U.K., 1950. Many editions of this work, one of the great masterpieces of world literature, are available, including inexpensive paperbacks. Though the character and deeds of *Don Quixote* are now stereotyped by the windmill episode, this wonderful novel has much, much more to offer.

Hill, D. *A History of Engineering in Classical and Medieval Times*. Barnes and Noble: New York, 1997. Chapter 9 provides information relating to the use of wind as an energy source, covering the period from 600 BCE to about 1450.

Ramelli, A. *The Various and Ingenious Machines of Agostino Ramelli*. Dover: New York, 1987. This relatively inexpensive reprint of a book first published in 1588 provides some examples of devices operated by wind. Excellent illustrations clearly indicate how each device functioned.

Smil, V. *Energies*. The MIT Press: Cambridge, U.K., 1999. Chapter 1 contains a brief discussion of the origin of winds; Chapter 4 discusses, also briefly, windmills and waterwheels.

Smil, V. *Energy in Nature and Society*. The MIT Press: Cambridge, U.K., 2008. Chapter 7 includes a detailed discussion of the uses of water and wind in pre-industrial societies, with many interesting diagrams of early mills.

White, M. *Leonardo: The First Scientist*. St. Martin's: New York, 2000. This new biography of Leonardo da Vinci is perhaps the first one that provides extensive treatment of his inventions and investigations of science.

10 Steam Engine

The highest industrial development of the medieval world was based entirely on three ENERGY sources: draft animals, watermills, and windmills. The transition from the medieval to the modern world—in terms of ENERGY and its use—is marked by a shift to reliance on new energy sources, especially coal, and on new devices for doing WORK. The first major invention, and the most crucial, was the steam engine.

The steam engine derives from the confluence of two totally separate issues, which will appear at first to be completely unrelated to each other and furthermore even completely unrelated to the steam engine. One was the problem how to remove underground water that had seeped into mines. The other, to be discussed in Chapter 11, is the work of the early scientists studying the properties and behavior of gases.

"PREHISTORY" OF STEAM

The first demonstrations of the possible motive power of steam are due to Hero of Alexandria (around AD 50). Besides the water-powered bellows mentioned in Chapter 8, Hero also invented devices that were operated by steam. One such device, probably a toy never used for doing WORK, was the aeolipile (Figure 10.1).

The central sphere contained water, which, when heated, turned to steam. The steam escaped through two opposite pipes. The force of the escaping steam caused the sphere to rotate. Leonardo da Vinci recorded in his notebooks ideas for a steam-operated gun. Around 1680, Denis Papin,* a professor of mathematics at the University of Marburg (a city in what is now the state of Hesse, in west-central Germany) designed and built a "steam digester," the first pressure cooker. Perhaps not coincidentally, he is also credited with inventing the safety valve, which would release steam if the internal pressure got too high and would prevent the pressure cooker from blowing up. Papin demonstrated his device by cooking some meals with it. Much more importantly, he also conceived and demonstrated how a piston could be moved upward by low-pressure steam and returned downward by atmospheric pressure.

---

* Denis Papin (1647–1712), who sometimes used the first name Dionysius, was born in France, but, by virtue of being a Huguenot (a French Protestant), was forced to spend much of his life outside his native country, a victim of the religious intolerance of Louis XIV, who had revoked the Edict of Nantes, which promised religious freedom in France. Much of Papin's best work was done in England and Germany. An excellent scientist and inventor, Papin developed not only the first pressure cooker, but also early steam engines, the diving bell, and a design for the first steam-powered boat.

FIGURE 10.1 The aeolipile designed by Hero (or Heron) of Alexandria, around AD 50. In a sense, it was the world's first steam engine, though used only for an interesting demonstration, and not for doing WORK. (From http://en.wikipedia.org/wiki/File:Aeolipile_illustration.JPG.)

OTTO VON GUERICKE AND THE FORCE OF THE ATMOSPHERE

The person who established the foundation for the eventual development of steam as an energy source was Otto von Guericke, the mayor of Magdeburg, a city on the Elbe River in Germany. Though much of his life was devoted to political and diplomatic endeavors, von Guericke maintained a second parallel career as an amateur scientist, including the study of vacuums and their effects.

Von Guericke created an apparatus consisting of two hemispheres that were very carefully fitted together. The apparatus is sometimes named in honor of von Guericke's home town, "the hemispheres of Magdeburg" (Figure 10.2).

Using an air pump (Figure 10.3), von Guericke could create a vacuum inside the hemispheres. In a public exhibition in 1672, von Guericke hitched a team of eight horses to each hemisphere. The two eight-horse teams were unable to pull the hemispheres apart. (Some modern observers believe that, considering the relatively poor status of precision machining and the crude nature of air pumps in von Guericke's time, these must have been very weak horses indeed, but no matter, the essential point was demonstrated.) von Guericke's public exhibition showed that the air itself can exert a tremendous amount of force.

In a related experiment, von Guericke showed—in another public demonstration—that when a partial vacuum was created below a large piston working in a cylinder, the combined strengths of fifty men could not prevent

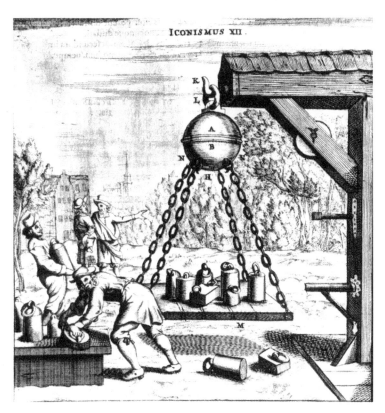

FIGURE 10.2 A public demonstration of heavy weights (on the platform) to overcome the force of the atmosphere pushing against a partial vacuum in the hemispheres of Magdeburg.

atmospheric pressure from driving the piston into the cylinder. Another, lesser known experiment provided remarkable evidence of the potential effects of the pressure of the atmosphere. Von Guericke connected his air pump to a spherical copper container. Two men worked the pump, to remove as much of the air as possible. The sphere suddenly crumpled, in what was the world's first recorded "implosion." Reports of the imploding copper sphere, and the hemispheres of Magdeburg, very likely inspired other scientists to recognize that air pressure, pushing against a vacuum, had considerable force and could be harnessed somehow to do useful WORK. A variation of this demonstration remains a staple of science classes some 350 years later. A small amount of water is put inside a can—an aluminum soft drink can works very well—and heated to form steam. When the interior of the can is filled with steam, the opening is closed up, and the can is allowed to cool. The steam condenses back to liquid water, creating a partial vacuum inside the can. The pressure of the atmosphere outside the can does the rest, with a result looking like Figure 10.4. The can is crushed by the force of the atmosphere pushing against the vacuum inside the can.

Two key observations directed the attention of scientists and engineers to the possibility of using the properties of steam to do WORK: first, the understanding that

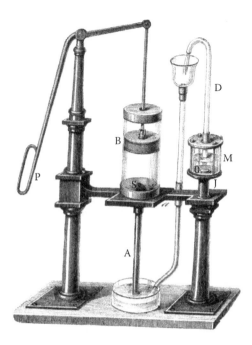

FIGURE 10.3   The air pump was a convenient way to create partial vacuums and to experiment with the force of the atmosphere. Though it has long been superseded by mechanical pumps, the air pump was an important piece of apparatus in the early days of science. (From *Elementary Treatise on Physics Experimental and Applied*, 17th edn., New York, William Wood and Company, 1906.)

the atmosphere possessed weight, as indicated by its ability to push against pistons or crumple evacuated containers; and, second, the discovery that it was possible to create a partial vacuum, either by using an air pump or by condensing steam in an enclosed vessel. As Figure 10.4 demonstrates, the weight of the atmosphere pushing against a partial vacuum can cause some remarkable effects. Von Guericke's demonstration led other scientists to recognize that somehow the air itself could be harnessed to do WORK, if only there were some convenient way to create a vacuum, as von Guericke had done inside the hemispheres.

ATMOSPHERIC ENGINE

At the Académie des Sciences in Paris, Denis Papin (Figure 10.5) had the opportunity to work with the great Dutch scientist Christiaan Huygens,* who encouraged Papin in his efforts. Huygens may have been inspired by an idea attributed to Leonardo da Vinci: If gunpowder could drive a cannon ball down the bore of a cannon, then why

---

* Christiaan Huygens (1629–1695) made brilliant contributions to mathematics, astronomy, and physics. These include the notion that light is transmitted by waves, and the concept of centrifugal force. He spent fifteen very productive years working in Paris. He returned home to the Netherlands in 1681 to recover from illness, and was prevented from returning to France when the Edict of Nantes was revoked in 1685.

FIGURE 10.4 Creating a partial vacuum inside a soft drink can make a dramatic demonstration of the force of the atmosphere working against the partial vacuum. (From http://science-mattersblog.blogspot.com/2010/03/air-pressure-crush-soda-can.html.)

couldn't it be used to operate an engine? Papin was familiar with the devices used in those days to assess the quality of a batch of gunpowder. He reasoned that if the amount of gunpowder and its explosion within the cylinder could somehow be controlled or regulated, then the movement up and down of the cylinder head (with each succeeding gunpowder explosion) could be used as the driving force for an engine.* Perhaps fortunately for posterity, Papin soon realized that a gunpowder engine was rather impractical and turned to steam instead.

Around 1690, Papin had the idea that one way, possibly a very useful way, of making use of steam would be to use a piston and cylinder (Figure 10.6). Papin was also the first to suggest the use of steam in this device. He reasoned that water could be boiled inside the cylinder. The piston would be pushed upward, rising on the column of steam made from the boiling water. Then, if the fire is removed, the steam will condense, creating a partial vacuum inside the cylinder. The pressure of the atmosphere will force the piston down, just as it can crumple a can. If the piston is connected to something, then work can be done, for example, raising a heavy load.

---

* In retrospect, of course, the idea of operating machinery or driving a vehicle using a "gunpowder engine" may seem more than a little hair-raising. However, we should consider that Huygens, or perhaps more appropriately, Leonardo had hit upon the concept that we now call the internal combustion engine. We will see more about the concept of explosions inside engines when we discuss engine knock and the diesel engine in later chapters.

FIGURE 10.5   Denis Papin (1647–1712) developed the idea of a movable piston inside a cylinder, an idea that survives today in gasoline and diesel engines. (From http://chestofbooks.com/.)

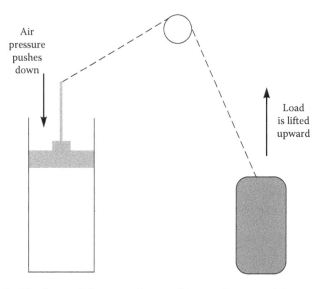

FIGURE 10.6   The force of the atmosphere pushing against a partial vacuum can lift a heavy weight.

Since it is a property of water that a small quantity of it turned into vapour by heat has an elastic force like that of air, but upon cold supervening is again resolved into water, so that no trace of the said elastic force remains, I concluded that machines could be constructed wherein water, by the help of no very intense heat, and at little cost, could produce that perfect vacuum which could by no means be obtained by gunpowder.

—Papin[1]

In 1690, an engine was built that used steam to raise a piston to the top of a vertical cylinder. At the top of its stroke, the piston was held in place by a latch. After the steam had condensed in the cylinder, the latch was released and atmospheric pressure forced the piston down. Historians of technology differ on whether Papin himself built the engine; some accounts suggest that though Papin had conceived the design for what eventually became the steam engine, he never actually built one. Papin died poor and unknown, unrecognized for his work.

In late seventeenth-century England, coal mines had already been dug so deeply into the Earth that they faced serious problems of water getting into the mines and flooding them. Some was surface water that ran down the mine shaft after storms, but much of it was water from underground streams seeping into the mine. Water could be hauled out in buckets, or it could be pumped, over very short vertical distances. The bucket system or the crude pumps could have been worked by draft animals or by the miners themselves.

Around 1698, the English inventor Thomas Savery* introduced the idea of using a separate boiler to create the steam, rather than boiling the water right in the cylinder itself. He also recommended condensing the steam more thoroughly and more quickly by pouring cold water onto the cylinder. Soon it was realized that the ENERGY in steam could be used to obtain the desired reciprocal motion. As steam is forced into the cylinder of Savery's device, a modification of Papin's piston-in-cylinder design, it pushes the piston up, against the weight of the atmosphere. Now, the supply of steam is shut off, and the steam is made to condense back to liquid water. The volume of the liquid water produced will be, roughly, about one-thousandth the volume of the steam. Condensing the steam thus creates a near-vacuum in the cylinder. The weight of the atmosphere on the piston then pushes it down. By connecting this device to a pump, the ENERGY in the steam could be used for WORK, to operate the pump, and thus to pump water out of the mines. When the piston is pushed up, it pushes—via the connection in the lever—the pump handle down. When the piston falls, it raises the pump handle.

Savery received a patent for a steam-operated pump in 1698. In the following year, he demonstrated a working model of an engine to the Royal Society. Savery advertised his machine as the "Miner's Friend" (Figure 10.7).

---

* Somewhat like von Guericke, a politician and diplomat with a second career in science, Captain Thomas Savery (1650–1715) served as a military engineer but with a keen interest in inventions. In addition to the engine for which he is famous, Savery's inventions included a glass-polishing machine and a rather bizarre paddle wheel boat in which the paddles were operated by the boat's crew, trudging around and around turning a capstan (a device roughly the size and shape of a large barrel and rotating around a vertical axis, most commonly used for raising the anchor).

FIGURE 10.7 Thomas Savery's engine, from about 1698. (From http://en.wikipedia.org/wiki/File:Savery-engine.jpg.)

It was inefficient, expensive to buy and operate, and, at times, downright dangerous. The Miner's Friend was at best a first faltering step on the way to modern machinery and engines. Unlike Papin's pressure cooker, it didn't have a safety valve. With the comparatively crude metal-working technology of the day, the high temperature of the engine would melt solder and cause the joints to split. Even so, Savery's Miner's Friend deserves recognition as a significant achievement, because, with all its faults, it was still the first real working device to go beyond the medieval technology of waterwheels and windmills.

I am very sensible a great many among you do as yet look on my invention of raising water by the impellent force of fire a useless sort of project that never can answer my designs or pretensions; and that it is altogether impossible that such an engine as this can be wrought underground and succeed in the raising of water, and draining your mines, so as to deserve any encouragement from you. I am not very fond of lying under the scandal of a bare projector, and therefore present you here with a draught of my machine, and lay before you the uses of it, and leave it to your consideration whether it be worth your while to make use of it or no....

For draining of mines and coal-pits, the use of the engine will sufficiently recommend itself in raising water so easie and cheap, and I do not doubt but that in a few years it will be a means of making our mining trade, which is no small part of the wealth of this kingdome, double if not treble what it now is. And if such vast quantities of lead, tin and coals are now yearly exported, under the difficulties of such an immense charge and pains as the miners, etc. are now at to discharge their water, how much more may be hereafter exported when the charge will be very much lessen'd by the use of this engine every way fitted for the use of the mines?

—**Savery**[2]

The first truly successful engine is attributed to another English inventor, Thomas Newcomen,* a hardware dealer with little formal education. His friend and partner was a plumber, John Calley. Newcomen and Calley tinkered for years with Savery's pump. They developed what is, in essence, a totally different device, but one that derived almost entirely from existing techniques and parts. Their engine worked at low pressure, so it was safer than Savery's, and cheaper to build. The design, which dates to about 1712, involves using a very heavy counterweight to raise the piston (Figure 10.8). He also developed a better way of condensing the steam—a spray of cold water *into* the cylinder. (Some historians attribute this invention to Newcomen's observing a Savery engine in which the cylinder had cracked, due to the poor quality of metal in those days, and cold water leaked inside.) Newcomen also realized that it is uneconomic to condense *all* of the steam. It's better, he decided, to condense only part of the steam, keeping the cylinder fairly warm, and letting the engine work more quickly. Since POWER = WORK/time, decreasing the amount of time required to do a given amount of WORK makes the engine more *powerful*. A major defect of the Savery design was that the iron cylinder where the steam was condensed back to liquid water then had to be reheated with every cycle of operation (i.e., with every stroke), which is very wasteful of fuel.

The Newcomen design was made fully usable in the mid-1720s. Newcomen's engine was the first new practical way of doing WORK since the extensive adoption of waterwheels some seven hundred years earlier. In the course of the next half-century, Newcomen's engine became very widespread throughout Europe, for operating pumps and other kinds of machinery. Nevertheless, the engine had a number of flaws. It was cumbersome and very slow. It sometimes required several minutes for the piston to go up and down, and the cylinder had to be heated and cooled on each stroke. Because of the comparatively crude manufacturing methods and metallurgical skills of the time, the engine leaked badly. The ability to do WORK was limited by the fact that the real effort was supplied by the pressure of the atmosphere. The engine was extremely wasteful of energy because, at one point in its operating cycle, the cylinder has to be hot (full of steam), but at another point in the cycle the cylinder has to be cold (to condense the steam). The engines were of enormous size and yet produced about three kilowatts. The engine was notorious for its high fuel consumption. Probably no more than about one or two percent of the chemical potential ENERGY in the coal was eventually transformed into useful WORK. A contemporary criticism of the Newcomen engine was that it took an iron mine to build one and a coal mine to keep it going. Fortunately, most

---

* Thomas Newcomen (1664–1729) conducted his engineering and business affairs through a company having the remarkable name Proprietors of the Invention for Raising Water by Fire. By the time he died, 75 of his engines were working in mines throughout England. Like von Guericke with a second parallel career, Newcomen served for many years as a layman preacher in the Baptist church.

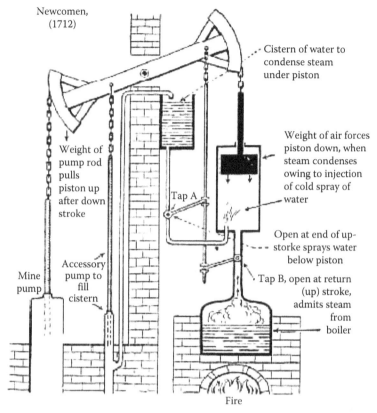

Diagrammatic view of Newcomen's atmospheric on
fire engine (1712)

FIGURE 10.8 The Newcomen engine, from about 1712. (From Hogben, L., *Science for the Citizen: A Self-Educator Based on the Social Background of Scientific Discovery*, Illustrated by J.F. Horrabin, p. 555, Figure 247, London, U.K., George Allen and Unwin Ltd., 1938.)

of these engines were used at coal mines, where the coal was essentially free. And yet despite all these flaws, Newcomen's engine was an immediate commercial success. Forty years later, more than a hundred were in use, mostly in pumping water from coal mines, as well as copper, lead, and tin mines.

Not only did Newcomen's engine become an immediate commercial success, it was a long-lived one as well. A Newcomen engine at a coal mine in Derbyshire (England) operated from 1791 to 1918. In fact, the last surviving Newcomen engine (in Yorkshire, England) was not scrapped until 1934! It had operated for more than a hundred years without needing extensive maintenance. Despite the notorious inefficiency of the Newcomen engine, it was better than anything else then available. In 1775, a Newcomen engine, built by the English engineer John Smeaton, was sold to Russia for the purpose of pumping out the naval dry docks at Kronstadt. The new engine replaced two windmills, each thirty meters high, which had been installed by

Dutch engineers some sixty years earlier. The windmills took a full *year* to pump the water out of the dry dock; the Newcomen engine did it in two weeks.

In the Newcomen engine, the steam functions only as an intermediary agent. Steam is used to create the necessary vacuum. The pressure of the atmosphere pushing against the vacuum is the source of WORK. For that reason, the Newcomen engine is more appropriately called an atmospheric steam engine, or an atmospheric engine.

JAMES WATT AND THE STEAM ENGINE

Despite its inefficiencies, Newcomen's engine steadily gained applications throughout Britain and Europe. Even universities began to acquire models of the Newcomen engine to study. One such model was at the University of Glasgow. By the 1760s, the university's model had been broken for some time, and finally someone took the broken model engine to the university's scientific instrument maker to have him repair it. On Easter Sunday of 1765, he was thinking about the working of this model engine while walking on Glasgow Green. Suddenly he realized the major problem of the Newcomen engine—the one cylinder must alternate between being hot and being cold. This very inefficient operation wastes considerable amounts of fuel. He recognized that it would be much better to have two chambers—one, the cylinder with piston, that is always hot and one that is always cold (Figure 10.9).

> *...as steam was an elastic body, it would rush into a vacuum, and if a communication was made between the Cylinder and the exhausted vessel, it would rush into it and might be there condensed, without cooling the Cylinder.*
>
> **—Watt[3]**

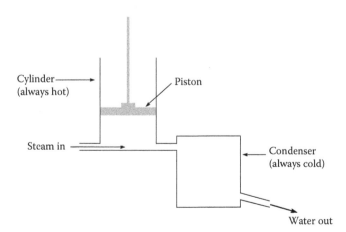

FIGURE 10.9 A schematic drawing of Watt's engine, showing his key contribution—a condensing chamber that is always cold, allowing the cylinder to remain hot.

FIGURE 10.10 James Watt (1736–1819), the father of the true steam engine. A portrait from about 1792 by the Swedish painter Carl Frederik von Breda. (From http://en.wikipedia.org/wiki/File:Watt_James_von_Breda.jpg.)

This modification incorporated two major advances: First eliminating temperature cycling (hot–cold–hot–cold…) in the single cylinder greatly increases the efficiency of operation. Second in the Papin, Savery, and Newcomen designs, the pressure of the air moves the piston. The function of the steam is simply to provide a convenient way of making a vacuum. The new device actually uses the expansion of the steam to do the WORK of the engine so, strictly speaking, is the first true steam engine. The instrument maker, the great Scot who came up with these ideas, was James Watt* (Figure 10.10). A steam engine is an apparatus or device for converting heat (in steam) into kinetic energy. A steam engine is one representative of a more general class of devices called *heat engines*. We will examine the concept of heat engines in more detail later.

---

* James Watt (1736–1819) was the son of a man who repaired ships and made nautical instruments. Watt was sickly as a boy, so had the great good fortune of not having to go to school. Instead he was schooled at home. His father made his workshop the "school." At the comparatively young age of 21, he had become the "mathematical instrument maker" to the University of Glasgow. In the mid-to-late eighteenth century, the Scottish universities, Edinburgh and Glasgow, were possibly the best in the world. Watt's employment at Glasgow let him meet some of the world's leading scientists and to learn their ideas. Today Watt seems seldom remembered, yet his development of the true steam engine must certainly rank him as one of the most influential scientists and inventors of modern times—in fact, it can be argued that the steam engine helped create "modern times." His name lives on as the accepted unit of POWER, the watt.

FIGURE 10.11 A sketch of a sun-and-planet gear system, a key invention that allowed the conversion of reciprocating motion, as in a steam engine, into rotary motion. (From http://en.wikipedia.org/wiki/File:Sun_and_planet_gears.gif.)

Watt made a series of improvements, both to the earlier engine of Newcomen and to his own original design for the steam engine. One of the crucial improvements was a separate chamber to condense the steam. A second very important contribution of Watt was the invention of a system of so-called sun-and-planet gears (Figure 10.11) that converted the reciprocating up-and-down motion of the piston into rotary motion of a shaft or axle.* The ability of the Watt engine to drive rotating machines made it attractive to the British cotton industry, the main user of automatic machinery at that time. In turn, this freed industry from a reliance on waterwheels for operating factories. Consequently, manufacturers could build larger factories and could locate them nearer to sources of raw materials and population, rather than near sources of water.

---

* The gear systems described in Chapter 8, in connection with waterwheel systems, are the simplest application of gears. One gear meshes with the other. One is driven by something (e.g., the waterwheel) and the other is caused to rotate because it meshes directly with the first. Suppose, however, that we had a different system, in which one of the gears was held stationary. The remaining gear would then rotate around it. If the sun-and-planet gears are attached to a casing, the rotation of the planet gear around the sun gear results in a rotary motion. Connecting the casing to the reciprocating piston rod of a steam engine converts the reciprocating motion into rotary motion. This is one method of converting reciprocating motion (as from a steam engine) into rotary motion. Sun-and-planet gear systems of much greater complexity are in everyday use, in, for example, the automatic transmissions of automobiles.

FIGURE 10.12 A patent model of an early governor, this one from about 1856, for regulating the speed of engines. (From http://www.patentmodel.org/patent-models/14967; Rothschild Petersen Patent Model Museum, All rights reserved, Web Site by Bentley & Hoke LLC, Copyright 2010.)

I intend in many cases to employ the expansive force of steam to press on the pistons, or whatever may be used instead of them, in the same manner in which the pressure of the atmosphere is now employed in common [Newcomen] engines.

—Watt[4]

Watt worked hard on further improving his invention. He invented a governor that automatically controlled the engine's output of steam (Figure 10.12), and, later, a gauge for determining pressure in the cylinder and an "indicator" for following changes in steam pressure during the operating cycle of the engine.* Watt's invention

---

* The "indicator" of a steam engine, one of the many improvements introduced by Watt, is a device intended to measure the "indicated work." One common form of the indicator consists of a very small cylinder with piston that, like the main engine itself, is operated by the pressure of the steam. Through a system of springs and levers, the other end of the piston is connected to a marking pen. The pen rests against a vertical drum, to which is attached a sheet of paper. The drum itself is connected to some part of the engine and is caused to rotate as the engine operates. The paper moves horizontally as the drum rotates, while the marking pen moves vertically. The extent of horizontal motion of the drum is proportional to the stroke of the engine; the extent of vertical motion of the marker is proportional to the steam pressure in the engine. Because both motions occur simultaneously, at the end of test, the pen will have traced out a closed curved shape. This curve is called an indicator diagram. The area enclosed by the closed curve determines the WORK done by the steam in driving the piston. The characteristic shape or form of the closed curve shows how the steam pressure in the cylinder varied from point to point throughout the stroke of the piston and is useful in checking the operation of various valves in the engine.

of the governor made the easy adaptation of the steam engine to rotary machinery possible.* Watt's governor adjusted the amount of steam passing into the cylinders according to how rapidly the machinery was going. This is the first example of real feedback control of a machine. The Greek word for "governor" is *kybernetes*. It has come into our language as "cybernetics." That is, fundamentally, what cybernetics means—governorship. Watt's governor and Papin's safety valve were the two essential devices that made steam engine operation safe, in terms of preventing engines from running out of control or possibly exploding.

By 1790, Watt's engines, steadily improved by his stream of inventions, had almost totally replaced the older Newcomen engines. To improve efficiency even more, Watt installed a steam jacket around the cylinder, to keep it as hot as the steam itself. These improved engines had fuel consumption one-third that of a typical Newcomen engine for the same amount of WORK. This gain of efficiency, from the 1%–2% of the Newcomen engine to 6%–7% in the Watt engine, made all the difference in accounting for Watt's engine displacing Newcomen's. The Newcomen engine was so inefficient that it could only be used in coal mines, where the coal cost nothing. By the time of the development of Watt's engine, there were other mines that needed to be pumped, such as the tin and copper mines in Cornwall. The Newcomen engine could not compete in that application because coal had to be *purchased* at the Cornish tin and copper mines. The extremely low efficiency of the Newcomen engine made the cost of purchased coal prohibitive.

Engineers of that era spoke of the "duty" of an engine. The term probably derives from the practice of having humans or animals perform WORK (operating machinery) by serving a turn of duty in a treadmill. The duty was essentially the WORK done multiplied by the time it took to do it. For steam engine operation, the duty of the engine was expressed in terms of the amount of water pumped from a mine for a given quantity of coal that was burned. James Watt measured the amount of work a horse could do by making it pull a known weight, lifted over a pulley, up to a certain height. He conceived the idea of work being the product of force and distance and of power being the rate of doing work. The name of the unit of POWER developed by Watt is the horsepower. This unit is still in quite common use in the United States and lingers in other nations that once used the British system of units.

Watt's steam engine rapidly became the crucial device for powering all sorts of factory machinery in virtually every industry in Britain, Europe, and America. Over a thousand steam engines of Watt's design were installed in factories in the 1790s. This is more than the total number of engines of all kinds sold in the period 1700–1790. Partly this is because the Newcomen engine had largely been restricted to

---

* The governor of an engine, another of Watt's inventions, is a device for automatically regulating engine speed. A simple form consists of two weighted balls connected to a hinge at the top of a rotating spindle. The spindle is connected to the engine and is rotated as the engine operates. As the speed of the engine increases, the balls tend to fly outward; in doing so, they partially close the valve that admits steam to the cylinder. As the engine slows, the balls will drop down toward the spindle; in this case, they will open the steam valve more widely. As the engine speeds up beyond a desired value, the governor restricts steam flow, and the speed drops. If the speed drops below the desired value, the governor opens the valve and speed increases.

use in operating pumps, because they provided WORK only on the downward stroke of the piston. Essentially, this meant that the engine was limited to unidirectional motion. This unidirectional motion for doing WORK meant that there was no easy way to apply the Newcomen engine to operate most kinds of factory machinery, which usually required a continuous rotary motion—in fact, just like that produced by a waterwheel. Watt invented the so-called double-acting cylinder, in which steam alternately admitted and released from both sides of an enclosed piston provides WORK in both directions.

An important step in spreading the steam engine away from the mines and into factories was due to the enterprise of Matthew Boulton (Figure 10.13).

I sell here, Sir, what all the world desires to have—Power.

—Boulton[5]

He owned the largest button factory in Birmingham, operated by water. Unfortunately, Birmingham is not a hilly area, so there's not much gravitational potential energy, nor is it a very wet place. The millpond used to supply water to the waterwheels dried up in the summer. It was suggested to Watt that his engine could be used to pump the water up from the millpond into the millrace and so work the wheel. In other words, a steam engine would be used to pump water to run a waterwheel, which in turn ran the factory.

FIGURE 10.13 Matthew Boulton (1728–1809), James Watt's business partner. Like Figure 10.10, this is another 1792 portrait by Carl Frederik von Breda. (From http://en.wikipedia.org/wiki/File:Matthew_Boulton_-_Carl_Frederik_von_Breda.jpg.)

The people in London, Manchester and Birmingham are steam mill mad. I don't mean to hurry you, but I think in the course of a month or two, we should determine to take out a patent for certain methods of producing rotative motion from the fire engine.

—Boulton[6]

Essentially, James Watt's invention revolutionized industry at the end of the eighteenth century. The engine quickly became the principal means of doing WORK in factories.

Our road turned to the right, and we saw, at the distance of less than a mile, a tall upright building of grey stone, with several men standing upon the roof, as if they were looking out over battlements. It stood beyond the village, upon higher ground, as if presiding over it—a kind of enchanter's castle, which it might have been, a place where Don Quixote would have gloried in. When we drew nearer we saw, coming out of the side of the building, a large machine or lever, in appearance like a great forge-hammer, as we supposed for taking water out of the mines.* It heaved upwards once in half a minute with a slow motion, and seemed to rest to take a breath at the bottom, its motion being accompanied with a sound between a groan and a "jike." There would have been something in this object very striking in any place, as it was impossible not to invest the machine with some faculty of intellect; it seemed to have made the first step from brute matter to life and purpose, showing its progress by great power. William* made a remark to this effect, and Coleridge* observed that it was like a giant with one idea. At all events, the object produced a striking effect in that place, where everything was in unison with it...

—Wordsworth[7]

Humankind was no longer dependent on draft animals, rivers, or the wind. It should not diminish Watt's luster in any way to point out that Watt did not invent the steam engine essentially out of nothing. Watt's tremendous achievement was to improve upon the earlier existing designs of men such as Papin, Savery, and Newcomen to develop the first truly *practical* steam engine. We will meet other cases of how a brilliant inventor, scientist, or engineer built upon earlier, perhaps flawed efforts of others finally to create a practical, useful device. A particular example that we'll see is in the story of the light bulb (Chapter 14).

RICHARD TREVITHICK: STEAM ON WHEELS

The next major development, another crucial one for society, came as a result of returning to Savery's seventeenth-century use of high-pressure steam. A drawback of the Miner's Friend was that it could be dangerous, because it used high-pressure steam. Watt stayed away from high-pressure steam and its potential explosion

---

* The device referred to here was the second Watt engine to be built in Scotland (1788). "William" was the author's brother, William Wordsworth, who is generally considered to have been the greatest of the English Romantic poets. Examples of his work can be found in most anthologies of great poetry, and certainly in anthologies dealing with the poetry of the Romantic era (i.e., the late eighteenth and early nineteenth centuries), as well as in collections devoted entirely to his work. "Coleridge" was Samuel Taylor Coleridge, another of the great English poets of the Romantic era. Perhaps his work best known nowadays is "The Rime of the Ancient Mariner." He also had a vigorous and productive career thinking and writing about religion and about politics, as well as being a literary critic.

hazards. He never built engines with steam pressures much greater than that of the atmospheric pressure, because he was afraid of boiler explosions and their potentially terrible consequences. (In this regard, Watt's fears were quite justifiable. As but one example, an 1850 high-pressure steam boiler explosion in a New York City factory instantaneously demolished the entire factory, with 67 deaths and another 50 injuries.)

My predecessors put their boilers in the fire; I have put the fire in the boiler.

—**Trevithick**[8]

In fact, Watt had essentially a patent monopoly on the use of low-pressure steam. Richard Trevithick* may have had the insight to realize that high-pressure operation could offer significant advantages (not the least of which is the fact that high-pressure engines would do the same WORK as a low-pressure engine but could be considerably smaller and lighter, and therefore cheaper), or perhaps he was looking for ways to circumvent the patents. If a high-pressure engine could be made smaller, and hence lighter, than a low-pressure engine doing the same amount of WORK, then perhaps it could be made so small, and so light, that it would be portable. The Watt engines were so heavy that it is quite doubtful that they could have generated enough ENERGY to move themselves, let alone pull a train behind. Yet another of a long line of exceptionally talented British engineers, Trevithick realized that the moving piston of the steam engine could be made, with appropriate mechanical linkages, to turn wheels. Trevithick built his first steam "road carriage" (Figure 10.14) in 1801, making the first trial run on Christmas Eve.

Captain Dick got up steam, out in the high-road, just outside the shop at the Weath. When we get see'd that Captain Dick was agoing to turn on steam, we jumped up as many as could; maybe seven or eight of us. "Twas a stiffish hill going from the Weight up to Camborne Beacon, but she went off like a little bird...

When she had gone about a quarter of a mile, there was a roughish piece of road covered with loose stones; she didn't go quite so fast...She was going faster than I could walk, and went on up the hill about a quarter or half a mile farther, when they turned her and came back again to the shop...

—**Williams**[9]

Two years later, he was driving through London at the princely rate of 16 kilometers per hour. In February 1804, he mounted a steam carriage on rails (Figure 10.15) and hauled a load of ten tonnes, along with seventy men, on a 16 km trip in Wales. This was the first successful operation of a locomotive.

---

* Richard Trevithick (1771–1833), labeled disobedient, slow, and obstinate by his schoolmaster, somehow managed to invent the steam locomotive anyway. His one talent as a boy seemed to be finding unique or unconventional solutions to mathematical problems. Trevithick became a mining engineer whose work took him to many countries. He started his career at the Ding Dong mine in Cornwall, England, and began experimenting with high-pressure steam there. Trevithick's career seemed to alternate between major professional and financial successes and failures, at the low points often on the brink of ruin.

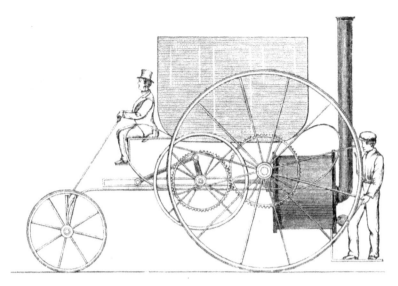

FIGURE 10.14 Trevithick's steam-operated road carriage, demonstrated in London in about 1803. (From http://en.wikipedia.org/wiki/File:Trevithicks_Dampfwagen.jpg.)

By the 1850s, the steam engine had become the dominant source of WORK in most of its possible areas of application. It dominated in the application for which it had first been developed, the pumping of water from mines. It played a major role in the operation of machinery in factories. It was the source of locomotion on all the world's railroads. It was beginning to displace the sail as the means of propelling ships.

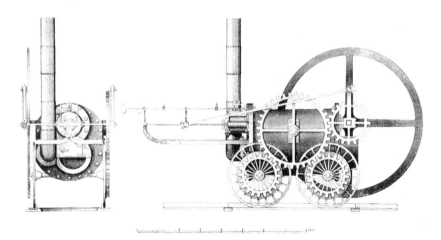

FIGURE 10.15 Trevithick's locomotive, demonstrated in Coalbrookdale, around 1803. (From http://fr.wikipedia.org/wiki/Fichier:Trevithick%27s_Coalbrookdale_locomotive,_1803_%28British_Railway_Locomotives_1803-1853%29.jpg.)

STEAM: ENERGY FOR THE INDUSTRIAL REVOLUTION

Steam was *the* ENERGY source of the Industrial Revolution. Consider: Watt's steam engine operated the pumping machinery to get water out of mines and operated the hoisting machinery to lift coal and ores from the mines. This made it possible to get coal and ore cheaply from deeper and deeper seams. The steam locomotive made it possible to transport the coal and ore—and other raw materials—cheaply. The steam engine, when applied to the blast furnace, allowed continuous operation and the production of cheap iron and steel. The iron and steel could be used to make a wide variety of industrial machinery that could be operated cheaply and on a large scale by the steam engine. The products of these manufactories could in turn be transported cheaply on land by the steam locomotive and internationally in steam-worked iron ships.

Steam engines were readily adopted in nineteenth-century Britain. However, in the United States, even up until about the time of the Civil War, waterwheels provided far more WORK than steam engines, and wood supplied much more heat than coal. It was not until about 1875 that steam became dominant in American industry. Nevertheless, steam was instrumental in the growth of industrial cities in the East. Worcester and Fall River, Massachusetts, are examples. These cities did not have access to fast-flowing rivers to take advantage of waterwheels. Instead, factories in those cities relied on steam engines, fuelled by coal from Pennsylvania.

A water mill is necessarily located in the country afar from the cities, the markets and magazines of labor, upon which it must be dependent....A man sets down his steam engine where he pleases—that is, where it is most to his interest to plant it, in the midst of the industry and markets, both for supply and consumption of a great city—where he is sure of always having hands near him, without loss of time in seeking for them, and where he can buy his raw materials and sell his goods without adding the expense of a double transportation.

—Scientific American[10]

A steam engine capable of generating as much POWER as a typical lawnmower engine would fill a two-story building. But complaining about the inefficiency of the early steam engine overlooks its absolutely critical contribution to society. The steam engine freed human society from reliance on ENERGY sources supplied by nature (wind, water, or animal muscles). The steam engine is the first device that provides useful WORK from an ENERGY source (steam) that could be created on demand and as needed by humans. Too, the steam engine was the first step in being liberated from the constraints of geography. That is, it was no longer necessary to build factories on the banks of fast-flowing rivers or to dam a river for a water supply. It was no longer necessary to find locations that offered a reasonably steady wind. A steam engine could be erected any place that WORK was needed.

It is in these things of iron and steel that the national genius most freely speaks.

—Howells[11]

Another aspect of the dominance of the steam engine in society is the way in which it entered everyday speech. Some of the colloquialisms are still heard occasionally. For example, a busy and vigorous worker was someone who "got up a head of steam," meaning that he or she had, in essence, developed a high pressure of steam and was ready to do some job. Many of us are annoyed from time to time by "high-pressure" salesmen. One reason these people are so obnoxious and unpleasant is that they tend to use "steamroller tactics" to convince you to buy something. In fact, if you are really annoyed by such a person, it wouldn't be unreasonable if you got "steamed up." When you're feeling angry like that, it's a good idea to find some safe way to relax and "let off a little steam." Otherwise, you might just "blow a gasket," or, worse, you could actually "explode."

The steam engine is only the first step in being freed from the constraints of geography. There is still one constraint in using the steam engine: it provides WORK by moving a piston up and down, or making a shaft or axle turn. This means that the engine itself still must be very close to the device that it will operate. Whatever mechanism is being operated by the steam engine has to be directly physically connected to it. One consequence of this limitation is that every factory with steam-engine-driven machinery had to have its own engine, in the basement or yard, and its own supply of fuel, usually coal. A little over a century after Watt's great invention, the second liberating step began to emerge from research in a completely different field of science—the study of electricity. This we will explore in future chapters.

REFERENCES

1. Papin, D., cited in Derry, T.K.; Williams, T.I. *A Short History of Technology.* Dover: New York, 1993, Chapter 11.
2. Savery, T., cited in Bernal, J.D. *A History of Classical Physics.* Barnes and Noble: New York, 1972, p. 260.
3. Watt, J., cited in Derry, T.K.; Williams, T.I. *A Short History of Technology.* Dover: New York, 1993, Chapter 11.
4. Watt, J., cited in Kirby, R.S.; Withington, S.; Darling, A.B.; Kilgour, F.G. *Engineering in History.* Dover: New York, 1990, Chapter 7.
5. Boulton, M., cited in Morton, O. *Eating the Sun.* HarperCollins: New York, 2008, p. 329.
6. Boulton, M., cited in Pool, R. *Beyond Engineering: How Society Shapes Technology.* Oxford: New York, 1997, p. 122.
7. Wordsworth, D. *Recollections of a Tour Made in Scotland.* Yale: New Haven, 1997, p. 50.
8. Trevithick, R., quoted in Rosen, W. *The Most Powerful Idea in the World.* Random House: New York, 2010, p. 296.
9. Williams, S. (blacksmith) quoted in Rosen, W. *The Most Powerful Idea in the World.* Random House: New York, 2010, p. 291.
10. From *Scientific American* magazine, cited in Nye, D.E. *Consuming Power.* MIT Press: Cambridge, U.K., 1998, p. 71.
11. Howells, W.D., quoted in Burton, O.V. *The Age of Lincoln.* Hill and Wang: New York, 2007, p. 257.

FURTHER READINGS

Bernal, J.D. *A History of Classical Physics*. Barnes and Noble: New York, 1997. A well-written survey of physics from the ancient Greeks through the end of the nineteenth century, easily readable by the layperson. Chapter 10 discusses the development of the steam engine, and its relationship to the development of theories of heat.

Laidler, K.J. *To Light Such a Candle*. Oxford University: Oxford, U.K., 1998. The focus of this book is the interrelationship between science and technology; how in some cases advances in basic science spark technological advances, and in others a step forward in technology stimulates new work in science. Chapter 2 deals with Watt and the steam engine, and how the steam engine led to developments in the science of heat.

Nye, D.E. *Consuming Power*. MIT: Cambridge, U.K., 1998. A social history of the use of energy in the United States. Chapter 3, "Cities of Steam," discusses the impact of the steam engine.

Parker, K. *How a Steam Locomotive Works*. TLC Publishing: Forest, VA, 2008. Steam locomotives have been gone from major railway systems for a half-century, but to those who love them they will never die. Literally thousands of photo-illustrated books have been published in many countries. This book goes beyond being a catalog of photographs for the railway enthusiast, and actually explains how these magnificent machines functioned.

Petroski, H. *Remaking the World*. Knopf: New York, 1997. This collection of essays on engineering includes one ("Harnessing Steam") on the development of the steam engine.

Rosen, W. *The Most Powerful Idea in the World*. Random House: New York, 2010. An excellent history of the development of steam engines and their impact on industrial development.

11 Heat and Thermal Efficiency

James Watt's invention of the steam engine was the most important development in the use of ENERGY in a millennium. Initially applied to manufacturing, it provided a source of ENERGY that could replace human or animal muscles, did not depend on having suitable rivers available, and did not depend on whether the wind was blowing. Richard Trevithick's extension of Watt's steam engine to develop the steam locomotive provided an entirely new form of transportation. These two applications of the steam engine—to manufacturing and to transportation—place a sharp dividing line between medieval and modern technologies.

NOTION OF EFFICIENCY

There is a second dividing line as well. An advantage of water and wind as energy sources is that they are essentially free. After the initial investment in the windmill or waterwheel (and dam, if needed), there is no cost for the water or wind itself. Not so with the steam engine! The heat needed to generate the steam was obtained by burning a fuel, most commonly coal. This coal had to be purchased.* Buying coal represented a continuous expense for the owner of a steam engine. For this reason, interest quickly developed in determining how much WORK an engine could do per amount of coal consumed. An engine that required less coal to accomplish a set amount of WORK (pumping a certain quantity of water, for example) represented a lower cost to operate. The necessity of having to buy coal to "feed" the engine, and the early applications of these engines for pumping water from mines led early engineers to define the "duty" of a steam engine: the amount of water that could be raised from the mine per bushel† of coal burned. In modern terminology, we would refer to the amount of WORK done per amount of fuel consumed as the *efficiency* of the engine.

In broad terms, engineering efficiency is defined as the ratio of the output to input:

$$\text{Efficiency} = \text{output/input}$$

---

* It might be argued that, if steam engines were being used to pump water from coal mines, the coal would be free too, because it was available right there on the job site. However, coal consumed by the engines operating the pumps represented coal that could not be sold to customers. Such internal consumption of coal also represented a financial burden to the mining company, because it reduced the income from the sale of coal.

† In British units, a bushel is a measure of volume, sometimes still used for selling farm products, such as a bushel of apples. A bushel is equivalent to about 36 liters. The weight of coal that could be packed into a bushel container depends on the size of coal particles, so that a large weight of pebble-sized pieces of coal could make up a bushel, but a lesser weight of football-sized pieces.

The ratio can be expressed in terms of any convenient engineering parameter, such as ENERGY, raw materials and products, money flow, or the use of time.

At first the steam engine, and earlier atmospheric engines, had been simply a means of raising water from coal mines. These early engines were being continually improved by hunches, trial and error, and tinkering. The science of steam engines was developed only later. (It comes, as we shall see in the following text, in part from the study of the behavior of gases by the French physicists Charles and Gay-Lussac.) The relationship between science and technological development is complicated. Sometimes useful, functioning devices are invented and improved with little understanding of their underlying scientific principles. Only later, after the practical success of these devices had been established, did other workers determine the relevant scientific principles.

In every department of human affairs, Practice long precedes Science: systematic enquiry into the modes of action of the powers of nature, is the tardy product of a long course of efforts to use those powers for practical ends.

—Mill[1]

Or put a bit differently,

...the practical mechanics of the engineer arrives at the concept of work and forces it on the theoreticians.

—Engels[2]

The steam engine provides an excellent example. In other cases, though, the fundamental science is established first, and from that scientific basis, engineers then design and build devices having practical commercial applications. An example of this kind of situation is provided by the development of the science and technology of electricity will be discussed in Chapter 12.

The steam engine has done much more for science than science has done for the steam engine.

—William Thomson[3]

In the early days of the steam engine and Newcomen's atmospheric engine, nothing was known as to what—if anything—limited the amount of WORK that could be done by these engines for the amount of ENERGY provided in fixed amount of fuel. Chapter 10 introduced the notion of engine efficiency and showed that the Newcomen atmospheric engine was singularly inefficient at converting the chemical potential ENERGY present in coal into useful mechanical WORK. In some especially poor cases, less than 1% conversion was achieved. Watt's steam engine was an enormous improvement compared to the Newcomen engine. Nevertheless, by modern standards, it too was rather inefficient. Only some 5%–7% of the ENERGY in coal was converted to useful WORK.

TEMPERATURE AND THERMAL POTENTIAL ENERGY

Consider some common observations: A cup of hot coffee or tea, or a bowl of hot soup, allowed to stand for some time, will become much cooler. That is, it has become closer to the temperature of the room. A dish of ice cream, or a cold beverage, allowed to stand for a while, will become warmer. Again, it has become closer to the temperature of the room. No one has *ever* observed the reverse; that is, a glass of water sitting on a table would spontaneously begin to boil or spontaneously freeze.

The first two observations—common experience for everyone—result from the same phenomenon. Heat has moved from one object to another and has moved in the direction from the hotter object (the cup of coffee in the first case, or the air in the room in the second case) to the colder object (the air in the room in the first case, or the ice cream in the second). What measurement allows us to determine the relative hotness or coldness of an object? Temperature. Thus the first two observations generalize as follows:

Heat spontaneously flows from an object at high temperature to one at low temperature.

The third observation illustrates another important point: heat *never* spontaneously flows from low temperature to high temperature. This is now the third situation in which a spontaneous change in one direction never occurs in reverse. Water never spontaneously flows uphill or jumps up a waterfall. A mixture of carbon dioxide and water vapor never spontaneously converts itself back into methane and oxygen. The waterfall and the burning methane cases were illustrated by simple energy diagrams (Figures 5.5 and 8.6, respectively). An analogous diagram (Figure 11.1) can help to explain the flow of heat.

To be consistent with the previous energy diagrams, heat must be flowing spontaneously from high potential energy to low potential energy. To emphasize the point that, in these SYSTEMS, heat is the kind of ENERGY of interest, we refer to *thermal* potential energy. That is,

Temperature is a measure of thermal potential ENERGY.

Therefore, the energy diagram can be drawn as shown in Figure 11.2.

The adjective *thermal* serves to indicate a SYSTEM in which heat is moving from high to low potential. In a waterfall, water at high potential falls down the waterfall under the influence of gravity. In that case, it would be appropriate to speak of *gravitational*

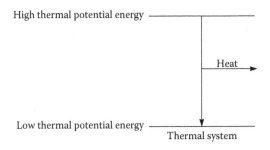

FIGURE 11.1 The energy diagram for thermal systems.

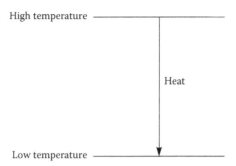

FIGURE 11.2   Thermal energy diagram with the thermal potential energy labeled as *temperature*.

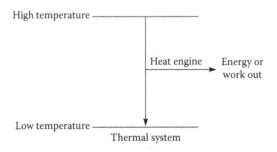

FIGURE 11.3   Energy diagram for a thermal system, showing extraction of WORK.

potential energy. ENERGY stored in the chemical bonds in molecules of methane and oxygen is a more subtle concept, but represents a form of *chemical* potential energy. In all three SYSTEMS, the key concept is that ENERGY flows spontaneously from high potential (regardless of whether thermal, gravitational, or chemical) to low potential.

A second important analogy with the waterfall comes from the way that ancient humans learned to design waterwheels to capture some of the ENERGY of the moving water and convert it into useful WORK. That is, a device, a piece of hardware, is inserted into the SYSTEM to extract ENERGY or WORK. We can do exactly the same thing with a thermal energy system (Figure 11.3). In the thermal system, the kind of device that we use to extract useful WORK from the SYSTEM is called a heat engine. A steam engine is one kind of *heat engine*.*

The point at which technology really took off is with the invention of the heat engine...

—Benson[4]

Obtaining 100% conversion of kinetic energy to WORK in a windmill or waterwheel, if it were possible, would require that the kinetic energy of the wind or water to be

---

* Though the term is very rarely used, we could, by analogy, refer to a waterwheel as a *gravitational engine*. Regardless of how noisy, unsafe, or improbable it might be, we could design a wheel to operate by dropping bowling balls on it. Just as water falls from high to low gravitational potential, so too does a bowling ball when we drop it. A gravitational engine would be some kind of device designed to extract useful WORK when something spontaneously falls from high to low gravitational potential.

zero when it leaves the mill. For example, the water just downstream of a waterwheel would have to come to a dead stop. A kinetic energy of zero would correspond to the complete conversion of the kinetic energy of the moving wind or water into WORK. A kinetic energy of zero is also the lowest possible value of kinetic energy, since there is no conceivable way that an object can have a negative kinetic energy. In a thermal system, therefore, it should be possible to obtain 100% conversion of thermal energy (i.e., heat) into WORK, provided that we can establish a point of zero thermal energy. Because temperature is the measure of thermal energy, this "zero point" of thermal energy would be, therefore, the lowest temperature that could theoretically be attained. In other words, to find out how to achieve 100% conversion of thermal energy to WORK, we first need to find out what the lowest possible temperature is.

GAS LAWS AND THE QUEST FOR ABSOLUTE ZERO

The first detailed study of the effect of temperature changes on gases was done by the French physicist Jacques Alexandre César Charles* (Figure 11.4) in 1787, but these studies were carelessly carried out, the quality of the data was poor, and the work was never properly published in the scientific literature.

The law governing the thermal expansion and contraction of gases should correctly be attributed to Joseph Louis Gay-Lussac[†] (Figure 11.5), though it is known in many textbooks as Charles's law.

Charles had remarked on these properties of gases 15 years ago, but not having published the results, I have the great good luck to make them known.

—**Gay-Lussac**[5]

Gay-Lussac studied many gases, but found that, regardless of the specific gas, every gas expands by about 1/300 in volume for each degree Celsius when it is heated and contracts by the same fraction when it is cooled. However, this observation depends critically on two restrictions: The pressure of the gas must remain constant, and the amount of the gas being studied must also remain constant.

Exploring how to find the lowest possible temperature, or the point of zero thermal potential energy, begins with this observation by Charles and Gay-Lussac: If a sample of gas is kept at constant pressure, its volume is directly proportional to its temperature. This relationship is now usually known as Charles's law. Table 11.1 shows some data that are consistent with this law. (These are not original experimental

---

* Jacques Charles (1746–1823) was a French scientist and balloon ascensionist, who, in 1783, made the first manned flight in a balloon filled with hydrogen. Not only was this an epochal moment in the early history of flight, it also represents the first practical application of hydrogen energy. One of the persons in the crowd watching this flight was the American scientist and diplomat Benjamin Franklin. Hydrogen-filled balloons became known as "Charlière." Charles made the hydrogen by reacting sulfuric acid with iron. Nowadays the early balloon flights of the Montgolfier brothers are perhaps better known, but the Montgolfier balloons relied on hot air for lift.
† Joseph Gay-Lussac (1778–1850) made numerous contributions to the development of chemistry and physics in the early nineteenth century. In 1805, Gay-Lussac and a copilot reached an altitude of 7000 meters in a hot air balloon, more than ten times higher than the pioneering flight of Charles in his hydrogen balloon. Today there is a street and a hotel in Paris named in his honor.

FIGURE 11.4   Jacques Charles (1746–1823), a French scientist who contributed to our knowledge of the behavior of gases. (From http://en.wikipedia.org/wiki/File:Jacques_Alexandre_C%C3%A9sar_Charles.jpg.)

data; these data are an example that is consistent with the law and that may make it easier to see the important effect that is occurring as the gas is cooled.) These data show that for every 1°C drop in temperature, a gas contracts by 1/273 of its volume. The extension of this observation suggests that reducing the temperature to −273°C would cause the gas to contract by the fraction 273/273, which is equal to 1. In other words, the gas would lose all its volume and vanish. This is physically impossible, since matter cannot be destroyed. Therefore, −273°C represents the lowest possible temperature. (The argument we are making here ignores the fact that all real gases would turn to liquids at some temperature above −273°C.*)

---

* At any given temperature, the molecules of any gas have a characteristic kinetic energy. The kinetic energy is directly proportional to the temperature, so that if the temperature is reduced, the kinetic energy of the gas molecules is reduced also. In addition, the molecules of any gas exert weak attractive forces among one another. The strength of these attractive forces depends on the chemical nature of the gas molecules. At ordinary conditions of temperature and pressure, the kinetic energy of the molecules is far stronger than the weak attractive forces, and the gas molecules whiz around in their container largely independently of each other. As the temperature—and consequently the kinetic energy—is reduced, however, a temperature will be reached at which the attractive forces overwhelm the kinetic energy of the molecules. At that point, the gas molecules will condense into the liquid state. Ammonia, a gas that dissolves in water to make a common household cleanser, and neon, a gas used in colored electric lights, illustrate the importance of the intermolecular attractive forces. These forces are exceedingly feeble in neon, which does not liquefy until a temperature of −246°C is reached. A molecule of ammonia, which has about the same mass as an atom of neon, experiences fairly strong attractive forces, so ammonia liquefies at −33°C.

FIGURE 11.5 Joseph Gay-Lussac (1778–1850), another of the brilliant French scientists of the late eighteenth and early nineteenth centuries, and another contributor to our understanding of gases. (From http://en.wikipedia.org/wiki/File:Gaylussac_2.jpg.)

TABLE 11.1

Hypothetical Data Showing the Relationship between Temperature and Volume according to the Gas Law of Charles and Gay-Lussac

Temperature, °C	Volume Contracted	Volume Remaining
0[a]	0[a] (0/273)	1[a] (273/273)
−1	1/273	272/273
−10	10/273	263/273
−50	50/273	223/273
−100	100/273	173/273
−273	273/273 (1)	0/273 (0)

[a] These are the values at the beginning of the experiment.

TABLE 11.2

Comparison of Basic Temperatures on the Fahrenheit, Celsius, and Kelvin Scales

	Degrees Fahrenheit	Degrees Celsius	Kelvins
Water boils	212	100	373
Water freezes	32	0	273
Absolute zero	−459	−273	0

Two temperature scales are in common use, the Celsius scale, in which water freezes at 0°C and boils at 100°C and used worldwide; and the Fahrenheit scale, in which water freezes at 32°F and boils at 212°F, used mostly in the United States. It should seem possible, then, to define any sort of temperature scale we'd like, just by assigning numbers to the melting and boiling points of water, or of some other convenient substance. If that's the case, the relationship of temperature to a certain amount of thermal energy could be quite arbitrary. This is not a very satisfactory state of affairs for measuring quantitatively the amounts of thermal energy. It is necessary to have a definite, immutable fixed point on which a temperature scale could be based. Since −273°C represents the lowest possible temperature, it represents the point of zero thermal energy and can be called *absolute zero*.

Absolute zero is the lowest possible temperature (the point of zero thermal ENERGY).

This allows resetting the temperature scale so that the zero point of temperature is at absolute zero. This temperature scale is called the Kelvin scale.* Its units are called kelvins (*not* "degrees Kelvin") and have the symbol K (*not* °K). In comparison with the Fahrenheit and Celsius scales, we find the relationships shown in Table 11.2. Comparing the last two columns of Table 11.2 shows that the temperature in kelvins can be found from

$$\text{Kelvins} = \text{degrees Celsius} + 273$$

QUANTIFYING EFFICIENCY

In the early nineteenth century, steam engines "took off." Their use in operating factory machinery, and the freedom they provided from the geographic constraints

---

* Named in honor of William Thomson, 1st Baron Kelvin (1824–1907), a very gifted and accomplished physicist and engineer. Kelvin served as a professor at the University of Glasgow for more than 50 years. He was the first scientist in the history of the United Kingdom to be elevated to serve in the House of Lords. Unquestionably, he ranks among the greatest scientists of the nineteenth century, even with the caveat that bestowing such recognition is akin to the interminable arguments of selecting who was the greatest athlete in his or her sport.

FIGURE 11.6   Sadi Carnot (1796–1832), a French artillery lieutenant who single-handedly established the foundations of the science of thermodynamics. (From http://en.wikipedia.org/wiki/Nicolas_L%C3%A9onard_Sadi_Carnot.)

of wind and water use, allowed industrial productivity to soar. However, the details of building, operating, and improving these devices were still entirely pragmatic and empirical. There was no theoretical scientific understanding of the relationship between heat and mechanical WORK, or indeed even much understanding of the nature of heat itself. The steam engine developed largely as a result of trial-and-error tinkering and experimenting.

The person who established the scientific understanding of the steam engine and who single-handedly laid the foundation of the science of thermodynamics (i.e., movement of heat) was the French investigator, Nicholas-Léonard Sadi Carnot* (Figure 11.6). Carnot, the first scientist to analyze quantitatively the relationship between heat and WORK, recognized that an engine functions because heat flows from a hot body to a colder one. In 1824, he wrote a paper on the "power" of heat.

---

* Nicholas-Léonard Sadi Carnot (1796–1832) usually went by the name of Sadi, given to him in honor of the great medieval Persian poet Sadi of Shiraz. As a young man, Carnot served in the French army. After the defeat of Napoleon, he spent most of his time in scientific research. His book, *Reflections on the Motive Power of Fire*, which helped establish the foundations of thermodynamics, went almost completely unrecognized during his lifetime. Carnot died during a cholera epidemic when he was only 36. Cholera was, and is, a highly infectious disease. In Carnot's time, when nothing was known of disinfectants, antiseptics, or the germ theory of disease, a standard approach to fighting the spread of cholera was to burn everything the victim had been in contact with recently. Virtually all of Carnot's belongings were burned, including his writings. We can only speculate as to what other scientific advances went up in smoke that day.

He did not use the term "power" as we have defined it, but in speaking of the way in which heat causes motion in steam engines. Carnot considered that this "power" comes from the *one-way-only* transfer of heat, from a hotter to a colder body; during this transfer, heat can be made to do WORK.

For Carnot, heat is *the* "moving force" of the world:

> *It causes the agitations of the atmosphere, the ascension of clouds, the fall of rains and of meteors, the currents of water which channel the surface of the globe, and of which man has thus far only employed but a small portion. Even earthquakes and volcanic eruptions are the result of heat. From this immense reservoir we may draw the moving force necessary for our purposes.*

—Carnot[6]

Wherever there is a temperature difference in any SYSTEM, the possibility exists for generating useful WORK (or, in Carnot's terminology, "motive power"). A temperature difference is like a column of water: the greater the temperature difference, the greater the WORK for the same quantity of heat, just as water falling 200 meters will do twice the WORK of the same amount of water falling 100 meters.

Carnot analyzed the workings of a steam engine in this same way. For a given volume of water that flows over a waterfall, the amount of ENERGY that can be converted to WORK depends on the distance through which the water drops.* The critical information is the height of the pool of water at the bottom of the waterfall subtracted from the height of the cliff that is the top of the waterfall—the difference in gravitational potential energies of the two locations. There are two important points related to this analysis: First, it is possible, in principle, to measure the two heights from any agreed-upon reference point or "zero point." For measurements made on Earth, the uniform zero point for height would be distance from the Earth's center. Second, the absolute values of the two heights are not important in deciding the amount of WORK we can extract from the ENERGY of the falling water—what counts is the difference between them.

In the operation of a heat engine (Figure 11.7), only some fraction of the thermal energy is converted to WORK, because the system still has some thermal energy left (i.e., it's not at absolute zero). In the waterfall analogy, water drops to the bottom of the waterfall and forms a pool there. The pool by itself is not capable of turning a waterwheel. A distinction must be drawn between the total energy content of the water and available ENERGY that we can actually use. Yes, the pool at the bottom of the waterfall is not capable of turning a waterwheel, but it still contains much potential energy. Imagine that someone dug a hole that would let the water in the pool run out. If that person placed a second waterwheel at the bottom of the hole, an additional amount of the ENERGY in the water could be converted to WORK. Disregarding the practicality of doing this, imagine that someone could dig a hole all the way to the center of the Earth. Then (and only then!) *all* the gravitational potential energy

---

* In fact, this is true of any form of matter falling in a gravitational potential system. If a bowling ball slips out of your hand and hits you on the foot, it might hurt a little bit, aside from the embarrassment. The same bowling ball falling from the roof of a very tall building and hitting you on the head would probably kill you.

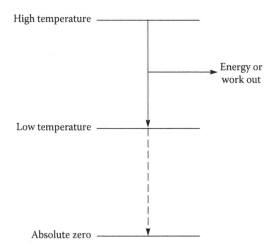

FIGURE 11.7 When absolute zero is added to the thermal energy diagram, we can see that even at the "low temperature" side, some thermal energy is remaining in the SYSTEM, which has not been converted to WORK.

of the water could be used, conceivably to run a whole series of waterwheels. Of course we do not have the engineering technology to dig a 6400-km-deep hole clear to the center of the Earth. Therefore, in practical reality, only the ENERGY of the falling water of the original waterfall is used. That ENERGY is available to us for doing WORK, and it represents only a portion of the total ENERGY of the water. The remaining potential energy of the water, which still "counts" as part of the total ENERGY of the water, is not available to us by any known technology.

In principle, we can get more WORK out of a waterfall by letting the water fall a greater distance. We ought to be able to do exactly the same thing with a heat engine by allowing the temperature to fall a greater distance, which means that we make the difference between the high temperature and low temperature larger—creating a greater difference in thermal potential energy. So, the bigger the difference between the high and low potential (whether gravitational or thermal), the more the WORK we can extract from the system.

Following the same line of thinking, in a heat engine, it is not the value of either the high or low temperature that dictates the amount of ENERGY that can be converted to WORK, but rather the temperature difference. The available ENERGY (not the total energy!) can be expressed in terms of the temperature difference within the heat engine. In 1851, a professor of physics at the University of Glasgow, William Thomson, later and better known as Lord Kelvin (Figure 11.8), recognized that there must be a temperature below which no further WORK could be produced.

This concept can be expressed conveniently in terms of the concept of the absolute zero of temperature. On the basis of this analogy, available ENERGY can be expressed as $T_h - T_l$. The subscripts h and l stand for *high* and *low*, respectively.

The pool of water at the bottom of the waterfall still had some ENERGY. Analogously, the cold side of the steam engine still contains heat. Conceptually, more WORK could be extracted from the water in the pool until all the available gravitational potential

FIGURE 11.8 Sir William Thomson, Lord Kelvin (1824–1907), one of the foremost physicists of the nineteenth century. Kelvin developed most of his theories while a professor at the University of Glasgow. The unit of the absolute temperature scale is named in his honor. (From http://commons.wikimedia.org/wiki/FileSir_William_Thomson_Baron_Kelvin_1824_-_1907_Scientist_resting_on_a_binnacle_and_holding_a_marine_azimuth_mirror.jpg.)

energy of the water was used up—until we had reached the center of the Earth, which is the zero point of gravitational energy. Quite similarly, suppose that the condenser on a steam engine is at, say, 25°C. The water produced in the condenser could, in principle, be cooled further and further, until we reached a point at which no more work could be extracted from it. This point would be the temperature analog of the center of the Earth, the zero point of potential energy below which it is impossible to go. In a thermal system, this zero point of energy is the temperature called absolute zero. The total energy of the thermal system would be represented by the difference between the temperature of the hot side and absolute zero: $T_h - 0$, or, simply T_h.

The heat engine efficiency is a measure of the ENERGY that can be converted to WORK as a fraction of the total ENERGY. This concept can be expressed mathematically in an equation for calculating efficiency:

$$\eta = \left(\frac{T_h - T_l}{T_h} \right)$$

Using this equation, the numerical value of efficiency will be a fraction having a value between 0 and 1. Often it is more convenient or easier to express efficiency as a percentage, rather than as a fraction. This is done readily by using the comparable formula:

$$\%\eta = 100\left(\frac{T_h - T_l}{T_h}\right)$$

In this case, efficiency has a value between 0% and 100%. The equations are equivalent and provide the same information; the difference is whether the calculated efficiency is expressed as a fraction or as a percentage. However, there is one crucial fact in the use of efficiency calculations: *these equations only provide the correct answer if the temperature is expressed in kelvins.*

The ratio of the available ENERGY to the total ENERGY represents the *maximum* efficiency of a heat engine. This efficiency value, sometimes called the *Carnot efficiency*, is reached only if the engine is mechanically perfect, for example, if there are no losses of energy through friction or losses of heat to the outside world. In any real heat engine, the efficiency is less—sometimes considerably so—than the predicted maximum (Carnot) efficiency, simply because of such problems as friction among the parts and losses of heat to the surroundings.

Carnot readily recognized that as heat "falls" from a higher to a lower temperature, the "duty" of the engine will be greater as the difference between the two temperatures becomes greater. This is analogous to the notion that, in a waterfall, the greater the drop of water (i.e., the greater the change in gravitational potential energy), the more WORK will be done by the waterwheel. Carnot also found that for a given drop in temperature (which we would write as $T_h - T_l$), more WORK is done for lower values of T_h. For example, an engine working between the temperatures of 400 and 300 K will do more work than one working between 1000 and 900 K, even though the temperature drop (100 K) is the same in both cases. Indeed, the efficiency of the first engine is 0.25 and, of the second, only 0.10.

Suppose a waterfall has a height of 100 meters, with its base at sea level, or a nominal elevation of 0 meter, and the top at an elevation of 100 meters. Suppose further—leaving aside the possibly thorny details of the construction—that a waterwheel was mounted in this waterfall at the 90 m mark. This wheel would take advantage of a 10-m drop of water (i.e., from 100 to 90 meters). However, much gravitational potential energy of the water (corresponding to the drop from 90 to 0 meters) is "thrown away" in this arrangement. Consider now a second waterfall, some 30 meters high, with a waterwheel mounted at the 20-m level. The same 10 m drop of water (i.e., now from 30 to 20 meters) occurs for this wheel. But much less potential energy is "thrown away"—in this case, only the amount corresponding to a drop from 20 to 0 meters. If a heat engine operates between 1000 and 900 K, it "throws away" all the rest of the thermal potential energy from 900 to 0 K. A comparable heat engine running between 300 and 200 K only discards the thermal energy between 200 and 0. Since efficiency depends on the amount of thermal energy *not* thrown away, the engine running at lower temperatures will be the more efficient, exactly in accord with the calculation in the previous paragraph.

The value of these equations, in addition to their obvious utility in calculating efficiency, is that they provide an understanding of what it would take to achieve a complete conversion of ENERGY to WORK. This complete conversion would represent an efficiency of 1.000 (or 100%). This value of the efficiency can be achieved for the special case when both the numerator and the denominator of the efficiency equation are equal (to each other). The denominator is always the value of the high temperature, T_h. Regardless of the specific value of T_h, the numerator $(T_h - T_l)$ can *only* be equal to T_h if the numerical value of T_l is zero. In other words, to achieve an efficiency of 1, or a percent efficiency of 100%, the value of T_l must be zero. This tells us that to achieve a complete conversion of thermal ENERGY to WORK (an efficiency of 1.000 or 100%), we must let the low temperature be absolute zero (0 K).

From Charles's law, at absolute zero, a gas would, in principle, have zero volume; that is, it would vanish completely. Put another way, Charles's law suggests that to reach absolute zero, we would have to destroy matter. But an enormous abundance of information collected over many centuries in laboratories all over the world shows that matter cannot be destroyed. In other words,

It is impossible to attain a temperature of absolute zero (0 K).

Let us explore the consequence of this unfortunate fact. The mathematical expressions for efficiency show that to achieve a complete conversion of thermal ENERGY to WORK, the low temperature of the thermal energy SYSTEM must be at absolute zero. We have also just concluded that it is impossible to achieve a temperature of absolute zero. Since we need a temperature of absolute zero to have an efficiency of 1, but we can't get to absolute zero, we are brought to the appalling conclusion that

It is impossible to convert thermal ENERGY to work completely (i.e., with 100% efficiency).

Evidence hinting at this conclusion has already surfaced in other SYSTEMS. Our bodies get hot when we exercise or work vigorously; some of the chemical potential energy in food is lost as this heat, rather than being converted to WORK. We know that the wind or water leaving a mill still has some kinetic energy left after it's done the WORK of the mill. Now we have found a SYSTEM—a thermal system—where we can directly show, and indeed calculate, the fact that the complete conversion of this kind of ENERGY to WORK is impossible.

It is impossible to extract heat from the hot side of a heat engine and convert it entirely into WORK, because it is impossible to attain a temperature of absolute zero. Some heat must be passed to the cold side. This fundamental limitation means that achieving a high efficiency from the heat engine cannot be done by lowering the temperature of the cold side arbitrarily (because we "hit a wall" at absolute zero).

Therefore, the only practical alternative is to raise the temperature of the hot side as much as possible. In doing this, the limitation is the high-temperature strength of the materials of the engine and, in some ultimate sense, the melting point of the material used to make the engine.

EQUIVALENCE OF HEAT AND WORK

The idea that heat is not a fluid but, rather, is a form of motion was occasionally advanced by scientists, many of whom are among the great names of science: Francis Bacon, Robert Boyle, and Isaac Newton as examples.

Heat is a very brisk agitation of the insensible parts of the object (i.e., the atoms), which produces in us that sensation from which we denominate the object hot; so that what in our sensation is heat, in the object is nothing but motion.

—Locke[7]

John Locke is now best known as a philosopher, but he was active as a physician and did a certain amount of work in experimental chemistry.

The person credited with establishing a link between heat and motion was Benjamin Thompson (Count Rumford, Figure 11.9), certainly one of the most colorful characters in the history of science.* In the late eighteenth century, it was observed that drilling the "bore" into cannon barrels makes them very hot. The process generates so much heat that, if the barrel and the drill were covered with water, the water would get hot enough to boil. As the boring was continued, there were two further observations: The water still got hot even when the drill was so dull that it was no longer cutting metal. And even if the water boiled away completely, any newly added batch of water would soon boil, and if that second batch of water boiled away and was replaced, this new batch of water would boil also.

Rumford argued that if heat were some kind of physical entity (that the scientists of the time called *caloric*) present in the metal of the cannon barrel, then surely a point would have to be reached sooner or later when all of the caloric originally in the barrel was used up. This should be observable at some late stage of the drilling operation, by the drilling producing less heat and boiling away less water. Rumford tested this by an experiment in which he submerged a cannon barrel in a tank of water, began boring the cannon, and measured the time it took the water to boil.

---

* Thompson (1753–1814) was indentured to a merchant when he was 13. He learned science on his own, became a schoolmaster, published scientific papers, married a wealthy widow, became a major in the New Hampshire militia, and, at the same time, also became a secret agent for the British. Fleeing to England, he contributed to the design of weapons, winning a Fellowship of the Royal Society, to this day one of the highest honors to which a scientist can aspire. Thompson was knighted by King George III. Moving to Bavaria, he reorganized the army; worked to improve living conditions of the poor; was the inventor of thermal underwear, the drip coffee maker, and the pencil eraser. He was awarded the title of Count of the Holy Roman Empire by the Bavarian government and chose the name of one of his hometowns, Rumford (now Concord), New Hampshire, to go with the title. In a final twist, Rumford moved in 1804 to Paris, where he married the widow of Antoine Lavoisier, the brilliant scientist who established the modern field of chemistry. Nothing in the historical record suggests that this was a happy marriage.

FIGURE 11.9   Benjamin Thompson, Count Rumford (1753–1814), spy, scientist, Bavarian minister of war, Fellow of the Royal Society, and second husband of Lavoisier's widow. One of the most colorful characters in the history of science. (From http://commons.wikimedia. org/wiki/FilePSM_V73_D044_Count_rumford_at_forty_five_years_of_age.png.)

> *It would be difficult to describe the surprise and astonishment expressed in the countenances of the bystanders, on seeing so large a quantity of cool water heated, and actually made to boil, without any fire.*

—Rumford[8]

Water would boil for as long as the drill could be turned. There was never a point reached when it seemed as if the caloric was completely consumed. But then where was the heat coming from?

> *Anything which any insulated body, or system of bodies, can continue to furnish without limitation, cannot possibly be a material substance, and it appears to me to be extremely difficult, if not quite impossible, to form any distinct idea of any thing, capable of being excited and communicated in the manner the Heat was excited and communicated in these experiments, except it be Motion.*

—Rumford[8]

The only thing going on in Rumford's SYSTEM (cannon barrel, drill bit, and water) is the turning motion of the drill bit inside the barrel. As a result, this led to the hypothesis that

Heat is a form of motion.

This hypothesis received substantiation from a simple experiment performed by Humphry Davy.*

"Sir Humphry Davy?" said Mr. Brooke, over the soup, in his easy smiling way, taking up Sir James Chettam's remark that he was studying Davy's Agricultural Chemistry. "Well, now, Sir Humphry Davy: I dined with him years ago at Cartwright's and Wordsworth was there too—the poet Wordsworth, you know.... But Davy was there; he was a poet "too".

—Eliot[9]

He went outdoors on a day when the temperature was below freezing and rubbed two ice blocks together. The ice melted. Since the temperature outdoors was too cold to melt ice, the only possible source of heat in this experiment was the motion of the blocks against each other. This very simple, but very profound, experiment gave immense credence to the new concept of heat as a form of motion.

Heat is a form of ENERGY. The ENERGY associated with motion is kinetic energy. Therefore, heat is a special kind of kinetic energy—the kinetic energy of atoms or molecules. Temperature scales actually measure the average thermal kinetic energy per atom in an object. The more kinetic energy each atom has, the more vigorous the thermal motion is. The atomic-scale WORK that one vibrating atom does on another is what actually passes heat through an object, or from one object to another. The object with more average thermal kinetic energy per atom will pass heat to an object with less. The temperature at which all thermal energy has been removed from an object would be absolute zero.

We have defined ENERGY as the capacity for doing WORK. We have also defined WORK as moving a body against a resistance or a force, one common example being lifting an object against gravity. A simple weight-and-pulley system with the concept for doing WORK is shown in Figure 11.10. If the heavy weight falls, it will make the light object rise against gravity, that is, it will do WORK. The heat released from fuel, or from food, can be measured with a calorimeter (as we saw in Chapter 3). James Prescott Joule[†] (Figure 11.11), conducted the key experiment in relating heat

---

* Humphry Davy (1778–1829) was first to isolate the elements sodium, potassium, boron, calcium, magnesium, and barium from their compounds. He also showed that chlorine and iodine, originally isolated by others, were elements and not compounds as most of the scientists of the time thought. Davy's most important invention was the Davy lamp, or miner's safety lamp, which had a major role in reducing gas explosions in coal mines. Davy claimed that his most important "discovery" was Michael Faraday, whom he had hired as a laboratory assistant and valet, and who eventually was recognized as an even greater scientist than Davy himself. Davy's contemporaries considered him an excellent poet; Samuel Taylor Coleridge is said to have claimed that "if he had not been the first [i.e., first-ranked] chemist, he would have been the first poet of his age."

† James Prescott Joule (1818–1889) started life as a brewer, managing the business and having science as a hobby. His insightful discoveries at first failed to be recognized by the scientific establishment, perhaps because he was perceived as an amateur and from the comparatively provincial environment of Manchester. Today, though, his greatness as a scientist is recognized by the use of his name for a unit of energy in the international system of units.

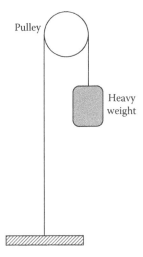

FIGURE 11.10   A simple SYSTEM for the concept of doing WORK. As the weight falls, it will cause the piston to rise.

FIGURE 11.11   James Joule (1818–1889), an English scientist who made numerous contributions to physics in the nineteenth century. A unit of energy is named in his honor. (From http://en.wikipedia.org/wiki/FileJoule_James_sitting.jpg.)

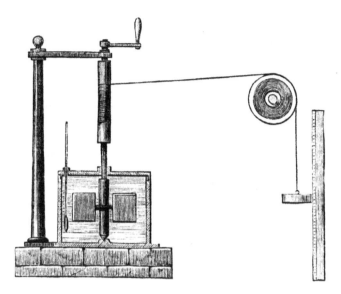

FIGURE 11.12 Joule's apparatus for measuring the mechanical equivalent of heat, around 1869. The falling weight caused the paddle wheel to revolve, adding heat to the water. (From http://en.wikipedia.org/wiki/File:Joule%27s_Apparatus_%28Harper%27s_Scan%29.png.)

to work by connecting these two pieces of equipment together. He used an apparatus in which a paddle is turned by the action of a falling weight (Figure 11.12). Releasing the weight causes a vigorous stirring of the water, and the temperature rises. This experiment established the *mechanical equivalent of heat*, the amount of mechanical energy equal to one unit of thermal energy.

As this simple concept was refined, there emerged some more rigorous defi-nitions: Thermal ENERGY is a kind of energy due to the random motion of the constituent atoms or molecules of an object (or of a SYSTEM). Heat is the thermal energy that is in the process of being transferred from one object to another (or from one SYSTEMS to another). Therefore, heat is not a physical fluid, like air or water, but is "energy in motion" between objects or systems having differ-ent amounts of thermal energy. Temperature provides the quantitative measure of thermal energy.

If the energy from a mechanical system, such as the one in Joule's experiment, is somehow equivalent to thermal energy (and indeed Joule proved this to be the case), then the conclusion reached about the efficiency of a thermal energy system should extend to mechanical systems as well. Indeed it does. Subsequently, other investigators were able to determine, for example, the electrical equivalent of heat, and to amass a growing body of evidence that the various forms of ENERGY can be interconverted. Joule himself measured the heat produced by an electric current, by the friction of water against glass, and by compressing gas. So the conclusion we had arrived at regarding efficiencies and conversion of ENERGY to WORK in heat engines is even worse than it seemed at first. Actually

It is impossible to convert any kind of ENERGY to WORK completely (with 100% efficiency).

This is a dreadful situation. It means that no matter what we do to convert ENERGY into WORK in any SYSTEM, some of the energy is going to be wasted. Very often (but not always), the wasted energy leaves the system in the form of heat.

All of the investigators who worked on this problem found that a fixed amount of one kind of ENERGY was converted into a fixed amount of another kind of energy. Furthermore, if ENERGY in all its varieties was considered, then when these transformations take place, no energy was either lost or created. "Extra" ENERGY did not suddenly appear in the experiment, nor did any ENERGY unaccountably vanish. By *very* careful measurements (as, e.g., in making sure to account for possible heat losses to the room in the Joule experiment), it was shown that the ENERGY present in one form could be converted completely to another, *but* that no "extra" ENERGY appeared in the SYSTEM nor disappeared from it. This growing body of carefully obtained experimental data finally led to an important statement summarizing the results:

The total amount of ENERGY in the universe is constant. It can be converted from one form to another, but is neither created nor destroyed.

This statement was formulated by the German scientist Hermann von Helmholtz* (Figure 11.13), at the time he was serving as a doctor in the German army. Helmholtz's summary statement shows us that the various forms of ENERGY are equivalent and that one form can be converted into another.

Not all of the kinetic energy of water is transferred to a waterwheel, because the water is still flowing (i.e., it still has some kinetic energy) after it leaves the wheel. Therefore, the amount of WORK taken out of this SYSTEM is less than the total ENERGY in the SYSTEM. From our perspective, some of the ENERGY has been wasted (because it's in the water flowing away from the wheel). Could we get all of the ENERGY transferred into WORK? In principle, yes. All we need to do is stop the water absolutely dead at the instant it hits a blade of the wheel, so that *all* of the kinetic energy of the water is transferred to the wheel. No one has ever succeeded in designing a waterwheel that would function in this way. This suggests that the maximum amount of WORK we can get from a SYSTEM is limited by the amount of available ENERGY in it.

Consider two different ways that we could position a waterwheel under a waterfall (Figure 11.14). Intuitively, it should seem that the SYSTEM on the left is not very efficient. The wheel is positioned at a point where not very much of the potential energy of the water has been converted to kinetic energy. On the other hand, by putting the wheel as far down the waterfall as we can get it, shown on the right-hand side,

---

* Hermann von Helmholtz (1821–1894) made important contributions to mathematics, physics, and other branches of science. Helmholtz surely must be counted as one of the leading scientists of the nineteenth century. His research encompassed topics ranging from the origin of the solar system to the physiology of nerves. His other accomplishments included the ability to speak at least nine languages.

FIGURE 11.13   Hermann von Helmholtz (1821–1894), a medical doctor who made important contributions to mathematics and physics, and who spoke at least nine languages. (From http://commons.wikimedia.org/wike/File:Hermann_von_Helmholtz_by_Ludwig_Knaus.jpg.)

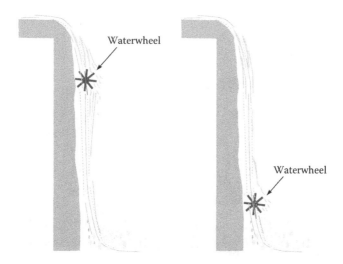

FIGURE 11.14   The picture on the left shows a SYSTEM that is not very efficient because it has rather limited conversion of potential energy to kinetic energy. The SYSTEM on the right takes much better advantage of available potential energy.

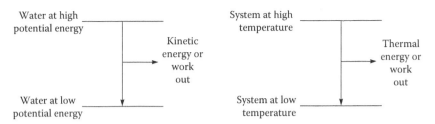

FIGURE 11.15 These energy diagrams show the similarities between gravitational and thermal systems.

we should have a much better SYSTEM, one that takes advantage of maximum conversion of potential energy to kinetic energy.

Joule showed that mechanical energy is equivalent to thermal energy. Figure 11.15 suggests an analogy between the two systems. With the waterwheel, the lower the position of the wheel, the better; to get complete conversion of ENERGY to WORK, we would have to stop the water dead, so to speak. By analogy, the lower we would make the low-temperature of the thermal system, the better. So how low can we go? The answer: absolute zero. We can sum up by saying that WORK is the result of using ENERGY to cause something to happen in the real world—it is the physical manifestation of ENERGY. In the real world, we are never able to convert ENERGY completely into WORK.

Suppose that some clever inventor could find a way to make a device that could do WORK essentially forever, without needing any input of fresh ENERGY to replace the amount consumed or wasted. Devices of this sort are known as perpetual motion machines. The possibility of building a perpetual motion machine has captured the imagination of scientists, engineers, inventors, tinkerers, backyard mechanics, cranks, and outright nutcases for well over a millennium. Figure 11.16 illustrates a design promoted in 1630 by the English physician and astrologer Robert Fludd.

FIGURE 11.16 One of the innumerable ideas for a perpetual motion machine, none of which actually work. This one was developed in the seventeenth century by Roger Fludd as a perpetually operating watermill. (From http://www.kilty.com/pmotion.htm.)

While some remarkable designs of *apparent* perpetual motion machines have been developed, on careful analysis, it inevitably turns out that the machine is obtaining energy from some source, often a very subtle or obscure one. There is general agreement that it is impossible to construct a *true* perpetual motion machine, because the machine would violate the law of conservation of energy, which is one of the fundamental underpinnings of science.

HEAT AND HOW IT IS TRANSFERRED

Just as gravitational potential energy tells us which way matter will flow (or, put crudely, whether or not it will fall down) in a gravitational system, temperature indicates which way (if any) thermal energy will flow. If no heat flows when two objects touch, then those objects are said to be in thermal equilibrium and, by definition, their temperatures are equal. But if heat flows from, say, the first object to the second, then, by definition, the first object is hotter than the second. Another way of saying essentially the same thing is that a hotter object will always transfer heat (thermal energy) to a colder object. The behaviors of the coffee, tea, soup, ice cream, or cold beverage used as examples earlier in this chapter are manifestations of this general rule.

Heat can move in a SYSTEM by three mechanisms: conduction, convection, and radiation (Figure 11.17). *Conduction* occurs when heat is transferred directly through the body of an object. An example would be the observation you'd make by holding a metal object in your hand and sticking the other end of it into a flame. All of us realize that sooner or later, the end being held in your hand will get hot enough to cause a burn. In this case, the heat of the flame is transferred directly to your hand through the motion of the heated atoms in the metal object. As a rule, conduction is particularly important for heat transfer through solids.

As heat is conducted through a solid, one end of the object gets hot (by being thrust into a fire, for example), and, as it does so, the atoms in that portion of the

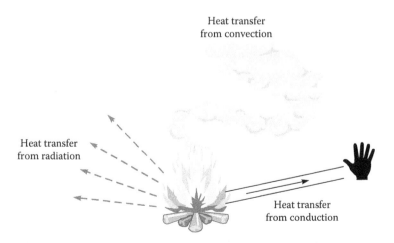

Heat transfer
from convection

Heat transfer
from radiation

Heat transfer
from conduction

FIGURE 11.17   Heat transfer can occur in three ways: radiation, convection, or conduction.

object increase in kinetic energy. Assuming that the object doesn't melt or catch on fire, the *average* position of each atom in the solid remains fixed, in some crystalline array that is characteristic of that particular solid. However, at any temperature above absolute zero, each of these atoms that has a fixed average position can, and indeed does, vibrate around that position. As the atoms gain kinetic energy by being heated, these atomic vibrations become more and more rapid, and the movements of the atoms extend further and further from their fixed "equilibrium" position. The atoms at the end inserted into the flame, which will be the hottest portion, will vibrate the most because they have the most kinetic energy. As they do this "heat dance," they will jostle neighboring atoms in the slightly cooler portion. Those atoms in turn will acquire kinetic energy as a result of these impacts. Therefore, they vibrate more energetically themselves and jostle atoms in an adjoining cooler portion of the object. This goes on and on until the kinetic energy bumps its way, atom by atom, through the whole length of the object, and we feel the heat at the other end, for example, in the palm of our hand.

Convection is a process of heat transfer that can occur in gases or liquids. We can see convection, for example, when we're cooking and watch a pot of water come to a "rolling boil." We can feel convection, when we're warmed by a hot summer breeze or cooled by a cold draft.

Heat can also be transferred by electromagnetic *radiation*. We can sense the heat from an object that is glowing red hot, even though we are not (let's hope!) in direct contact with it for conduction, nor feeling currents of heated air from convection. Unquestionably, the most vital aspect of heat transfer by radiation is that radiation is the process that brings heat from the sun to the Earth.

Regardless of the method by which the heat is actually transferred—conduction, convection, or radiation—heat will flow from a hot object to a cool object until the temperatures of different portions of the system are equal. Essentially, the temperature scale classifies or "orders" objects according to the direction in which heat (thermal energy) will flow between them. This lets us create a straightforward definition of the concept of heat: energy that flows from one object to another because of a difference in their temperatures.

HOW MUCH HEAT FLOWS?

A final question on this topic is to determine the amount of heat that actually flows. The answer to this question again relies on some common observations. Some substances, such as water, can remain perceptibly warm for a long period of time. (For example, people who go swimming at night after a very hot day may find the water still quite warm, even though the air has cooled.) Other materials, such as silver, lose heat remarkably quickly. Thus water has the capacity to store a lot of heat, and silver does not. This is measured by a fundamental property of materials called the *heat capacity*. The amount of material involved is important. A single drop of boiling water splashed on your skin might cause a momentary irritation or tiny burn, but a bowl full of boiling water could cause a very severe burn. Relative temperatures are also important. A red-hot ingot of iron can be felt to put out a lot of heat, while a block of iron just a few degrees warmer than the room might not even be perceived to be warm. Thus the amount of heat that flows is a function of the amount of material (the mass) in the system, the heat capacity of the system, and the difference in temperature.

Heat flow = Mass of material × temperature difference × heat capacity

It is possible to look up values of heat capacity for different materials in handbooks or on reliable websites. Then, knowing two of the other three terms—heat flow, mass, or temperature difference—makes it possible to calculate either the heat flow or the temperature change.

REFERENCES

1. Mill, J.S.; quoted in Heilbroner, R. *Teachings from the Worldly Philosophers*. Norton: New York, 1996, p. 130.
2. Engels, F.; quoted in Bernal, J.D. *Science and Industry in the Nineteenth Century*. University of Indiana Press: Bloomington, 1953, p. 75.
3. Thomson, W.; (Lord Kelvin), quoted in Rosen, W. *The Most Powerful Idea in the World*. Random House: New York, 2010, p. 67.
4. Benson, R.; quoted in Brand, S. *The Clock of the Long Now*. Basic Books: New York, 1999, p. 101.
5. Gay-Lussac, J.; quoted in Glashow, S.L. *From Alchemy to Quarks*. Brooks-Cole: Pacific Grove, CA, 1994, p. 177.
6. Carnot, S.; quoted in von Baeyer, H. *Maxwell's Demon*. Random House: New York, 1998, p. 38.
7. Locke, J.; quoted in Laidler, K.J. *To Light Such a Candle*. Oxford University Press: Oxford, U.K., 1998, p. 263.
8. Rumford, C.; quoted in Glashow, S.L. *From Alchemy to Quarks*. Brooks-Cole: Pacific Grove, CA, 1994, p. 207.
9. Eliot, G. *Middlemarch*. Numerous editions of this classic novel are always in print; the quotation is from p. 27 in the edition published by The Folio Society: London, U.K.

FURTHER READINGS

Atkins, P. *Four Laws*. Oxford University Press: Oxford, U.K., 2007. A superb introduction to thermodynamics, requiring little prior scientific background. Highly recommended.

von Baeyer, H.C. *Maxwell's Demon*. Random House: New York, 1998. An account, written for readers of limited scientific background, of the behavior of heat. The work of Rumford, Joule, and Carnot, among others is discussed. An excellent introduction to the topic.

Bloomfield, L.A. *How Things Work*. Wiley: New York, 2010. This introductory physics textbook addresses the subject by analyzing the ways in which common devices that we encounter in everyday life function. Chapters 7 and 8 treat heat and thermodynamics, including analyses of wood stoves, air conditioners, and automobiles.

Carnot, S. *Reflections on the Motive Power of Heat*. Dover: New York, 2005. A relatively inexpensive paperback reprint of the English translation of Carnot's seminal work.

Coopersmith, J. *Energy: The Subtle Concept*. Oxford University Press: Oxford, U.K., 2010. An extensive look at thermodynamics through the works of, among others, Carnot, Davy, Helmholtz, Rumford, delving much deeper than this present chapter but requiring little mathematics.

Cutnell, J.D.; Johnson, K.W. *Physics*. Wiley: New York, 2007. All introductory physics text-books will have material on temperature and heat, heat transfer, and thermodynamics. The book by Cutnell and Johnson is an example of a good, reasonably modern text. Chapters 12, 13, and 15 are particularly relevant. Requires more mathematical sophistication than this book, or the others listed here.

Rosen, W. *The Most Powerful Idea in the World*. Random House: New York, 2010. A very fine history of the development of the steam engine and its role in the Industrial Revolution.

12 Introduction to Electricity

Electricity has become a mighty kingdom. We perceive it in a thousand places where we had no proof of its existence before... the domain of electricity extends over the whole of nature.

—Hertz[1]

"PREHISTORY" OF ELECTRICITY

Some 2500 years ago, Thales of Miletus* observed that rubbing a piece of amber would give the amber the ability to attract light objects. We can repeat Thales's experiment today with an ordinary comb, making it attract small pieces of paper. This phenomenon we now call static electricity. It's the "static cling" of clothes as they come out of the dryer. It's the spark or shock from reaching for a metal object after shuffling across a carpet. It's the maddening tendency of Styrofoam packing "peanuts" to stick to just about anything. It makes people's hair stand on end (Figure 12.1). Like so many ideas and observations of the ancient Greeks, little or nothing was done to develop the practical consequences. The study of electrical phenomena languished for the next two millennia.

EARLY "ELECTRICIANS"

In the beginning of the modern scientific era—in the seventeenth century—many investigators in Britain and in Europe began to build up a fund of experimental observations on static electricity. One of the first significant contributors was William Gilbert.† In 1570, he developed the term "electrics" to describe substances that would generate these feeble attractive forces when rubbed. A substance that showed such attractive forces was said to be "electrified." The electrified substance was thought to have gained some sort of electrical fluid that—once it was actually in the substance—remained stationary, hence the term *static electricity*. (About that

---

* Thales (circa 625–545 BCE) is regarded as having been the first philosopher in the long and glorious line of ancient Greek thinkers. Like most who came after him, Thales thought about a remarkably broad range of subjects, geometry, electricity, reasons for occurrence of earthquakes and eclipses, and early theories of how the mind works. He was active at various times in politics and in business. Thales's most important contribution was his determination to explain natural events, such as earthquakes, without falling back on mythology or other supernatural explanations.

† William Gilbert (1544–1603) was the official court physician to Elizabeth I and, after she died (presumably not as a result of Gilbert's ministrations), to James I, who became the new monarch. Gilbert had ample free time to do his own scientific work and was interested in phenomena related to electricity and magnetism. His most enduring work is his book *De Magnete*, one of the first discussions of magnets and their behavior.

FIGURE 12.1 This girl's hair is standing on end because of the effects of static electricity. (Photograph courtesy of Cole and Arielle Forrester, Shakopee, Minnesota.)

time, too, those scientists who studied electricity came to be called "electricians." Of course, the word has a totally different connotation today.)

Around 1660, the famous mayor of Magdeburg, Otto von Guericke, built a machine that consisted of a large ball of sulfur that could be rotated at high speeds by means of a crank (Figure 12.2).

It was excited by friction by applying a cloth pad, or even his hand. von Guericke was able to transmit electricity a meter or more through a moistened string. He rubbed the spinning sulfur globe in the dark and saw it glow. He saw the glow eerily extend from the globe to his hand, some inches away. Unfortunately, von Guericke never followed up these observations. He had no way of realizing it at the time, but he had actually constructed the first electric generator. He not only generated electricity in greater amounts than ever before by means of his machine, but he also showed that it could be *transmitted* over distances.

Almost a century passed. Then, in the Boston summer of 1746, a visitor from Philadelphia decided to attend some public lectures on electrical phenomena. The visitor was Benjamin Franklin. He was so intrigued that he purchased various pieces of apparatus and began doing experiments on his own.

Franklin, the first American scientist to be recognized as "world-class," proposed in the mid-eighteenth century the concept that electricity is a fluid.

... The greatest man and ornament of the age and country in which he lived.

—Jefferson[2]

FIGURE 12.2 Otto von Guericke's static electricity generator, using a large rotating ball of sulfur. (From http://en.wikipedia.org/wiki/File:Guericke-electricaldevice.PNG#file.)

Nearly two centuries after Gilbert, Stephen Gray, an English electrician, learned that rubbing a piece of glass would cause its ends to attract light objects. Then, sticking a cork into the open end of a glass tube and rubbing the glass in the usual manner, he found that now the cork attracted small objects. The next experiment involved sticking a nail into the cork. Now the end of the nail attracted things. Replacing the nail with a metal rod with a knob on its end—that too attracted objects. After experimenting with the longest metal rod he could find, Gray next tried thread, stretched out longer and longer until it ran all around the garden of a friend's house—and it still attracted things. In the ultimate "scale-up" of the experiment, an electrified thread was able to cause a feather to stick to its other end, at a distance of about 240 meter.

Gray recognized that the electric fluid is not necessarily static or stationary, but can actually move through some kinds of objects. He distinguished between substances called "electrics," which could conduct electricity, and "nonelectrics," which electricity would not pass through. In fact, the electric fluid can move so easily through some bodies that they can't even retain any of the fluid. Based on Gray's observations, scientists recognized that substances could be divided into two classes: those that readily pass the electric fluid, which nowadays we call *conductors*; and those that retain the electric fluid, which we now call *insulators*. As a rule, good conductors are metals. Wood, rubber, and plastics are examples of insulators. Later in his career, Gray suspended an iron rod sharpened at both ends. When he brought an electrified object near the rod in a darkened room, he saw sparks of light shining from it. He also heard crackling noises that accompanied the sparks. The noise and the sparks reminded him of the roar of lightning bolts. That small observation of Gray was the first realization that lightning was somehow similar to electricity.

Following Gray's lead, the French electrician Charles François Du Fay discovered what seemed to be two "kinds" of electricity. One is generated by rubbing a piece of glass; the other, from rubbing a piece of resin. He referred to these as vitreous electricity and resinous electricity. He showed that the same kinds of electricity repelled each other, but opposite kinds attracted.

Franklin realized that the concept of two kinds of electricity—vitreous and resinous—was not necessary. The presence of one ensured the absence of the other. Franklin refined Du Fay's original concept to the idea of positive and negative electricity. He showed that matter under ordinary conditions was not electrified. If some electrical charge were accumulated on an object, it would be said to be positively charged. If charge were removed, then that object would be said to be negatively charged.* Franklin had the important insight that electricity was not created by rubbing a glass tube; it was only being transferred. In other words, when an "unelectrified" object was rubbed, either it could gain electric fluid and become positive, or it could lose some electric fluid and become negative. When a positively charged body is brought into contact with an uncharged or negatively charged body, the excess electric fluid that is in the positively charged body should flow into the uncharged or negatively charged one. Similarly, when an uncharged object is brought into contact with a negatively charged object (i.e., one that has a deficit of electric fluid), some electric fluid should flow from the uncharged to the negatively charged object.

Franklin extended his work by proposing that lightning is electric. He proved it in 1752. His proof came in one of the best-known scientific experiments of history: the kite-in-a-thunderstorm experiment (Figure 12.3). He proved it by hanging a key from the kite and charging a Leyden jar from a lightning discharge. (A Leyden jar, shown in Figure 12.4, is a laboratory device used to accumulate an electric charge.) Remarkably, the experiment did not kill him. The next person to try the experiment, a Russian "electrician" in St. Petersburg, was struck dead. And despite Franklin's unquestioned brilliance, not much was truly understood about electricity by the early nineteenth century:

I am wonderfully ignorant of the whole subject: is there yet a general theory of electricity—electricity, what it is? Not that I know of. Its effects can be seen and measured, but apart from that and some pretty wild unsubstantiated statements, I do not think we yet know the ABC.

—O'Brian[3]

---

* In Franklin's nomenclature, an excess of electric charge is positive, and deficit is negative. This is analogous to the way we think about money. If we deposit money in the bank, our balance is positive. If we have an overdraft, our balance is negative. As things turned out, Franklin's reasoning was exactly backward. We know now something that Franklin did not—that the carriers of electric charge are electrons, which themselves have a negative charge. The physical reality of the situation is that a material with an excess of electric charge is negative, and one with a deficit is positive. Franklin's legacy has been an awkward and often confusing system of defining the signs (i.e., positive or negative) of electrical connections and the direction of electricity flow that is different between practical electrical work and physics. This mix-up has persisted for two-and-a-half centuries. However, we must realize that Franklin had no way of knowing that he was making an assignment that was backward. In Franklin's time, the atomic theory had not yet been established, let alone the concept of the components of atoms and the charges of subatomic particles. The fact that Franklin happened to get it wrong-way-round should in no way diminish his standing as one of the most eminent scientists of his era.

THE PHILOSOPHER & HIS KITE.

Drawn expressly for the Columbian Magazine by J. L. Morton

FIGURE 12.3 Benjamin Franklin's experiment of flying a kite in a thunderstorm is one of the most famous, and most dangerous, experiments in the history of science. Franklin showed that lightning is a discharge of static electricity. (From http://www.loc.gov/pictures/item/2006691772/; http://www.loc.gov/pictures/resource/cph.3a52954/.)

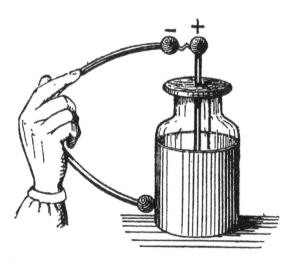

FIGURE 12.4 The Leyden jar was a common object in the laboratories of the early "electricians." These jars could accumulate an electric charge. Franklin was able to prove that lightning was electricity by storing some of its charge in his Leyden jar, where it remained until the jar was grounded. The process of grounding the Leyden jar is illustrated here.

With Franklin's gift for turning scientific observations into practical inventions, he immediately turned his kite experiment into the first practical—indeed, life-saving—application of the new understanding of electricity, the lightning rod.* In the Middle Ages, there was a custom of ringing church bells during lightning storms, on the ill-founded idea that the noise of the bells would somehow disperse the thunder and the storm. Unfortunately, a bell rope, once it's been wet by the rain, is a fairly good conductor of electricity. As a result, it was not uncommon for the bell ringers to be electrocuted during such storms. In mid-eighteenth century, more than a hundred people were killed in France alone at the ends of wet bell ropes.

Franklin reasoned that the lightning rod would work best with a sharp end. Scientists in England and Europe argued that a rounded end rod would work better. The matter was referred to the Royal Society in Britain to determine which one was the correct design.† The British settled for the round (i.e., the wrong) design, on the grounds that Franklin was a revolutionary. In fact, the argument of pointed vs. round argument was not settled by scientific trial (at the time)—it was settled in favor of the round design by King George III himself, in a rage against the American Revolution. It's said that the president of the Royal Society, Sir John Pringle, took the dangerous course of possibly offending the Society's royal patron, by replying "Sire, I cannot reverse the laws and operations of nature."[4] While this controversy over lightning rod design was raging, Franklin was in France negotiating the alliance between America and France that virtually ensured the success of the American Revolution.

> *The succession to Dr. Franklin, at the court of France, was an excellent school of humility. On being presented to any one as the minister of America, the commonplace question was "c'est vous, Monsieur, qui remplace le Docteur Franklin?" "it is you, Sir, who replace Dr. Franklin?" I generally answered, "no one can replace him, Sir; I am only his successor."*
>
> **—Jefferson[5]**

When he heard of the king's meddling, Franklin expressed the hope that King George III would just dispense with lightning rods altogether—and suffer the obvious result.

---

* When a severe earthquake hit Cape Ann, Massachusetts, in 1755, Reverend Thomas Prince, of Boston's South Church, blamed it on Franklin. Rev. Prince's "reasoning," such as it was, lay in the fact that, prior to Franklin's invention of lightning rods, God could use lightning to blast those who made him angry. Because of Franklin's invention, lightning was no longer an option, so God was now using earthquakes instead.

† A humorous example of the furor that can erupt over scientific arguments is provided by the great satirist Jonathan Swift in his immortal *Gulliver's Travels*. There he described how "the two great Empires of Lilliput and Blefuscu" actually went to war over the issue of whether it is best to open a boiled egg at the breakfast table at the sharper or at the more rounded end. This story of King George III and the lightning rod is probably not the first instance of political interference in science, and it surely was not the last. The twentieth century has witnessed tragic examples, most especially under Nazi and communist political regimes. Even now in the twenty-first century, political issues, particularly in the United States, affect the course of action for dealing with global climate change, to be discussed in Chapter 29.

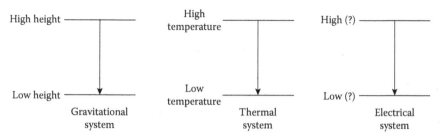

FIGURE 12.5 This set of energy diagrams shows the close relationships among gravitational, thermal, and electrical systems.

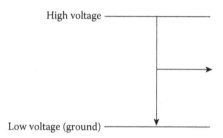

FIGURE 12.6 This is the energy diagram for an electrical system, complete with the appropriate labels for electrical potential energy.

ELECTRIC FLUID AND ITS POTENTIAL

Franklin's hypothesis about the flow of electric fluid suggests the analogy of a flow of water downhill. From the previous discussions of heat and waterfalls, if something flows, it must do so because of a difference in potential. It is possible to build on that analogy to describe an electrical SYSTEM (Figure 12.5). Electrical potential energy, or electrical potential, is measured in *volts*, in honor of the Italian "electrician" (physicist), His Excellency, Count Alessandro Giuseppe Antonio Volta. (Volta's work will be discussed in more detail in Chapter 13.) Thus the volt is the measure of electrical potential energy (Figure 12.6). In keeping with the analogy of a waterfall, the low-potential side of an electrical SYSTEM is sometimes referred to as the "ground."

ELECTRIC CURRENT

In a waterfall, water moves from high to low gravitational potential energy. In the more general case of a gravitational SYSTEM, the "thing" that is moving is *matter*. Water is simply one kind of matter. Not only is the difference in potential important, but so is the amount or quantity of matter that flows. As an extreme example, a single drop of water falling from the height of the world's tallest waterfall (the Angel Falls in Venezuela, at 980 meters) will be not nearly as effective in doing WORK as many thousands of liters of water flowing down just a few meters in a working waterwheel.

The same is true in a thermal system. Pouring a teacup of boiling water onto an icy sidewalk isn't going to melt much of the ice.

In a river or waterfall, the current of water can be measured as the volume of water flowing in a unit of time, for example, cubic meters per hour. Because a difference in electrical potential energy at two different points in space may result in the movement of electrical charge between them, the motion of charge is called an electric *current*. An electric current is measured by the number of charges going past a selected point in a unit time.

In a gravitational system, what flows is matter, measured in various units of mass. In a thermal system, what flows is heat, measured in units of joules. In an electrical system, the flow is the electric charge. However, it is more common and more convenient to express the flow as an electric current, which is the amount of charge flowing per unit time. That is,

$$Current = \frac{Change}{Time}$$

The units of electric current are named in honor of a French "electrician," a child prodigy who is said to have mastered *all* of what was then known of mathematics by the time he was 12—André Marie Ampère* (Figure 12.7). Thus the unit of electric current is the *ampère*, or, as often called, the *amp*.

RESISTANCE

Stephen Gray made the first observations that some substances (metals, such as copper, aluminum, or silver) conduct electricity easily, and others (e.g., wood, rubber, or plastic) do not. Within these two families of materials, not all are equally good conductors or insulators. Aluminum, for example, is a better conductor than iron, and silver is better than aluminum. The amount of electric current that will flow will depend not only on the potential difference (the voltage) but also on the ease with which the current flows. This latter property is called the *resistance*. About a century after Gray's work, a professor of mathematics in Cologne, Germany, Georg Simon Ohm (Figure 12.8),[†] worked out the fundamental laws of conductivity.

Ohm considered the flow of electric current in a wire as analogous to the flow of heat through some material that readily conducts heat. To develop this analogy, he needed a concept that would correspond to the role of temperature in the flow of heat. Ohm conceived a quantity that he called the electroscopic force (now called *electro-*

---

* André Marie Ampère (1775–1836) made major contributions to the field of electrodynamics, drawing on his extensive knowledge of the fundamentals of mathematics and physics. The importance of his work will be revisited in Chapter 13. Ampère spent much of his career at the École Polytechnique in Paris.

† Georg Simon Ohm (1789–1854), a Bavarian mathematician and physicist, taught mathematics and physics at the Jesuit Gymnasium in Cologne. In 1827, he published a book, *The Galvanic Current Investigated Mathematically*, which contained what was probably the best fundamental exposition of the theoretical basis of electricity published up to that time. The college administrators, however, didn't think much of it. Ohm resigned and moved on to the Polytechnic School in Nuremburg. In addition to his important work in electricity, Ohm made many other contributions, such as in the field of acoustics.

FIGURE 12.7 André Marie Ampère (1775–1836), the great French physicist in whose honor the unit of electrical current is named. (From http://en.wikipedia.org/wiki/File:Ampere_Andre_1825.jpg.)

motive force, essentially equivalent to the voltage), which pushed electricity along the wire in the direction of the current. The amount of heat that will flow via conduction depends in part on the nature of the material it's flowing through (contrast a metal cup with a Styrofoam cup, for example). In the same way, the flow of electric current should vary with the nature of the conductor. From that analogy, Ohm introduced the idea of a resistance to the current. Resistance is a property of whatever object the current is flowing through. Using a battery that would allow him to connect different materials to the terminals, Ohm showed that the current obtained was directly proportional to the voltage at the terminals and inversely proportional to the resistance of the conductor (Figure 12.9). This relationship is now known as Ohm's law.

It has become customary in electrical work to consider not the conductivity (i.e., the ease of flow of the electrical current) but rather the resistance (the opposition to the current flow). Ohm developed one of the important equations for dealing with electricity:

$$\text{Current} = \frac{\text{Potential difference}}{\text{Resistance}}$$

or, equivalently,

$$\text{Potential} = \text{Current} \times \text{Resistance}$$

FIGURE 12.8   Georg Simon Ohm (1789–1854), the German physicist for whom the unit of electrical resistance is named. (From http://en.wikipedia.org/wiki/File:Georg_Simon_Ohm3. jpg#file.)

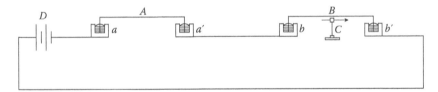

FIGURE 12.9   A diagram of Ohm's experiment. Battery *D* provides electricity to the entire circuit. Compass *C* measures the strength of the current. Specimens *A* and *B* were lengths of various test samples, placed into copper terminal pieces that dip into cups of mercury (*a*, *a'*, *b*, *b'*). By changing the materials used for *A* or *B*, Ohm could measure the change in current and determine the proportional resistance.

Electrical resistance has the units of ohms. Thus

$$\text{Volts} = \text{Amps} \times \text{Ohms}$$

Think of a hose used to water your lawn or garden. The pressure of water supplied to the hose is constant, comparable to an electrical system with a constant voltage. What happens if someone steps on the hose or drives a car over it? We can immediately see that the water flow diminishes when the hose is squashed. Compressing the hose increases the resistance to the flow of water. If someone puts a kink in the hose, the flow will decrease drastically. When the resistance increases and the pressure holds constant, the flow decreases. In exactly the same way, as can be seen from Ohm's law, increasing resistance in an electrical circuit and keeping the "pressure" (the voltage) constant will cause the current to drop. Even without kinks and crimps and fools driving their cars over your hose, there is still an inherent resistance to the flow of water. If you could lay the hose out in a perfectly straight line, it would still have some resistance. If we connected a second hose, to double the length, we'd find that the flow would drop. If we got a hose of a smaller diameter, we'd also see that the flow would drop. Electricity flowing through an electrical system will behave in the same way. The relationship between electricity flow and water flow is shown in Figure 12.10.

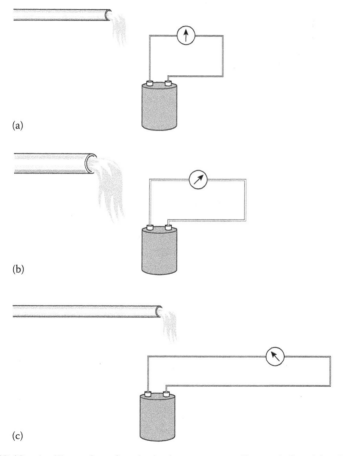

(a)

(b)

(c)

FIGURE 12.10   An illustration of analogies between water flow and electricity flow. In (a), water flowing through a pipe is comparable to electricity flowing through a circuit. In (b), increasing the diameter of the pipe allows more water to flow; increasing the diameter of the wire reduces resistance and allows more electricity to flow. In (c), lengthening the pipe reduces water flow; similarly, lengthening the wire increases resistance such that less electricity flows.

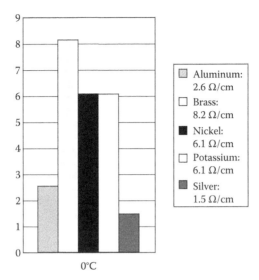

FIGURE 12.11 Conductors have various resistivities. However, the differences among them are minor compared to insulators, for which the resistivity could be ≈3500 Ω/cm.

Indeed, as Gray showed, different materials resist the flow of electrical current by different amounts. Some materials offer little resistance to the movement of electrons. Others offer a great deal of resistance. *Resistance*, as commonly measured, depends on the amount of the material, as well as on the specific substance. It would be useful to have a property that would characterize materials based only on their composition and nature, without having to take into account such factors as amount or size. *Resistivity* characterizes this property of materials in that way, that is, depending only on the nature of the material (Figure 12.11). Thus, for example, all samples of copper have the same resistivity, but a copper wire of a given diameter and, say, ten meters long has twice the *resistance* of a copper wire of the same diameter but only five meters long. This relationship is expressed by the equation

$$\text{Resistance} = \frac{\text{Length} \times \text{Resistivity}}{\text{Area}}$$

In terms of fundamental properties of the materials, iron, for example, has a higher resistivity than copper. However, we cannot say that a particular piece of iron wire has a higher resistance than some other piece of copper wire, unless we know the lengths and cross-sectional areas of each, so that we can appropriately calculate the resistances.

When a wire breaks, the current in the circuit stops because air has so high a resistance (or, in other words, is so good an insulator) that the voltage is not sufficient to overcome it. The electricity in household circuits stays in the wires and does not leak out through the walls because the copper or aluminum in the wires has a much lower resistance (i.e., a much better conductor) than the plastic or rubber insulation

FIGURE 12.12   The phenomenon of superconductivity allows a magnet to be levitated above a superconductor. (From http://en.wikipedia.org/wiki/File:Meissner_effect_p1390048.jpg.)

around the wires. Since it is easier for current to flow through the low-resistance path—the wire—than through a route of higher resistance, electricity follows the circuits we have created for it.

For all practical purposes, every material exhibits some amount of resistance to current. However, a very few materials, *superconductors*, have no resistance whatsoever. Superconductors began as curiosities in the physics laboratory (Figure 12.12). They now have some applications, though usually very specialized ones in, for example, scientific apparatus.* Superconductivity is under intense investigation in many laboratories around the world; it is hoped that future commercial applications might include long-distance transmission of electricity without losses and high-speed magnetically levitated ("maglev") trains to replace conventional railroads. So far, the materials known to be superconductors show this remarkable property only at low temperatures (e.g., so far the highest temperature at which superconductivity has been observed is −135°C), and this need—to keep them insulated and cold—has restricted their application from electrical transmission lines, high-speed transport, and ordinary household electrical products. A challenge of science in the twenty-first century is to develop materials that would be superconducting either at ordinary temperatures or at least at temperatures that

---

* As one example, chemists rely heavily on the technique called nuclear magnetic resonance (NMR) spectroscopy to determine the structures of complex molecules. (The method is akin to magnetic resonance imaging used in medicine for the diagnosis of injuries or disease.) NMR is especially powerful for determining the arrangements of carbon and hydrogen atoms in organic compounds. A scientist skilled in designing NMR experiments and interpreting the data can often unequivocally work out the molecular structure of a newly discovered substance, even from a sample of a few milligrams. Most modern NMR instruments rely on superconducting magnets, which the users must vigilantly keep cooled to liquid helium temperatures.

could be maintained with common refrigerating equipment.* Such a development would represent one of the most revolutionary advances in our use of energy in all of history.

ELECTRIC POWER

The definition of electric current has introduced the units of time. We have previously identified POWER as the rate of doing WORK or the rate of using ENERGY. Electrical POWER is determined by

$$\text{POWER} = \left(\frac{\text{Charge}}{\text{Time}} \right) \times \text{Potential}$$

$$\text{POWER} = \text{Current} \times \text{Potential}$$

The units of electrical power are *watts*. Thus

$$\text{Watts} = \text{Amps} \times \text{Volts}$$

In many parts of the world, consumers using electricity are charged a certain amount of money per kilowatt-hour (kWh). A kilowatt (1000 watts) is a unit of POWER, and of course an hour is a unit of time. A kilowatt-hour, therefore, is a unit of (POWER × time), which is ENERGY. When you pay your electric bill, you are *not* purchasing POWER, you are in fact purchasing ENERGY. So the local "power" company that produces electricity in "power" plants is not in the POWER business at all. It is producing and selling ENERGY.

The work of these early scientists and electricians, in establishing exact relationships among the amp, volt, watt, and ohm, made the buying and selling of electrical ENERGY possible. In turn, that made it possible for the electrical supply industry of every nation to come into being, a story that we will follow in subsequent chapters.

Most of us encounter the watt when we're purchasing lamp bulbs. The—POWER the "wattage"—of the bulb is the product of voltage times current. From Ohm's law,

---

* In the late 1980s, physicists discovered materials that display superconducting properties at the temperature of liquid nitrogen, 77 K (−196°C). While to our everyday experience this temperature is still extraordinarily cold (indeed, life as we know it could not exist at such temperatures), large-scale production of liquid nitrogen is easily within today's technology. Therefore, a device relying on superconductors operating at 77 K would be much easier to build and operate, and far cheaper to maintain at operating temperature, than ones requiring temperatures near that of liquid helium, 4 K. Today the record is up to about −135°C, or 138 K. At present, however, many so-called high-temperature superconductors (i.e., materials that display superconducting properties at or above liquid nitrogen temperatures) are nonmetallic. Fabricating them into wires might be difficult. Further, there seems as yet to be no theoretical understanding of high-temperature superconductivity that is accepted widely in the scientific community. This lack of a generally accepted theory of high-temperature superconductivity may also be impeding progress (though it is useful to recall how the steam engine was developed before there was a satisfactory understanding of heat and efficiency).

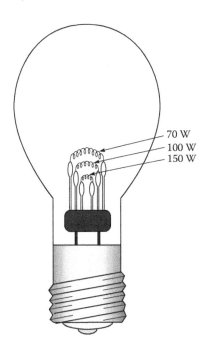

70 W
100 W
150 W

FIGURE 12.13 This cutaway image of a three-way light bulb allows us to see how differ-ent wattages are produced. In this case, rather than changing the cross-sectional area of the tungsten filament, the filaments have various lengths that affect their resistance to electricity, and, hence, their brightness.

current depends inversely on the resistance of the filament in the bulb (Figure 12.13). That is, the greater the resistance, the smaller the current. And the smaller the current, the less POWER consumed, and the less light will be produced. In other words, we will perceive that a 60–watt bulb is "dimmer" than a 100–watt bulb.

Generally, most incandescent bulbs use the same material, tungsten, as the material for the filament. If the material is the same from one bulb to another, then so is the *resistivity* of the filament, since all samples of the same material have the same resistivity. What has to be done is to change the *resistance* of the filament. The easiest way to make the resistance higher is to make the cross-sectional area smaller, that is, to make a filament of smaller diameter. So the following relation-ships apply: the thinner the filament, the higher the resistance; the higher the resis-tance, the lower the current; the lower the current, the lower the POWER (wattage); the lower the wattage, the less light is produced. Therefore, a 60–watt bulb has a thinner filament than one rated at, say, 100–watts.

We can imagine, and probably intuitively understand, that a stream of water flowing through a hose at a pressure of 10 MPa (about 100 times atmospheric pressure) is more powerful than an equal-sized stream at 1 MPa. A high-pressure stream of water can knock a person down or do considerable damage to structures. In the same way, a current of electricity provides more power at the relatively high "pressure" of 100–volts than the same current would at 10–volts.

The relationship among power, current, and resistance is that

$$\text{POWER} = \text{Resistance} \times (\text{Current})^2$$

$$\text{Watts} = \text{Ohms} \times (\text{Amps})^2$$

Any electric generating facility wants to have the smallest possible losses of electricity while it is being distributed from the point of generation to the individual consumers. In other words, the electricity supplier wants to get as much of the electricity generated delivered to you as possible. Even though the wires in electrical distribution system are made of excellent conductors, there is still some resistance to the flow of current. Sending electricity through wires requires that we do WORK to overcome the natural resistance of the wires. When that WORK is performed, some ENERGY is inevitably wasted. Much of that ENERGY appears as heat and indeed is sometimes referred to as resistance heating.*

For a distribution system having a given resistance, the losses are proportional to the current *squared*, as can be seen from the equations provided earlier. Significant reductions in losses can be achieved by making the current as small as possible. But if the current is made very small to reduce losses during transmission, then the voltage must be made very large to maintain a given value of the power. Consequently, electric generating plants transmit electricity at extremely high voltages—thousands or even tens of thousands of volts. The higher the voltage for a given amount of POWER, the lower will be the current, and, therefore, the lower will be the heating losses.

I've found out so much about electricity that I've reached the point where I understand nothing and can explain nothing.

—**Musschenbroek**[6]

REFERENCES

1. Hertz, H. Quoted in Schlesinger, H., *The Battery*. Smithsonian Books: New York, 2010, p. 197.
2. Jefferson, T. Letter to Samuel Smith, August 22, 1798. *Jefferson*. The Library of America: New York, 1984, pp. 1052–1055.
3. O'Brian, P. *Blue at the Mizzen*. Norton: New York, 1999, p. 201.
4. Pringle, J. Quoted in Derry, T.K.; Williams, T.I. *A Short History of Technology*. Dover Publications: New York, 1960, p. 609.
5. Jefferson, T. Letter to Reverend William Smith, February 10, 1791. *Jefferson*. The Library of America: New York, 1984, pp. 973–975.
6. Musschenbroek, P.V. (professor at the University of Leiden). Quoted in Schlesinger, H., *The Battery*. Smithsonian Books: New York, 2010, p. 26.

---

* We should recognize that resistance heating is not always a bad thing. Incandescent bulbs give off light because the resistance heating of the filament is so great that the filament becomes hot enough to glow. The heating coils on an electric stove rely on resistance heating; on many such stoves, turning the control to "high" generates enough heat for the coil to become red hot. Many other electrically operated devices that we use to produce useful heat make use of resistance heating.

FURTHER READINGS

Bloomfield, L.A. *How Things Work*. Wiley: New York, 2010. Subtitled *The Physics of Everyday Life*, this book provides good discussions of physical principles in the context of devices that many of us frequently encounter or use. Chapter 10 is relevant to the present discussion of electricity, and introduces the principles through such devices as photocopy machines and flashlights.

Bureau of Naval Personnel. *Basic Electricity*. Dover: New York, 1970. An inexpensive paperback reprint—still available—of a training course for Navy personnel in the theories and applications of electricity. First published by the U.S. Government in 1960, it remains an excellent introductory book on electricity, despite being more than a half-century old. After all these years, Ohm's Law still is valid.

Cutnell, J.D.; Johnson, K.W. *Physics*. Wiley: New York, 2007. All physics textbooks contain information on electricity, and on the physical and mathematical relationships that govern its generation and use. The book by Cutnell and Johnson is a good, modern example. Chapters 18 and 19 are particularly relevant.

Gribbin, J. *History of Western Science*. Folio Society: London, U.K., 2006. A survey of the development of science since the mid-1500s. Chapter 8 discusses the contributions of the early electricians.

13 How Electricity Is Generated

Any of us can sometimes observe a flow of electricity by accumulating static electricity on our bodies (shuffling your feet on a carpeted floor often works) and seeing a spark jump between a fingertip and an object we're reaching for. In dim light, we can see a momentary flash of electricity as the "electric fluid" flows from us to the object we're about to touch. But note that the spark happens only once, and it appears to be nearly instantaneous. Similarly, lightning, which is a flow of electricity driven by enormous potential differences, literally occurs in a "flash." Until the end of the eighteenth century, only these momentary flows of electricity could be produced. To put electricity to work for us, we need a continuous flow of electricity, to provide for the continuous operation of the many electrical devices that form part of everyday life.

LUIGI GALVANI'S FROGS

Until 1786, there was no way to make an electric current under controlled conditions. All electrical experimentation related to phenomena of static electricity. The first key observation leading to production of electric currents was made by an Italian professor of anatomy, Luigi Galvani.* While experimenting with frogs' legs, Galvani showed that the muscles would twitch when an electric spark was applied to them, even though the tissue was no longer living (Figure 13.1). Very likely other scientists had also made this observation, but Galvani added two very important new observations. First, the "twitch" was even more pronounced if he happened to be touching the muscle with a metal scalpel at the time. Galvani thought that perhaps the effect— that is, the greater response of the muscle tissue—was due to some sort of induced electric charge in the scalpel.

Franklin's studies of electricity, including the famous kite-flying experiment, were well-known among the European scientific community. Galvani reasoned that

---

* Luigi Galvani (1737–1798) spent all of his life in his hometown of Bologna, Italy. He graduated from the University of Bologna with degrees in both medicine and philosophy. He started his academic career as a lecturer in surgery, and then became a professor of anatomy. Some of his early studies involved the phenomena of hearing. Later, he became interested in studying the effects of electricity on organisms, in the course of which he made the famous observations on frogs. Galvani continued the study of "animal electricity" for the remainder of his career. When the French army took control of northern Italy in 1797, they founded the Cisalpine Republic, requiring a loyalty oath from all university professors. Galvani refused, and was stripped of his professorship and all sources of income. He died in poverty a year later.

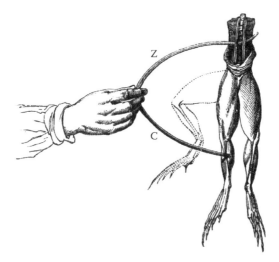

FIGURE 13.1 Luigi Galvani found that placing two different metals (e.g., copper, C, and zinc, Z,) on a frog so that one end touched the nerves and vertebrae while the other touched the leg muscles, a substantial twitch could be observed.

there ought to be a marked effect on the muscle tissue if he exposed samples to a thunderstorm. This reasoning led to the second important observation, which may have been accidental. Galvani used brass or silver hooks to attach the leg muscles to an iron railing. Indeed, just as Galvani had hoped, there was a significant effect on the "twitch" of the leg muscles. Then Galvani discovered that the same effect occurred even if there was no thunderstorm.

ALESSANDRO VOLTA AND THE EARLY "BATTERY"

For some time scientists thought that the effects observed by Galvani belonged to the study of animal physiology, not the physics of electricity. These observations with frogs' legs seemed not to fit in with the laws of electricity. Du Fay, for example, had shown that a wet string was a better conductor of electricity than a dry one. That is, the wet string could not maintain a difference in electrical potential energy from one end to the other; instead, electricity was conducted easily along the string. In Galvani's experiments, the tissues of the frogs' legs were wet, with the bodily fluids that permeate the tissues. Therefore, it would have been reasoned, the legs themselves could not sustain an electrical potential difference any more than could a wet string. In the scientific thinking of Galvani's era, this twitching of frogs' legs was a new phenomenon, one apparently different from Du Fay's demonstration of the easy way in which a wet string conducted electricity. For that reason, Galvani's discovery was at first called "animal electricity." Galvani's experiments were taken as proof (quite wrongly, in fact) that there was some kind of "organic" or animal electricity in the realm of physiology that was different from the "inorganic" electricity studied by physicists.

FIGURE 13.2   Alessandro Volta (1745–1827). His creation of the Voltaic pile, as the first practical continuous source of electricity, stimulated research in many laboratories in Europe. The unit of electrical potential energy is named in his honor.

This infuriated Galvani's countryman, Alessandro Volta (Figure 13.2).* Volta rejected the idea of animal electricity. He recognized that the effects observed in Galvani's experiments came from the use of two different metals. Volta had already established the existence of a "contact potential" between two different metals when they were put together. He wondered whether the so-called animal electricity resulted from the contact potential of different metals touching. What had really produced the effect Galvani observed in his experiments was not the thunderstorm, but rather the two different metals (the brass hooks and the iron railing) in contact.

Galvani's discovery remained largely a curiosity until Volta reasoned that two different metals in contact with a solution capable of conducting an electric current would generate an electric current (Figure 13.3). Volta produced devices for generating electric current consisting of two metals, zinc and copper for instance, in contact with a medium that would conduct electricity, such as a piece of blotting

---

* Alessandro Volta [1745–1827] was a physics professor who spent much of his career at the University of Pavia. Like numerous other brilliant scientists of the era, Volta made contributions in many areas. For example, he spent several years studying the chemistry of methane (natural gas) and how to ignite it with an electric spark. Volta's demonstration of the Voltaic pile to Napoleon Bonaparte so impressed the Emperor that he made Volta Count of Lombardy. Those who enjoy "what if" speculations might consider how the course of European history would have changed if Volta had overdone it and accidentally zapped the little guy.

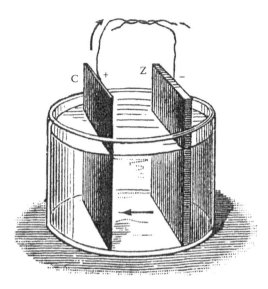

FIGURE 13.3 A voltaic cell often uses a liquid medium as the electrolyte, to allow passage of current between the electrodes.

paper soaked in a solution of salt. Volta found that it was possible to draw electricity continuously from the ends of this "stack," which came to be called a Voltaic pile (Figure 13.4). In Galvani's frogs' legs, the fluid in the muscle tissues acted to conduct electricity just like salt water. Volta argued that the generation of electricity had nothing to do with the frog at all; the frog's body parts were not necessary (a fortunate thing for untold generations of frogs yet to be born). The Voltaic pile and the remarkable effects of the electricity it generated became a popular scientific demonstration, as in this fictional account of the effect on Jeremiah Dixon.*

> *So forcefully that his Queue-Tie breaks with a loud Snap, Dixon's Hair springs erect, each Strand a right Line pointing outward along a perfect Radius from the Center of his Head. What might be call'd a Smile, is yet asymmetrick, and a-drool. His Eyeballs, upon inspection, are seen to rotate in opposite Senses, and at differing Speeds.*

> **—Pynchon**[1]

In modern terminology, the voltaic pile would be called an electrochemical cell. Two or more cells connected together constitute a "battery." However, in informal usage, we nowadays use the term "battery" to refer even to a single cell. In essence, the voltaic pile is the forerunner of our battery. A pile, in the literal sense, was necessary because Volta's only way of measuring electricity was by getting enough potential to

---

* Jeremiah Dixon (1733–1779) was an English surveyor and astronomer best known today for his collaboration with Charles Mason in surveying the Mason–Dixon Line in the United States, which at the time resolved the border between Pennsylvania and Maryland. The Mason–Dixon line is nowadays often taken as the unofficial dividing line between northern and southern states in the United States. Dixon's astronomical work included observations of the transit of Venus (the passage of Venus across the face of the Sun), for which he made measurements in South Africa and in Norway.

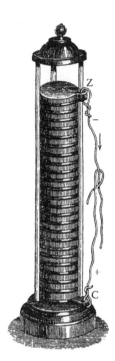

FIGURE 13.4 Many early Voltaic piles used blotting paper, moistened with salt water, between the copper and zinc electrodes. In this pile, electrodes are stacked on a wooden base with copper on the bottom and zinc on the top, held in place by glass rods.

raise a spark. In those days, there were no sensitive electrical instruments for measuring voltage or current. Volta had to stack twenty to thirty of the metal disks on top of one another to get his first "battery."

The voltaic pile made it possible for scientists to experiment with electric currents under controlled conditions. Humphry Davy, for example, using a voltaic pile, became the first scientist to isolate several elements—such as sodium, magnesium, and calcium—from their compounds. Stacking many alternating zinc and copper plates, with salt-soaked blotting paper between, made it possible to produce significant amounts of electric current. Despite the advance that the voltaic pile represented, it was still far from ideal as a steady source of electricity. Just like our modern-day batteries, the pile eventually "runs down." It is rather messy to work with. Generating a steady current for a long period of time requires devoting lots of space to setting up the pile.

BATTERIES

The modern descendant of the Voltaic pile, the battery,* creates an electric potential that is available when needed to do WORK. In this SYSTEM, it is necessary to *create*

---

* Strictly speaking, one of the devices that we commonly refer to as a "battery" is called a cell. A collection of two or more cells connected together would be a battery. However, there is an almost universal tendency to eschew the pedantic and speak of individual cells as batteries. We will do so here.

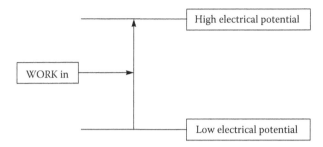

FIGURE 13.5 The reverse energy diagram shows that WORK must be put into the SYSTEM for the electrical cell to generate high potential energy.

a state of high potential energy. For us to extract WORK *from* a SYSTEM, we take advantage of a spontaneous change from high to low potential energy. Now, we want to cause a nonspontaneous change from low to high potential. To do this, we are required to put WORK *into* the SYSTEM. In any SYSTEM, providing a high potential energy—whether electricity or any other form—requires a "reverse energy diagram" (Figure 13.5).

We know that the fall of matter in a gravitational system is spontaneous, from abundant everyday experiences. We can also be assured that the reverse, that is, matter falling upward, is nonspontaneous, because, for instance, we could stand observing a waterfall until doomsday or until boredom sets in (whichever comes first) and be absolutely confident that we would never observe the water suddenly start to flow *up* the waterfall. But it's not hard to cause water to flow uphill—we just need a pump. And for the pump to operate, we must supply it with ENERGY, using our muscles, or a steam engine, or an electric motor.

The spontaneous direction of change is for electricity to flow from the high voltage, or high potential, to the low voltage. Since the battery carries out a nonspontaneous change, WORK or ENERGY is required to effect this change. A battery "pumps" electrical charge uphill to establish the high electrical potential energy (high voltage).

> *For every natural event, there is an allowed direction and a forbidden direction. Apples fall; they don't jump up to their branches. Sugar lumps dissolve in your coffee; they don't form by themselves in a cup of sweetened coffee. Hydrogen combines with oxygen to form water; water does not dissociate spontaneously into a mixture of hydrogen and oxygen. All nature's streets are one way. You can take them in the wrong direction, of course, but you have to work for it: lift the apple, extract the sugar, wrench the hydrogen off the water molecules, for example, with electricity.*
>
> *This is the absolutely fundamental way of nature, expressed by what scientists call the second law of thermodynamics.... If you have to work for it, you are going in the forbidden direction. In the allowed direction, on the other hand, the phenomenon, properly harnessed, may work for you, though it would never give you as much work as you would have to carry out yourself to reverse it....A convenient image to describe the two directions is downhill for what is allowed and uphill for what is forbidden.*

—de Duve[2]

How does the battery supply the WORK to create the high electrical potential? Chemical potential energy is stored in the chemical components of the battery—in the case of the Voltaic pile, in the copper and zinc and the salt solution. In a *primary cell*, the chemicals necessary for production of the electricity are sealed inside. Once they have been consumed, the battery is useless. The nearly ubiquitous AA batteries are an example of primary cells. A *secondary cell*, however, must be charged with electricity before it can be used. That initial charging puts the chemicals in the battery into a state from which they can then react to generate electricity for us when we need it. In a sense, we have put electricity into the battery before using it and withdraw it later as we need it. For that reason, many secondary cells are also referred to as *storage batteries*. The battery in an automobile is an example.

> *The storage battery is, in my opinion, a catch penny, a sensation, a mechanism for swindling the public by stock companies.*
>
> **—Edison**[3]

When we use a battery to run a flashlight, or calculator, or other device, the battery has to create the electrical potential energy (voltage) by consuming some of its chemical potential energy to "pump" the electric charges uphill. Each time it does this, some chemical potential energy is used up. Eventually, not enough of the original chemical components of the battery remain to supply the needed chemical potential. When that happens, we can no longer obtain an electrical potential from the battery. We speak of this effect by saying that the battery has "run down" or has "gone dead."

Because the secondary cell is charged with electricity to get it to operate in the first place, most such batteries can be recharged. An automobile battery, for example, supplies electricity to start the car, but then is recharged while the motor is running. Cordless power tools often have rechargeable batteries; they can be removed from the tool and recharged using household electricity. But no process of any sort operates with 100% efficiency. Therefore, each recharge cycle does not fully restore the battery to its original condition. After a large number of recharge cycles, even this kind of battery will reach the point at which it no longer functions adequately. In other words, even rechargeable batteries will "go dead" in the sense of reaching a point from which they cannot be restored. Modern battery technology is such that good rechargeable batteries will last a very long time before they are truly "dead."

THE GREAT DANE

In 1820, the Danish scientist, Hans Christian Ørsted (Figure 13.6), performed a ground-breaking experiment in physics.*

---

* Hans Christian Ørsted (1777–1851) was a friend of the better-known Hans Christian (Andersen). Ørsted was a physicist, chemist, poet, and philosopher. In the same year as his crucial discovery in physics, he also made an important discovery in chemistry. Ørsted discovered the nature of the active poison (a chemical called piperidine) in poison hemlock—the drink that Socrates was required to take to commit suicide in 399 BCE, as described by Plato in his dialogue *Phaedo*. In 1825, Ørsted was the first scientist to isolate aluminum in pure metallic form.

FIGURE 13.6 Hans Christian Ørsted (1777–1851), a professor at the University of Copenhagen, established the connection between electricity and magnetism.

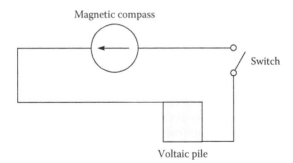

FIGURE 13.7 A sketch of Ørsted's experiment. When the switch was closed, allowing the electrical current to flow from the Voltaic pile, the compass needle moved, indicating the existence of a magnetic field.

Apparently it is not known what actually motivated Ørsted to try this experiment. Ørsted's apparatus is shown schematically in Figure 13.7. During a lecture demonstration, Ørsted observed that when the switch was closed—that is, the current was flowing—the compass needle responded just as if a magnet were brought near it. That is, the needle would swing violently around when the current was flowing. Since it is well known that a compass responds to a magnetic field, Ørsted deduced that

A flowing electrical current generates a magnetic field.

Within a *week* of learning of Ørsted's experiment, Ampère published an extensive theoretical treatment of the subject, showing that an electric current causes a compass needle to deflect. Ampère likely had been thinking along the same lines before Ørsted produced the necessary experimental demonstration.

MICHAEL FARADAY AND THE INVENTION OF THE GENERATOR

Ørsted's experimental observation and his deduction remained mainly a curiosity for about ten years. Then, in 1831, Michael Faraday (Figure 13.8), the great English chemist and physicist,* asked an extremely profound question: Will this experiment work backward? In other words, Faraday's question was: Does a moving magnetic field generate an electric current? A simple schematic diagram of Faraday's

FIGURE 13.8   Michael Faraday (1791–1867), one of the greatest experimentalists of the nineteenth century, showed that a moving magnetic field generates an electric current.

---

* Michael Faraday (1791–1867) ranks among the greatest of nineteenth-century scientists, making important contributions in physics and in chemistry, despite (or possibly because of) having little formal education. In chemistry, e.g., he discovered benzene and tetrachloroethylene (a widely used dry-cleaning solvent) and isolated magnesium from its compounds. Faraday was a very popular lecturer, and his series of Christmas lectures at the Royal Institution was eventually published as *The Chemical History of a Candle*, a classic of science that remains in print today (Dover Publications).

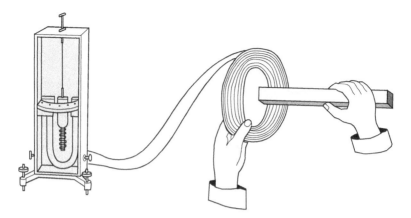

FIGURE 13.9 A schematic diagram of Faraday's demonstration that a moving magnetic field generates an electric current. The device on the left is an early instrument for detecting current flow.

experiment is shown in Figure 13.9. By passing the magnet in and out of the loop of wire, Faraday showed that a current is generated as the magnetic field passes through the conductor (the wire). Faraday's very simple experiment answered his original question in the affirmative. Faraday showed that mechanical kinetic energy (pushing the magnet back and forth) can be converted into electrical energy. This simple observation is responsible for *all* electricity generation today, with the exceptions of the small amounts of electricity we obtain from batteries and from the direct conversion of solar energy into electricity (photovoltaics, Chapter 33). Electric heating and lighting at home, all of our electrical appliances, electric motors and other electric devices in manufacturing industries, electrical railway locomotives and streetcars, or subways all come from this simple experiment. By showing how mechanical work can be transformed into electricity, Faraday brought electricity into the scope of the Industrial Revolution.*

A moving magnetic field generates an electric current.

During the time between Ørsted's observations and Faraday's discovery, other scientists had simply placed magnets near the wires carrying currents. They found that nothing happened. The idea that it was necessary to move the magnets, or the wire, was entirely due to Faraday. This discovery was not some serendipitous accident, but rather had been carefully thought through and is recorded in Faraday's laboratory

---

* There are two versions of a story about Faraday and his coil-and-magnet experiment. Perhaps both actually occurred. Sometime in 1831, Faraday was demonstrating the experiment and was asked by a woman, "But, Mr. Faraday, of what use is this?" Supposedly he replied, "Madam, of what use is a new-born baby?" In the other version, Faraday was asked the same question by a Member of Parliament, William Gladstone (who later served four terms as prime minister of Great Britain). In this version, Faraday is said to have replied, "Sir, in twenty years, you will be taxing it."

books. In Faraday's experiment, WORK is supplied into the SYSTEM to generate the high electrical potential energy. In fact, in the original experiment, Faraday supplied the WORK himself, holding the magnet and moving it through the coil of wire.

As long as the magnetic field was in motion, an electric current could be generated essentially forever. However, the reciprocating (back-and-forth) motion of Faraday's original experiment is not desirable for large-scale generation of electricity. Mechanically, it is not easy to produce smooth reciprocating motion on a large scale. Also, there are always two "dead spots" when the magnet is stopped at each end of its motion. Faraday soon refined his original experiment with two improvements. First, it's generally easier to sustain rotary motion than reciprocating motion. Second, he realized that the key is having the magnetic field and electrical conductor moving relative to each other. It doesn't matter, in principle, which one of the two is moving and which is stationary. Moving a magnetic field past a stationary electrical conductor has exactly the same effect as moving the conductor through a stationary magnetic field. Thus Faraday refined his original bar magnet and coil of wire experiment into the concept of an electrical generator or *dynamo* (Figure 13.10).

All of the electricity we use, again excepting the little from batteries and from solar energy, comes from this device. Therefore, the essence of electricity generation on a large scale is to keep the generator turning as cheaply and reliably as possible. An electrical generator is a device that converts kinetic energy (usually rotary) into electrical energy. One cheap, reliable way of spinning the generator is to take advantage of the kinetic energy of falling water. As we'll see, all the other approaches make use of steam to turn the generator.

The first full-scale electricity-generating machines were built in the 1840s and 1850s. In those early devices, the magnetic fields were produced by permanent magnets. Using permanent magnets causes a disadvantage in the construction of these machines, because a very large array of horseshoe magnets was required to provide a sufficient magnetic field. In the 1850s, many designers tried to solve this problem by replacing the permanent magnets by electromagnets. In the ultimate stage of development, a part of the current produced by the generator can be used for the energy needed in the magnetic field windings of the electromagnets. Essentially, this provides a self-generating magnetic field. This specific type of generator constitutes a dynamo. In common usage, the terms *dynamo* and *generator* are often used interchangeably.

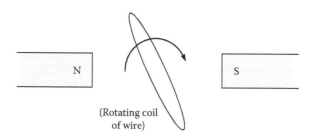

FIGURE 13.10   The basic principle of a dynamo generator is a coil of wire rotating inside a magnetic field.

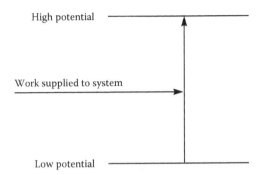

FIGURE 13.11 When a SYSTEM goes from low potential to high potential energy due to WORK being put into it, the reverse energy diagram applies.

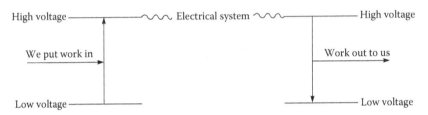

FIGURE 13.12 WORK put into this SYSTEM results in high electrical potential energy that can be transmitted to a different location before we convert it back to WORK.

In an electrical generator, *we* must supply WORK to the system, that is, the WORK of spinning the generator, as illustrated in the "reverse energy diagram" (Figure 13.11). WORK put into spinning the generator can be taken advantage of by physically separating the site of electricity generation (even by tens or hundreds of kilometers) from the place where the electric ENERGY is used, as shown in Figure 13.12 and implied by the long-distance electric transmission lines in Figure 13.13. This provides a way of taking kinetic energy that might be available in one place, and easily and conveniently making it available for use someplace else.

Earlier chapters showed that the energy diagrams for various kinds of SYSTEMS, such as thermal, gravitational, and chemical, are *exactly the same* in concept. This chapter has introduced the idea of a "reverse energy diagram" in which we supply WORK to a SYSTEM to cause a change from low potential to high potential (and note that this would be a nonspontaneous change). Does that mean that there are "reverse energy diagrams" for all of these systems? Yes. As examples: If we supply WORK, we can cause water to flow uphill. We can do this with a pump. If we supply WORK, we can cause heat to flow from a cold object to a warmer object. One kind of device we can use for this process is a refrigerator. Indeed, if we supply WORK (in the form of chemical energy), we can even convert carbon dioxide and water back to methane and oxygen. We'll see some of these examples in detail later. The crucial point is that regardless of what kind of SYSTEM we're considering, the principles of the energy diagram and the reverse energy diagram are *exactly the same* (Figure 13.14).

FIGURE 13.13 The principle of the energy diagrams of Figure 13.12 is put into practice through numerous complex electrical transmission lines across the world. (Reprinted from first edition, permission from Electric Power Generation Association, Harrisburg, Pennsylvania.)

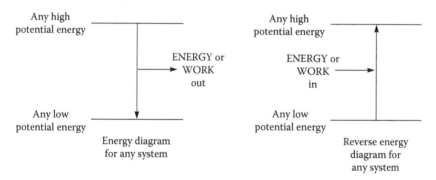

FIGURE 13.14 Any SYSTEM whatsoever can be described by one of these diagrams: spontaneous flow from high to low potential with ENERGY (or WORK) out (on the left) or nonspontaneous flow from low to high potential, requiring that ENERGY (or WORK) be put in.

REFERENCES

1. Pynchon, T. *Mason and Dixon*. Holt: New York, 1997, p. 765.
2. de Duve, C. *Vital Dust*. Basic Books: New York, 1995, p. 36.
3. Edison, T. Quoted in Schlesinger, H. *The Battery*. Smithsonian Books: New York, 2010. For once, Edison was wrong.

FURTHER READINGS

Bloomfield, L.A. *How Things Work*. Wiley: New York, 2010. Chapter 11 of this book discusses electrodynamics and some of its applications in everyday life.

Faraday, M. *Experimental Researches in Electricity*. Green Lion Press: Santa Fe, NM, 2000. An account, in three volumes, of most of Faraday's work on electricity.

Franklin, B. *Experiments and Observations on Electricity*. Eighteenth Century Collections Online (ECCO), 2010. Other print and electronic versions are available. Here are Franklin's contributions, as described by himself.

Gribbin, J. *History of Western Science*. The Folio Society: London, U.K., 2006. Chapter 8 provides a fine discussion of the historical development of our knowledge of electricity, involving most of the people discussed in these past two chapters.

Hopkins, G.M. *Experimental Science*. Lindsay Publications: Bradley, IL, 1987. This is an inexpensive paperback reprint of the 1906 edition of this book, which is a trove of information on relatively simple experiments for those who like to learn by trying things out. Chapters XVII and XVIII of Volume One provide many potentially useful experiments in static and dynamic electricity.

Schlesinger, H. *The Battery*. HarperCollins: New York, 2010. A splendid and very readable discussion of batteries, from the earliest experiments to very recent developments in battery technology.

14 Impacts of Electricity on Society

Ørsted showed that a moving electric current produces a magnetic field. Faraday then turned this observation around and showed that a moving magnetic field produces an electric current. Following Faraday's fundamental discovery in physics, two engineering design modifications of the original experiment incorporated the advantages of using rotary instead of reciprocating motion, and of moving the coil of wire while keeping the magnet stationary. Many people contributed to the development of the truly practical large-scale generator. Credit is usually awarded to the Belgian engineer Zénobe Gramme* (Figure 14.1). Gramme's achievement has considerable similarity with James Watt's. Neither can truly claim to have originated or invented the device with which they are associated, but each genuinely deserves credit and praise for bringing the device to a state of reliable practical operation and a scale large enough for industrial and commercial application. Perfection of the generator by Gramme in 1870 provided for rapid expansion of the electricity industry. In this chapter, we examine some of the ramifications of this on society.

BREAKING THE GEOGRAPHIC BARRIER

In the early days of using electricity, it was assumed that it would be generated on-site, that is, generated on the same spot where it was used. One of the first such applications occurred in 1878, when Wanamaker's department store in Philadelphia was illuminated with arc lights, supplied with electricity from a generator in the basement. The 1870s and 1880s saw a growing realization that the future of electricity was not in local on-site generation, but rather in distributing electricity produced in large central power stations. It is important to realize that electricity is different from other kinds of energy because, with electricity, the energy source does not have to be in the same location (e.g., in the same building) as where the energy is used. A factory or industrial community using electricity did not have to be located on a river for a source of water power or built around a big, heavy steam engine. The beauty of electricity as an energy source is that any electrical device can be operated at any place whatsoever, as long as wires can be run to it from a source of electricity.

---

* Zénobe Gramme (1826–1901), like Faraday, had very little formal education and minimal skills in higher mathematics. He got his start by working for a French company that manufactured electrical equipment; Gramme's job was to build models of the company's equipment. From that somewhat humble beginning, Gramme went on to perfect a dynamo that, in the early 1870s, was possibly the best in the world. For a time, dynamos were even known as "Gramme machines."

FIGURE 14.1 Zénobe Gramme (1826–1901) is one of several nineteenth-century engineers whose work resulted in the development of practical electrical generators. (From http://commons.wikimedia.org/wiki/File:Z%C3%A9nobe_Th%C3%A9ophile_Gramm%C3%A9.jpg.)

Because of the relatively low cost and high efficiency of transmitting ENERGY over long distances via electricity, large quantities of ENERGY can be produced in central generating stations ("power" plants) and transmitted to locations where it's needed. Though the economics and structuring of the electric industry vary from one country or region to another, in most cases, it is probably cheaper to have a small number of very large generating stations than a larger number of smaller stations scattered around an area.* An additional benefit of electricity as an ENERGY source is that it can be generated either from fuel—coal, natural gas, biomass, or oil—that can be transported to the power plant, or from the kinetic energy of water or wind, or from nuclear reactions, or from the direct conversion of sunshine into electricity. In the chapters that follow, we will examine how these kinds of generating plants operate and what some of the advantages and disadvantages are for each.

---

* This notion is being increasingly reexamined and questioned. Large generating stations offer the advantages of economy of scale—it is cheaper to build one big plant than two plants each half the size of the big one—and that it is easier to capture potential pollutants from one large centralized source than from many small sources. But with the continuous improvement in small relatively nonpolluting sources of electricity, such as direct conversion of solar energy, the twenty-first century may see a move back to small-scale distributed generation of electricity, possibly even down to the level of the individual business or homeowner.

ELECTRIC LIGHTING

Volta developed the voltaic pile around 1800. Within two years, Humphry Davy had shown that a spark could pass between two carbon rods connected to a voltaic pile. Davy's observation was responsible for the development of the first practical form of electric lighting, the arc light (Figure 14.2). In this device, electricity flows through two rods of carbon that are touching at the tips. When the rods are gradually pulled apart, electricity continues to flow across the gap, as an arc. The tips of the carbon rods become heated to incandescence, creating a bright light.

Arc lights (Figure 14.2) first came into practical use in 1844. Initially they had two applications, for street lighting and for lighting harbors and railway shunting yards. Lighting of streets certainly improved safety, since a person now could see where he or she was going. It also contributed to a reduction in street crime, by reducing the number of darkened places where criminals could lurk. Lighting of harbors and railway yards increased the number of hours that they could be in operation and thus increased the number of hours during which goods of all sorts could be loaded for shipment or unloaded for distribution.

The inherent glare and brilliance of the light from an arc lamp meant that an obvious application was in lighthouses. In 1858, the South Foreland lighthouse (Figure 14.3) and, in 1862, the Dungeness lighthouse in England were equipped with arc lights. The electricity supply for these lights came from large generators driven by steam engines. The first athletic event played under lights was a soccer match in Sheffield, England, in the late 1870s.

At first, most systems of arc lights were arranged in series (Figure 14.3). This is a significant drawback, because if one lamp fails, the entire system shuts down.

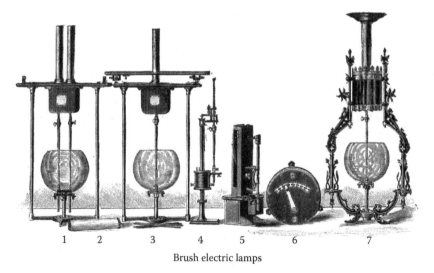

Brush electric lamps

FIGURE 14.2   Examples of early arc lamps, from Brush Electric Company. 1, Double lamp; 2, carbons; 3, single lamp; 4, focusing lamp; 5, head-light lamp; 6, dial attachment to machine; 7, ornamental lamp. (From http://en.wikipedia.org/wiki/FileArc_Lamp_Examples.jpg.)

FIGURE 14.3 The South Foreland lighthouse, one of the first lighthouses in the world to be equipped with arc lamps. (Photograph courtesy of U.K. National Trust/David Lewing.)

This problem can still be encountered with the decorative strings of lights used at the holiday season. Older sets of these lights are wired in series, and, if one bulb burns out, the whole string stops functioning until we have made a frustrating and time-consuming search, bulb by bulb, for the defective one.

Despite these successes, arc lights have a number of disadvantages, particularly for applications in which a competent trained electrician may not be available (such as in most of our homes). Arc lights require dangerously high voltages of electricity. As the light "burns," the carbon electrodes are consumed, so the lamp requires occasional adjustment. It was, mechanically, a complicated piece of equipment. Additional devices were needed to keep the tips of the carbon rods positioned at the correct distance for producing an arc. The carbon rods had to be replaced from time to time. In operation, the lamp emitted a hissing noise. All these limitations might be acceptable in a lighthouse, where the keeper could attend to these duties, or in a commercial enterprise, with maintenance workers on the staff, but it would hardly be acceptable in the average home, which is where the potential for an enormous market lay.

Several criteria have to be met for electric lighting to be successful in the home. First, it had to be reasonably safe, operating at potentials of no more than about 240 volts. Second, it had to be simple and easy to use, even for persons with no special training; for example, there had to be an easy way to change bulbs. Third, it needed to produce a "soft" light—that is, light that is not glaring to the eyes nor casts shadows.

In an incandescent lamp, the electric current heats a filament that glows to give off light. The concept of an incandescent lamp originated in 1820. For several decades, many talented scientists in America and Europe experimented with the development of a *practical* incandescent electric lamp, that is, one that could be relied upon to operate for long periods of time at relatively safe voltages. By 1877, the consensus of the international scientific community was that no practical result would ever come of all these experiments. And then a partially deaf former paperboy and candy seller came up with two crucial ideas. First, all of the air had to be removed from the bulb, to avoid burning up the filament. Second, out of thousands of candidate materials tried, a very thin filament of carbon was superior to all others, including thicker filaments of carbon and special materials like platinum. The first test of an incandescent bulb that incorporated these new ideas occurred on October 21, 1879. The bulb operated continuously for two days and was the first truly successful demonstration of an incandescent lamp.

> *The hunt was a long, tedious one. Many materials which at first seemed promising fell down under later tests and had to be laid aside. Every experiment was recorded methodically in the notebooks. In many there was simply the name of the fiber and after it the initials 'T.A.,' meaning 'Try Again.'*
>
> *Literally hundreds of experiments were made on different sorts of fiber.... Threads of cotton, flax, jute silks, cords, manila hemp and even hard woods were tried.*
>
> *Chinese and Italian raw silk both boiled out and otherwise treated were among those used. Others included horsehair, fish line, teak, spruce, boxwood, vulcanized rubber, cork, celluloid, grass fibres from everywhere, linen twine, tar paper, wrapping paper, cardboard, tissue paper, parchment, holly wood, absorbent cotton, rattan, California redwood, raw jute fiber, corn silk, and New Zealand flax.*
>
> *The most interesting material of all... was the hair from the luxurious beards of some of the men about the laboratory.*

—Jehl[1]

AND THEN THERE WAS EDISON...

Thomas Edison (Figure 14.4) was a deaf, former paperboy and self-taught inventor. Contrary to popular mythology, Edison did not invent the light bulb, just as Watt did not invent the steam engine, nor Gramme the dynamo.* Several persons had had the idea of an incandescent light before Edison and had designs that more-or-less worked. What Edison *did do*, and for which he deserves enormous credit, was invent the first incandescent lamp that could be made in large quantities, would operate reliably, and could be easily sold.

> *At last, on 21 October 1879, Edison made a bulb that did not burn out. Its filament was of carbonized cotton sewing thread.... The first commercial bulb, which followed swiftly, had a horseshoe filament of carbonized paper.*

---

* Very, very few scientists, inventors, or engineers ever achieve celebrity status in the way that entertainers or athletes do. Edison was one. Thomas Alva Edison (1847–1931) was born in Milan, Ohio. His first contact with electrical devices was working, as a young man, as a telegraph operator. Around 1868, Edison purchased a set of books describing Faraday's work, which may have inspired him to begin a serious career as an inventor of electrical equipment. His first triumph came in 1870, the invention of the stock ticker. This was a critical device in financial markets for nearly a century, until the development of instantaneous electronic communications. It was a modification of the telegraph that could print rapidly, rather than requiring a human operator to listen to electrical dots and dashes. Edison sold the invention for $40,000, an enormous sum in those days and likely worth many millions today. He used the money to set himself up in business.

FIGURE 14.4   Thomas Edison (1847–1931), whose genius lay in understanding the need for an entire electrical system to provide electricity to customers—generators, wires and cables, switches, meters, and finally the bulbs. He is shown here working on chemical experiments, sometime around 1905. (From http://www.loc.gov/pictures/item/2005687120/.)

> *Three years later the Pearl Street Central Power Station was completed in New York—the first of the world's great cities to be electrically lit. The coming of electric light was widely seen as banishing the fear and superstition that darkness had bred.*
>
> *The same point is made, though less poetically, in Conan Doyle's The Hound of the Baskervilles (1902) when the new heir to Baskerville Hall arrives from North America and remarks, on viewing his spooky ancestral home:*
>
> *It's enough to scare any man. I'll have a row of electric lamps up here inside of six months, and you won't know it again with a thousand-candlepower Swan and Edison right here in front of the hall door.*
>
> —Carey[2]

In the late 1870s, Edison began to think in detail about a *system* of electric lighting. One of his critical contributions—far more important than the mere invention of incandescent light bulb itself—was the recognition that the bulb is useless as an isolated device by itself. Other inventors perfected the incandescent light at just about the same time that Edison did. Edison's true brilliance was to recognize, and conceive the details of, an entire electric lighting system rather than a new object (the bulb). He recognized that an electric lamp needed to be considered as part of a system along a central electricity-generating station and all the necessary distribution equipment.

> *It was not only necessary that the lamps should give light and the dynamos generate electric current, but the lamps must be adapted to the current of the dynamos, and the dynamos must be constructed to give the character of current required by the lamps, and likewise all parts of the system must be constructed with reference to all*

other parts, since, in one sense, all the parts form one machine, and the connections between the parts being electrical instead of mechanical. Like any other machine the failure of one part to cooperate properly with the other part disorganizes the whole and renders it inoperative for the purposes intended.

The problem then that I undertook to solve was stated generally, the production of the multifarious apparatus, methods and devices, each adapted for use with every other, and all forming a comprehensive system.

—Edison[3]

Despite the scattered successes of arc lamps, the major source of illumination through the 1870s was gaslight. Most gaslights functioned by burning a combustible gas made by heating coal under conditions at which it would decompose but not burn.

What use of wealth so luxurious and delightful as to light your house with gas? What folly to have a diamond necklace or a Correggio, and not to light your house with gas? How pitiful to submit to a farthing-candle existence, when science puts such intense gratification within your reach! Better to eat dry bread by the splendor of gas, than to dine on wild beef with wax candles.

—Murray[4]

Edison began electrical power generation with the goal of competing with gas lamp illumination in New York City. By this time, Edison had developed a friendship with a prominent attorney, Grosvenor Lowrey, who set up contacts for Edison with Wall Street financiers. He helped get financial backing for Edison from Drexel, Morgan and Company. By one of those interesting confluences of events in history, Edison stood on the brink of international fame for his lighting system just at the same time that Drexel and Morgan came into the hands of the young J.P. Morgan, who became likely the most famous (or notorious) of the Wall Street financiers of the late nineteenth century.*

The widespread public acceptance of electric lighting in the home was a "chicken and egg" story, since of course electric lighting can't be used without a source of electricity. Edison and his associates not only developed the first practical bulb, but also developed the components of the entire system for electric lighting: improvements in the generator; cables for transmitting electricity to the consumer; schemes for electric wiring in the home; electric meters, fuses, and switches; and even the sockets or other fittings for the bulbs.

One of Edison's crucial concepts was the idea of parallel wiring rather than series wiring (compare the wiring schemes in Figures 14.5a and 14.5b). If lights are wired in parallel, one bulb burning out does not affect the operation of the others. Those

---

* John Pierpont Morgan (1837–1913) was the dominant financier in the United States, and possibly the world, in the late nineteenth and early twentieth centuries. Morgan and his partners were instrumental in putting together the huge corporations that became United States Steel and General Electric. He was a hero and role model to the political right, and a villain to the left. Morgan was scheduled to sail on the maiden (and only) voyage of the *Titanic* in 1912, but canceled his plans at the last minute. He died very peacefully in his sleep while visiting Rome the following year.

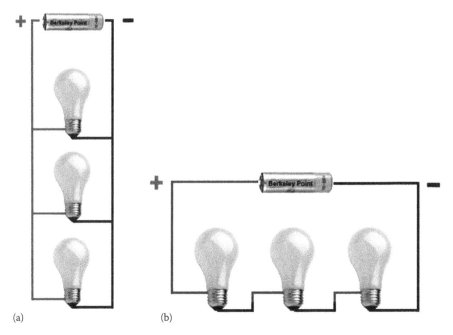

(a)   (b)

FIGURE 14.5   (a) These lights are wired in parallel. In this case (compare Figure 14.5b), when one bulb burns out, the current is not interrupted, and all of the remaining bulbs stay lit. Much less frustrating for the homeowner than the series wiring shown in (b) (From http://www.berkeleypoint.com/learning/parallel_circuit.html.)

who have experienced the slow, trial-and-error hunting of one burned-out bulb in a holiday decoration can imagine the incredible frustration of having an entire house wired in series and having to search in darkness through the house to find the defective bulb.

Edison opened an electricity-generating station on Pearl Street, New York, in 1882. On its first day, Edison had 52 customers. While it was not the first electricity station of any kind, it was the first dedicated especially to supplying electricity for incandescent lighting. The Pearl Street plant was only a tiny cog in the vast system of Edison's enterprises.

There were also the Edison Electric Light Company, the Edison Machine Works to build dynamos, the Edison Electric Tube Company to make the underground conductors, and the Edison Lamp Works to manufacture incandescent lamps. The Pearl Street station would have been useless without all of the other components of this system. By 1890, manufacturing companies and electric utilities owned by Edison and his backers had spread all over the United States and were getting established in Europe.

This single event—the opening of the Pearl Street station—was an absolute disaster for the gaslighting industry, and investment in shares in gaslight companies plummeted.

A sign in hotel rooms in 1892 read, 'This room is equipped with Edison Electric Light.
Do not attempt to light with a match. Simply turn key on the wall by the door.'

—Reynolds[5]

Information about Edison's new system spread remarkably quickly. Within a few months, people were applying to Edison for licenses to build electricity-generating plants all over the country. Here is where Edison's vision of electricity as an entire system really paid off, because it was the system concept that allowed the very rapid expansion of the electrical network. Within twenty years of Pearl Street, there were 2250 generating stations in the United States and, by 1920, almost 4000. By then, about a third of the nation's homes were wired for electricity. Just about the time that the Pearl Street station opened, a group of investors in Japan established the Tokyo Lighting Company, which began supplying electricity in 1887. In the same year as Pearl Street, the first electric lighting company was established in Chile. The Korean government ordered a generating plant from Edison Electric in 1884, with electricity becoming publicly available in early 1887. In Britain, electricity was used for street lighting in 1881, marking the first public application of electricity. By 1920, England and Wales together had 480 suppliers of electricity.

The formation and development of large firms—such as General Electric in the United States—essentially marked the end of an era for electricity. By 1900, electricity had become an advanced technology. It was no longer the domain of the lone inventor, the empirical tinkerer, or the "electrician" of the days of men like Gray and Franklin. Now electricity was big business, with further research and development in the hands of the electrical engineer.

Despite the extreme cost of early electric rates, possibly five to ten dollars per kilowatt hour, the access to electricity changed living conditions drastically. In 1889, the first domestic electric appliance, the clothes iron (Figure 14.6), began to eliminate some of the laundry drudgery of heating irons on a stove and periodically reheating them as they cooled. Electric fans (Figure 14.7) followed around 1890; electric refrigerators came into fairly wide use in the 1930s (Figure 14.8), followed by electric washing machines in the 1940s (Figure 14.9). Edison himself remarked that he had made electric lighting so cheap that the day would come when only rich people dined by candlelight.

Early electric rates were so high because a centralized electricity station requires a major investment of capital. It also used very large equipment, such as steam boilers, that is most efficiently and economically utilized if it can be operated all the time. The problem with generating electricity only for lighting is that most people use electric lights only six to twelve hours per day, even in the winter when daylight hours are short. To improve the economics of centralized electricity generation, other uses were needed, especially applications of electricity that could potentially operate round the clock.

ELECTRIC MOTORS

As early as 1837, the inventor Thomas Davenport (of Brandon, Vermont) used primitive motors to do industrial work, such as drilling iron and steel and turning wood in his workshop. These motors were not very satisfactory because they relied

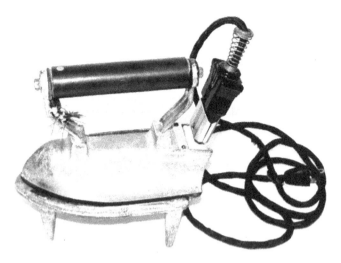

FIGURE 14.6 An early electric clothes iron. (Photograph courtesy of Larry Meeker, Somerset, California, From http://www.Patented-Antiques.com.)

FIGURE 14.7 An early electric fan. (Photograph by Donald Audette, courtesy of Gary Lillemoen, Hatton, North Dakota.)

FIGURE 14.8   An electric refrigerator from the 1940s. (Photograph by Donald Audette, courtesy of Gary Lillemoen, Hatton-Eielson Museum, Hatton, North Dakota.)

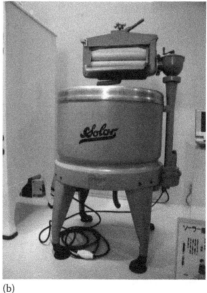

(a)   (b)

FIGURE 14.9   (a) This antique hand-operated washing machine was the sort that homeowners were delighted to replace with an automatic electric appliance. (Photograph by Donald Audette, courtesy of Gary Lillemoen, Hatton-Eielson Museum, Hatton, North Dakota.) (b) The first electric washing machine in Japan from about 1930. (From http://commons. wikimedia.org/wiki/File:TOSHIBA_Solar.JPG.)

on voltaic piles that eventually "ran down."* In 1851, the English inventor Charles Babbage† recognized that there were some applications that cannot be fulfilled by steam engines because such engines are simply too big and too costly. These are applications that require devices that can start and stop at a moment's notice, require no time for maintenance or management (e.g., for shoveling coal), have a moderate initial cost, and a low daily operating expense. In the 1870s, Friedrich Engels,‡ the German-born owner of a textile factory in Manchester, England, predicted that if electricity could be transmitted over long distances, it would revolutionize industry.

Then, in 1873, Zénobe Gramme showed that the electric generator could be operated "backward." That is, if current were supplied to the generator, the armature would spin. This was the first practical electric motor. By some accounts, this discovery was accidental. Working with a colleague, Hippolyte Fontaine, and getting ready to show off his generator at the Vienna Exhibition, somehow the connections got reversed, and electricity was supplied *to* the generator. When this happened, the generator shaft began to spin.

The first factory to be completely electrified in the United States was a cotton mill, built in 1894. Initially, factory designers thought that large steam engines or large waterwheels would be replaced by equally large electric motors. However, it soon became clear that there are more advantages to having a relatively small electric motor on each machine. This eliminates the complicated drive train of belts, pulleys, and gears needed to operate each machine from a centralized source. Having individual motors allows much more precise control of the speeds of individual machines, in turn providing the ability to make new kinds of precision-machined goods. Each machine can be controlled directly by its operator. The independent operation of each machine can also speed up the circulation of raw materials and partially assembled or partially finished goods.

The replacement of steam engines by electric motors brought with it major shifts in factory design and location. First, it was no longer necessary to build factories that were several stories high to accommodate a huge steam engine and the complex system of shafts and belts used to operate the individual machines. Second, it was no longer necessary to locate factories near water sources or coal fields.

---

* Nowadays battery-operated tools, such as portable drills or saws, seem to be enjoying steady growth in popularity. Commonly, these tools have rechargeable battery packs. A steady user of such tools, a construction worker for example, usually keeps at least one battery pack recharging while another is operating the tool. A homeowner who may have only occasional need, such as for a simple repair project, can get enough life from the batteries to get small jobs done and then let the battery pack recharge until the next calamity strikes.

† Charles Babbage (1791–1871) also invented a mechanical device, called the "analytical engine," to help him with his calculations. Today we operate analytical engines with electricity—but we call them "computers."

‡ Friedrich Engels [1820–1895] certainly knew something about revolutionizing—he was the friend, benefactor, and collaborator of Karl Marx. Nowadays Engels is probably best known as the coauthor of *The Communist Manifesto*. Based on contemporary accounts of Engels being an outgoing generous person who hosted Sunday dinner parties that lasted until 3 a.m. on Monday, interested in hunting and in poetry, it is likely that he would have been appalled by the monstrous activities of Josef Stalin and his henchmen.

Neither Marx nor any other devotee of the economic interpretation of history was able to foresee the coming of electrical power, one of the consequences of which was that factories no longer needed to be built near coalfields; moreover the properties of electrical power are such as to make possible the springing up of numerous small factories containing machine-tools individually powered by their own electrical outlets.

—Medawar[6]

The electric motor meant that factory owners no longer had to be in the business of generating their own POWER. From the other perspective, utilities could achieve economies of scale by supplying electricity to many customers from centralized electricity-generating plants. No single manufacturer could hope to achieve such economies. The significant advantages of electric factory machinery combined with electric lighting (which allowed for safe, round-the-clock operation of factories) led to great improvements in industrial output, which in turn helped contribute to a generally rising standard of living.

In 1889, the Deptford Power Station, on the Thames estuary in England, began supplying electricity to customers. The Deptford station is the prototype for the type of large, centralized, coal-fired power plants that we will describe in a later chapter. The location of the plant was not dictated by where the market was, but rather by convenience of the coal supply. For this plant, coal was brought by ship from the north of England. The Deptford station supplied alternating current (AC) at 10,000 volts to substations in London some ten to fifteen kilometers away.

The development of the large centralized power station, as at Deptford, raised speculation as to the possibilities of using electric locomotives on railways (Figure 14.10), operated by electric motors, usually supplied with electricity from overhead wires. Electric railways were being designed in Germany in 1879, initially limited because

FIGURE 14.10   An electric locomotive, class VL60 of the former Soviet Union railways, circa 1960. (From http://en.wikipedia.org/wiki/File:%DO%AD%DO%BB%DO%B5%DO%B A%D1%80%D0%BE%D0%B2%D0%BE%D0%B7_%D)%92%D0%9B60%D0%B!-1196.jpg.)

FIGURE 14.11 A modern electric commuter train, this example operating in Istanbul, Turkey. (From http://commons.wikimedia.org/wiki/File:E14000s_at_Goztepe-1.JPG.)

of a lack of long-distance transmission of electricity. The beginning of the electric railroad, a system of streetcars or trams, occurred in Richmond, Virginia, and in Berlin in 1881, followed two years later by developments in Britain and Austria. By the 1930s, it was apparent that a train hauled by an electric locomotive had the advantage of more rapid acceleration relative to a train with a steam locomotive. In those days, a steam train starting from rest could accelerate at about a half to one kilometer per hour every second, but an electric train could accelerate at twice that rate and, at the end of half a minute, would be traveling at 50–65 km/h. This quick acceleration is exactly what was needed for passenger traffic in urban and suburban areas, where stops are frequent and traffic heavy. The electric locomotive provided a higher average speed and therefore shorter intervals between successive trains arriving and departing a given station. In turn, this meant a great increase in the passenger-carrying capacity of the line. Further, electric locomotives do not emit the plumes of smoke, cinders, and soot that steam locomotives do. Electric locomotives operating on dense commuter rail networks around large cities resulted in far less air pollution and less soot on passengers' clothing.

Suburban electric trains were developed that did not need a separate locomotive (Figure 14.11). The driving motors were distributed among the coaches. This so-called multiple-unit system allowed any number of coaches to be coupled together without reducing train speed, since each additional coach contributed its own share of the power. This system provided great flexibility. During the course of a day or week, passenger trains could be lengthened or shortened as needed to cope with demand, without any effect on the speed of the trains or the time interval between them.

AC OR DC?

The story was told at about this time [1890] that a Cambridge student, asked in his final viva voce examination what electricity was, hesitated and then stammered that he had known this once but had forgotten. The examiner sighed: 'What a pity, what a pity! Only one man in the world who knew what electricity is, and now he has forgotten!'

—**Gratzer**[7]

As electric systems began to spread in the 1880s and 1890s, a ferocious controversy erupted regarding the better way of transmitting and using electricity: AC or direct current (DC).* The early generators of Gramme and his contemporaries provided DC, around which Edison built his system. Edison's system was based on a series of small generating plants located near their customers, rather than a few very large plants transmitting electricity to large numbers of customers over a wide geographical area. After the Pearl Street station began operating successfully, it was quickly followed by about sixty more small plants in New York City alone.

However, for long-distance transmission, AC has numerous advantages relative to DC. A DC transmission system uses the same voltage throughout. An AC system can take advantage of a transformer to "step up" the voltage to very high values, transmitting the electricity at high voltage and low amperage, to minimize losses during transmission. Then, at the consumer's end, a second transformer "steps down" the voltage and increases the amperage to the values normally used in household or commercial devices. For one of the few times in his life, Edison was on what proved to be the wrong side of the argument. Having started with DC, he resisted AC to the last ditch and had a prominent role in a bitter, acrimonious, and even ignominious dispute on the superiority of AC vs. DC.

George Westinghouse,† whose start in business came from his invention of the air brake for trains, had the vision to take AC seriously. The Westinghouse Company began, around 1885, to make equipment for high-voltage AC. Edison tried to discredit AC in the public's mind, but really succeeded in discrediting only himself. Edison and his associates showed public demonstrations of how stray dogs and cats, pushed onto sheets of metal connected to AC, were promptly electrocuted by the high voltage. Based on those demonstrations, which were dreadful enough, Edison then proposed using an electric chair to replace the standard (at the time) method of hanging condemned murderers. Of course, for this awful application, Edison enthusiastically recommended using AC and equipment supplied by Westinghouse. The first such execution by AC occurred in 1890, using Westinghouse equipment. The lowest and ugliest point of this squabble was reached

---

* In an AC system, the direction of current flow reverses periodically, 60 times per second (60 hertz, Hz) in the United States and 50 Hz in most other countries. In a DC system, the direction of current flow is constant. Nowadays AC is produced in large generating plants and transmitted to consumers, so that most lighting, household appliances, and industrial equipment operates with AC. Batteries and solar cells produce DC, which is used in nearly all electronic devices. When a battery charger or laptop computer is connected to a household electric circuit, the connection includes a block- or brick-like object called a rectifier, which converts AC from the wall socket into the DC needed by the device. Some very large-scale industrial processes also rely on DC; making aluminum from its ore is an example.

† George Westinghouse (1846–1914) is probably the only person who served in the National Guard, the Army, and the Navy during the American Civil War. He was inspired to invent the air brake by having witnessed a railway accident involving two trains that could not be stopped in time to avoid colliding. The Westinghouse air brake is one of the first inventions employing the "failsafe" concept; any disconnection or accidental breakage of the air supply anywhere in the train will bring the train to a stop.

by the suggestion from Edison and his allies that, instead of saying that a person had been electrocuted, we should say that he or she had been "Westinghoused." None of these negative campaign tactics had the slightest effect in counteracting the growing acceptance of AC.

The chief source of scientific and engineering expertise for Westinghouse and the commercial development of AC was a brilliant, colorful, somewhat bizarre (and possibly crazy) Croatian immigrant, Nikola Tesla* (Figure 14.12). As a student at Graz Polytechnic Institute in Austria, Tesla saw a class demonstration of a generator running "backward" as a DC motor. Parts of the motor were giving off sparks, and it seemed obvious that it would soon wear out. Five years later, walking with a friend in the Budapest city park, Tesla envisioned a simple way to produce a rotating magnetic field through a particular way of using multiple phase currents. This discovery was the heart of his AC motor.

Edison enjoyed many technical successes despite having little formal education and no skills in higher mathematics. Tesla did have an excellent education and mathematical ability, which may have been a decisive advantage in the development of AC systems.

> *If Edison had a needle to find in a haystack, he would proceed at once with the diligence of the bee to examine straw after straw until he found the object of his search. I was a sorry witness to such doings, knowing that a little theory and calculations would have saved him ninety percent of his labor.*
>
> **—Tesla**[9]

In 1876, Sir William Siemens[†] had suggested that the energy of the falling water at Niagara Falls could be used to operate generators that would supply electricity economically to New York, Toronto, Philadelphia, and Boston. In the 1890s, Tesla, with Westinghouse's financial backing, developed the key components of an AC electricity system. By 1895, the designs were in place for a complete generating system, and Westinghouse was awarded the contract for the new generating station to be installed at Niagara Falls. The plant transmitted electricity to Buffalo, about 35 kilometers away.

---

* Nikola Tesla (1856–1943) had the unusual ability to visualize a complete invention, right down to its dimensions, in his head before building it; he could actually make a device from the information in his head, without first preparing drawings or blueprints. Westinghouse offered to buy Tesla's patents for $60,000 plus a royalty of $2.50 per horsepower. Some years later, when Westinghouse had suffered serious financial setbacks and was in dire straits, Tesla agreed to accept a cash settlement of $216,000 instead of royalties. The lost royalties likely cost Tesla some $12 million at the time and possibly $50 million over his lifetime. Depending on one's point of view, Tesla's settlement was either an act of extreme loyalty and generosity to the man who helped him get started or, alternatively, one of the dumbest financial transactions ever.

† Born in Germany as Carl Wilhelm Siemens (1823–1883), he emigrated to England, changing his name to Charles William Siemens, and, upon receiving a knighthood late in life, became known as Sir William Siemens. Like many other great inventors of the nineteenth century, Siemens was both prolific and able to turn his talents to many different fields. In addition to numerous inventions pertaining to electricity, he is also the coinventor of the Siemens–Martin process for making steel.

FIGURE 14.12 Nikola Tesla (1856–1943), calmly sitting in his laboratory in Colorado Springs, Colorado, in 1899, nearly surrounded by electrical discharges. (From http:// en.wikipedia.org/wiki/File:Tesla_colorado_adjusted.jpg.)

The use of energy in cities has been marked by a constant displacement of burning fuels by electricity. Gaslights were replaced by electric lamps. Steam-powered factories were modernized with machines having individual electric motors. Steam locomotives were replaced by electric locomotives. Now, early in the twenty-first century, we see the beginning of a trend to replace gasoline-powered autos with electric cars. The incandescent bulb, which dominated lighting technology for about 150 years, is giving way to compact fluorescent bulbs and light-emitting diodes for illumination. These current trends are motivated by increasing awareness of the need to conserve energy and of the connections between conserving energy and protecting the environment. We will explore these themes in later chapters.

REFERENCES

1. Jehl, F. *Menlo Park Reminiscences*. Edison Institution: Dearborn, MI, 1937.
2. Carey, J. *The Faber Book of Science*. Faber and Faber: London, U.K., 1995, pp. 169–172.
3. Edison, T. Quoted in Jehl, F. *Menlo Park Reminiscences*. Edison Institution: Dearborn, MI, 1937, p. 854.
4. Murray, V. *An Elegant Madness*. Viking: New York, 1998, p. 278.
5. Reynolds, F. *Crackpot or Genius?* Chicago Review Press: Chicago, IL, 1993, p. 14.
6. Medawar, P. *The Strange Case of the Spotted Mice*. Oxford University Press: Oxford, U.K., 1996, p. 184.
7. Gratzer, W. *A Bedside Nature*. W.H. Freeman: New York, 1998, p. 91.
8. Tesla, N. Quoted in Davis, L.J. *Fleet Fire*. Arcade Publishing: New York, 2003, p. 199.

FURTHER READINGS

Adair, G. *Thomas Alva Edison: Inventing the Electric Age*. Oxford University Press: New York, 1996; Baldwin, N. *Edison: Inventing the Century*. Hyperion: New York, 1995; Israel, P. *Edison: A Life of Invention*. Wiley: New York, 1998. Edison biographies appear to come in waves. Any of these three books will provide details of Edison's life and inventions, and are easily readable with little or no technical background. The book by Paul Israel has numerous interesting illustrations from the original laboratory notebooks of Edison and his associates.

Cheney, M. *Tesla: Man Out of Time*. Barnes and Noble: New York, 1981. Probably the most popular biography of Tesla, written with a somewhat melodramatic flair. It tends to read much like a novel, and requires no scientific background.

Davis, L.J. *Fleet Fire*. Arcade Publishing: New York, 2003. This book provides a history of the so-called electric revolution, sweeping from Franklin to Marconi and radio transmission. Edison is a central figure, but Samuel Morse and the telegraph, the trans-Atlantic cable, and of course Nikola Tesla all figure into this fascinating story.

Glenn, J. *The Complete Patents of Nikola Tesla*. Barnes and Noble: New York, 1994; Martin, T.C. *The Inventions, Researches, and Writings of Nikola Tesla*. Lindsay: Bradley, IL, 1988; Tesla, N. *Experiments with Alternate Currents*. Lindsay: Bradley, IL, 1986; Tesla, N. *My Inventions*. Barnes and Noble: New York, 1995; Tesla, N.; Childress, D. *The Fantastic Inventions of Nikola Tesla*. Adventures Unlimited: Steele, IL, 1993. Most of these are inexpensive paperback reprints of books originally published more than a century ago. They are anthologies of Tesla's writings, including patents. A background in electricity would be needed to understand some of the material in depth. Nevertheless, these books provide overviews of some of the many concepts and devices developed by Tesla, an appreciation of which can be obtained even by glossing over the more technical passages.

Verne, J. 'In the twenty-ninth century: The day of an American journalist in 2889.' This short story can be found in various anthologies of Verne's work, of science fiction, or of French literature. One useful source is The Folio Society edition of *French Short Stories*. The city in which the journalist works is run totally by electricity; Verne's predictions of the state of technology some 900 years from now are interesting.

15 Electricity from Falling Water

THE ROTARY GENERATOR

Michael Faraday's discovery of the concept of continuous generation of electricity was the first crucial step on the way to large-scale availability of electricity. Converting Faraday's discovery to a practical working device is credited to several engineers, among them Zénobe Gramme. The generator is, in essence, as shown in Figure 15.1. In this simple generator, the conductor moves through the magnetic field because we turn it. In other words, we must do WORK on (or supply ENERGY to) the generator, for which we therefore have the "reverse energy diagram" shown in Figure 15.2. (Two points to keep in mind regarding these "reverse energy diagrams" are, first, as the name implies, it is "backward" from the energy diagrams introduced earlier, and, second, *both* arrows in the diagram have to change direction from one type of energy diagram to the other.) The entire essence of large-scale electric energy generation hinges on finding ways to spin the generator reliably and economically.

THE WATER TURBINE

Turning the generator reliably and, ideally, at low cost requires a source of ENERGY that itself is reliable and inexpensive. One source of cheap, reliable rotary motion is the waterwheel. In principle, WORK provided by the waterwheel could be used to operate a generator, as shown in Figure 15.3. Two changes occur in this SYSTEM. In the waterwheel, water flows spontaneously from high to low gravitational potential energy. When this happens, some ENERGY becomes available. The waterwheel is the device used to capture or extract a portion of this energy as useful WORK. The generator causes a nonspontaneous movement of electrical charge from low to high electrical potential energy—from low voltage to high voltage. This nonspontaneous change will happen only if we supply WORK *to* the generator. The WORK being supplied is that provided by the waterwheel.

Waterwheels used in medieval times were not very efficient devices for converting kinetic energy into WORK, because the water still possesses significant kinetic energy even after it leaves the wheel. The quest for more efficient waterwheels began around the time of the blossoming of the Renaissance. Leonardo da Vinci recognized that water flowing through *curved* blades bears against those blades with a force or impulse, illustrated schematically in Figure 15.4. The device that Leonardo invented was at first called an "impulse wheel." Nowadays it is generally

FIGURE 15.1   (a) The field magnet and (b) drum armature of a direct current generator.
(c) Armature and (d) field magnet of an alternating current generator.

known as a *turbine*. A turbine is a device that converts the kinetic energy of a
moving fluid into rotating kinetic energy. The fluid used in the turbine is called the
working fluid. For the kinds of turbine discussed in this chapter, the working fluid
is water. Later we will meet an extremely important family of turbines in which the
working fluid is steam.

In a turbine operated by water, the water flows through pipes or ducts over curved
vanes. To provide the impulse to the vanes, the turbine has to be enclosed in a casing.
The higher the pressure of water supplied to the turbine (essentially, the higher the

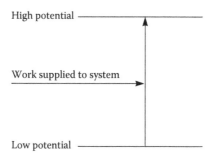

FIGURE 15.2   The generic reverse energy diagram, which is applicable to all SYSTEMS.

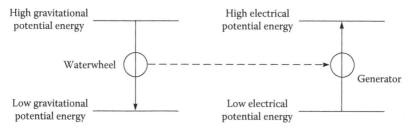

FIGURE 15.3   Hydroelectricity is based on the relationship between gravitational and electrical SYSTEMS as shown in this pair of energy diagrams.

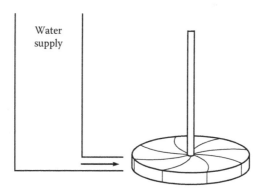

FIGURE 15.4   A basic water-operated turbine relies on a stream of water directed against the curved blades of the turbine.

column of water, or its "head"), the faster the turbine will spin. The rotational speed of a waterwheel is limited by the speed of the stream. This is not so with turbines, where it is easy to get rotational speeds as high as 30,000 revolutions per minute (rpm).

The development of the turbine during the mid-nineteenth century was a major engineering advance in providing a source of WORK for the electrical generator. The combination of a turbine and a generator can be represented as shown in

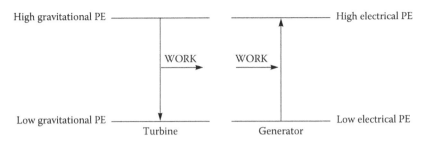

FIGURE 15.5 This pair of energy diagrams expresses the basic turbine–generator relationship.

Figure 15.5. This diagram is actually a generic scheme for generating electricity, regardless of the working fluid in the turbine. In the special case where the kinetic energy of water drives the turbine/generator set, the technology is referred to as *hydroelectricity* or hydropower. (The term hydroelectricity does not imply that the electricity is somehow different from electricity generated in other ways. The prefix "hydro-" is used simply to remind us that water is the fluid used to operate the turbine.)

The essential design feature of the turbine that allows it to capture the kinetic energy of the water is the flow of the water over curved vanes. ENERGY is captured by the "impulses" or "reactions" between the fast-flowing water and the passages created by the curvature of the vanes. These turbine designs will work only if water is confined, and therefore the passageways for the water created by the vanes are enclosed in an external closely fitting casing.

The Pelton turbine (Figure 15.6), suitable for large heads of water that have a low to moderate flow rate, goes back to the California gold rush in the middle of the nineteenth century. In those days, a common way of extracting gold was hydraulic mining. A jet of water at high pressure was used to flush the sand and gravel away. The water, from reservoirs upriver, was brought in pipes to the mining location. When the gold finally ran out, all of this system of reservoirs, pipes, hoses, and other paraphernalia was still in place. Rather than waste all of the effort that had been expended in setting up these water systems, piped water was used in some places to drive waterwheels. Lester Pelton, a mining engineer who had worked in the gold fields, designed and built a waterwheel with curved blades. One day, the water jet happened—possibly by accident—to strike the outer edges of the blades. So much kinetic energy was transferred to the wheel that its rotational speed increased to the point at which the wheel disintegrated. Pelton recognized that the kinetic energy of the jet could be best utilized when its direction was almost completely reversed by the blade.

The Pelton wheel hydroelectric plant at Fully, Switzerland, illustrates some of the remarkable design features of these installations. At the Fully plant, water is led downward from a lake through a pipe almost 5 km long. In that 5 km distance, the pipe drops in elevation by nearly 2 km. This very steep gradient results in the water, at the lower end of the pipe, having a pressure of 14 MPa—that is, a pressure of a

FIGURE 15.6   The Pelton wheel. This view emphasizes the curvature of the blades. (From http://en.wikipedia.org/wili/File:Peltonturbine-1.jpg.)

140 kilograms per square centimeter. The water leaves the nozzle at a speed of nearly 650 kilometers per hour. (To indicate some of the extraordinary properties of water under these conditions, it is estimated that a water jet coming out of a 7.5 cm pipe at a pressure of a quarter of the value in the Fully plant, 35 kg/cm^2, could not be cut through or deflected even by hitting it with a steel crowbar.) The Pelton wheels are 3.5 meters in diameter. Water at this pressure and velocity spins the wheels at a rate of 500 revolutions per minute.

The other major turbine design uses horizontal wheels. Like the Pelton turbines, they move at high speeds and are made of curved vanes mounted on a central shaft. The pressure of water flowing through these vanes causes the wheel to rotate. A well-designed and operated version of this turbine can convert more than 90% of the kinetic energy of the water into rotary motion, at high rates of revolution.

In the late 1820s, the French engineer Bénoit Fourneyron developed a small turbine (Figure 15.7) capable of producing about 4.5 kW. Fourneyron's design featured two concentric horizontal wheels. The fixed inner wheel had curved guide vanes that directed the water against runner blades on the outer wheel, the rotor. Over the next 10 years, he had refined his design to the size of a 45 kW unit turning at 2300 rpm. Remarkably, it was only about 30 cm in diameter and weighed about 18 kg. And, its efficiency of converting was about 80%, that is, it captured about 80% of the kinetic energy of the water for operating a generator.

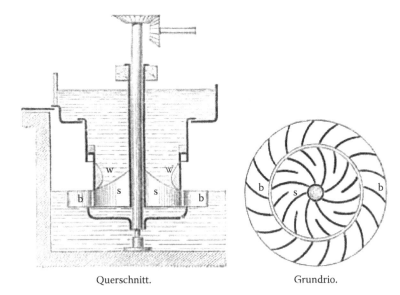

Querschnitt.   Grundrio.

FIGURE 15.7 A Fourneyron turbine. (From http://nl.wikipedia.org/wiki/Bestand:
Fourneyron_Turbine01.jpg.)

Fourneyron turbines were installed in the hydroelectric plant at Niagara Falls, on
the U.S.–Canadian border, in 1895.

An Anglo-American engineer, James Francis,* compared the performance of a
Fourneyron turbine with one that he had designed himself. To try to ensure that the
results obtained from the two turbines could reasonably be compared, both were
installed at the same site—a factory in Lowell, Massachusetts. The information
obtained on turbine design and performance from this series of head-to-head trials
led to the design of what has come to be called the Francis turbine. It became, by
the early twentieth century, the most commonly used turbine in hydroelectric plants.

Nowadays most of the world's large hydroelectric plants use Francis turbines. The
world's largest hydroelectric plant, the Three Gorges plant in China, represents probably the largest engineering project of any kind ever undertaken anywhere. It has an
intended generating capacity of 22,500 MW. To put this figure into perspective, the
output of the Three Gorges hydroelectric plant is some twenty times larger than that
of a large coal-fired or nuclear electrical plant.

---

* James Francis (1815–1892) was born in England and emigrated to the United States at age 18. One
of his first contributions to water engineering was the invention of the sprinkler system now widely
used in buildings for fire protection. Francis's design involved a series of perforated pipes running
throughout the building. It had two defects: it had to be turned on manually, and it had only *one* valve.
Once the system was activated by opening the valve, water would flow out everywhere. If the building
did not burn down, it would certainly be completely flooded. Only some years later, when other engineers perfected the kind of sprinkler heads in use nowadays—that turned on automatically and were
activated only where actually needed—did the concept become popular. Later in life, Francis put his
accumulated knowledge of water systems to use in his service on the committee that investigated the
catastrophic Johnstown (Pennsylvania) flood that killed more than 2000 people.

At the other extreme of scale from plants like Fully and Three Gorges are many small-scale hydroelectric plants. Such installations have various names, intended to convey a notion of their small size. Customarily, small hydro plants generate less than 10 MW; mini-hydro, less than 1 MW; and micro-hydro, less than 100 kW. All use the same basic concepts of turbine design and operation. Micro-hydro projects are popular because their output is sufficient for a small community or small commercial installation. At the small end of the micro-hydro range are units that would serve one home or a few homes. Small-scale projects that don't require constructing dams to form large reservoirs of water represent an aspect of hydropower that is growing in popularity around the world. (As discussed later, large-scale hydropower projects seem to have lost most of their luster and are under attack in many places.) Small hydro plants that can feed electricity into a regional or national grid represent a source of relatively inexpensive, renewable energy. This can be attractive in regions or countries that now have mandates requiring the generation of a set fraction of electricity from renewable resources. Some countries in the developing world find the installation of "mini-hydro" plants to be an attractive alternative, economically, to expanding nationwide electricity grids to connect a few large, expensive electricity-generating stations. Micro-hydro units can be set up to generate electricity in rural regions that have no connections to an electric grid. They can also be installed by those who choose to disconnect from the electric grid. Vendors in several countries now offer virtually complete commercial packages for those who want a source of electricity or who want to live "off-grid." Many can be installed as a do-it-yourself project.*

HYDROPOWER

Worldwide, hydropower provides about 16% of electricity demand, equivalent to about 4% of *total* energy demand. World electricity production from hydropower plants is about 3400 billion kWh. In terms of total production of hydroelectricity, China easily leads the world, producing almost as much as the next two countries, Canada and Brazil, combined. But in terms of the percentage of total electricity supplied by hydro, Paraguay is the world leader. Hydroelectricity supplies 100% of Paraguay's electricity demand. Paraguay is well endowed with hydro resources and has a relatively small demand, due to its small population. Consequently, Paraguay exports almost 90% of the electricity generated within the country to its neighbors. Norway, with about 98% of its electricity from hydro sources, runs a very close second to Paraguay.

Hydropower has a number of advantages. Compared with fossil fuels, hydropower is renewable, that is, highly unlikely to be depleted on any human time scale. While we will likely someday experience shortages of fuels, barring the most extreme predictions for the effects of global climate change, there is little likelihood that rivers will dry up. Water will continue to flow through turbines. Hydropower is highly efficient. About 90% of the kinetic energy of water is converted to electrical energy

---

* A search of the web will turn up firms offering micro-hydro packages that can be installed as do-it-yourself, or DIY, projects. As a cautionary note, it should always be remembered that DIY has an alternative: HAP—hire a pro.

(as compared, e.g., to 30%–35% when electricity is generated from chemical energy in coal-fired power plants). Hydropower is highly reliable mechanically, since the only moving parts are the turbine and the generator. As a result, maintenance costs are low. Without doubt, hydropower plants are proven technology. Unlike fossil fuel or nuclear plants, there are no waste products and virtually no pollution, including no carbon dioxide emissions. Since there are no waste products, there are also no recurring costs for waste disposal. Another contrast to fossil fuel or nuclear plants is that the energy source, the water, is free—there is no recurring cost to buy fuel. If opportunities exist for the development of hydropower, it is still possible for regions or countries that do not have their own indigenous fossil fuel resources to develop heavy industries on a large scale. From the plant operator's perspective, an important advantage of hydropower is that it can easily be ramped up and down, or even switched on and off entirely (by diverting water onto or away from the turbines), making it an excellent choice for providing extra electricity at times of peak demand or in energy emergencies.

Another virtue of hydropower is it is economic. In most places, hydropower has always been one of the least expensive sources of electricity. For example, some of the large hydropower installations built in the United States during the middle of the twentieth century today generate electricity at a consumer cost of about 4.5 cents/kWh. This is roughly half of the national average cost of electricity. The initial capital investment required to build a new hydropower installation can be very high. Large plants may cost about \$86/MWh of capacity. This is slightly less than a typical coal-fired plant, which might be about \$95/MWh. However, once the plant is built and running, the recurring costs for operations and for maintenance are usually quite low. An alternative to building an entirely new plant is to augment the capacity of an existing one (sometimes referred to as "uprating"); where possible, this is a cheaper option than building a new plant.

Estimates suggest the potential of producing five times the total output of all present power stations. While this may sound very attractive and promise a great future for hydro, many of the potential sites for new development are places that are inaccessible, far from potential consumers and markets for electricity, and often both. In many of the developed countries, there is growing public concern about the environmental effects of constructing dams to provide hydroelectric sites. In some Third World nations, construction of dams and their large reservoirs could cause many people to have to be displaced from their homes. With these issues taken into consideration, it is likely to be the case that no more than about half of the world's potential hydro capacity will ever really be developed.

The first-ever hydroelectric plant was installed in Cragside, England, the home of the industrialist William Armstrong.* It provided electricity for exactly *one* lamp,

---

* William Armstrong (1810–1900) made a fortune and earned a baronetcy, as a manufacturer of armaments, notably various kinds of cannons. His firm sold weapons to both sides in the American Civil War of 1861–1865. Though Armstrong himself was gone from the scene, his firm also helped to arm the Japanese fleet that won what was likely the most lop-sided naval battle in history, the 1905 battle of Tsushima. A Russian battle fleet spent months laboriously steaming some 33,000 km from the Baltic Sea to the Strait of Tsushima, only to be annihilated in two days by a Japanese fleet having superior weapons, superior seamanship, and superior coal in its ships' bunkers.

but soon extended through the entire home. Practical hydropower for regional or on-site generation results from the confluence of two critical developments: the generator, such as the Gramme machines; and the turbines of Pelton and Francis. Developments came quickly in the late nineteenth century: An installation at Lisleby Brug, Norway, in 1877; the Schelkopf plant in Niagara Falls in 1881 (Figure 15.8); the Minnesota Brush Electric plant in Minneapolis (Figure 15.9), providing electricity

FIGURE 15.8 A view of the Schelkopf electric generating plant at the base of Niagara Falls, taken from the Canadian side of the falls. (From *The Niagara Falls Electrical Handbook: Being a Guide for Visitors from Abroad Attending the International Electrical Congress*, St. Louis, MO, September, 1904. Niagara Falls, NY: Published under the auspices of the American Institute of Electrical Engineers, 1904. p. 36; https://archive.org/details/niagarafallsele00engigoog.)

FIGURE 15.9 The Minnesota Brush electric plant. (From http://nokohaha.com/category/sky-blue-waters/page/9/.)

to the Pillsbury flour mill in 1882; and three weeks later, Edison's first hydro plant, in Appleton Wisconsin. Edison's Appleton plant was the first installation in the United States supplying hydroelectricity for domestic use. By 1885, a hydro plant in Canada was supplying electricity to Quebec City; in the same year, one in New Zealand was operating machinery at a mine. The third key piece was Tesla's invention of the AC generator, because AC is much more suitable for transmission over long distances. Within a year, Tesla's 1890 invention was followed by the first long-distance (180 km), high-voltage (8 kV) AC transmission line, in Germany. By then, the technology was in place for the development of the major electrical systems of the twentieth century.

By 1895, three 3725 kW generators had been installed at Niagara Falls, producing electricity at 2200 volts. The local business community and civic boosters made this "electrification" of the falls a major theme of the 1901 Pan-American Exposition in Buffalo. By 1905, the two generating stations installed at Niagara Falls produced one-tenth of all the electric energy generated in the United States. Much of this electricity was consumed by the industries in Buffalo, only about 50 km away. The early days of electricity generation at Niagara Falls are illustrated in Figure 15.10.

As a portent of things to come later in the century (and discussed in the next section), not everyone saw the hydro plants at Niagara as an unqualified benefit. The technical success of the plants soon led to plans to install even more

FIGURE 15.10   The generating capacity of the Niagara Falls plant is evident in this long row of generators run by the turbines. (Reprinted from first edition, courtesy of Hagley Museum and Library, Wilmington, Delaware.)

generating capacity there. Soon some segments of the press were predicting the possibility of the falls essentially drying up, because of the diversion of so much water to the hydro plants. Remarkably, by 1900, New York's state government had granted permits to companies that would, in principle at least, use essentially *all* the water flowing over the falls. Some developers went so far as to propose simply converting the entire falls to economic uses. The potential of using all the water was a real one. For example, by 1908, one company was pulling 225 cubic meters of water *per second* from the river. This enormous diversion of water had the effect of lowering the water level in the river by about 15 cm. If all of the other companies had actually been able to proceed as planned, the likely effect would have been to reduce the crest of the beautiful Horseshoe Falls (Figure 15.11) from 900 meters to 490 meters, a drastic change that would have shifted these falls totally to Canada.

In the United States, the first transmission system intended to carry electricity from a hydroelectric plant was built by the Westinghouse Company in 1891. It carried electricity from the generating plant at Willamette Falls, Oregon, to the nearby city of Portland. The main transmission line from the plant to the city operated at 3300 volts. In Portland, transformers reduced the voltage to 1100 for distribution within the city. In those days, voltage standards had not yet been developed, so depending on the requirements of the local householders, voltages were reduced further, either to 50 or 100 volts. The previous chapter mentioned the "battle of the currents" between Edison, favoring DC, and Westinghouse and Tesla, who supported AC. In fact, it was Tesla's AC system that made it possible for Westinghouse to construct the transmission line between Willamette Falls and Portland; at the

FIGURE 15.11 The Horseshoe Falls, at Niagara Falls, one of the world's best-known waterfalls. (From http://en.wikipedia.org/wiki/File:Horseshoe_Falls_2006.jpg.)

time, there was no economical way to transmit DC that distance (21 km). The victory for Westinghouse and Tesla in this "battle" greatly facilitated the growth of hydroelectricity. AC transmission systems made it possible for electricity to be generated at hydro plants, which might be some distance from large markets for electricity and yet have the electricity transmitted reasonably economically to those markets.

In the 1910s and 1920s, electric companies in several countries interconnected their generating plants by means of transmission lines. Of course, the motivation was for the companies each to extend their service areas. We now use the terms "electric grid" or "power grid" to describe this interconnection of transmission lines. The development of the grid, which by now essentially interconnects entire nations, provided numerous benefits. First of all, grids provide a way for the most efficient (and therefore most economical) plants to generate most of the electricity. Second, an interconnected grid increases the chances of supplying energy to consumers even if one plant happens to shut down unexpectedly. The grid allows incorporating the electricity-generating capacity of hydro plants, even if they are located far from places of high demand for electricity. On the other hand, if some technical problem, major weather event, or terrorist attack brought down a nation's entire electric grid, the effects would be truly calamitous.

These facts lead to the question: If hydropower is so good, why not convert all of our electricity-generating capacity to hydro? As one problem, most of the suitable hydropower sites have already been developed. Though it may seem that the contribution of "only" 16% of the world's electricity by hydropower is a small fraction, that number represents the worldwide generation. At least sixty countries obtain more than 50% of their electricity from hydropower, and a few, such as Norway, get close to 100%.

Second, many countries face growing public concern over the environmental effects of building dams for water reservoirs for hydropower generation. Dams can have very negative impacts on some species of endangered fish and other wildlife. The Swiss, for example, have expressed concern about the environmental impact of dam construction on some of the pristine valleys in the Alps. Opposition to the proposed Narmada Sarovar hydroelectric project in India has sparked what may have been the largest demonstrations in human history. This issue of dams will be explored in the next section. Because of the public concerns, in many countries it does not seem likely that there will be lessening of regulatory or licensing provisions affecting hydroelectricity. In fact, the total hydro capacity could *drop* in the future. There will likely be increasingly fierce competition for use of water, with needs for drinking water and agricultural irrigation contesting with its use in hydro plants. This makes a potentially explosive issue in regions where a major river flows through more than one country, and questions of water rights become international squabbles. Some serious observers believe that wars over water will replace wars over oil as a major source of conflict in the world. These same concerns are evident in other nations. In Switzerland, for example, about 15% of per capita energy consumption comes from hydro plants. If global climate change (Chapter 29) causes climate shifts to reduce rainfall, then water resources could become even more scarce. A decrease in hydro capacity due to global climate change would be a sadly ironic situation, because hydroelectricity generation is one of the technologies that does *not* produce emissions thought to contribute to climate change in the first place.

DAMS: PROS AND CONS

Not many rivers have the fast currents or significant waterfalls needed to provide an adequate head for a turbine. We can solve that problem, though, by creating an artificial head, by building a dam. Dams are by no means new. The documentary history of dams can be traced at least to 1177. Even in the Middle Ages there were arguments and lawsuits over dams, for illegally raising the height of a dam (and thus reducing the water flow downstream) and for interfering with navigation. The building of dams has long been a subject of literature:

Daily they would vainly storm,
Pick and shovel, stroke for stroke;
Where the flames would nightly swarm
Was a dam when we awoke.

—Goethe[1]

Dams provide a number of significant benefits: They create a source for hydropower generation. They control seasonal fluctuations in water flow, helping navigation and downstream flood control. The newly created artificial lake can be used as a source of recreation, or the water in it can be used for agricultural irrigation. However, barring some very major changes in public perception and public policy, it remains questionable whether hydropower will expand into a significantly larger share of the world's electricity generation.

In the 1930s and 1940s, dam building was a highly praised activity, in countries as different as the United States and the Soviet Union. Dam-building projects were seen as a way of bringing electricity to many rural communities and farms, and the availability of cheap electricity would foster the growth of industries and create jobs. This was especially important during the Great Depression of the 1930s. American folk singers of that era, such as Woodie Guthrie and Pete Seeger, who were politically on the left, had several songs about—and in favor of—building dams.

And on up the river is Grand Coulee Dam
The mightiest thing ever built by a man
To run the great factories and water the land
It's roll on, Columbia, roll on

—Guthrie[2]

This particular example concerns the Grand Coulee Dam on the Columbia River (Figure 15.12).

Today there has been a swing of opinion. Environmentalists, many of whom are also politically on the left, now vehemently oppose dam construction. There are

FIGURE 15.12   This early picture of the Grand Coulee Dam, taken before it was upgraded in the 1980s, represents a large, centralized hydroelectric plant.

numerous counterarguments to the proclaimed benefits of building a dam: The artificial lake floods valuable farmland, or pristine wilderness. Dams do not last forever. Silt can accumulate behind the dam, eventually filling in the reservoir. If the dam fails, there is great potential for a catastrophic accident with large loss of life and property damage. For example, a dam failure in the Hunan province of China in 1975 cost the lives of 20,000 people.

The breaking of a dam is an event that can cause terror to anyone. Consider the quintessential American cowboy story:

> *"What's that?" Hopalong suddenly demanded, drawing rein and listening. A dull roar came from the dam and he instinctively felt that something was radically wrong.*
> *"Water, of course," Red replied, impatiently. "This is a storm," he explained.*
> *Hopalong rode out along the dam, followed by Red, peering ahead.*
> *"She's busted! Look there!"*
> *A turbulent flood poured through a cut ten feet wide, and roared down the other side of the embankment, roiled and yellow.*
> *"God almighty!" Red shouted. "The dam's broke!"*

> **—Mulford**[3]

Dams can be vulnerable to acts of war or terrorism. Likely the most spectacular example was the 1943 attack of the Royal Air Force on three dams on the Ruhr River, in the industrial heartland of Germany. Two of the dams were breached, with an estimated death toll of about 1600.

The turbines and generators themselves have a negligible effect on the environment. That is, the actual generation of hydroelectricity is a very environmentally friendly technology. However, the damming of a river or stream, which, in many locations, is a necessary prerequisite to provide the needed head of water, has a potentially major impact on the local ecosystem. Dams create an artificial reservoir—a pond or lake—where a flowing stream or river used to exist. Those aquatic plants and animals that are best adapted to free-flowing water may not be able to flourish in the relatively stagnant conditions of the reservoir. Similarly, new species, that *are* adapted to still or sluggish water, may move in. The relatively sluggish or stagnant water in the reservoir can potentially serve as a breeding ground for mosquitoes. Dams can block fish migration and destroy spawning grounds, affecting both commercial and sport fishing. For this reason, in some parts of the United States, such as New England and the Pacific Northwest, there is a growing effort to dismantle small hydroelectric plants to restore trout and salmon to the streams. Not only do the reservoirs potentially change the ecosystem in the river itself, but also, depending on the specific conditions where the dam is built, reservoirs can flood large areas of forest and wildlife habitat. In some cases, they can even flood valuable farmland or towns.

Dams allow accumulation of silt in the reservoir. These sediments, which may carry vital nutrients, accumulate in the reservoir instead of being carried downstream. This results in a potential impact on ecosystems downstream of the dam, as well as those affected by the upstream reservoir. The downstream situation is further aggravated because the water flowing out of the hydro plant may be warmer, have lower oxygen content, or both, compared to the normal river conditions.

The resulting pond or lake behind a dam also usually raises the water table behind the dam (as a result of seepage) and lowers it below the dam. Evaporation of water from the reservoir increases the amount of dissolved salt and minerals in the water. If the river has a high concentration of dissolved minerals to begin with—an example being the Colorado River in the United States—then this can become a serious problem.

As an example, the salt cedar or tamarisk tree (Figure 15.13) is a native of Asia, but was introduced to the United States in the late eighteenth century.* It spread only slowly during the nineteenth century. In the early years of the twentieth century, it began to move rapidly through the American southwest. Dams may have given the tamarisk a crucial advantage over competitor trees by suppressing the natural flooding of rivers. The tamarisk produces and releases its seeds on a different cycle than those of the native cottonwoods and willows. Natural flooding favors the seed cycle of the native trees; when this is suppressed, the tamarisk has a chance to take over. What had been brought to the United States as a pleasant tree for shade and for firewood has now become the dominant tree over more than 400,000 hectares. As a result, these tamarisks now soak up about twice the amount of water every year as

---

* There are about ninety different species of tamarisks. Most grow in semiarid regions including seashores and salt deserts. They are noteworthy for being able to resist drought and salinity in the soil. For example, the salt cedar (also called the French tamarisk) is actually planted along seacoasts to provide shelter or windbreaks. Tamarisks have a long association with human affairs, being mentioned at least three times in Homer's *Iliad*; in one case, the Trojan warrior Adrestos was captured by Menelaos when the horses pulling his war chariot ran into a tamarisk tree.

FIGURE 15.13 The tamarisk, or salt cedar, trees have become a common sight in the American southwest. The spread of range of these trees may be an unintended consequence of hydroelectric plants. (Reprinted from first edition, courtesy of Steven Dewey, Department of Plants, Soils, and Biometerology, Utah State University, Logan, Utah.)

used by all the cities of southern California. Not only did the tamarisk displace the native cottonwoods and willows, the dominance of the tamarisks created another problem as well. As they drop in the autumn, the tamarisk leaves release so much salt that the soil may not support other vegetation.

Since most—but not all—of the environmental problems associated with dam building derive from the creation of the reservoir, then ideally we should maximize the amount of electricity generated while minimizing the size of the necessary reservoir. One measure of estimating the environmental impact of a hydroelectric plant is the power produced per hectare of surface covered by the reservoir. The larger this indicator number, the smaller the likelihood of a serious impact on the environment; conversely, the smaller the number, the worse are likely to be the problems. The Balbina hydroelectric plant in Brazil is an example of a major environmental disaster; there, only about one kilowatt is generated for each hectare of reservoir surface. To put this number in context, another plant in the same country has an indicator value of 588 kW/ha, which indicates a substantial generation of electricity for a relatively small area covered by the reservoir.

But, it is fair to say that in most cases, when the construction of new dams was not perceived to be a threat to popular locations such as Niagara Falls, then there was actually little opposition. Hundreds of new dams were built in the early part of the twentieth century in many countries of the world. These dams were, in many cases, symbols of technological progress and of economic prosperity. During the Great Depression of the 1930s, the single project that received most public acclaim was the construction of the Hoover Dam on the Colorado River (Figure 15.14). In late 1930, Elwood Mead, then the Commissioner of Reclamation,

Hoover Dam
C45-300-021094

FIGURE 15.14 Hoover Dam, the best known of the numerous hydroelectric plants constructed in the United States in the 1930s. (From http://www.usbr.gov/lc/hooverdam/ gallery/damviews.html.)

noted that the difficult national economic conditions had caused "pressure for action on this matter, as a means of furnishing employment and encouraging a revival of business." When the dam was completed in 1936, President Franklin Roosevelt told the nation how badly these projects were needed to "create a new world of abundance."

The Hoover Dam was the engineering and construction marvel of the 1930s. Even the construction site became a tourist attraction. The pouring of the concrete began in June 1933 and continued round the clock for two years. In 1934 and 1935, the Hoover Dam construction site was as popular a tourist destination as the Grand Canyon. At the time, the Hoover Dam was the tallest dam anywhere in the world, at 220 meters above bedrock. The base of the dam is about 190 meters thick, and the top can accommodate three lanes of traffic. The four water intake towers are as tall as 33-story buildings. The finished dam used 2,300,000 cubic meters of concrete. The new reservoir covered an area of 600 square kilometers. The turbine and generator installation covers about 4 ha. The 17 turbines generate five billion kilowatt-hours a year. (The indicator value for Hoover Dam is about 10.) The dam was finished in March 1936, two years ahead of schedule. Since then, more than 27 million people have visited the dam. The American Society of Civil Engineers has declared it to be one of the country's Seven Modern Civil Engineering Wonders.

FIGURE 15.15   A portion of the Abu Simbel Temple on the Nile River, parts of which were salvaged by an international archeological effort prior to construction of the Aswan High Dam. (From http://en.wikipedia.org/wiki/File:Abu_Simbel_Temple_May_30_2007.jpg.)

In the 1960s, attention turned to the Aswan High Dam,* on the Nile River in Egypt. The motivations for constructing this dam were to help control the annual flooding of the lower Nile and to provide electricity for increasing industrialization and the standard of living in Egypt. The Aswan High Dam became a major geopolitical issue, with the Egyptian government playing off the United States and United Kingdom against the Soviet Union. This situation was further complicated by the eternal rivalries among the Arab states, as well as the poisonous Arab–Israeli antagonisms. Eventually, construction began with financial help from the Soviet Union. This touched off a second issue, because it became apparent that the new reservoir would flood priceless antiquities of the ancient Egyptian kingdoms, such as the Great Temple of Abu Simbel (Figure 15.15). A major international rescue effort was launched to relocate or save some of these relics.

The Aswan High Dam has indeed provided many benefits. These include the increased electricity production and improved navigation below the dam. Also, the dam helps protect the downstream regions from flooding. Overall, it has also contributed to improved agricultural production. On the other hand, accommodating the reservoir required the resettlement of over 120,000 people. Silt accumulating behind the dam reduces the amount of water that can be stored in the reservoir. Soil salinity increases. Worse, some of the irrigation system that utilizes water from the reservoir

---

* The name reflects the fact that there was an Aswan Low Dam, built by the British around 1900. In 1946, a major Nile flood nearly overtopped this dam.

has become a breeding ground for snails that carry and spread schistosomiasis. This is a chronic, albeit seldom fatal, disease that leads to fatigue, fever, diarrhea, and possibly may trigger cancer of the bladder.

After the Aswan High Dam had become a bone of contention between the West and the Soviet Union, Ghana found it much easier to get funding from the United States for the Volta River power project. Initially, this dam provided cheap electricity to an aluminum company that imported ore from elsewhere (rather than using Ghanaian resources) and shipped the raw aluminum to Europe for finishing. Rather little benefit accrued to Ghana.

The most monumental project (in more ways than one) is the Three Gorges Dam, on the Yangtze River in Hubei Province, China. It is the world's largest electric generating station (Figure 15.16). The main portion of the dam itself was completed in 2006. To compare with the Hoover Dam, the Three Gorges Dam used 27 million cubic meters of concrete. Once again, the motivating factors were the need for increased electricity production, improving downstream navigation, and flood control. An additional advantage comes from the production of large amounts of electricity (22,500 MW when operating at full capacity) with no carbon dioxide emissions. China has frequently been criticized, in the West, for its rapid industrial and commercial development with scant regard for the environment. The Three Gorges Dam provides a substantial contribution to China's energy economy with no increase in carbon dioxide or other emissions that would degrade air quality. Not only does the dam itself have no carbon dioxide emission, by increasing the ability of the downstream length of the river to accommodate more barge traffic (barges having no engines, but being towed by a ship), there is a further reduction in carbon dioxide owning to the increased barge traffic.

FIGURE 15.16 The Three Gorges Dam in China, the largest engineering project in human history. (From http://en.wikipedia.org/wiki/File:ThreeGorgesDam-China2009.jpg.)

Like the Aswan High Dam, the Three Gorges Dam has required relocation of significant numbers of people, in this case at least 1,300,000, which some suspect to be a low estimate. Like Aswan, the Three Gorges Dam has led to loss of antiquities, or a need for rescue operations to salvage some of this heritage of ancient China. As with other large dams, Three Gorges is having, and will continue to have, impacts on ecology. Of particular concern is the fate of endangered species such as the Chinese river dolphin and the Siberian crane. But possibly the most worrisome aspect is the fact that the dam is built in a seismically active region. Hairline cracks are already evident in the dam. If a major earthquake should strike the region, especially when the reservoir is full and holding a huge weight of water pressing against the dam, perhaps—one can only hope not—the dam would rupture. The resulting flood could be a catastrophe of unimaginable proportion.

Some observers have written off the construction of large hydroelectric projects as "yesterday's technology." This is because of the record of the huge dams for damage to the environment, for massive cost-overruns, failure to meet the promised amounts of electricity and water, and social disruption. Yet, on the other hand, there is a long list of projects in progress. About 100,000 MW of hydro capacity are in various stages of construction worldwide. Areas experiencing rapid development of hydroelectricity include the Mekong River basin in Southeast Asia, Iran, and South Africa, including the kingdom of Lesotho. The construction companies argue that their industry is more socially, environmentally, and financially responsible than indicated by its past record.

On the other hand, the international news is not all positive. Worldwide environmental protests resulted in (or at least contributed to) stopping the Narmada project in India. Similarly, the Cree people in Quebec organized an international campaign to halt a hydroelectric scheme on the James Bay. The Ilisu Dam is planned to be built in eastern Turkey in the heart of territory whose ownership is disputed by the Kurds. The dam will be built in the headwaters of the Tigris River just before it flows over the Turkish border into Syria and Iraq. This construction is contrary at least to the spirit, if not the letter, of the United Nations Convention on the Non-Navigable Uses of Transboundary Waterways. A dam project in Brazil was canceled due to worker riots.

As the future for hydroelectricity unfolds, it will be interesting to see if there will be a "Goldilocks solution"—one that is not too big, so would be a time bomb for environmental disaster; one that is not too small, so that thousands or millions would have to be constructed, but one that is "just right."

REFERENCES

1. Goethe, J.W. *Faust. Part Two.* Many editions of this work are available. The cited passage is from lines 11123–11126.
2. Guthrie, W. *Roll on Columbia.* Woodie Guthrie Publications, Inc. the Richmond Organization: New York, 1963.
3. Mulford, C. *Hopalong Cassidy.* Tom Doherty Associates: New York, Chapter XX.

FURTHER READINGS

Boyle, G. *Renewable Energy*. Oxford University Press: Oxford, U.K., 2004. This book is an edited collection of chapters dealing with various aspects of renewable energy, each written by one or more experts. Chapter 5 deals with hydroelectricity.

Klare, M. *Resource Wars*. Metropolitan Books: New York, 2001. This book offers an important and well-reasoned argument that international conflicts in the twenty-first century may involve access to scarce resources, including, but not limited to, oil. Chapters 6 and 7 relate to possible conflicts over water.

Macaulay, D. *Building Big*. Houghton Mifflin: Boston, MA, 2000. All of Macaulay's books are treasures, splendidly illustrated. Building Big has a section on dams, including Hoover Dam and the huge Itaipu station in South America.

16 Electricity from Steam

Prudence and justice tell me that there is more love for mankind in electricity and steam than in chastity and abstention from meat.

—Chekhov[1]

Bringing together the development of the water turbine with the dynamo allowed harnessing the kinetic energy of water to operate an electricity generator. Adding Tesla's AC generator paved the way to long-distance transmission of electricity. Together, these provided the foundation for today's regional, national, and international electricity grids. Hydroelectricity has many advantages, not the least of which is its being virtually pollution free. Yet there remain concerns about environmental and ecological effects of dam building; even if these concerns were mitigated, many parts of the world lack good sites for potential hydroelectric generation. In other words, it is imperative to find an alternative working fluid to use in the turbine, to avoid relying exclusively on hydro for large-scale electricity generation.

Such a working fluid has to satisfy a number of stringent criteria:

- It needs to be cheap and readily available.
- It needs to be easy to produce or generate, and easy to handle on a large scale.
- It should be reasonably safe to work with, and safe in the environment.
- It should have enough mass to provide impulse to the turbine.
- It should be easy to generate a sufficient head to run the turbine.

This is a very tough list of criteria to satisfy. The only material that meets these criteria is water. But having ruled out any further significant application of *liquid* water, that is, disregarding hydro, we must recognize that there is another useful form of water—steam. Steam is a form of water that can be made to flow easily and that can be practically anywhere where enough fuel is available. It satisfies the other criteria listed earlier. Essentially all of the world's electricity except for the amount generated by hydropower derives from steam-driven turbines.*

RECIPROCATING STEAM ENGINES IN ELECTRICITY GENERATION

By the late nineteenth century, the steam engine had been brought to a high state of development. The early solution of the problem of reliable operation of electricity

---

* Small quantities of electricity are generated from batteries, fuel cells, and from the direct conversion of solar energy. Though there is much hope for direct conversion of solar energy to become a major source of electricity—perhaps even the major source—in the future, at the present time, the contribution of these "nonsteam" sources to the world electricity supply is very tiny.

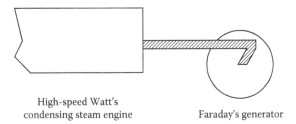

High-speed Watt's
condensing steam engine   Faraday's generator

FIGURE 16.1 A schematic diagram illustrating how a rotary generator could be operated by a reciprocating steam engine.

generators was to connect the steam engine to the generator. This is illustrated schematically in Figure 16.1.

Unfortunately, for this application, steam engines have several disadvantages: The reciprocating motion of the piston in the cylinder needs to be converted to the rotary motion of the generator. This is inefficient; for one thing, it usually produces significant vibration. Reciprocating steam engines are too slow. The best steam engine technology of Edison's day provided operation at about 350 revolutions per minute (rpm). This meant that the steam engine had to be "geared up" to operate a generator at, for example, the 1000–1500 rpm typically required for the equipment of that era. When the first central generating stations were built, Edison had a difficult time finding a firm that could build a reciprocating steam engine capable of running at 350 rpm. As the demand for electricity continued to grow, bigger and bigger and more and more complex reciprocating steam engines would have been needed to operate the generators.

As the continual improvement of electricity-generating and distribution systems made a public electricity supply increasingly practical, great efforts were made to design high-speed steam engines to drive generators. As the nineteenth century drew to a close, a major need of the rising electrical industry was for an engine that could both run at very high speeds and provide a smooth, vibration-free operation. A turbine in which steam is the working fluid—the steam turbine—meets both requirements. The steam turbine soon took over the job of driving electricity generators. (At about the same time, the steam turbine found a second important application: propelling high-speed ships.) By about 1910, the reciprocating steam engine had been almost totally displaced by the steam turbine in electricity generation application.

EARLY HISTORY OF THE STEAM TURBINE

In 1769, James Watt experimented with a ring-shaped "cylinder" in which a piston was to be driven around by steam. He could not get the device to operate, so abandoned the idea. However, in 1784, Watt was forced to reconsider the concept, since the Hungarian Baron Wolfgang von Kempelen patented a steam turbine, claiming that his engine was superior to Watt's and would easily push it off the market. Von Kempelen patented what was, in his words, a "reaction machine set in motion by Fire, Air, Water, or any other Fluid." Despite this grandiose all-encompassing claim (which is by no means unique in the patent literature), von Kempelen's machine was mainly intended to be driven by "boiling waters or rather the vapor proceeding therefrom [i.e., steam]." Watt said little about von Kempelen's engine on the presumption that even discussing the device might suggest to von Kempelen some needed improvements. Once Watt set to work analyzing this competitor and made a thorough examination of the steam consumption, the speed of rotation, and their relationship to the output of WORK, he concluded that von Kempelen's design was technically impossible. Indeed, von Kempelen is now best remembered for another of his remarkable inventions, the "chess-playing Turk."*

About thirty years after von Kempelen's attempt at a steam turbine, Richard Trevithick made a more serious and successful effort. He built a so-called whirling-engine that consisted of a pair of hollow arms mounted on a wheel set on a shaft. The wheel was about 4.5 meters in diameter. Steam at a pressure of 700 kPa escaped at a tangent from a small hole at the end of each arm, causing them to whirl around. The weakness of Trevithick's design was that the rotational speed of 250–300 rpm produced only about one-fifth of the WORK potentially available from the steam.

The first person to use successfully the energy of steam to produce rotary motion was Carl Gustaf Patrik de Laval, who developed his first invention, the milk separator, in the 1870s. To accomplish the separation of cream from milk, containers of milk had to be rotated at very high speeds, typically 6,000–10,000 rpm. These speeds cannot be achieved by a reciprocating steam engine. De Laval wanted a device that could be connected directly to the separator without the complexity of an intervening gear mechanism.

Fully convinced that speed is a gift from the gods, I dared, in 1876, to believe in the direct exploitation of steam on a power-generating wheel. It was a bold venture. At that time, only low speeds were employed. The speeds which were considered normal in later separators were never used. In contemporary handbooks they wrote about steam: "Unfortunately, the density of steam is too low to permit any thought of exploiting it in a power-generating wheel." In spite of this I succeeded.

—de Laval[2]

THE PARSONS TURBINE

The English engineer Charles Parsons (later Sir Charles, Figure 16.2) set about developing a steam turbine specifically for driving an electricity generator. The basic concept for his turbine was taken from the water turbine designed by Fourneyron.

---

* Toward the end of the 1760s, von Kempelen had built an automaton in the form of a male in Turkish dress, which sat beside a large, boxy chest, which had a chessboard on its top. Von Kempelen challenged famous chess players to play against the "Turk" and for many years toured Europe and the United States with his show. The android could also talk, though with a two-word vocabulary: "Check!" and "Gardez!". Von Kempelen claimed that in the android, he had managed to copy human intelligence and ability to reason. After attending a demonstration of the chess-playing Turk in London, Edgar Allan Poe wrote an essay denouncing any suggestion that the device had an ability to think. The fraud was eventually revealed—von Kempelen had hidden a dwarf, who was a skillful chess player, in the chest, and the android's machinery consisted simply of mechanisms that would allow the hidden dwarf to move the chessmen on the board.

FIGURE 16.2   Charles Parsons (1854–1931) invented a practical steam turbine and demonstrated its potential in his own yacht, the *Turbina*. (Reprinted from first edition, courtesy of Robert Hazen, Geophysical Laboratory, Carnegie Institution, Washington, D.C.)

Parsons' steam turbine used a sequence of fixed "guide" blades and moving blades. The steam flows between fixed guide vanes, so that it meets the blades of the wheel at right angles. The steam was admitted at the middle of the axle and flowed parallel to it in both directions, while delivering its kinetic energy to the turbine and expanding to atmospheric pressure. The electricity generator was directly connected to the turbine axle.

As steam passed through the series of rotors and fixed blades in Parsons' machine, the size of both the moving and fixed parts increased along the length of the turbine to correspond with the reduction in the pressure of the steam as it expanded. The flow of steam through the turbine was assisted by a condenser that drew off the exhaust steam. In the condenser, the steam, after doing WORK on the turbine (i.e., keeping it spinning), was condensed at atmospheric pressure. In earlier designs, the low-pressure, low-temperature "spent" steam had always been exhausted simply to the atmosphere. In Parsons' design, the condensed steam (i.e., liquid water) could be pumped away or pumped back to the boiler. This *condensing turbine* had smaller steam consumption than a conventional steam engine of equal POWER. It also provided advantages of space saving, greater reliability, and much lower vibration.

The first Parsons turbine, built in 1884, drove a generator that produced 7.5 kilowatts at 100 volts. A much larger Parsons turbine/generator set, working at 4800 rpm, was installed in 1888 at the Forth Banks (England) generating station. It had an initial capacity of 75 kW. The development of this machine for commercial application in the burgeoning electricity supply industry took place remarkably quickly, so that steam turbines were coming into widespread use in electricity supply stations by the end of the century. The general style of the Parsons turbine has not been significantly changed since the 1880s, though there are now many variants of this type. One noticeable change that has been made is a remarkable improvement in efficiency. The first Parsons turbine of 1884 had a steam consumption of 60 kilograms per kilowatt-hour. Nowadays the steam consumption per kilowatt-hour would be about 3.5 kg. In the course of time, the small Parsons unit of 1884, generating 7.5 kW, has evolved into the gigantic units in modern electricity stations that develop 450 MW at a fraction of the fuel consumption of earlier steam engines.

The turbine can be built on a larger scale than a reciprocating engine. The limits of the reciprocating engine were not important when it was used simply to meet the needs of a single factory installation. However, development of large centralized generating stations in the electrical industry created demands for WORK and POWER that the reciprocating engine could not meet. By 1900, the best marine engine had a maximum capacity of 6000 kW, reaching the limit of development of the reciprocating engine. In contrast, steam turbines for luxury ocean liners, discussed in the next section, were already at some 13,000 kW.

DIGRESSION: STEAM-TURBINE APPLICATIONS IN SHIP PROPULSION

Parsons was quick to recognize that, with adequate gearing *down*, his turbine could be a very effective method of propulsion for ships. In the eternal tradition of bureaucracies the world over, the British Admiralty was reluctant to admit the potential of ship propulsion by steam turbines. Parsons convinced them—and marine engineers in other parts of the world—by an unsolicited but sensational demonstration at the Jubilee Naval Review in 1897. Parsons' own experimental turbine-propelled launch, the *Turbinia*, sped past the comparatively slow-moving naval ships at the then-incredible speed of 34 knots.

Parsons crowned his achievement of providing a turbine suitable for the rapidly growing electrical industry by making the first successful application of the steam turbine to ship propulsion. Within a decade, steam turbines were being extensively adopted for both naval and merchant ships. The luxury liner *Mauretania*, launched in 1906, developed a speed of 27 knots. The *Mauretania* and her ill-fated sister ship *Lusitania*,* built in 1907, had turbines directly connected to their screws. There

---

* The *Lusitania*, a British passenger liner, was en route from New York to Liverpool in May, 1915. The ship was carrying a load of munitions for the British war effort in addition to her crew and the passengers. On the night of May 7, she was struck by a torpedo from the German submarine U-20. Although the incident occurred only a few miles off the coast of Ireland, the *Lusitania* sank so quickly that only about 800 people of the nearly 2000 passengers and crew could be rescued. The casualties included 128 American citizens. The incident led President Woodrow Wilson to insist that Germany should not attack passenger or merchant ships without making some kind of provision for the safety of the people on board. The resumption of unrestricted submarine attacks by Germany in early 1917 ultimately led to American involvement in the First World War. The sinking of the *Lusitania* was the beginning of a two-year-long process that eventually dragged the United States into the war.

were four turbine units, of about 13,000 kW each, in each vessel. The success of the turbines in these two gigantic, luxurious ships paved the way for the general installation of turbines in other vessels. The *Mauretania* was one of the last large hand-fired ships, requiring a crew of 324 firemen and "trimmers" to cart and shovel a thousand tons of coal per day into her boilers. The *Lusitania* and the *Mauretania* reduced the transatlantic travel schedules from seven or eight to five days per crossing.

Most large ships for the next generation of ocean liners and naval vessels were equipped with turbines. Steam turbines were eventually displaced by diesel engines for most purposes of marine propulsion, but they remain in use in some larger vessels. However, the steam turbine is still supreme in generating much of the world's electricity. Those turbines use inlet steam at about 600°C and 30 MPa pressure, which expands in twenty or more stages of expansion to generate nearly 56,000 kW. The efficient, high-speed operation of a steam turbine in any application still requires a way to generate a "head" of steam.

ROBERT BOYLE AND THE BEHAVIOR OF GASES

One of the great scientific developments of the sixteenth and seventeenth centuries was that scientists began to understand the properties of gases. Charles and Gay-Lussac showed that, at constant pressure, the volume of a given quantity of gas is directly proportional to its temperature. In the early 1660s, two English scientists, Henry Power and Richard Towneley, observed that the pressure and volume of a gas are inversely related, provided the temperature is held constant. This relationship can be expressed in mathematical terms as

$$\text{Pressure} \times \text{Volume} = \text{Constant}$$

Their work was known to several of the eminent scientists of the time, including Robert Boyle, Robert Hooke, and Isaac Newton. In France, Edmé Mariotte, presumably unaware of Boyle's publications, published this same relationship without giving credit to Boyle, let alone to Power and Towneley. By a quirk of history, this law of gas behavior is known by Boyle's name in English-speaking countries, and by Mariotte's name in France, but it really should be known as Power and Towneley's law.

Robert Boyle (Figure 16.3) enjoyed a vigorous scientific career during the seventeenth century, among other things promoting the idea that the scientific study of nature was actually a religious duty. Boyle poured mercury into a long open tube whose end was shaped like the letter J, as illustrated in Figure 16.4. By doing so, the mercury in the tube trapped some of the air in the short closed side. By adding more mercury, he raised the pressure on the trapped air. The trapped gas was always able to support the column of mercury on the long open side once its volume had been reduced by the appropriate amount, so that the pressure it exerted was equal to the pressure exerted upon it. It was relatively easy to measure the pressure increase, by measuring the difference between the lengths of the two columns of mercury in the open and closed sides. He found that doubling the pressure on the trapped gas roughly halved its volume, tripling the pressure reduced the volume to a third, and so on. Furthermore, this effect could be reversed: A gas whose volume has been

ROBERTVS BOYLE NOBILIS ANGLVS

FIGURE 16.3   Robert Boyle (1627–1691) was one of the leading figures of the scientific revolution of the seventeenth century, emphasizing the importance of experimental work. He is best known now for the law showing an inverse relationship between volume and pressure of a gas held at constant temperature. (From http://sceti.library.upenn.edu/sceti/smith/ scientist.cfm?PictureID=1988&ScientistID=52.)

compressed by increasing the pressure can be made to expand again by lowering the pressure, and vice versa. This is why we can say that, for a given quantity of gas, the pressure is inversely related to the volume, so as one goes up, the other goes down. Therefore, the product of the two remains constant.

Charles and Gay-Lussac showed that a volume of gas will expand if heated and contract if cooled. So, in studying the effect of changes in pressure on volume, one would have to be sure to keep the gas at constant temperature. Otherwise, volume changes would take place as a result of two effects—one of pressure, that is, the parameter under investigation, and the other caused inadvertently by changes in temperature. The specific effect that one was trying to isolate for study (in this case, the role of pressure in changing the volume of a gas) would be confounded by a second effect—the effect of temperature. In other words, one would likely measure volume changes for which pressure was not solely responsible. Boyle did not, apparently, take note of this fact and therefore may not have appreciated its significance. Mariotte did

FIGURE 16.4 A more modern version of Boyle's J-tube experiment that helped establish the gas law named in his honor. (Image originally appeared in Elementary Treatise on Physics: Experimental and Applied, 17th edition, edited by A.W. Reinhold, William Wood and Company, New York, 1905.)

draw attention to the importance of making measurements at constant temperature. So, calling the relationship of pressure and volume Mariotte's law rather than Boyle's law is not simply an act of European, or French, chauvinism, but actually there is some justice in our doing this. But as for Power and Towneley—well, they're history.

Why does Boyle's law "work"? The explanation derives from the molecular nature of gases. Think of operating a small hand-pump of the kind used to inflate bicycle tires, sports balls, or air mattresses. If you give the piston a quick shove inward, and let go, you will notice that it jumps back at you. This seemingly odd behavior is a manifestation of what Boyle called "the spring of air." In the early days of science, in Boyle's era, no one knew of the existence of atoms or molecules and their behavior. Some of the early scientists thought of the particles of air as having actual little springs on them (illustrated by the diagram in Figure 16.5), and when measuring the spring of air, they thought they were measuring the springs that held the particles apart in the air.

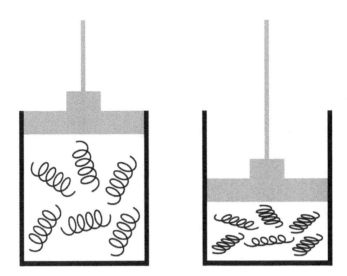

FIGURE 16.5 A cartoon diagram illustrating the concept of gas molecules being tiny springs, the idea that suggested an explanation for the relationship between pressure and volume.

The first valid explanation was formulated by James Clerk Maxwell, a brilliant theoretical physicist, in the 1870s.* Pressure exerted by a gas confined in some container is actually due to the bouncing of gas particles off the walls of the container. As it collides and bounces, each particle subjects the wall to a tiny force. The total force exerted by all the gas molecules colliding over a unit area is the pressure. That is, what we observe macroscopically as a "pressure" is the result of an enormous number of separate "pushes" against the walls of the container by each of the gas molecules that collide with the walls. There are so many of these "pushes" in an instant of time, and each separate push is so tiny that the total effect is sensed on a macroscopic scale as a smooth even pressure. Since the gas molecules are able to move freely and randomly in all directions, the observed pressure is equal in all directions.

Suppose that the sample of gas is in a container that is enclosed at the top by movable frictionless piston. (For our purposes, we will assume that the piston itself has no weight, and we will neglect any possible friction between the walls of the container and the piston.) The pressure is caused by the force of the particles bouncing against the walls of the container and the underside of the piston. Assume that the top of the piston has been stacked with just enough weights resting on it to balance exactly the gas pressure. If one of those weights is removed, the external force pressing down on

---

* It is as difficult to rank scientists as it is to choose the all-time greatest athlete in a particular sport. But by any reckoning, James Clerk Maxwell is among the elite of the nineteenth-century physicists. In fact, he is probably the greatest of theoretical physicists in the era after Newton and before Einstein. Among his numerous contributions include Maxwell's equations, a set of four equations that form a complete description of the relationships between electric and magnetic fields; Maxwell's relations, another set of four equations that are central to the science of thermodynamics; and the imaginary creature called Maxwell's demon, which, by segregating fast-moving molecules from slow ones in a container of gas, could bring about a flow of heat.

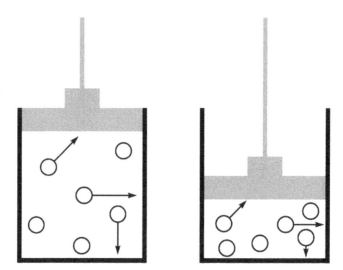

FIGURE 16.6 Individual gas molecules are able to travel further before colliding with the walls or with the movable piston when the volume is increased.

the top of the piston is necessarily decreased. Now, the upward force of the bouncing gas molecules is greater than the downward force of the piston, and the piston moves upward. However, the piston moving upward causes the volume of the container to increase. As the volume of the container increases, each molecule of gas has to travel a greater distance to reach the underside of the piston. Because the molecules must travel further to collide with the bottom of the piston, the number of collisions in any given instant of time must necessarily decrease, as shown in Figure 16.6.

This decrease in the number of collisions per instant is a direct consequence of the fact that each gas molecule has to spend more time traveling and, therefore, less time colliding. Since the measured pressure is the result of the collisions of the gas molecules against the container walls and the bottom of the piston, we observe that the pressure decreases. In fact, the pressure will drop to the point at which it is exactly balanced by the remaining weights on the piston, and when that happens, the piston will no longer rise. The volume has increased, and the pressure has decreased. This inverse relationship of pressure and volume is exactly as described by Boyle's law.

In the same way, adding weights to those originally present on the piston will cause the piston to move downward against the force of the particle collisions. As the piston moves downward, the volume of the container decreases. Now, each gas molecule has a smaller distance to travel to reach the bottom of the piston. Less time is spent traveling and more time is spent colliding; the number of collisions in any given instant rises, and we observe that the pressure increases. Eventually, the pressure will increase to a value at which the additional weight placed on the piston is exactly balanced by the gas pressure. In this example, volume has decreased and pressure has increased. Once again, the inverse relationship of pressure and volume holds, exactly in the manner described by Boyle's law.

Boyle gets credit for establishing the first *quantitative* law describing the measurable properties of matter. He gave to science and engineering a mathematical formulation that allows predictions of how the volume of a gas will change to compensate for a change in its pressure, or how the pressure will change to compensate for a change in volume. This legacy is remarkably ironic in that Boyle himself expressed serious reservations about mathematical formulations or expressions of physical phenomena, and his written records of his own experimental work contain few, if any, such mathematical representations—including the "law" relating the pressure and the volume of air for which he is best known. This is a "law" that Boyle never called a law and to which he never personally gave symbolic mathematical expression.

Boyle and Mariotte showed that for a fixed quantity of gas at constant temperature, the product of pressure and volume is a constant. From the work of Charles and Gay-Lussac, we know also that the volume of a fixed quantity of gas held at constant pressure is proportional to its temperature. By combining the discoveries of Boyle and Charles, we find that an increase in temperature of a fixed volume of gas must increase its pressure. Thus Boyle gives us the key to creating a head of steam—raising a fixed volume of steam to a high temperature. The high-pressure head of steam entering the turbine therefore will also be at high temperature.

TURBINES AS HEAT ENGINES

To obtain high pressures of steam to operate a turbine, we heat it to high temperatures. As the high-pressure steam goes through the turbine, it expands, that is, increases in volume. And if the volume increases, the pressure must drop. The essence of a steam turbine is illustrated conceptually in Figure 16.7. We have already seen that when a SYSTEM changes temperature, we can—by inserting the right device—extract WORK, as in the energy diagram of Figure 16.8. The generic device for extracting WORK from this spontaneous change in temperature (i.e., thermal potential energy) is a *heat engine*. The steam turbine represents one example of a heat engine.

The *efficiency* of operation of a heat engine is given by the equations

$$\text{Efficiency} = \frac{T_{\text{HIGH}} - T_{\text{LOW}}}{T_{\text{HIGH}}}$$

or

$$\% \text{Efficiency} = 100 \times \frac{T_{\text{HIGH}} - T_{\text{LOW}}}{T_{\text{HIGH}}}$$

where the temperatures are *always* expressed in kelvins. Ideally, of course, we would like the efficiency of any device to be as close to 1 (or 100%) as possible. But from the discussion in Chapter 11, it is impossible to attain a temperature of 0 K (absolute zero), so that

It is impossible to attain an efficiency of 100% from any heat engine.

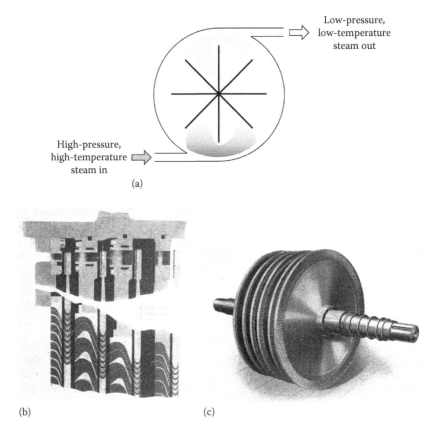

FIGURE 16.7 As steam passes through a turbine, its pressure and temperature drop, creating an expanded volume. (a) is a cartoon diagram of a simple steam turbine; (b) is a cutaway drawing of a turbine; and (c) shows the rotating blades and shafts of an actual turbine.

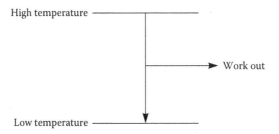

FIGURE 16.8 The energy diagram for a thermal SYSTEM also applies to the steam turbine.

In addition, it is generally much simpler and less costly to heat something than to cool it below ordinary temperatures. (As a quick example, compare the prices of a stove and a refrigerator of comparable quality of materials and workmanship.) Therefore, the approach to achieving high efficiencies is to make T_{HIGH} as high as possible. In other words, the higher the high-temperature side of the turbine, the greater will be its efficiency.

This concept of efficiency derives from the work of Sadi Carnot. Efficiencies calculated in this way represent the best efficiencies for a heat engine. That is, Carnot's efficiency represents an ideal efficiency; the efficiency of real-world devices is often less than the Carnot efficiency because of friction, leaks, and other mechanical imperfections. In a modern electricity-generating plant, the temperature entering the turbine is 540°C or 813 K. The heat eventually winds up in a body of water such as a river. In the summertime, its low temperature might be about 25°C or 298 K. The Carnot efficiency would be 63% for a device operating between 813 and 298 K. In other words, if the electricity plant were operating at this efficiency, then 37% of the chemical potential energy in the fuel gets lost into the "cold side," usually a river or an ocean. In reality, efficiencies of fossil-fuel-fired generating plants are less than the ideal Carnot value, usually less than 40%. Consequently, about two-thirds of the chemical potential energy of the fuel simply winds up in the low-temperature reservoir.

Why do we put up with this? Why can't we turn all the heat liberated from the fuel straight into electricity? It is because we are stuck. Since the electricity production extracts ENERGY from the generator, something has to supply WORK to the generator to keep it turning. The WORK comes from the steam's thermal energy (i.e., heat). Since heat can't be converted directly into WORK at 100% efficiency, the steam turbine, operating as a heat engine, converts a limited amount of heat into WORK as it flows from a hotter object to a colder object. In the steam turbine, the hotter object is the high-pressure steam and the colder object is the outside water or air. If we run the plant as a heat engine, we absolutely must have a cold side to the engine (otherwise there is no heat engine). And since we cannot operate a heat engine at 100% efficiency, the plant must dump heat—a large amount of heat, as it turns out—into the cold reservoir.

Because turbines avoid many of the mechanical problems (friction and leaks, for example) that reduce the efficiency of reciprocating steam engines, they also make better use of steam at much higher temperatures and pressures. This lets turbines achieve much higher efficiencies than were ever possible with the reciprocating steam engines. If we disregarded all of the other advantages of the steam turbine relative to the reciprocating engine (its speed of operation, freedom from vibration, quietness, etc.), the steam turbine would still have to be regarded as an invention of absolutely first-class importance, solely on the basis of the extraordinary improvement in efficiency—which translates directly into much less fuel that must be extracted from the earth, shipped, processed, and burned.

THE TURBINE/GENERATOR SET

The steam turbine has become the standard technology for operating the generator in all the world's electricity-generating plants (except hydroelectric plants), and to use the turbine most effectively, we need to supply it with a head of steam, as sketched

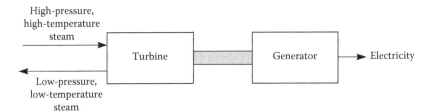

FIGURE 16.9 Most of the electricity used in most of the countries of the world comes from generators turned by turbines that move as a result of high-pressure, high-temperature steam.

in Figure 16.9. Generating electricity from steam involves a turbine/generator set in which high-pressure, high-temperature steam is used as the working fluid in the turbine, and electrical energy is provided from the generator. Previously, the central problem in electricity generation was said to be that of finding a cheap and reliable way to turn the generator. Now, taking a figurative "step back" from the generator, the practical problem to be solved is to find a cheap and reliable way to generate high-pressure, high-temperature steam.

HOW TO GET UP A HEAD OF STEAM

The short answer to the question of finding a way to generate large amounts of high-pressure, high-temperature steam as cheaply and reliably as possible is this: boil water! This requires raising its temperature, as shown in Figure 16.10, where the "ENERGY in" is supplied as heat.

In modern electric generating plants, the steam rate, that is, the amount of steam required to be generated in a given time, is often in excess of 500,000 kg of steam per hour. The ability to boil water on this prodigious scale is governed by two factors. The first is the rate at which the necessary heat can be produced. In steam-turbine plants (except nuclear plants, which will be discussed later), the necessary heat comes from burning a fuel. Therefore, the rate at which heat is released relates directly to the rate at which the fuel can be burned. The second factor is the rate at which this heat can be transferred into the water or steam. The rate of heat transfer

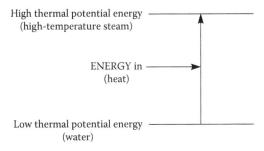

FIGURE 16.10 To generate electricity in steam plants, ENERGY in the form of heat must first be put into the SYSTEM.

depends on several factors, including the temperature difference between the hot flame and the relatively cool steam, the thickness of the material through which the heat must travel, the ability of that material to conduct heat, and the available surface area across which that heat is transferred. Since in most steam-generating equipment the first three factors are, very roughly, the same, surface area is the key to heat transfer.

A steam-generating system consists of two major components: the *furnace* where heat is liberated by burning the fuel and is transferred to the *boiler* where water is heated for generating steam. In modern systems, the furnace and the boiler are physically the same piece of equipment. Because of that, we often use the term "boiler" to mean *both* the place where fuel is burned and the place where steam is generated.

HEAT TRANSFER

Heat is transferred by three different mechanisms (Figure 11.17). In some SYSTEMS, all three may be operating at once; in others, one may dominate. One mechanism is conduction, the movement of heat directly through a material object (often, but not necessarily, a solid). A second mechanism is convection, the transfer of heat as a result of movement of fluid, often, for example, hot air. A hot stove warms a room chiefly through the convection resulting from movement of heated air from the stove. The third mechanism is radiation. The heat that Earth receives from the sun arrives in the form of radiation.

We have already seen how the work of Rumford and Davy led to the understanding that heat is a form of motion. In most materials, conduction is a consequence of the motion of the molecules or atoms of the object conducting the heat. (The average positions of the atoms themselves do not change in conduction—only their energy of motion.)

There are wide differences in the ability of various substances to conduct heat. Metals have by far the greatest ability to conduct heat. Convection is much simpler than conduction, since it consists of the actual motion of a volume of hot fluid (which can be either a gas or liquid) from one place to another. Convection may be either natural or forced. In natural convection, the buoyancy of a heated fluid leads to its motion. When a portion of the fluid is heated, it expands, becoming lower in density than the surrounding cooler fluid, and therefore moves upward. This is the basis of the common expression that "hot air rises." In forced convection, a blower or pump directs the heated fluid to wherever we would like it to deliver its heat.

Heat transfer by radiation involves transport of ENERGY by electromagnetic waves. These waves travel at the speed of light and require no material medium for their passage. All objects radiate such energy continuously, whatever their temperature. However, any object can also absorb electromagnetic energy from its surroundings. If the object and its surroundings are at the same temperature, the rates of emission and absorption are the same. An object at higher temperature than its surroundings radiates more energy than it absorbs. We detect this excess of radiation as heat. The rate of radiation of energy from an object increases rapidly as its temperature increases.

THE BOILER

A *boiler* is a device for heating water to convert it to steam. Conceptually, the simplest possible coal-fired boiler would look something like that sketched in Figure 16.11. We already know that heat will flow spontaneously from high temperatures (high thermal potential energy) to low temperatures. Thus heat will flow from the hot gases produced by the burning coal—that is, from the fire—to the relatively cool water. The transfer of heat takes place, in this crude cartoon design, across the surface of the water container that is exposed to the heat. Figure 16.11 shows that coal is used as the fuel. The very same concept would apply with other fuels, such as oil, natural gas, or biomass fuels such as wood.

In any simple boiler, the rate of steam generation is limited by the amount of surface across which heat can be transferred, that is, the heat-transfer surface. The simple design shown in Figure 16.11 is adequate for generating very small amounts of steam (such as in a tea kettle) or modest amounts of low-pressure, low-temperature steam (e.g., for home heating), but is hopeless for producing the large quantities needed by typical modern electric generating plants. An immense flat-bottomed boiler would be needed to produce 500,000 kg of steam per hour, and great difficulty would be experienced in distributing the fuel evenly in a furnace of such immense size.

The first design improvement in increasing the ability to generate large quantities of steam involved sending hot combustion gases (using metal pipes or tubes to contain them) right through the water reservoir. This design was called a fire-tube boiler. It places the heat in the midst of the water. As the demand for steam (i.e., the amount of steam generated per unit time) continued to increase, even the fire-tube boiler could not keep up. The limitation of heat-transfer surface was addressed by, in essence, turning the fire-tube boiler inside out, so that the water or steam is completely surrounded by the high-temperature gases from the burning fuel. This design is called a water-tube boiler or steam-tube boiler. The water-tube boiler, with numerous water tubes, is about as good as can be done in increasing the effective heat transfer.

A water-tube boiler consists of a number of tubes in which the water circulates, with the flames and hot gases acting on the exterior surface of the tubes.

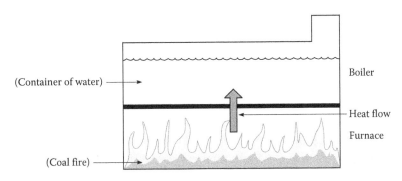

FIGURE 16.11 In the simple boiler, heat is transferred across the bottom surface of the water container. A tea kettle on a stove functions in exactly this way.

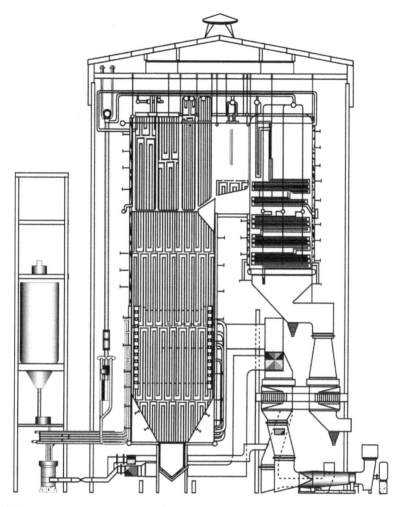

FIGURE 16.12   A modern water-tube boiler involves an immense amount of piping for the water and steam tubes.

Nowadays, water-tube boilers have become almost universally adopted for supplying steam to turbines for generating electric power. Because of the large heating surface provided by the many tubes, they are able to produce steam very quickly. The large surface area inherent to the water-tube design is further increased by a convoluted system of piping, illustrated in Figure 16.12.

HOW TO BURN COAL

The limitations to high rates of steam generation are the heat-transfer rate (determined by heat-transfer surface) and the heat release rate. The water-tube boiler solves the first problem. To discuss the second problem, we consider coal as the fuel to be used in the plant. Coal is the dominant fuel for electricity generation worldwide, with about 40% of total world electricity generation coming from plants burning coal. (For comparison, natural gas is the second most popular energy source, accounting for about 20% of world electricity generation; nuclear and hydro each contributes 15%–16%.) South Africa and Poland are the world leaders, producing about 90%–93% of their electricity from coal. China and Australia also have heavy reliance on coal, which supplies some 75%–80% of their electricity. Furthermore, burning coal in electric generating plants is by far the dominant use of coal worldwide, with 60% of coal production going to electric plants. Although there are numerous environmental issues associated with the mining and burning of coal, as will be discussed later, it is fair to say that without coal, the electricity systems in many major industrialized countries would be severely crippled.

Think of the internet as being a coal-fired information transfer system.

—Dutton[3]

Burning coal is the first step in raising the steam that eventually drives the turbine. Burning releases some of the chemical potential energy that was stored in the chemical bonds in the coal.

He... composed in the grate a pyre of crosslaid resintipped sticks and various coloured papers and irregular polygons of best Abram coal at twentyone shillings a ton from the yard of Messrs Flower and M'Donald of 14 D'Olier street, kindled it at three projecting points of paper with one ignited lucifer match, thereby releasing the potential energy contained in the fuel by allowing its carbon and hydrogen elements to enter into free union with the oxygen of the air.

—Joyce[4]

For the moment, coal can be considered to be pure carbon. (Unfortunately, it isn't. As we'll see later in this chapter, the fact that coal really isn't pure carbon causes considerable technical problems in plant operation.) Using this simplifying assumption for the time being, the complete combustion of coal can be represented by the equation

$$C + O_2 \rightarrow CO_2$$

Because coal is a solid, the reaction of carbon with oxygen molecules will occur where the oxygen molecules have access to the carbon—on the surface of the coal.

For a given amount of coal, the rate of combustion will be limited by the available surface area. That is, the heat release rate will be governed by the surface area of the coal (via the combustion rate). To increase the heat release rate, the surface area available for reaction must be increased. This is done by pulverizing the coal to extremely fine particle sizes (of diameters between 0.02 and 0.05 mm). Using finely pulverized coal provides greatly increased heat release rates. Unfortunately, it also introduces a mechanical problem. In furnaces that burn lumps of coal, the coal

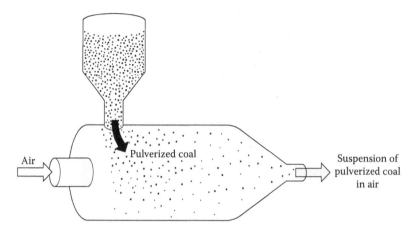

FIGURE 16.13 A conceptual diagram of a coal feeder and burner for pulverized coal firing. The finely pulverized coal is suspended in air and blown into the combustion chamber.

is supported on a grate that holds the burning coal off the floor of the furnace and allows air to circulate through the burning coal bed. With pulverized coal, there's no way to support it on a grate. Either the powdered coal would sift through the grate, or it would be so tightly packed as to limit the access of air and negate the benefits of pulverizing it in the first place. The solution is to blow the pulverized coal, with air, into the boiler. This approach is called suspension firing or, often, *pulverized coal firing*, illustrated in the sketch in Figure 16.13.

Roughly, a pulverized coal-fired steam-turbine generating plant will operate at about 30%–35% efficiency for conversion of the chemical potential energy of the coal into the electrical potential energy (high-voltage electricity) leaving the plant.* The loss of some 70% of the heat energy of the coal is high, but still a major improvement over early plants. Edison's Pearl Street Station in 1882 used about 5 kg of coal per kilowatt-hour of electrical energy generated. This requirement dropped to about 1 kg by the 1920s. Nowadays, a kilowatt-hour can be generated with less than 0.5 kg of coal. Nevertheless, there obviously remains plenty of room for improvement. One reason for the relatively low efficiencies of approximately 30% is that our present technology takes such a roundabout route to converting the chemical potential energy stored in the coal to electrical energy. Consider Figure 16.14: First the chemical potential energy is released as heat; then the heat energy is used to produce high-temperature (high thermal potential energy), high-pressure steam; next the thermal potential energy of the steam is converted into mechanical work in the turbine; finally, the work of the turbine is used in the generator to produce the high electrical potential. With this long sequence of operations, it is inevitable that numerous losses and

---

* This figure of 30% efficiency represents an approximate average of the best modern plants, which may approach 40%, and wheezing antique plants that may be somewhere in the low 20% range.

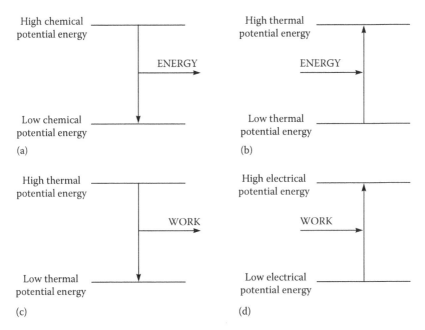

FIGURE 16.14 This quadruple energy diagram traces the conversion of chemical potential energy in the fuel (e.g., coal) into electrical potential energy. (a) Furnace, (b) boiler, (c) turbine, and (d) generator.

inefficiencies occur. There ought to be better, more direct ways of producing electricity from a primary energy source; perhaps in the course of this new century, we will see some of them realized.

THE COAL-FIRED ELECTRIC GENERATING PLANT

In the middle of the previous century, power plants were fairly small—on the order of 5 MW, for example—and served a town or city in the immediate vicinity. After the Second World War, many countries began constructing new plants to larger and larger sizes. Nowadays, a typical midsized unit might generate (in round numbers) 400 MW of electricity and consume 10,000 tonnes of coal a day. To put this into perspective, a single, large railroad car of coal—about a hundred tonnes of coal—supplies either enough coal to heat a modest two-story wood-frame house during the winters for 10 *years* or enough coal to operate a typical modern power plant for 15 *minutes*. Rather than just providing electricity to people in the local area, new plants provide electricity to regional or national power grids, so may serve customers hundreds of kilometers away. Today some very large plants have outputs over 1000 MW. However, in those cases, plants are built with two or more generating units of, say, 500 MW capacity in parallel to generate the total plant output, rather than relying on a single gigantic turbine and generator.

It's important not to waste the heat generated inside the boiler (but a lot of it is lost anyway, even in the best of designs). For this reason, boilers are configured

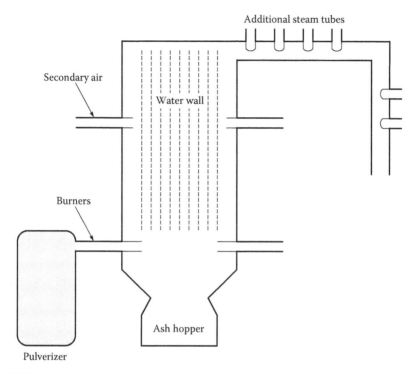

FIGURE 16.15 A sketch of the full boiler system.

to enclose as large a volume as possible while presenting a relatively small surface to the outside world. This helps minimize the heat loss from the boiler to its surroundings. The large heat-transfer surface is achieved by the water-tube design. The boiler, usually a big rectangular box, could be ten to twenty stories tall, depending on the size of the plant (its generating capacity) and the type of coal used.

All coals will contain some amount of incombustible materials that remain behind as an ash. Some provision is made to catch and collect the ash residue from the coal in an ash hopper. The basic components are laid out as sketched in Figure 16.15.

Complexities in design and operation of the plant come from the fact that coal is not really pure carbon. As coal comes from the ground, it consists of (1) a carbonaceous portion, which may contain 70%–95% carbon, 2%–6% hydrogen, 2%–20% oxygen, and some nitrogen and sulfur; (2) a noncombustible inorganic portion, consisting of minerals that accumulated with the carbonaceous portion, and any rocks or minerals that might accidentally have been mixed with the coal during mining; and (3) some amount of moisture, much of which is removed by drying during the pulverizing process. One way of remembering the constituents of coal is by the mnemonic NO CASH, in which the letters stand for nitrogen, oxygen, carbon, ash (which is, strictly speaking, not a constituent, but rather a product of heating the inorganic portion of coal), sulfur, and hydrogen.

We can treat the combustion of coal as if it were a simple mixture of the elements. Thus, for complete combustion

$$C + O_2 \rightarrow CO_2$$

$$4H + O_2 \rightarrow 2H_2O$$

$$S + O_2 \rightarrow SO_2$$

$$2S + 3O_2 \rightarrow 2SO_3$$

$$2N + O_2 \rightarrow 2NO$$

$$N + O_2 \rightarrow NO_2$$

$$\text{Minerals} \rightarrow \text{Ash}$$

Both sulfur and nitrogen form more than one oxide. A convenient simplification that avoids worrying about which specific oxide has formed lumps all the oxides together as SO_x (pronounced "socks") and NO_x (pronounced "knocks"). The noncombustible minerals in the coal undergo a variety of chemical and physical transformations to form the ash residue. The ash is sometimes humorously referred to as RO_x ("rocks").

If the products of coal combustion could simply be released into the air and thoroughly diluted by perfect mixing with the total volume of the atmosphere, the potential harm they cause might not be very apparent. (This gives rise to the somewhat facetious—and totally unacceptable—statement, "The solution to pollution is dilution.") But being able to "dilute" pollutants to levels at which they appear to be harmless is seldom possible. In many cases, these combustion products are generated and released in or near large centers of population. In many such cases, the local geography and the prevailing weather patterns make it difficult to achieve an extensive dilution of emissions. Consequently, cities or regions may suffer severely from various kinds of atmospheric pollution. In British and Australian slang, large cities such as London used to be called "the big smoke."*

SO_x and NO_x are of concern because of their role in the formation of acid rain (Chapter 27). Highly acidic rain has been falling for decades in the industrial regions of the United States, particularly in the northeast; in the neighboring parts of Canada, where the smelting of sulfide ores may be an important contributor to the problem; in England, especially in the industrialized north; and in heavily industrialized parts of Europe. This rain can be ten times more acidic than ordinary rainfall in unpolluted areas. If the pollution is particularly severe, the rain may be a thousand times more acidic than normal rain. Acid rain harms crops and forests, rivers, and lakes.

---

* As early as 1860, British tramps used the term "the smoke" in their argot to refer to London. By 1900, "the great smoke" or the "the big smoke" had become colloquial in Britain. For example, one might speak of having to "go up the smoke" as meaning to have to go to London. Meanwhile, Australians adopted "the big smoke" as a slang term for any city during the mid-nineteenth century; by about 1920, it was usually taken to refer to Sydney. This term derives from the very high concentration of coal-burning equipment, especially household stoves and fireplaces, in the comparatively small area of a city. Many of these devices did not burn coal very efficiently or effectively, producing a smoky fire; when thousands upon thousands of these smoke producers were concentrated in a city, the total smoke emission was enormous.

As it accumulates in lakes and causes them to become increasingly acidic, it severely disrupts aquatic ecosystems. It damages buildings, corrodes statues and monuments, and attacks metal.

There are several approaches to mitigating SO_x emissions. One, *coal cleaning*, removes some (but usually not all) of the sulfur from the coal. A second is to switch to burning a coal that has less sulfur than the one presently being used in the plant. This approach has some possible problems, because many boilers have been built to fire coals with a narrow range of specifications. A third approach is to switch the plant to burn an entirely different kind of fuel of lower sulfur content, such as natural gas. This option could be extremely expensive, since it may involve rebuilding a considerable section of the boiler. All three of these strategies involve doing something to reduce SO_x formation before the coal is burned. The alternative is to allow the SO_x to form and then deal with it.

The best available technology to do that is *flue gas desulfurization*, often called simply FGD. The hardware or device in which FGD is accomplished is called a *scrubber* (Figure 16.16). In the scrubber, the combustion gases pass through a spray of wet alkaline solution or slurry, usually of lime or limestone. The scrubber takes advantage of the fact that SO_x dissolves in water to produce sulfurous and sulfuric acids. Because lime and limestone are alkaline, they will react with the acid. Lime and limestone offer two other advantages. First, they are about the cheapest alkaline substances available. Second, the reaction products, calcium sulfite and calcium sulfate, are not soluble in water and so form a precipitate sometimes called "scrubber sludge."

Scrubbers work—they work very well indeed. In the last quarter of the twentieth century, when the use of coal in the United States roughly doubled (so it might not be unreasonable to expect that sulfur emissions would also double), sulfur emissions actually *dropped* by 30%. But the other important message about scrubbers is that they do not *destroy* pollution. The sulfur isn't gone. It has only been converted from a very dilute form hard to handle—SO_x in the combustion gases—to a much more concentrated form that is easier to handle—scrubber sludge. That results in a new problem: what to do with the sludge.* Furthermore, scrubbers use a lot of water. Using this same 1000 MW hypothetical plant as an example, the water consumption in the scrubbers would be about 4000 liters per *minute*. This will be of increasing concern in countries or regions facing water supply problems.

NO_x is more difficult to deal with than SO_x. First, there are actually two sources of NO_x in a combustion system. *Fuel NO_x* is NO_x generated from the reaction of nitrogen atoms chemically incorporated in the fuel molecule; *thermal NO_x* is generated from the air used for combustion when molecules of nitrogen and oxygen react at the high temperatures of the combustion system. Air is about 80% nitrogen. At flame

---

* There is a partial solution to the problem of what to do with scrubber sludge. The calcium sulfate that forms is usually associated with two molecules of water ($CaSO_4 \cdot 2H_2O$) as the material commonly known as gypsum. This material can be used in the manufacture of sheet rock or wallboard used in building construction; scrubber sludge has been used in this application. Gypsum also has other uses, in cement manufacture and in soil treatment, as examples. The income from sale of some of the by-product gypsum could offset some of the costs of operating the scrubber. However, doing so means that the generating plant owners must take on some aspects of being a chemical manufacturer, something they may be reluctant or unable to do.

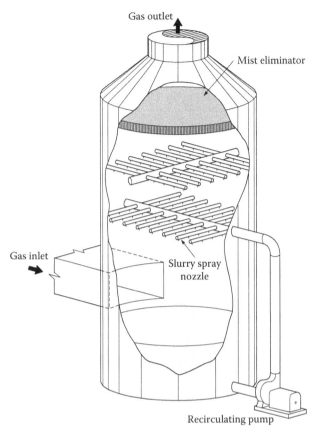

FIGURE 16.16 A cutaway diagram of a scrubber for flue gas desulfurization. Units like this can do an excellent job in reducing the amount of sulfur oxides emitted to the environment.

temperatures above about 1500°C, the nitrogen in the air burns to form nitrogen oxide (NO), as shown earlier. In the presence of more air, NO is quickly converted into nitrogen dioxide:

$$NO + \frac{1}{2}O_2 \rightarrow NO_2$$

Nitrogen dioxide is toxic. In the presence of water droplets, it dissolves to produce nitric acid, the other component of acid rain. Nitrogen dioxide is brown; its presence in air that is badly polluted gives the sky a sickly brownish color. Because of the formation of thermal NO_x in this way, even if it were possible to reduce the nitrogen content of a fuel to 0.000%, there would still be NO_x emissions from a combustion system. (In contrast, if the sulfur content of a fuel were reduced to 0.000%, there would be no SO_x emissions at all.)

NO_x is hard to deal with firstly because no easy, cheap way exists to remove nitrogen from coal. Much of the sulfur in coal is present as the mineral pyrite; when coal is pulverized, the pyrite is relatively easy to separate. Nitrogen, on the other hand, is chemically incorporated in the molecular structure of the coal. Second, scrubbers work well in part because of the calcium sulfate, which is not soluble in water and forms a "sludge." An analogous system for "flue gas denitrogenation" could not function this way, because all nitrate salts are soluble in water. Third, thermal NO_x formation is almost inevitable, and the higher the combustion temperature (which we want, both to drive heat transfer and to increase the steam temperature to the turbine), the more thermal NO_x is formed.

Today, there is a steady development of approaches to NO_x reduction or removal. In Europe and Japan, many boilers are equipped with a selective catalytic nitrogen oxide reduction system, where ammonia is mixed with the nitrogen oxides in the stack. This process produces harmless nitrogen and water vapor from the chemical reaction of ammonia with the nitrogen oxides. American efforts to reduce NO_x have also focused on the combustion process itself. New, so-called low-NO_x burners are being designed in which the fuel and air mixture and the temperatures attained are more carefully controlled to lower thermal NO_x production.

Ideally, the ash residue from the coal should fall to the bottom of the boiler, where it could be removed via the ash hopper. Some 80% or more of the ash does that. However, the extremely turbulent flame results in some very small ash particles being carried upward and out of the boiler with the hot gaseous products of combustion. Such ash particles are called *fly ash*. If they escape into the environment, they are also termed *particulate emissions*.

Particles in the stack gases of coal combustion vary in size from 0.01 to 10 microns (μm) in diameter. (A micron is 1-millionth of a meter.) The smallest particles, in the range 0.01–1 μm, are the most dangerous to health. If they are inhaled, they can be trapped in fine passages of the respiratory system and add to pollution-related respiratory distress. Particulates lodged in the lungs can irritate and damage the lungs. Particulates have large surface areas with many tiny cracks and crevices and therefore can absorb other pollutants on their surface or inside these crevices. When these particles enter the lungs, the pollutants are released at a higher concentration to the surrounding tissue than if they were simply inhaled directly from the atmosphere. This could exacerbate a health problem, such as emphysema, in the sick individual. Particulate emissions, which can include soot (from incomplete combustion) and coal dust as well, are undesirable for several reasons, including the aesthetics of having fine ash particles deposit on objects outdoors.

Particulates are released when coal burns, unless they are trapped or removed from the effluent smokestack gases. The basic approaches to reducing particulate emissions are a baghouse, which uses a fine fabric filter to trap the ash particles (crudely, working like a gigantic vacuum cleaner) and an electrostatic precipitator (ESP), which uses a high-voltage electric field to give an electric charge to the fly ash particles, trapping them on high-voltage plates. A third alternative is the cyclone collector, which whirls the combustion gas around like a miniature cyclone or tornado, forcing the entrained particles to the walls by centrifugal force. ESPs have long been used to remove suspended particulates from coal smoke. The particles become ionized by a strong electric field between two electrodes of an ESP. Once ionized, the particles are drawn to and captured on the electrodes. Periodically, the particles are removed from the electrodes and disposed of.

Every coal-burning plant produces ash as a waste product (both fly ash from the flue and "bottom ash" from the ash hopper on the boiler). Somehow this ash has to be disposed of. Though ash is only a small percentage of the coal burned, the absolute amounts of ash produced are huge because of the massive quantities of coal used. If a plant burns 10,000 tonnes of coal a day, a not unreasonable figure for a large plant, if the ash yield from the coal is 10%, then 1,000 tonnes of ash have to be disposed of every day. One solution is to send the ash to a landfill, but this requires a significant amount of space, either adjacent to the plant or somewhere within the range of low-cost transportation for the waste. Landfills are becoming increasingly difficult to find, especially in densely populated regions. Some ashes can be returned to the mine for use in mined-land reclamation. Fly ash can also be used to make concrete. If appropriate processes could be developed to sell ash and scrubber sludge into useful by-products and to make money doing so, rather than paying a cost for disposal of these wastes, the net financial gain could perhaps reduce environmental control costs overall or even pay the costs of environmental compliance.

OVERALL PLANT LAYOUT

The overall layout of a coal-fired electric generating plant might look like Figure 16.17 (neglecting the turbine and generator portion of the plant). The combination of scrubber and ESP or baghouse, together an appropriate NO_x control strategy, would make it seem that we have removed or reduced all of the harmful pollutants.

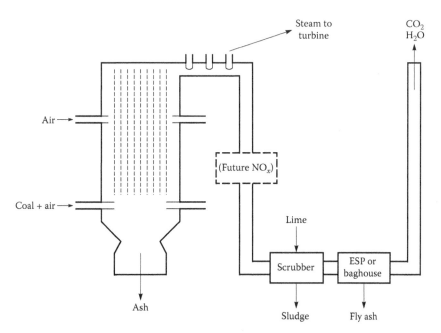

FIGURE 16.17 A sketch illustrating the overall plant layout, including the hardware used to help reduce emissions of pollutants.

Only carbon dioxide and water vapor go "up the stack." In recent years, however, there has been growing concern about carbon dioxide emissions and their role in the environmental problem of global climate change. We will return to this issue later (Chapter 29).

After doing its WORK, the low-pressure, low-temperature steam leaving the turbine must be condensed back to water, using a device called a *condenser*. To handle the large amounts of steam used in today's plants, many a condenser has several thousand square meters of cooling surfaces. Cold water from a natural source, a river, a lake, or the ocean, is brought in to extract heat from the steam. The loss of heat from the steam causes the cold condenser water to warm up or even get hot because the cooling water that passes through the condenser absorbs the heat given off when the steam condenses back to water. Hot water from the condenser cannot be discharged back to the environment as is. Doing so could, potentially, upset the local environment by causing an unnatural growth of aquatic plants or some kinds of aquatic animals, a problem sometimes known as *thermal pollution*. The hot water has to be cooled back to near-ambient temperature before it can be discharged. This is done in a *cooling tower*. The full system resembles the sketch in Figure 16.18. Nowadays power plants are generally built close to rivers, lakes, and oceans, which provide a ready source of water for cooling.

The efficiency of conversion of chemical potential ENERGY in the coal to electrical potential ENERGY leaving the plant is about 35%. This figure depends on the specific plant and may get close to 40% in a very modern and well-run plant and dip into the 20% range in old plants. Given an overall efficiency of about 35%, where does the rest of the ENERGY go? If the generator is running well, the loss there is only about 1%. The turbine may have an efficiency of 45%. About 10% of the ENERGY in the coal goes

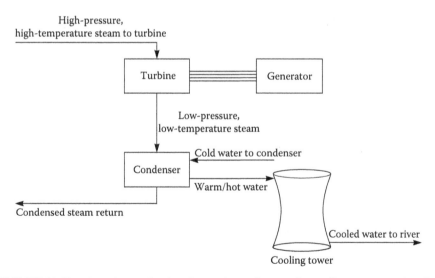

FIGURE 16.18 Heated water leaving the condenser flows to the cooling tower to reduce its temperature before being released back to the environment. Fresh cold water taken from a source such as a river or lake is used as the input to the condenser.

straight up the stack in the form of heat in the hot combustion products. Some heat is lost from the steam as it moves through the plant, from the boiler to the turbine and from the turbine to the condenser. More heat is wasted in warming cooling water in the condensers. Fans are needed to provide the air for combustion; pumps are needed to circulate water through the condensers and cooling towers; scrubbers must be operated. All of these consume more ENERGY. A plant that is highly efficient by today's standards will throw away the equivalent of 2 MW in wasted ENERGY for every 1 MW of electrical ENERGY put into the distribution system. (Another way of putting it is that two-thirds of all the energy extracted from the ground by coal miners is thrown away as hot air and hot water.) By the time the electricity actually reaches the consumer, more losses have been encountered in the wires and transformers. Consequently, only about 30% of the chemical potential ENERGY in the coal is available to the consumer in the form of electrical potential ENERGY at the outlet in the wall.

REFERENCES

1. Chekhov, A.; Quoted in Keizer, G. *The Unwanted Sound of Everything We Want.* Public Affairs: New York, 2010, p. 218.
2. de Laval, G.; Quoted in Strandh, S. *The History of the Machine.* Dorest Press: New York, 1979, p. 131.
3. I owe this comment to John A. Dutton, former dean of the College of Earth and Mineral Sciences, The Pennsylvania State University, University Park, PA.
4. Joyce, J. *Ulysses.* Folio Society: London, U.K., 1998, pp. 626–627. Many other editions of this extraordinary novel are available.

FURTHER READINGS

Freese, B. *Coal: A Human History.* Perseus Publishing: Cambridge, MA, 2003. This book documents many aspects of coal use, from an environmental, political, and social perspective.

Gebelein, C.G. *Chemistry and Our World.* Brown: Dubuque, IA, 1997. Most introductory chemistry texts include a discussion of Boyle's law and related aspects of the behavior of gases. This is a well-illustrated book with abundant examples of the relevance of chemistry to daily life. Chapter 2 includes a discussion of Boyle's law and development of the applications of steam. Chapter 18, "Air Pollution, Energy, and Fuels," relates to the present discussion.

McNeill, J.R. *Something New under the Sun.* Norton: New York, 2000. This remarkable book, subtitled "an environmental history of the twentieth-century world," surveys the history of the last century's technological developments, politics, and international relations in terms of humankind's dramatic impact on the environment. Chapters 3 and 4 discuss the atmosphere and what we have done, and may still be doing, to it.

Miller, B. *Clean Coal Engineering Technology.* Elsevier: Amsterdam, the Netherlands, 2011. An excellent book covering many aspects of production of electricity from coal, including emissions controls and research and development activities for generating electricity with near-zero emissions. The book assumes that the reader has some technical background.

17 Energy for Transportation

Most of us make direct personal utilization of ENERGY mainly for two reasons. The first is domestic comfort, including lighting, cooking, heating, air conditioning, and entertainment. The second is to transport ourselves and to move our goods, from place to place. A reliable and efficient transportation system is vital in a modern industrialized society, because it lets us move ourselves and our belongings and lets others move to us the food and consumer goods we need, easily, at reasonable cost, providing us with remarkable convenience.

HUMANS AND OTHER ANIMALS

For almost all of human history, transportation of ourselves, our belongings, or other cargo has depended either on the use of human muscles or on the muscles of certain animals (horses, camels, oxen, and occasionally more exotic animals such as elephants). From the dawn of early hominids several million years ago, the obvious form of "transportation" was to use the energy of our own bodies for walking or running. We can also carry or pull things. Photographs of some of the primitive peoples of the world show them carrying animal carcasses or other burdens slung from a pole carried along by two or more people (as in Figure 17.1).

Animals were being used for transportation by about 3000 BCE. Over the years, an extraordinary variety of animals has been pressed into service as draft animals: horses, oxen, asses, camels, reindeer, llamas, elephants, water buffalo, and yaks. Humans in northern Europe had been pulling loads on sledges since about 5000 BCE. No doubt the idea spread that one could just as easily hitch an animal to the sledge rather than packing the load onto the animal itself. Then some unknown inventor mounted a sledge on wheels. Drawings of these first vehicles have been found on tablets excavated in Mesopotamia, dating from about 3500 BCE (Figure 17.2). Actual models of vehicles, sometimes even the vehicles themselves, have been found in Middle Eastern tombs that date from 3000 to 2000 BCE.

Nowadays only very few of us can appreciate, or even imagine, what life must have been like before the development of modern transportation systems. Almost everyone lived out their lives in the immediate vicinity of the village or farm where they had been born. There might be occasional visits to the local market or fair, or, rarely, to the regional capital for some unusual event. For a few privileged people, there might have been a once-in-a-lifetime religious pilgrimage. But for almost all the people almost all the time, life was confined to a region limited by the distance that could be traversed on foot or horseback. Even as recently as two hundred years ago, people tended to remain settled and live close to where they were born. A major reason for this state of affairs was the sheer physical difficulty of moving oneself and one's belongings in the absence of reliable means of transport.

FIGURE 17.1   These Bushmen show their reliance on human ENERGY for transportation. (From www.horstklemm.com.)

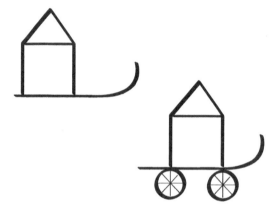

FIGURE 17.2   An artist's representation of wheeled vehicles in use in ancient Mesopotamia.

It's also likely that few of us can appreciate what life must have been like, especially in cities, when reliance on horses represented a major aspect of transportation. And yet, certain aspects of the horse-drawn transportation system sound remarkably modern. They include gridlock on city streets, excessive noise, fatal traffic accidents, and air pollution. There were even concerns about diverting food crops to provide "fuel" (i.e., fodder) for the horses, somewhat like today's food vs. fuel debate concerning production of ethanol from corn (Chapter 31) as fuel for automobiles. Around 1900, there were 200,000 horses in New York City alone. No doubt most of the world's metropolitan areas were equally clogged with horses. The death toll from accidents involving horses in 1900 was 1 per 17,000 citizens. A century later, with streets clogged by automobiles, the death toll in car accidents in New York City was 1 per 30,000, nearly half that of the "good old days" when life was supposedly simpler.

It's an immutable fact of biology that what goes in one end of a horse sooner or later emerges, in considerably altered form, from the other end. New York City's 200,000 horses collaborated to produce about 2,300 tonnes of manure and kept on doing it every single day, whether they were in harness or not. There had been a time, around mid-nineteenth century, when cities were still small, countryside farms were not far from city center, and the population of people and horses was relatively small. Then, the solution to this problem was easy, because the manure could be hauled away to farms as a natural organic fertilizer. But, 200,000 horses overwhelmed this system. The manure was piled in vacant lots, twenty meters deep. It was banked along the sides of streets, much as freshly fallen snow is banked up by ploughs. The odor must have been horrific, and the flies must have numbered in the millions, possibly billions.

When a horse could work no longer, it was "put down," a polite euphemism for "killed." To minimize insurance fraud, many life insurance policies on horses required that the horse be put down by an authorized third party, such as a veterinarian. Imagine a horse that has dropped in its traces, too exhausted, or too ill to do more. No mobile phones in those days. A message had to be sent to fetch a person authorized to put the horse down. That person then has to get to the scene. Meanwhile, traffic is backing up, and up, and up… Whether the horse has been put down or has simply dropped dead, the next problem is what exactly can be done with a dead horse. Horses are heavy; depending on breed, age, and conditioning, a horse can top 500 kg. Shifting 500 kilos of (literally) dead weight is not a simple job. Supposedly in some cities, the carcass was just left on the spot, until it had decomposed enough that it could be cut up and hauled off piecemeal.

The transportation problem was particularly severe for people who did not have access to large bodies of water, such as the sea or large navigable rivers. Inland, transport depended on two factors: the availability of draft animals and a good system of roads. In fact, the latter was much the more difficult problem. In western Europe, the road system had been neglected ever since the collapse of the western half of Roman Empire in the early 400s. In the early 1800s, a Scot, working as a surveyor of road systems, thought up a better way of making roads than the rutted, eroded, muddy travesties that then existed. The idea was to begin with a bottom layer of stone or gravel to allow rain water to drain away and to put on top of the gravel layer some impervious material that would not turn to mud but would have some "give" (i.e., elasticity) to it so that the surface would not wear so quickly. The Scottish surveyor was John McAdam. His better way of building roads was quickly recognized, and his expertise soon became in demand throughout Britain. By the 1830s, he had been responsible for improving hundreds of miles of road in the British Isles. We remember him today when we speak of a "macadam" road.*

Once a good road system had been developed based on McAdam's ideas, it became possible to establish regular passenger transportation via fast stagecoach services (Figure 17.3). Improvements in the design of the coaches themselves, and setting up a system to allow for regular changes of horse teams, allowed the stagecoach companies to maintain unprecedented speeds over long distances.

---

* John Loudon McAdam (1756–1836) developed the idea of using a very thick, tarry material to help bind the road surface together. The tar could be applied hot, so that it could be mixed with, and worked into, the stones used in the road surface. Once the hot tar had cooled, it would be so thick as to provide a solid surface largely impermeable to water. McAdam's original approach has largely been displaced by other materials, such as asphalt or concrete. The term *tarmac*, still used especially in conjunction with air fields, is a condensation of the words tar and McAdam. In some parts of the world, any paved road, regardless of surface, is called a *tarred* road, another term that is part of McAdam's legacy.

FIGURE 17.3 The horse-drawn stagecoach represented, for its era, relatively rapid and regular transportation service. Its success depended on improved road quality. (Image courtesy of State Historical Society of North Dakota.)

It was not long before inventors got the idea of designing a "horseless" carriage, propelled by the major energy source of the time—steam.

SAILING SHIPS: MOVING WITH THE WIND

Early humans may have gotten the idea of building rafts by seeing logs floating down rivers. Boats emerged very early in human history as a technological innovation among people who lived along rivers. They provided effective transport for people and their goods. Primitive rafts were simple bundles of logs or reeds, but as early as 4000 BCE, some Egyptian rafts had an early form of sail. These designs were improved by the Cretans and Phoenicians into two kinds of boats: rowing galleys, used for warfare and powered by humans, and sailing ships, used for commerce and powered by the wind. The invention of improved sails, probably by Arabs, may have occurred around AD 200. The Islamic conquests, followed by extensive Chinese voyages of exploration, made the sailing vessel dominant by the 1400s.

In the West, the steerable rudder was invented by the Dutch, in about 1200. (The Chinese had already known about it for a thousand years.) Magnetic compasses were first used around 1000. With perfected sails, rudder, and compass, it was possible to navigate and sail accurately and to develop genuine sea routes, first in the Mediterranean and then in the great European voyages of discovery in the sixteenth century. Sea-going ships tended to be larger than those developed for river transport. Improvements in ship construction and the increasing mastery of sailing techniques and navigation, which collectively took centuries to occur, gave Western societies a form of transportation that required no fuel and that became capable of carrying a substantial cargo over the seas of the world. Sailing ships provided western Europe, and, later, America, with much of the wealth without which the subsequent processes of industrialization could likely never have started.

STEAM FOR TRANSPORTATION

LOCOMOTIVES

Richard Trevithick used the steam engine to turn wheels, creating the first steam locomotive. While Trevithick's designs certainly did work, they had plenty of room for improvement. Analogous to Newcomen's building of a workable engine that was later perfected by Watt, many engineers worked on the challenge of developing a practical steam locomotive. Credit is generally awarded to George Stephenson.*

The locomotive steam engine emerged as an application of *high-pressure* steam. Watt's low-pressure, stationary engine had dominated the market until 1800 and had totally swept aside the earlier Newcomen engine. The Watt engine was too big and too heavy to be adapted easily for locomotive purposes. Trevithick demonstrated his high-pressure machine, smaller and more compact than a Watt engine (and therefore also lighter), at the beginning of the nineteenth century. This successful demonstration led people to realize that the steam locomotive could be a viable idea. George Stephenson's role was to introduce further improvements. He put his engine, *Locomotion Number 1*, into commission on the Stockton and Darlington Railway in 1825. Four years later, the success of Stephenson and his son Robert at the Rainhill Trials of 1829 established their locomotive *Rocket* as the prototype for the Liverpool and Manchester Railway. When this railway opened in 1830, it was the first railway in the world in the sense that we would recognize now, that is, with passenger and freight trains operating on a regular timetable.

In 1829, Stephenson's locomotive *Rocket* (Figure 17.4) demonstrated the ability to haul a 12-tonne train 50 kilometers (from Liverpool to Manchester)—at an average speed of 22 km/h. Stephenson's demonstration set the stage for an enormous expansion of railways throughout the industrialized world. Throughout the nineteenth century, there was an immense sustained activity in building railways worldwide. Over the next century and a half, steam locomotives increased massively in size and power. In most countries, they were replaced by diesel or electric traction in the last half of the twentieth century, except in China, which was one of the last bastions of steam power on railways. It is remarkable that in the lengthy evolution from Stephenson's 7-tonne *Rocket* to the most recent steam locomotives of several hundred tonnes, there were no revolutionary changes in fundamental design. The line of development of steam locomotives was a steady increase in efficiency and in tractive effort (the capacity for hauling heavy trains). Since condensing of steam after it leaves the

---

* George Stephenson (1781–1848) was another stellar example of the many superb scientists and engineers in eighteenth-and early-nineteenth century Britain. Remarkably, he did not learn to read or write until he was 18. In addition to his contributions to railway engineering, Stephenson invented a safety lamp to prevent explosions of gas in coal mines, touching off an acrimonious dispute with Humphry Davy, who claimed that Stephenson had somehow stolen his ideas. Though it came to be established that Stephenson had arrived at the invention independently (a situation for which there are many, many other examples in the history of science and technology) and was innocent of any theft or plagiarism, nowadays Davy gets all the credit. Stephenson established the track gauge—i.e., distance between rails—of 1.435 m (4 ft 8½ in.), which has come to be standard on almost all the world's railways. This rather odd measurement (why not, say, 1.5 m?) may have come from the wheel gauge on stagecoaches of the time, which in turn may have derived from even earlier vehicles.

FIGURE 17.4 Stephenson's *Rocket* was one of the first successful steam locomotives for railway use. (From http://commons.wikimedia.org/wiki/File:Stephenson%27s_Rocket_drawing.jpg.)

cylinder is not practical on a locomotive, it is exhausted directly into the atmosphere still at a relatively high temperature.* Therefore the efficiencies of steam locomotives are no better than about 8%, and frequently less than 5%. Furthermore, there are practical upper limits to the height and width of locomotives, because they have to be able to pass through tunnels and fit between station platforms.

Throughout almost the entire steam locomotive era, coal was the fuel of choice for generating the steam. Wood was used on early locomotives in the United States and stayed in use in some countries that had ample wood supplies but few resources of coal or petroleum. Railways provided a new and extraordinary facility for personal transportation that was quickly adopted across the world. Development of the locomotives and tracks themselves created a need for "infrastructure," such as station buildings. By the end of the nineteenth century, railways had entered into the imaginative experience of more people in the world more generally and in a shorter space of time than any previous technological innovation in the entire history of the world. Though the rail system became very widespread in most industrialized nations by the end of the nineteenth century, both personal travel and shipping of goods were still limited to places linked by rails, and then only at the times scheduled and the rates established by the railway companies. As late as the turn of the twentieth century, most of the people in modern industrialized nations still depended on themselves or horses for personal transportation.

---

* A few steam locomotive designs have used so-called condensing tenders, which have some provision for condensing the exhaust steam and recirculating the water back to the boiler. Mostly, these have been used in places where it is necessary to haul trains across long distances where water is in limited supply. Many designs have been tested on railways around the world. Likely the most successful were two classes of large condensing locomotives put in service by South African Railways.

Ships

In 1685, Denis Papin suggested the possibility to using steam propulsion as a way of propelling ships. About a hundred years later (in 1783), a ship called the *Pyroscaphe* was propelled by steam on the Saóne River in France—for about fifteen minutes. The first practical steamship was the *Charlotte Dundas*, built in 1802 to tow canal barges on the Forth and Clyde Canal in Scotland. It was soon retired from service because the waves created by its paddle wheel blades caused severe erosion to the canal banks. In the United States, credit for the first practical steamship is given to Robert Fulton for *The Claremont*, launched in 1807. Although Robert Fulton did not invent the steamboat, for all practical purposes his was the first viable one.

At the beginning of the nineteenth century, the finest development of the sailing ship was represented by the East India merchant ships (Figure 17.5) and the principal ships of battle of the navies of the world. Those ships represented the ultimate in the technological development of wind as the energy source and in the building of ships from wood. Using steam to propel ships and converting their construction

FIGURE 17.5 The great sailing ships of the nineteenth century are the most powerful devices ever developed, even by today's standards, for the use of wind ENERGY. (From http:// commons.wikimedia.org/wiki/File:EastIndiaman.jpg.)

FIGURE 17.6 A close-up view of the paddle wheel on the *Belle of Louisville*. Steam-driven paddle wheel ships were used worldwide, but remain most famous for their use on the Mississippi River in the United States. (From http://www.luminosityquest.com/wp-content/uploads/2011/11/wheel-1.jpg.)

to iron or steel caused a profound transformation of maritime technology, in both its civilian and naval applications. Some five years after Fulton, Henry Bell began to operate a successful steamship service on the River Clyde (Scotland) with his paddle steamer *Comet*. All the early steamships were propelled by paddle wheels, which were adapted from waterwheel technology, quite familiar to engineers of the day (Figure 17.6). The steam engine is not a very efficient device for converting ENERGY to WORK. Consequently, a large ship was likely to require an immense fuel supply, which in turn posed formidable problems of how to store it on board the vessel. For this reason, early steamships were not thought to be practical except for journeys on lakes and rivers.

By virtue of geography, England and France approached steam propulsion on water from a different perspective than did America. Both England and France already had efficient well-run systems of inland waterways on rivers and canals and so thought of steam primarily in terms of ocean-going naval or merchant vessels. This meant large vessels and long voyages. In early America, however, the interest was in building smaller boats to navigate the rivers, especially large rivers like the Ohio, the Mississippi, and the Missouri. These trips would be short, quick, and cheap.

As early as 1787, the *John Fitch* (modestly named for its builder, John Fitch) was navigating the Delaware River at Philadelphia. In 1790, Fitch operated a summertime steamboat service between Pennsylvania and New Jersey. The steamboat logged some four thousand kilometers and achieved speeds as high as 12 km per hour. However, it was not until the efforts of Fulton that it was possible to produce an efficient, reliable steamboat that—most importantly—would actually turn a profit. His contribution, much like that of James Watt, was for building, logically and systematically, on the work of others, for understanding the principles involved, and learning from mistakes—his and others. By early 1807, Fulton's 45-meter vessel, *Steamboat*, was able to make the 240 km run from New York to Albany in 32 hours, compared to four days by sail. Commercial trips began in September of that year, and in the next year, the famous *Claremont* was launched. Fulton's first steamboats traveled 8 km per hour. As steamboats were improved, they became an extraordinarily successful business. Fulton's first New Orleans boat resulted in a net profit in its first year of $20,000 on an initial investment of $40,000.

By 1838, the Great Western Steamship Company began to operate trans-Atlantic service with the first large steamship, designed by the archetype of the nineteenth-century engineer, Isambard Kingdom Brunel. The *Great Western* was a wooden ship and paddle wheel propelled. Brunel had correctly calculated that the volume of space required to carry fuel, expressed as a fraction of the total volume of the ship, would decrease as the size of the ship increased. The *Great Western* arrived in New York fifteen days out of Bristol with plenty of coal remaining. Brunel's innovations were the first key steps that enabled steamships to take over from sailing ships most of the traffic in goods and passengers on the lucrative North Atlantic crossing.

When profits like those earned by Fulton are possible, competition develops quickly. Speed of travel provides an advantage in this competition. Getting higher speed means using higher-pressure engines, since the higher the pressure, the smaller the pistons and other engine parts could be for the same amount of POWER. However, owners and operators were tempted to operate even these engines at pressures above design levels and to reduce costs by compromising on maintenance and safety margins. The result was a rash of boiler explosions. A 1-tonne boiler linked to a 7.5 kW engine contains enough ENERGY to throw the boiler about six kilometers high with an initial velocity of projection of 330 meters per second. When the steamboat *Oronoko* exploded in 1838, steam swept through the whole length of the boat as if it were hit by a tornado, killing more than a hundred passengers in seconds.

Therefore, the steamboat was also the subject of legislation that first established the authority of the U.S. federal government to regulate an industry in the interest of public safety. In the 1840s, some seventy explosions killed 625 people. In December 1848, the commission of patents estimated that over the years 1816–1848, 233 steamboat explosions had killed 2563 people. In 1850 alone, 277 people were killed in boiler explosions, and a year later the number rose to 407. The steadily escalating death toll resulted in a public outcry. In 1852, Congress created the Joint Regulatory Agency of the Federal Government. This agency actually was successful in lowering fatalities by about a third up to 1860. Eventually, uniform codes and regulations for the construction and operation of high-pressure boilers were adopted, and boiler explosions on boats became a thing of the past.

Despite the success, tremendous controversy was associated with the creation of this agency, and bitter opposition came from those in the steamboat business. (Anyone following current events will realize that some things never change!) To free-market conservatives, government regulation of business was economically and philosophically distasteful. A Congressional inquiry had actually started in 1824, after an accident that may have killed 13 people. Six years later, another inquiry looked into an accident near Memphis that killed 50–60 people. (In the meantime, about 270 people had been wiped out in less spectacular blasts.) This 1830 inquiry led to legislation requiring inspection of steamboat boilers every three months, but it failed to win passage. Among the arguments raised against it was the idea that the Constitution did not include "ensuring the public safety" among the powers reserved to Congress. Seven years later, President Van Buren urged passage in his 1837 State of the Union address. Congress finally acted the following year, after a truly spectacular explosion near Charleston took the lives of 140 people.

This law, which had caused such furor and took so long to win passage, provided that each federal judge would appoint a boiler inspector, who was to examine every steamboat boiler in his jurisdiction twice a year. For this service, the owner of the boiler was to pay the inspector five dollars, in return for which his license to navigate would be certified. The law also provided that, in suits against boiler owners for damage to persons or property, the fact that a boiler had burst was to be considered evidence of negligence. Congress strengthened the law in 1852. Until the problem of so many tragic deaths resulting from bursting steamboat boilers, most Americans had believed that the federal government could not, indeed ought not, interfere with the rights of personal property. The several hundred boiler-related deaths finally convinced many Americans, and most congressmen, that there had to be a point at which property rights of some individuals had to give way to the civil rights of others. This boiler legislation established the precedent on which all succeeding federal safety-regulating legislation would be based.

Through most of the nineteenth century, sail competed with steam, but as the century progressed, steam became increasingly important for ocean-going cargo, passenger, and war ships. By the end of the nineteenth century, steam—generated most often by burning coal—was dominant in both land and water transportation. By the mid-twentieth century, there were four important general types of ship propulsion in service: reciprocating steam engines directly connected to the propeller shafts, steam turbines driving directly or through gears (recall Charles Parsons), diesel engines (discussed later and in Chapter 22) directly connected to the shaft, and steam turbines or diesel engines driving electricity generators that operated motors turning the shafts. Early steamships had reciprocating engines, as did all commercial steamships in the nineteenth century. The reciprocating steam engine was very popular, and it was not until the middle of the twentieth century that the combined tonnage of turbine- and diesel-operated ships exceeded the tonnage of ships with reciprocating steam engines. Steam turbines were eventually displaced by diesel engines for most purposes of marine propulsion, but they still remain in use in some larger vessels.

ROAD VEHICLES

The dominant forms of land transportation of the nineteenth century had some significant drawbacks. Railroads could certainly haul large amounts of freight, or large numbers of passengers, swiftly, but only where there were railway tracks. Horses required constant food and care (regardless of whether they were being used or not) and room for stables. There is also the generally unpleasant "by-product" that has to be collected and disposed of. During the nineteenth century, and even before, engineers dreamed of developing the "horseless carriage." All sorts of propulsion schemes were tried. Steam might be an obvious choice, because of its success with locomotives.

We've seen that Trevithick's idea was to rely on high-pressure steam to build a compact, and therefore light, engine that could be movable and propel a vehicle. On Christmas Eve, 1801, he was ready to test his steam-driven carriage (Figure 10.14). The first test was not successful, because the boiler could not produce enough steam. But three days later, he did successfully drive the vehicle up a hill.

Soon after Trevithick's successful trial of a steam carriage, the Philadelphia Board of Health asked the American inventor Oliver Evans* to construct a steam dredge to deepen Philadelphia's harbor. Since the dredge would have to be moved to the harbor in any case, Evans placed it on a wheeled frame; in doing so, Evans produced America's first self-propelled road vehicle, which he named the *Orukter Amphibolos* (the "amphibious digger," Figure 17.7). This dredge/boat/car used a steam engine to drive wheels on land and a paddle wheel when it was in the water, as well as a bucket

FIGURE 17.7 Oliver Evans's amphibious digger traveled on land and in water. Its mechanical shovel was also operated by steam. (Reprinted from first edition, from *Engineering in History*, by R.S. Kirby, S. Withington, A.B. Darling, and F.G. Kilgour, Dover Publications, Inc., Mineola, New York, 1990.)

---

* Though he is virtually unknown today, Oliver Evans (1755–1819) was a brilliant inventor. He developed the first automated flour mill, which processed grain into flour on a continuous basis. It became widely used during the nineteenth century. Evans's mill connected half a dozen separate operations via a series of conveyor belts, moving buckets, and feeders. It many have been an inspiration for the later development of assembly-line manufacturing. Evans also designed what was probably the first mechanical refrigerator, although the first working version of his design was not built until some years after his death.

chain dredge. On August 12, 1805, Evans drove it around the streets of Philadelphia at six kilometers per hour. This, of course, is just about the same speed that can be attained in any large congested city today with the most modern automotive technology of the twenty-first century. Unfortunately, the rough cobblestone streets broke the wheels of this remarkable contraption. Several months later, the Pennsylvania legislature banned "steam wagons" from turnpikes, partly because of the experiences with Evans's vehicle.

By the mid-nineteenth century, development of steam-propelled automobiles was mostly, though not completely, abandoned for further improvements in locomotive design. In the decades before 1890, the steam-propelled cars failed mostly because of governmental regulation and prohibitions (driven partly by public fears relating the spectacular explosions of engines on steamboats), and not because of perceived limitations in their mechanical efficiency. Compared to a horse-drawn vehicle of comparable capacity, a steam vehicle provided greater speed, lower operating costs (you only need to buy fuel when the vehicle is operating, but you have to feed a horse all the time, whether it is working or not), and less pollution than horses. None of those considerations mattered—the public still feared steam vehicles. The government regulators simply reflected public opinion when they banned steamers. The regulators cited the issues of their dizzying speed (maybe 15 kilometers per hour), the inevitable smoke and steam exhaust, and their potential for blowing themselves up (along with the driver and any hapless passengers).

Travelling was in a troubled state, and the minds of coachmen were unsettled. Let them look abroad and contemplate the scenes which were enacting around them. Stage coaches were upsetting in all directions, horses were bolting, boats were overturning, and boilers were bursting.

—Dickens[1]

Steam as an energy source for transportation has not yet totally disappeared, and of course, the use of steam turbines for generating electricity is of immense importance to our society. But in most of its direct applications to transport, steam has been almost entirely replaced by internal combustion engines and electricity.

THE COMING OF PETROLEUM

One of the first steps toward a practical new form of personal land transport occurred around 1850 in Scotland. James Young, a chemist, found that an oil that seeped out of the rocks in a coal mine in England could be separated into a paraffin wax, an oil that could be used for lubrication, and a liquid like kerosene. Later, Young found that by heating the oil, he could drive off—and subsequently collect—a liquid that we would recognize today as gasoline.

In 1855, Benjamin Silliman Jr., a professor of chemistry at Yale University, was asked to analyze some samples provided to him by the Pennsylvania Rock Oil Company. Silliman found that about 50% of the rock oil could be distilled to kerosene and light lubricating oil, called paraffin oil. The kerosene was especially useful as a fuel for lamps, replacing whale oil.* Silliman also noted that another 40% of the rock oil could be distilled into other products. On the basis of Silliman's one laboratory report, the Pennsylvania Rock Oil Company decided to commercialize their product. Rock oil is much better known to us in the Latinized form of its name, petroleum (*petro*, rock, and *oleum*, oil). The Pennsylvania Rock Oil Company drilled the first successful oil well, the Drake Well, near Oil Creek, Pennsylvania, in 1859.

Kerosene and lubricating oils were valuable products from petroleum. Some of the paraffin oil could also be used as a fuel. One by-product of distilling petroleum was a useless nuisance—gasoline. The temperature at which gasoline boils is very low, and therefore, even though it had no particular use, it was the first component to distill when the petroleum was treated. In other words, while distilling petroleum to make kerosene and paraffin oil, gasoline was an inevitable by-product.

When coal is heated in the absence of air (so that it doesn't catch on fire), various gaseous products are formed. The gas is combustible. Especially before the development of electric lighting, so-called gaslights found extensive use for domestic and public illumination. The gas was also useful as a fuel, for example, in stoves. As the nineteenth century drew to a close, small engines operating on the gas from coal were able to compete successfully with steam engines for many small and medium duties. Gas engines also provided the practical advantage of not needing a boiler to raise steam. There seemed to be no way that a gas engine would compete with a steam engine for transportation, because the engine had to be connected to the gas supply through pipes. But when it was recognized that an alternative was available—the easy-to-vaporize gasoline from petroleum—this situation was revolutionized. Gasoline could be easily carried in a tank to supply the engine.

In comparison to solids or gases, liquid fuels offered several advantages. For one thing, they are easy to transport and store. Second, they are generally easy to feed into the engine, sometimes just by gravity flow alone. When compared on an equal weight basis, the liquid fuels derived from petroleum provide more heat per unit weight than coal. At the time when the petroleum industry was developing as a supplier of illuminating fuel (kerosene for oil lamps), it was also producing quantities of the relatively useless by-product, gasoline. Therefore, as interest developed in using an easy-to-vaporize liquid fuel for vehicle engines, the petroleum industry just happened to be in position to offer such a fuel at an attractive price.

---

* The head of a sperm whale contains large cavities that accumulate oil—in the largest of whales, several tonnes of oil. Whale oil was a very useful product for lamps, in which it was burned to produce illumination. The popularity of whale oil for illumination led to the growth of a significant industry, dominated by the United States, dedicated to hunting and killing whales primarily to obtain this oil. Oil can also be extracted from the blubber of other species of whales, but sperm whale oil was generally considered to be the top quality. Had it not been for the coming availability of kerosene as a replacement, sperm whales might well have been hunted to extinction. Whale oil had other uses in addition to illumination and was a component of the fluid used in automatic transmissions until about 1970. Whaling was a hard, brutal, dangerous business. Herman Melville's *Moby-Dick, or The Whale*, is one of the greatest American novels of the nineteenth century.

In the steam engine, the fuel is burned outside the cylinders and is used to "raise" steam. The steam pushes the piston. This type of device is called an external combustion engine. Adapting an external combustion engine for a horseless carriage would require a firebox to burn the fuel and a separate boiler to raise steam and deliver it to the engine. The alternative to external combustion is the *internal combustion engine*. In this case, the fuel is burned *inside* the cylinder. The heat liberated by the burning fuel is transformed directly to mechanical energy, eliminating entirely the steam-generating step and also eliminating the hardware needed to produce the steam.

AUTOMOBILES

The idea of obtaining work from combustion or explosion in a confined space—the gunpowder engine—dates from the seventeenth century, when several inventors, including Denis Papin, independently attempted to make gunpowder-fueled pumps. In 1684, a French monk, Jean de Hautefeuille, fared no better with suggestions for a gunpowder-driven fountain (but it is interesting to imagine what such a thing might have been like in operation!). Beginning in the 1790s, many people worked on various designs of internal combustion engines, using coal gas, wood gas, or volatile hydrocarbons as fuel. It was not until the late 1860s, when the French inventor Joseph Etienne Lenoir produced small gas engines, that the internal combustion engine became a commercial success. The machine was noisy and cumbersome, but it worked, and with sufficient development, it could be transformed into a viable engine. Smooth action of the Lenoir engine was achieved by the adoption of the "Otto cycle," which we will examine in Chapter 19.

The German merchant Nikolaus August Otto* became so interested in the Lenoir gas engine that he had a couple made under license by a workshop in Cologne. They did not work very well, so Otto, who was convinced of the engine's great potential, sat down and redesigned the whole thing. In this endeavor, Otto teamed up with an engineer, Eugen Langen. The operating efficiency of the Otto–Langen engine was much better than that of Lenoir's gas engine. Otto and Langen carried on experimenting, and the ultimate outcome of their work, in 1876, was a great breakthrough, the first four-stroke engine.

Successful operation of Otto's internal combustion engine design (Figure 17.8) required very fast combustion inside the cylinder. The fuel needed to mix almost instantaneously with air, meaning that it had to be a material that would vaporize very easily. A cheap material that fitted this criterion very nicely was the useless by-product of distilling petroleum—gasoline. The problem of getting gasoline into the cylinders quickly and efficiently was solved in 1893 by a German engineer, Wilhelm Maybach. The inspiration for his device was a popular fashion accessory of that era, the perfume atomizer. Maybach and his colleague, Gottlieb Daimler, adapted the concept of the perfume atomizer to inject or spray gasoline into the cylinders of Otto's internal

---

* Nikolaus Otto (1832–1891) served an apprenticeship in business and then worked in businesses in Frankfurt and Cologne. He apparently had no formal training in engineering or related subjects. Fortunately, he partnered with Eugen Langen, who contributed the technical expertise to improve Lenoir's gas engine and eventually perfect the gasoline-fired engine using what has become known as the Otto cycle.

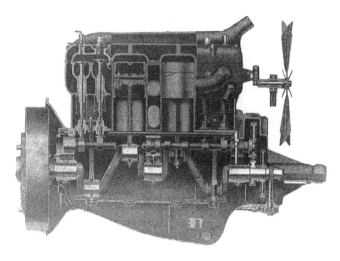

FIGURE 17.8   A cutaway view illustrating the interior of an early four-stroke Otto-cycle engine.

combustion engine. They had invented the carburetor (Figure 17.9). Otto's engine fitted with the carburetor was a new contraption that Maybach and Daimler decided to name in honor of the boss's daughter, a young woman named Mercedes.

About ten years after Otto developed the internal combustion engine, Karl Friedrich Benz used it to build the first practical automobile, a three-wheeled vehicle. Benz created the "horseless carriage," the first genuine motor car. Daimler built a motor truck in 1891, and Benz, the first motorbus in 1895. In 1901, Benz eliminated the last vestiges of carriage design by placing the engine under a hood in front of its car. Though Benz developed the first automobile around 1886, it was not until the early decades of the twentieth century that automobiles became cheap enough for most individuals to be able to obtain one. One day in 1876, a young man in rural Michigan watched a steam traction engine at work. The device was rather interesting, and the experience led him to believe that he might make some improvements and developments in the field of "road locomotion." His name was Henry Ford. He built his first motor car in 1896. In 1913, Ford dropped the price of his Model T (Figure 17.10) to $500. This action meant that the automobile was no longer mainly a plaything for the rich, but was within economic reach of the average working person.

It does not decrease the originality of these innovations, nor should it in any way diminish the accomplishments of those who made them, to realize that they represent inspired assemblies of preexisting parts. The internal combustion engine had been developing steadily since Lenoir's gas engine of 1859. As internal combustion depended on coal gas for fuel, it could be only for stationary engines, because it was directly attached to its source of supply—usually the town gasworks. The advent of petroleum-derived liquid fuels opened the way to a mobile internal combustion engine, carrying its supply in a tank on the vehicle. The gasoline engine, with a carburetor to atomize the fuel, became a versatile lightweight engine suitable for horseless carriages.

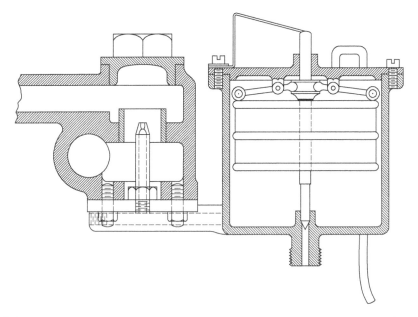

FIGURE 17.9 Maybach's carburetor, inspired by a perfume atomizer, was an effective way of introducing gasoline into engine cylinders. (Reprinted from first edition, with permission from *The Construction of a Gasoline Motor Vehicle*, by C.C. Bramwell, Lindsay Publications, Bradley, Illinois.)

FIGURE 17.10 A Model T Ford, circa 1910. This is one of the first successful automobiles that persons of ordinary financial means could purchase. (From http://en.wikipedia.org/wiki/File:1910Ford-T.jpg.)

By the end of the nineteenth century, there were three choices for powering the transportation device that the French called an "automobile": steam, gasoline, and electricity. The very first automobile speed record was set by an electric car at forty kilometers per hour. Those early electric cars were plagued by the same problems that have plagued electric car development in recent decades: a heavy weight, because of all the batteries needed, and a limited driving range. Steam-powered cars were somewhat successful, notably the Stanley "Steamer," which was probably faster than most other cars of its time, but had the disadvantage that the driver had to wait for the car to develop a head of steam before driving off. By about 1910, the gasoline-fueled internal combustion engine dominated, just in time for Henry Ford and the Model T. The internal combustion engine used ENERGY derived directly from burning fuel, that is, gasoline. All the apparatus connected with generating steam could be eliminated. The engine and its fuel supply were relatively compact, so that it could be economical to build a vehicle that would carry only a few people at a time—that is, the personal passenger car as we would recognize it today.

The automobile engine was almost invariably a four-stroke gasoline engine operating on the Otto cycle, though eventually diesel engines were manufactured small enough to be competitive in automobiles. Otto's engine was not only a major stride in the development of land transportation as we know it today, but also became the parent of powered aviation.

The 1890s' gasoline-powered internal combustion engine inventors chose that technology for more than just its greater potential. They were concerned about a possible revival both of legislative bans on, and of consumer resistance to, steam, though steam worked better than internal combustion engines for the automobiles of the 1890s. The switch of Locomobile (the leading American manufacturer at the time) and Bollée (a leading French manufacturer) from steam to internal combustion just after the turn of the twentieth century established the triumph of internal combustion. Nevertheless, in 1906, a steamer still held the land speed record, and as late as 1917 some steamers lingered on for heavy trucks. Another advantage for the internal combustion engine was that the driving rules of that era typically required that steam-car drivers hold a boiler operator's license, at a time when drivers of gasoline autos needed no license at all.

EARLY AVIATION

George Cayley,* in 1799, outlined the principles of aircraft design, with stiff wings and a body, and a stabilizing tail unit with control surfaces. These principles are in fact the ones on which aircraft design is based today, but Cayley's contemporaries did not even take him seriously. After all, Cayley needed a prime mover—a source of motive power for the airplane. A steam engine was obviously out of the question, being far too bulky and heavy. He even touched on the possibility of using a

---

* Like so many of his contemporaries, George Cayley (1773–1857) made significant contributions in many fields and is considered the founder of the science of aerodynamics. His design for a powered airplane never progressed beyond construction of a model, but he did build the first glider that could fly with a human passenger. A few of his many other contributions include the invention of seat belts, improvements in lifeboat design, and the caterpillar drive that was originally designed for tractors and has now become nearly universal in construction machinery and armored vehicles as well.

"gunpowder engine," but went no further than to suggest that a heat engine with air as its medium must be the best means of powering an aircraft.

In the latter years of the nineteenth century, many inventors took up the challenge of powered flight. The early years of the twentieth century saw the gasoline engine used to drive propellers in prototype aircraft. In this application, there was no real contest among different types of engines. The gasoline-fueled internal combustion engine was the only choice for a lightweight power source.*

By 1896, Professor Samuel Pierpont Langley, of the Smithsonian Institution, was experimenting on the Potomac River below Washington with model airplanes. Some of these, which had wingspreads of up to four meters, were powered with small steam engines. In 1903, he built a full-size one-man airplane and provided it with a remarkably light gasoline engine designed by Charles Manly. This engine was a five-cylinder water-cooled engine of 39 kW that weighed only 57 kg (1½ kg/kW). Langley's plane twice collapsed as it was being launched from the roof of his houseboat, nearly drowning Manly, the pilot. Langley became discouraged and gave up on further experiments. Manly could not have been too happy either. The Wright brothers' gasoline engine, which they made themselves, developed 9 kW and weighed 81 kg (9 kg/kW—compare this with Manly's engine). The whole airplane weighed 340 kg.

The Wright brothers, Wilbur (1867–1912) and Orville (1871–1948), sons of a bishop of the United Brethren Church, had a keen interest in mechanical inventions from boyhood. Their thoughts turned towards flying machines in June 1878 when their father gave them a toy helicopter designed by the Frenchman Alphonse Penaud, who first used rubber bands to power model aircraft. As young men, they experimented with kites and gliders, while running a business repairing and building bicycles. It was Orville who had the idea of constructing an aircraft wing with movable sections (ailerons), so that the pilot could vary their inclination. This was the original Wright Brothers patent. Their first powered machine was a 40 ft. wingspan biplane with a 16 h.p. four-cylinder motor. To reduce the risk of its falling on the pilot, the motor was mounted on the lower wing right of centre, and the pilot lay flat, left of centre, to balance it. It had two propellors, and sledge-like runners instead of wheels. For take-off it was put on a truck with wheels that fitted into the groove of a monorail track. It first flew on 17 December 1903 at Kill Devil Hill, a few miles south of the remote coastal hamlet of Kitty Hawk, North Carolina, watched by five locals from the Kill Devil Life Saving Station, two small boys and a dog. The brothers tossed a coin to decide who should be pilot first, and Wilbur won.

—Carey[2]

---

* In 1894, the American-born inventor Hiram Maxim proposed a high-pressure steam engine for *airplanes*, with the steam being generated on the ground. Though this concept was clearly a dud, one of Maxim's other inventions did take off (in a sense): the so-called captive flying machine, variations of which became, and still are, popular amusement park rides. Maxim was a prolific inventor, who spent years embroiled in a controversy with Thomas Edison over which one of them had really invented the incandescent light bulb (which was in fact first invented by Joseph Swan). Maxim rose to the stereotypical challenge presented to all inventors, that of building a better mousetrap, and supposedly did indeed invent one. Maxim's most enduring invention likely changed the course of human history in 1914–1918: the fully automatic machine gun. Untold thousands of these weapons were used by both sides in the war. Maxim himself became deaf in later life, presumably from the effects of the loud, repetitive muzzle blasts endured while perfecting and testing the design of the machine gun.

FIGURE 17.11 The Wright brothers' first flight, December 17, 1903. (Reprinted from first edition with permission from National Air and Space Museum, Smithsonian Institution, Washington, D.C.)

The Wright brothers first flew on December 17, 1903 (Figure 17.11).

Success four flights thursday morning all against twenty one mile wind started from Level with engine power alone average speed through air thirty one miles longest 57 seconds inform Press home Christmas.

—**O. Wright**[3]

This flight lasted only 12 seconds, but it was nevertheless the first in the history of the world in which a machine carrying a man had raised itself by its own power into the air in full flight, had sailed forward without reduction of speed, and had finally landed at a point as high as that from which it had started.

—**O. Wright**[4]

By 1905, they were carrying passengers.

Well, sir, we pulled that fool thing around over the ground of Huffman Prairie about thirty or forty times, hoisting it up on the derrick so it would get a good start, and we were all hot and sweaty and about played out. What was the use of our wasting our time over such a ridiculous thing any longer? But once more we pulled her up again and let her go. The old engine seemed to be working a little better than normal. Orville stuck his head and nodded to Wilbur and Wilbur turned her loose. And by God the damn thing flew.

—**Webbert**[5]

THE DIESEL

Internal combustion engines had one additional very significant difference from external combustion steam-engine technology. With steam engines, the technology developed first, and the scientific understanding of the engine followed later (e.g., Watt preceded Carnot). But in the case of the internal combustion engines, a much higher level of scientific understanding was required of both inventors and manufacturers virtually from the very beginning. The steam engine had been a product of the empirical tradition of practical millwrights and blacksmiths. Speculation about the fundamental behavior of the engine (the way in which it transformed heat into WORK) and the practical concern to measure the efficiency of the engine prompted the development of thermodynamics by Carnot and his successors. The new science then was available for, and became important to, the development of the internal combustion engine. In particular, the high-compression engine developed by Rudolf Diesel in 1892 was inspired by the scientific principle that it should be possible to induce self-ignition in the fuel by intense compression.

At this time, there was interest in developing internal combustion engines that would operate not on gasoline, but on fuels of poor quality, such as pulverized coal or the high boiling components of petroleum. The fuel that Diesel first intended to use in his engine was finely powdered coal, a fuel that many designers were then thinking of using for various kinds of engines. Diesel soon abandoned that idea and instead started to experiment with heavier petroleum fractions of much higher boiling point (less tendency to vaporize) than gasoline. Diesel also had the remarkable foresight to oils derived from plants—in particular, peanut oil—forerunners to today's biodiesel fuel. In 1892, Diesel patented an engine in which a mixture of air and heavy oil vapor could be ignited by compression alone.

Diesel had developed one form of his early high-compression oil-burning engine specifically for use in submarines, but it quickly proved to be an excellent engine for any small- or medium-sized ship. Marine diesel engines, like diesel engines used in locomotives, have higher thermal efficiency than steam engines. In 1903, five years after the diesel engine was first produced commercially, they were installed in two Russian tankers for service in the Caspian Sea. The first important ocean-going vessel to be diesel-powered was built in 1912. In the 1920s, oil became the dominant fuel for ships, thanks in large part to a decision made in 1911 by Winston Churchill, then serving as First Lord of the Admiralty, to convert the British navy from coal to oil. Eventually, the diesel engine challenged the steam turbine in even the largest marine applications. Virtually all large ships are now diesel powered.

By the second decade of the twentieth century, the diesel engine had been adopted for a wide variety of heavy-duty engines, in ships, tractors, and buses, and continues to enjoy enormous success today in virtually every country of the world. Railway locomotives with diesel engines were first used in the 1920s. They caught on slowly, because of the economic crisis of the Great Depression in the 1930s, followed by the tremendous burdens on manufacturing and logistics of the Second World War. During the 1950s, diesel locomotives totally displaced

FIGURE 17.12   An example of an early diesel railway locomotive. This was the first diesel locomotive in Brazil. (From http://commons.wikimedia.org/wiki/File:EE_VFFLB.jpg.)

the coal-burning steam locomotive from railways in the United States. Most other countries converted to diesel or electric traction for railways in the latter half of the twentieth century (Figure 17.12).

THE JET

In 1929, the Swedish company Bofors decided to develop a gas turbine engine for airplanes. They had in mind an "aeroplane without a propeller," where the propeller's powerful air stream would be replaced by the exhaust jet from a turbine. This kind of turbine would be operated not by steam, but by the high-temperature, high-pressure gases produced by burning a fuel and supplying the gaseous products of combustion as the working fluid to the turbine. Such engines became known as aviation gas turbine engines, commonly called jet engines. Bofors tried to interest the Royal Swedish Air Force in the project, but this happened at a time when the defense budget had been greatly cut, so the scheme was abandoned in 1935.

In the 1930s, Britain still possessed an empire that spanned the globe. Moving people, mail, or goods over the immense distances between London and such cities as Delhi, Johannesburg, or Sydney fostered a growing interest in relatively rapid transportation. High-speed aircraft provided the obvious solution. The English engineer Frank Whittle (Figure 17.13) began serious work on the jet engine.* Though it

---

* Sir Frank Whittle (1907–1996) was the son of a man who owned a machine shop, in which there was a simple gas engine. Whittle became quite familiar with its design and operation, and this experience may have been the motivation that led to his engineering career. After being rejected several times by the Royal Air Force, because of his small stature and poor health, Whittle eventually succeeded in joining and eventually became admitted to a special officers' engineering course. Completion of the course required that each student submit a thesis. Whittle's thesis, "Future Developments in Aircraft Design," basically laid out the foundations for jet aircraft development. Despite Whittle's efforts, great financial difficulties and bureaucratic bungling during the 1930s resulted in the British lead in jet aircraft development being overtaken by the Germans, who suffered their own bureaucratic bungling and an increasingly loony Adolf Hitler. The Germans deployed jet engine combat aircraft many months ahead of the British. Those who enjoy the "what ifs" of history can contemplate what might have happened if one side or the other had had operational jet combat aircraft at the outbreak of war in 1939.

FIGURE 17.13 Sir Frank Whittle (1907–1996), one of the inventors of the jet engine. The device he is holding is an early form of the calculator, made of wood and never requiring batteries. (From http://en.wikipedia.org/wiki/File:Frank_Whittle_CH_011867.jpg.)

is certainly possible to build and operate a jet engine using gasoline, at high altitudes too much of the gasoline can evaporate.* By this time, electric lighting had become widespread in homes, drastically reducing the demand for kerosene as a fuel for lamps. At the same time, gasoline was becoming in shorter supply due in part to building up stocks of fuel for the impending war. As a result, Whittle reworked his design of an engine to use kerosene as the fuel. Today's jet engines use a slightly modified and refined form of kerosene as jet fuel.

Designers in the German aviation industry had been trying since about 1930 to produce a gas turbine for use in airplanes. A team led by Ernst Heinkel was the first to reach a solution. A few days before the outbreak of the Second World War in 1939, the world's first jet plane, the Heinkel He 178, flew for the first time, at Warnemünde. The Junkers firm started to develop another gas turbine engine around the New Year of 1940, and the first flights took place two years later with the Jumo 004 engine. This later became the first jet engine to be mass-produced. A slightly modified version was installed in the German twin-engine fighter-bomber aircraft, Messerschmidt Me 262, which began to be produced in 1943.

After the war, jet engines began to be applied in civilian aircraft. The de Haviland Comet first entered service in 1952. France, the Soviet Union, and the United States

---

* The temperature at which any liquid boils depends on the prevailing atmospheric pressure. Pressure decreases at high altitudes. Gasoline is already a low-boiling, volatile liquid, even at normal pressures near sea level. At the reduced pressures of very high altitude flight, gasoline is even more volatile and evaporates all the more quickly. In some of the early jet fighters of the late 1940s and early 1950s with gasoline-fueled engines, pilots would reach operational altitude only to discover that they had only about a third of their fuel left—the rest had evaporated.

FIGURE 17.14 The Boeing 707 was the first successful civilian jet-propelled passenger airplane, and ushered in the "jet age" of air travel. (From http://en.wikipedia.org/wiki/ File:Boeing_707_138B_Qantas_Jett_Clipper_Ella_N707JT.jpg.)

all had jet airliners in service not long thereafter. Unfortunately, the early Comets suffered from problems of the metal becoming "fatigued" and breaking apart (the easy demonstration of this is to bend a paperclip back and forth until it breaks—surprisingly little effort is needed). Metal fatigue likely contributed to several spectacular crashes of the Comet, which made it difficult to interest airlines in buying Comets and to interest passengers in flying in them. The Boeing 707 (Figure 17.14) was the airplane that really ushered in the era of civilian use of jets.

> *So hug me and smile for me*
> *Tell me that you'll wait for me*
> *Hold me like you'll never let me go*
> *'Cause I'm leavin' on a jet plane*

—Denver[6]

ENERGY DEMANDS OF TRANSPORTATION

Just about every possible ENERGY source has been used, at one time or another, in transportation. Very likely, every one is in service someplace in the world right now. Yet in most industrialized nations transportation is dominated by liquid fuels from petroleum and by electricity. The transportation sector can account for 25% or more of a country's total ENERGY consumption.

Liquid fuels derived from petroleum operate internal combustion engines in gasoline- and diesel-fueled vehicles and jet engines in aircraft. This makes the economy of any country vulnerable to swings in the price of oil. Those countries that cannot supply their petroleum needs with indigenous resources are also vulnerable to the shifting geopolitical winds and to events in oil-producing countries, some of which are politically unstable, hostile to Western nations, or both. Such countries are very vulnerable to the availability of imported oil and its cost. In the last half-century, there have been numerous major shifts in oil pricing and availability: the oil price "shock" of 1973, the oil embargo and consequent price rise in 1979 and the early 1980s, the oil price "spike" during the Persian Gulf war of the early 1990s, and a major run-up in oil prices from the mid-2000s, thanks in part to global financial turmoil and wars in the Middle East.

Oil price hikes or embargoes cause a number of dislocations in society. A rise in oil prices usually kicks off a spurt of general inflation in the economy. When oil is expensive or is in short supply (or both), the public gets serious about such matters as energy conservation, including buying fuel-efficient cars, increasing domestic oil exploration and production, and seeking alternatives, such as making liquid fuels from coal. The United States has fought wars in the Persian Gulf region over oil (despite the official rhetoric about defending democracy in Kuwait or eliminating weapons of mass destruction from Iraq). With the terrible ethnic strife resulting from the breakup of Yugoslavia, some cynics observed that the best thing the people of war-torn Bosnia and Kosovo could have done to secure Western intervention and aid was to strike oil.

It's important to realize that hominids that we would recognize as our human-like ancestors existed about five million years ago. Modern humans (*Homo sapiens*) first evolved some 200,000 years ago. Up until the turn of the twenty-first century, there were still a few people alive who had been born before airplanes existed or before automobiles were readily available. The alternative to human or animal muscle for convenient personal transportation has existed for about a century (if we take the availability of the Model T Ford as the start of the "boom" in personal transportation). The phenomenal pace of technological advancement in the twentieth century is dramatically illustrated by a senior officer in the U.S. Air Force who commanded a unit of jet engine bombers loaded with nuclear weapons—he had taken flying lessons from the Wright brothers!* In many industrialized nations, most families either own a car or have ready access to a one through family members or friends. Those who don't own, or have access to, a car likely aspire to own one. For many students, a car very often becomes one of the first major purchases made after graduation. So, at the beginning of the twenty-first century, petroleum remains a dominant transportation energy source: gasoline for cars, light trucks, and piston-engine planes; kerosene (jet fuel) for jet aircraft; diesel fuel for heavy trucks, railway locomotives, and small ships; and fuel oil for large ships all derive from it.

There are ongoing efforts to examine alternatives to petroleum for transportation energy. Biofuels, primarily ethanol and biodiesel, are touted as replacements

---

* This is mentioned in the remarkable book *15 Minutes*, by L. Douglas Keeney (St. Martin's Press). This very well-written history of the U.S. Strategic Air Command during the Cold War has harrowing accounts of how close the world came to an all-out exchange of nuclear weapons and to the accidental explosion of nuclear bombs in aircraft accidents.

for gasoline and "petrodiesel." Synthetic liquid fuels can be produced from coal, a technology in which South Africa currently leads the world. Natural gas is a possible replacement for diesel engines in city buses. Methanol, which could be made from natural gas or from coal, is a possible automotive fuel.

A serious rise in the price of oil fuels could encourage the development of alternatives to the internal combustion engine as the form of propulsion, with various electric motors being the most obvious candidates. In many countries, railway locomotives are electric, especially true in many areas of Europe and in Japan. Electricity also operates streetcar and tramway systems (sometimes called "light rail"). Both electricity and steam provided strong competition to the gasoline engine at the beginning of the century, but the need to carry bulky batteries or steam-generating equipment weakened their challenge. Changing circumstances could provide new opportunities. Two circumstances in particular are important: First, continuing improvements in the performance of batteries and fuel cells for electric or hybrid (electric motors combined with an internal combustion engine) vehicles. Second, continuing environmental concerns about the emission of polluting gases, including carbon dioxide.

As an ENERGY source for transportation, electricity offers important advantages. Electricity can be generated from many different sources—hydroelectric plants, steam plants burning fossil fuels such as coal, steam plants using nuclear energy as the heat source, steam plants in which the steam is generated using solar heat, wind-operated turbines, and direct conversion of solar energy to electricity. Many of these sources are renewable, in that, in principle, we will never run out of them. This includes the ultimate source of renewable energy, the sun. A smart energy strategy for a country would combine several of these options, rather than relying on an "all the eggs in one basket" strategy. At its point of use, that is, in the vehicle, electricity is essentially nonpolluting. This is *not* necessarily true of the generating plants where the electricity is produced. But it is much easier to capture and deal with potentially harmful pollutants in a small number of large, centralized, stationary sources, than to try to do that with thousands or millions of comparatively tiny and highly mobile sources.

Despite the many virtues of electricity as an ENERGY source for transportation, it remains impractical in certain applications, particularly in aircraft and in ships. In at least some applications, liquid fuels will be crucial for transportation for decades to come.

REFERENCES

1. Dickens, C. *The Pickwick Papers*. Bantam Books: New York, 1983, p. 4.
2. Carey, J. *The Faber Book of Science*. Faber and Faber: London, U.K., 1995, pp. 236–240.
3. Wright, O. Text of telegram sent to his father, December 17, 1903. From Letters of Note, http://www.lettersofnote.com/2010/12/success.html.
4. Wright, O. *Flying* [magazine], December 1913. See also http://www.aero-web.org/history/wright/flight.htm.
5. Webbert, C. Quoted in Bernstein, M., *Grand Eccentrics*. Orange Frazer: Wilmington, OH, 1996, p. 91.
6. Lyrics by Denver, J., 1967. From http://www.lyricsfreak.com/j/john+denver/leaving+on+a+jet+plane_20073310.html.

FURTHER READINGS

Harvey, J.N. *Sharks of the Air*. Casemate: Philadelphia, PA, 2011. This is probably the most recent book on the development of the Me-262, the first operational jet fighter plane. Told from the German perspective, it includes much biographical information on Willi Messerschmidt and historical notes on the development of jet engines.

Levitt, S.D.; Dubner, S.J. *Superfreakonomics*. HarperCollins: New York, 2009. The first chapter of this interesting and entertaining book discusses, with additional detail, the problems created by burgeoning horse populations in the cities of the late 19th century.

Madrigal, A. *Powering the Dream*. Da Capo Press: Cambridge, MA, 2011. This interesting history of "green" energy technologies, including the electric streetcars in use before World War I, touches on the uncertainties, successes, and how some ideas make it into marketplace and others don't. So much of what is being discussed nowadays, in transportation and other energy fields, has been tried before.

O'Brian, P. *Master and Commander*. Norton: New York, 1970. There is no better way to learn the intricacies of the design, nomenclature, and operation of sailing ships than by following the adventures of Jack Aubrey of the Royal Navy, and his friend Dr. Stephen Maturin. Master and Commander is the first of a 20-volume series, all still in print in relatively inexpensive paperback editions.

Riley, C.J. *The Encyclopedia of Trains and Locomotives*. MetroBooks: New York, 2000. A nicely illustrated overview of steam, diesel, and electric locomotives and the history of their development.

Sperling, D.; Gordon, D. *Two Billion Cars*. Oxford: New York, 2009. Among the topics covered in this book is a history of automobile use in the United States, impacts of rapid increase in car ownership in China and India, and possible future innovations in efficiency, technology, transportation planning.

Suzuki, T. *The Romance of Engines*. Society of Automotive Engineers: Warrendale, PA, 1997. This book provides an idiosyncratic, but enjoyable, history of the development of engines, with the major focus on internal combustion engines. An extraordinary melange of engines, vehicles, and aircraft is discussed.

Twain, M. *Life on the Mississippi*. Usually in print in many editions. A classic of American literature that describes travel by steamboat on the Mississippi River in the mid-nineteenth century.

18 Petroleum and Its Products

FOSSIL FUELS AND THE GLOBAL CARBON CYCLE

Petroleum is one of three major ENERGY resources that collectively comprise the fossil fuels. The others are coal and natural gas.* We first begin by inquiring what the term *fossil fuels* actually means. A *fossil* is a remnant of formerly living organisms preserved in the Earth's crust. A *fuel* is a substance that we use as a source of ENERGY, usually by burning it to liberate heat. So, fossil fuels are ENERGY sources originating from the remains of once-living organisms that have accumulated inside the Earth. Collectively, petroleum, coal, and natural gas represent the most important source of ENERGY that we have, for electric power production, transportation, manufacturing, and use in the home.

At first sight, fossil fuels are very diverse materials: a colorless gas, a liquid that ranges in color from a pale yellow to black, and a brown or black solid. Despite these differences, fossil fuels as a group have two things in common: First, they are, by definition, fossils—remains of animals and plants. Second, their major chemical component is carbon. To begin the story of the origin of fossil fuels, we consider how carbon is dispersed throughout the nature. This can be represented by the *global carbon cycle* (Figure 18.1). The key step that starts the cycle going is photosynthesis. Photosynthesis occurs in green plants. It converts water and carbon dioxide from the atmosphere into sugars, which the growing plants then use as an ENERGY source to help produce all of the other chemicals needed for their life processes. Life processes operating on solar energy convert inorganic materials—carbon dioxide and water—into the organic compounds needed for life and growth. These compounds include cellulose, starch, proteins, and fats and oils. These are the compounds that will eventually transform into fossil fuels. The ENERGY source that drives the global carbon cycle, via photosynthesis, is sunlight (solar energy). Essentially,

The fossil fuels represent a stored-up supply of solar energy.

---

* Other fossil fuels are known. Tar sands, also known as oil sands, are a mixture of very thick, viscous oil with sand. This resource is steadily increasing in importance in Canada, where oil is extracted from immense deposits in Alberta. Much of the present and likely future production is exported to the United States. Oil shale is unfortunately named because there is no oil as such in this material, and often the associated rock is not shale. Enormous deposits exist in many places, including the Rocky Mountains of the United States. The oil shale has to be heated to transform the organic material into oil; the rock portion left over can be a serious disposal problem.

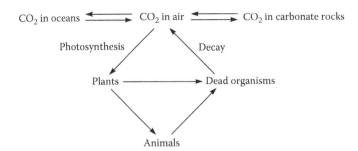

FIGURE 18.1   The global carbon cycle is a model for tracking the fate of carbon through the atmosphere into rocks, natural bodies of water, and living organisms, and its eventual return to the atmosphere.

Carbon dioxide provides the building block for the chemical substances in plants. The plants then either live out their life cycle and die or might be eaten by animals (including us, of course). Eventually, animals too die.* Normally, dead organisms, whether plant or animal, decay. As they decay, the carbon locked up in their bodies in a myriad of chemical compounds is converted back to carbon dioxide, which returns to the atmosphere. In principle, this cycle accounts for the fate of all the carbon in the world. Then where do fossil fuels come from? The preservation of plant and animal remains—the word *fossil* in the name fossil fuels—suggests that it is the decay process that must be interrupted.

Needing these vital ENERGY sources, we of course would like natural processes to maximize the amount of fossil fuels that are produced. Considering the formation of fossil fuels as a detour in the operation of the global carbon cycle, as can be seen in Figure 18.2, two options present themselves: The amount of fuel that can be produced can be affected by increasing preservation at the expense of the decay process or by increasing the production and input of the dead organisms. In fact, both happen in nature.

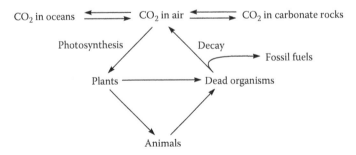

FIGURE 18.2   When some dead organisms are collected in the Earth and preserved against the natural decay processes, their carbon becomes part of fossil fuels, rather than immediately returning to the atmosphere as carbon dioxide.

---

* As we are reminded by the American author Damon Runyon (1880–1946), the odds on life are six to five against.

Maximizing the production of organisms requires abundant light, to drive the photosynthesis process important in the growth of plants, ample warmth (e.g., relatively few organisms live in the Arctic), and ample amounts of moisture (similarly, there are few living things in deserts). These criteria are satisfied by tropical or subtropical environments such as swamps, river deltas, lakes, and shallow seas. The importance of the watery environment derives from two reasons: The abundance of water stimulates growth of organisms. And, the water will eventually help in the preservation of these organisms. In fact, virtually all petroleum is found in what are clearly marine sediments that originally accumulated under the seas and oceans.

An observation that we probably never give much thought to is that there is usually not an extensive accumulation of dead plants on land. For example, a day's outing in a forest doesn't require clambering over the remains of trees that died and toppled decades or centuries ago. On land, exposure of dead organisms to the oxygen in the air causes virtually complete decay to carbon dioxide and water. Here, the main process in the decomposition of dead organisms is affected by aerobic (i.e., air-breathing) bacteria. Decay can be prevented by excluding the principal agent causing the decay—oxygen in the air needed by the aerobic bacteria. This can be accomplished by covering the dead organisms with some kind of sediment (e.g., mud) or having them accumulate in stagnant or slow-moving bodies of water, where the amount of oxygen dissolved in the water is not large. Particularly if the water is stagnant and contains little dissolved oxygen, it helps to protect accumulated dead organisms from the ravages of oxygen from the air.

The kinds of organisms that accumulate and gradually transform to fossil fuels include plankton (single-celled animals or plants that live in aquatic environments), algae, and the higher plants—grasses, shrubs, and trees. Chemically, these organisms are composed of carbohydrates (sugars, starches, and cellulose); fats, oils, and waxes; proteins; and—in the special case of higher plants—lignin, a special material that provides the mechanical structure and rigidity for these plants. As these various chemical components are decomposed, they gradually transform into a family of new substances called *kerogens*. These are brown or black solids that do not dissolve in ordinary chemical solvents and have large, complicated molecular structures (Figure 18.3). Kerogen is rich in carbon and hydrogen and contains some nitrogen, sulfur, and oxygen. A familiar form of kerogen is peat moss.

COOKING KEROGEN

Kerogen can be thought of as being the "halfway point" to petroleum. Eventually (on a long geological time scale) the accumulation of dead organisms and their conversion to kerogen stops. If the environment dries out, the accumulated kerogen could be exposed to the atmosphere and destroyed. Alternatively, the accumulated kerogen could be covered by sediments, such as mud, sand, or silt. Now a new process can begin. As the kerogen is buried more and more deeply inside the Earth, it becomes exposed to high temperatures, and sometimes high pressures, for thousands, hundreds of thousands, millions, or even hundreds of millions of years. The temperature increase inside the Earth is due to a natural phenomenon

FIGURE 18.3   This specimen of oil shale represents a rock rich in kerogen. This sample is from the Green River Formation in Colorado, USA. (From http://geosurvey.state.co.us/ energy/Oil%20Shale/Pages/Whatisit.aspx.)

called the geothermal gradient, an increase of temperature with depth, caused by the decay of naturally occurring radioactive materials (Chapter 23) in the Earth's crust. The long, slow "cooking" inside the Earth opens a new chapter in the formation of fossil fuels.

When kerogens have formed from algae, or plankton and algae, most of their chemical components have derived from fats, oils, and waxes. The molecules are characterized by long, continuous chains of carbon atoms. As the kerogen "cooks," some of the chemical bonds between these carbon atoms break apart. The effect is just like breaking any other sort of chain—smaller pieces of chain are produced. Ultimately, the fragments of chain become small enough that they are able to be liquids at ordinary temperatures. We recognize this liquid product of the cooking of kerogen as petroleum or crude oil.* (We will use these terms interchangeably.)

The amount of crude oil that forms from kerogen and the sizes of molecules in the oil depend on how much the kerogen has been "cooked" inside the Earth. This in turn depends on the depth to which the kerogen has been buried and the accumulated time during which it has been buried. This process is just like cooking at home: The final product depends on what ingredients were used, and how hot and for how long they were cooked. As the conversion of kerogen to petroleum occurs, some of the new molecules will be small enough to be gases. The continued breaking down of the chains of carbon atoms, with higher and

---

* If the kerogen contains abundant lignin, contributed by the higher plants, the chemical story becomes much different. Lignin is characterized by many rings of carbon atoms, interconnected by short chains of carbon atoms. The rings resist being broken apart chemically. As the kerogen undergoes transformation, the structure actually becomes more and more rigid, as rings unite to form larger, rather than smaller, molecular fragments. This lignin-rich kerogen transforms into a hard, carbon-rich solid, coal.

higher temperatures, at longer and longer times, or both, eventually leads to chain fragments so short that they are gaseous at ordinary conditions. This material we recognize as natural gas. Many deposits of petroleum around the world are accompanied by some natural gas.

As the petroleum and gas develop from kerogen, molecule by molecule, they migrate with the water out of the mud and shale strata into the more open strata. Usually they will migrate through the interior of the Earth, moving away from the source rocks through other porous rocks in the Earth. Three things might happen to the oil. It could migrate to the surface of the Earth, where it could be destroyed by contact with the air. Likely far more petroleum has disappeared in this manner than has been captured in the Earth. It could migrate through the porous rocks until it is completely dissipated. Or, it could migrate until it encounters a formation of an impervious layer of shale, salt, or other nonporous rock, which stops the migration and allows the oil to collect. This third option is the one that is desired. Oil accumulates in the pores of the porous rock, but it cannot migrate further because it's stopped by the nonporous rock structure. This trapped deposit of oil is called a *reservoir*. The oil reservoir contains an accumulation of oil held in place by nonporous rocks. This concept is shown schematically in Figure 18.4. If the accumulation is large enough—and if we can find it—it is worth drilling.

Three considerations make an oil reservoir valuable: its location, its size, and the quality of the oil. Factors affecting the location of the reservoir include its depth and the hardness of the rock intervening between us (on the surface) and the oil. These factors indicate how difficult it will be to drill to the reservoir. The size of the reservoir tells something about how much money might be made by extracting and selling the oil. The quality of the crude oil determines, first of all, how easy it is to get out of the ground. The ability of the oil to flow, or to be pumped, improves as the size of the molecules decreases. (The resistance to flow, the *viscosity*, is the property that is usually measured.) A second concern about quality is chemical composition of the oil, in particular its sulfur content. A third property of interest is the range of boiling points of its components. Boiling temperatures and viscosities are functions of the size of the molecules. As useful rules-of-thumb, the larger a molecule, the higher will be its boiling point and the higher will be its viscosity. (These rules work only as long as we confine comparisons to molecules within similar chemical families.) The sulfur content of oil depends on the extent to which the chemical bonds holding the sulfur atoms in the molecular components of the oil have been broken.

NATURAL GAS

Because natural gas is formed under conditions similar to those that formed crude oil, and usually from the same kinds of kerogen, gas and oil often occur together. At high pressures inside the Earth, the gas can even dissolve in the oil and is released when the oil is brought to the surface. A rough analogy is the fizzing when a soft drink bottle is opened, caused by the release of dissolved carbon dioxide that had been held under mild pressure. Alternatively, gas can exist as a separate material, usually over the pool of oil. In addition, the gas can also migrate away from the oil.

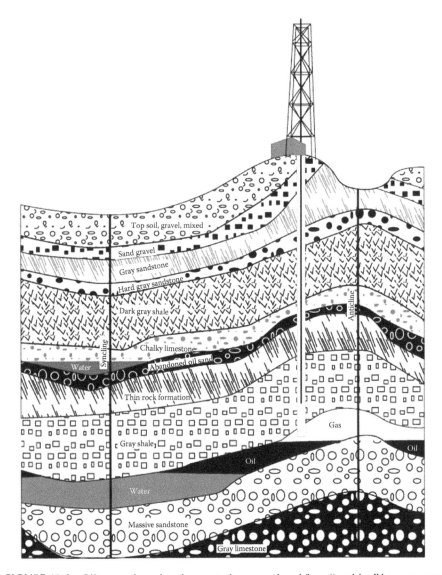

FIGURE 18.4 Oil reservoirs exist where petroleum, produced from "cooking" kerogen, can accumulate in porous rocks inside the Earth's crust. Finding these reservoirs can sometimes be challenging.

The only thing that can keep the gas inside the Earth—that is, keep it from being lost to the atmosphere—is a layer of impervious, nonporous rock that serves as a "cap" to prevent the further migration and escape of the gas. The gas itself is held in some sort of porous rock, such as sandstone.

Sometimes gas escapes and comes to the surface. Ancient seafarers could see the flames of a natural gas fire on mountains near Cape Gelidonya (near modern Finike, Turkey) and use them as a navigational aid in their travels.

Currently a new natural gas industry is developing in some parts of the world around the so-called tight gas. These deposits contain natural gas trapped inside impervious rocks, such as shale. Thus the term shale gas is also used for this resource. Worldwide, at least 32 countries have known deposits of shale gas. By far the largest amount is in China. Other major deposits occur in the United States, Argentina, Mexico, and South Africa. Smaller, but still significant, amounts are found in such countries as Australia, Canada, Libya, Algeria, Brazil, Poland, and France. Not all of these countries have been endowed with much recoverable petroleum or conventional natural gas. In such cases, their ability to exploit shale gas to meet domestic energy needs and possibly for exports could have great repercussions for the geopolitical balance of trade in energy supplies.

Releasing this gas requires drilling into the shale and breaking it apart. Breaking up the shale is often accomplished by pumping water at high pressures into the gas well, a process known as hydraulic fracturing or "fracking." The water used for fracking contains various chemicals, and the fracking process might release other materials, besides gas, from the shale. The so-called "frack water" has become an extremely contentious environmental issue. Concerned citizens believe that their drinking water supplies might become contaminated by frack water, leading to health problems; the gas industry argues otherwise.

The principal component of most natural gas deposits is methane. Many gas deposits contain the related compounds ethane, propane, and butane. Some deposits contain compounds, such as pentane and hexane, which would be liquids at ordinary temperatures, but are present as vapors in the gas because of the higher temperature inside the Earth. When such gases are brought up to the surface, pentane and hexane condense back to the liquid state, forming a product known as *natural gasoline*.

If the natural gas is cooled below ambient temperatures, butane and propane can be condensed to liquids. The condensed butane and propane are sometimes referred to as *liquefied petroleum gases*, more familiarly known by the initials LPG. In areas that have no easy access to natural gas distribution systems, LPG makes a very convenient fuel gas for home heating and cooking. Butane and propane can also be used by the chemical industry to produce many useful materials, such as polypropylene or synthetic rubber.

Some deposits of natural gas also contain hydrogen sulfide, H_2S. Gases that contain hydrogen sulfide are called *sour gas*. Hydrogen sulfide is a very undesirable component of natural gas. It has a vile odor, giving rotten eggs their characteristic odor, and is poisonous. It dissolves in water to form a mild acid that could corrode metal components of systems for gas distribution and use. If the gas were burned, the H_2S would be converted to the oxides of sulfur, which are health hazards and contribute to the serious environmental problem of acid rain (Chapter 27). Hydrogen sulfide, a mild acid, can be removed from natural gas by absorbing it in a chemical that is a mild base. This process is called *sweetening*.

After natural gas has been sweetened, and the butane and propane have been removed, the gas that remains is mostly methane, also containing some ethane. This is the product that actually comes to us as consumers of natural gas. Most commercial natural gas also contains a very small amount of sulfur compounds added to the gas deliberately to serve as odorants. These odorants function as a warning signal to our noses to indicate a gas leak. It may seem odd, even self-defeating, that we go to the trouble to sweeten natural gas and then add back a sulfur compound. However, the sulfur compounds that are used as odorants have such extraordinarily horrendous odors that we can detect even one part of odorant mixed with a million parts of natural gas. The contribution to acid rain is miniscule and is certainly outweighed by their value as a warning that a gas leak is occurring. This is important because methane itself is odorless, but certain proportions of methane in air are explosive. People need to know if the gas is leaking.

> *There was a gas explosion at the home of Larrieux, in Bordeaux. He was injured. His mother-in-law's hair caught on fire. The ceiling caved in.*
>
> —Fénénon[1]

Natural gas is truly a premium fuel. On an equal weight basis it provides the most heat of any of the fossil fuels, about 56 MJ/kg. Natural gas heating systems are easy to control from a thermostat. These systems can essentially be turned on instantaneously (as, e.g., compared to the time it takes to get a wood fire or coal fire burning well). Natural gas is "clean" heat—it leaves no ashes and produces no mess. No space in the home is required for storage of gas, in contrast to space needed for oil tanks, a coal bin, or a wood pile. There are no delivery problems—the furnace draws gas from the distribution system as needed, and there is no necessity to call a dealer to arrange for delivery.

PETROLEUM AND ITS AGE–DEPTH CLASSIFICATION

Petroleum, as a rule, contains 82%–85% carbon and 12%–14% hydrogen. These numbers do not sum exactly to 100%; small amounts of oxygen, nitrogen, and sulfur account for the balance. The range and complexity of naturally occurring hydrocarbons is extremely large, and the variation in molecular composition from one deposit to another is enormous. All crude oils differ in the fractions of the various hydrocarbons (described below) they contain. The specific chemical compounds in petroleum have molecules of different shapes and sizes. Physically, crude oils may be black, heavy, and thick, like tar, or brown, green, to nearly clear straw color, which flow almost as easily as water. Most crude oils are less dense than water.

The temperatures inside the Earth break down the long chains of carbon atoms in kerogen into shorter chains. Most of the compounds that are produced belong to a family called, in the terminology of the petroleum industry, paraffins. The physical state in which a paraffin exists at a given temperature and pressure depends on the number of carbon atoms in its molecule. When the carbon atoms are connected in a linear chain, that relationship is as summarized in Table 18.1.

A crude oil is likely to contain a mixture of paraffins having five or more carbon atoms. That alone would make oils chemically complex substances. However, the situation is made even more complicated by the many ways in which carbon atoms can join together. As a simple example, the molecular structure of the compound with four carbon atoms, butane, could be drawn as shown in Figure 18.5. However, there is another way in which four carbon atoms can be joined together (Figure 18.6).

TABLE 18.1

Relationship of the Number of Carbon Atoms to Physical State of Oil

Number of Carbon Atoms in the Chain	Physical State at Room Temperature and Pressure
<5	Gaseous
5–15	Liquid
>15	Highly viscous liquids to waxy solids

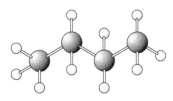

FIGURE 18.5 The molecular structure of butane illustrates a linear arrangement of four carbon atoms.

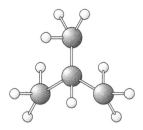

FIGURE 18.6 In comparison to Figure 18.5, the molecular structure of isobutene illustrates a branched arrangement of four carbon atoms.

As the number of carbon atoms in the molecule increases, so does the number of possible ways for linking them together. The five-carbon-atom compound, pentane, can have three possible arrangements (Figure 18.7). Notice that in the first structure, the carbon atoms link together in a chain that is more-or-less straight.* Neither of

---

* The apparent straightness of a chain of carbon atoms is an illusion caused by the fact that we are forced to represent a three-dimensional molecule on a two-dimensional sheet of paper. In the paraffins, each carbon atom forms chemical bonds with four other atoms (hydrogen or carbon). These four chemical bonds are at angles of 109.5° to each other, directed toward the corners of a tetrahedron. Therefore, there is no such thing as a truly straight chain of carbon atoms in these compounds. Rather, the chain of carbon atoms proceeds in a somewhat zig–zag fashion. Chemists have developed conventions for representing specific three-dimensional arrangements on paper for those situations where it is very important to attend to, and keep track of, the exact three-dimensional structure of a molecule. For our purposes, it is not necessary to do so; consequently we can write molecular structures *as if* the four bonds that a carbon atom forms with its neighbors were at 90° angles to each other.

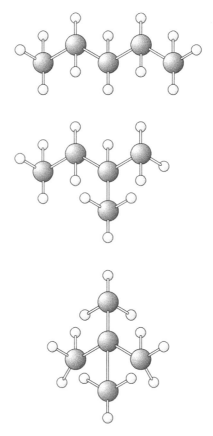

FIGURE 18.7 Five carbon atoms can be linked in a linear arrangement, or in two different branched structures. From top to bottom, these compounds are pentane, isopentane, and neopentane.

the two others has a single chain containing all five carbon atoms. Rather, these structures can be visualized as consisting of chains having branches. The number of possible molecular structures increases drastically as the number of carbon atoms increases, as summarized in Table 18.2. To distinguish between the two broad categories of linking carbon atoms, a straight chain and a chain with branches, we use the terms straight-chain paraffins and branched-chain paraffins. Both kinds can occur in petroleum, though usually the straight-chain form dominates.

A further possible structural arrangement of carbon atoms is a ring (e.g., Figure 18.8). This family is called cycloparaffins (the prefix "cyclo-" indicating the ring arrangement) or naphthenes. The naphthenes constitute a distinct family from the paraffins, because there are slightly fewer hydrogen atoms in the naphthenes. As an example, the paraffin with six carbon atoms (hexane) contains fourteen hydrogen atoms. The naphthene with six carbon atoms (cyclohexane) contains only twelve hydrogen atoms.

TABLE 18.2
**Relationship of the Number
of Carbon Atoms to Possible
Paraffin Structures**

Number of Carbon Atoms	Number of Possible Structures
1–3	1
4	2
5	3
7	9
10	75
20	366,319

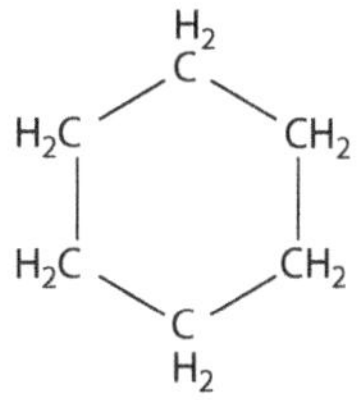

FIGURE 18.8 The molecular structure of cyclohexane shows a cyclic arrangement of carbon atoms.

Finally, a very special family of cyclic compounds is the aromatic hydrocarbons.* The parent compound in this family is benzene (Figure 18.9; notice that benzene has only six hydrogen atoms, compared to twelve in cyclohexane). In principle, naphthenes can have any number of carbon atoms greater than three in a ring. Aromatics are built only of hexagonal rings. Aromatics as a family have some very special chemical properties that make them quite distinct from the

* Benzene was discovered in 1825 by Michael Faraday. It is the parent of the family of aromatic compounds, all of which are built of joined hexagonal rings. The characteristic properties of these compounds derive from the nature of the bonds in the molecules. In any molecule, a carbon atom always contributes four electrons. In the paraffins and naphthenes, these four electrons are used to form bonds with four other carbon or hydrogen atoms. In aromatic compounds, a carbon atom forms bonds with only three other atoms. The fourth, "extra" electron from each carbon atom contributes to the formation of a type of bond that is spread over the entire molecule, rather than being only between two adjacent atoms. These so-called "delocalized" electrons make aromatic molecules particularly stable in many chemical reactions. The stability of aromatic molecules results in less heat being evolved when they burn than would be predicted from the simple molecular composition.

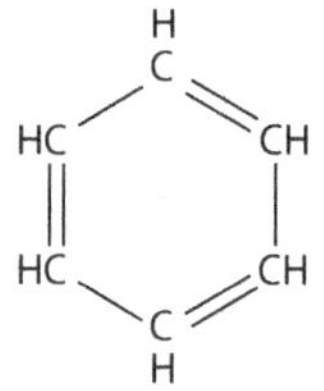

FIGURE 18.9 Aromatic hydrocarbons are a special class of cyclic carbon atoms with unusual stability. This molecule, benzene, is the parent compound of the whole family or aromatic hydrocarbons.

paraffins and naphthenes, and this distinction has important consequences when they are present in petroleum or its products.*

In addition to the hydrocarbons, which, strictly speaking, are compounds that contain only carbon and hydrogen, there can be many molecules that contain one or more atoms of nitrogen, sulfur, or oxygen. These are sometimes grouped together in the jargon of the petroleum business as NSOs. For the most part, we will not be too concerned about the NSOs. Sulfur-containing compounds are an exception. Burning them contributes to the environmental problem of acid rain. In addition, many sulfur compounds have stupendously awful odors. Reading their descriptions uncovers a litany of comments such as "odor of skunk," "rotten onions or leeks," "repulsive garlic," and even just plain "repulsive." Petroleum containing sulfur compounds is called a high-sulfur crude, or sour crude.† This sulfur will need to be removed at some point in the refining process.

The properties of a given sample of crude oil are determined by its constituent molecules. These in turn are determined by the kinds of organisms that originally accumulated, the chemical changes experienced during the formation of the original kerogen, and the depth and duration of burial of the kerogen. Thus various crude oils from around the world vary greatly in their molecular compositions. The chemical complexity of one sample of oil is illustrated in Figure 18.10.

Both the molecular size and the sulfur content depend on the extent to which the kerogen has been "cooked" inside the Earth. The extent of cooking is determined by how much the kerogen has been heated—which is determined by its depth of burial inside the Earth—and by how long it has been heated, which is determined by its

---

* Aromatics are usually not desirable components of petroleum products. In jet fuel and diesel fuel, they can lead to formation of soot as the fuel burns. Soot contributes to air pollution, as has been experienced by anyone exposed to the billowing cloud of soot in the exhaust of a bus or large truck with a diesel engine. Many of the components of soot are large aromatic molecules, which are also suspect carcinogens. Steady exposure to soot could possibly lead to various forms of cancer. Small aromatic molecules, such as benzene, used to be added to gasoline to enhance the octane number (Chapter 19). However, they too have come to be recognized as suspect carcinogens. Chronic exposure to benzene vapor may lead to leukemia. Since many crude oils contain some aromatic compounds, refiners are faced with the challenge of removing them to meet concerns for air quality and human health.

† Strictly speaking, sour crude contains dissolved hydrogen sulfide, while high-sulfur crude contains larger sulfur-containing molecules that would be liquids at ordinary temperatures and dissolve in, or mix with, the hydrocarbon molecules. Informally, the terms *sour* and *high-sulfur* are used interchangeably.

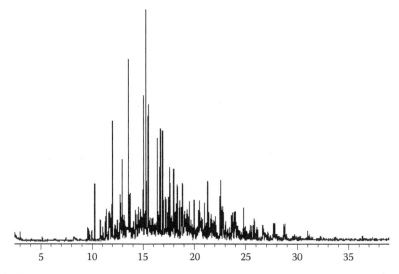

FIGURE 18.10   This is an analysis of a petroleum sample done by a technique called gas chromatography. The important information is that each vertical "spike" on this diagram represents a different chemical component of the petroleum, illustrating the highly complex composition. (Reprinted from first edition, courtesy of The EMS Energy Institute, Penn State University, University Park, Pennsylvania.)

geological age. In other words, the properties of a particular crude oil are determined by its age and its depth. This concept provides a way of classifying or categorizing crude oils, and relating these categories to oil properties.

In petroleum that is just at the beginning of the oil formation process, long chains of carbon atoms have not been broken down very much. First, the kerogen has likely not been buried very deeply and, consequently, has not been exposed to high temperatures inside the Earth. Second, the kerogen has likely not been buried for a long time. The dominance of long-chain compounds gives this oil some distinctive properties: It is likely to be fairly dense. The long chains of carbon atoms can tangle with each other as the oil tries to flow, so the oil is likely very viscous. It has a high percentage of high-boiling components. In addition, there has been little chance for any sulfur compounds to break down. Thus the oil likely has high sulfur content. Because of the short time (geologically) and shallow depth of burial, these oils are classified as *young–shallow* crudes. Young–shallow crudes are usually heavy, viscous, and sour. This oil will require considerable refining before use, and therefore is not a very desirable material for input to an oil refinery if alternatives are available.

Deeper burial brings exposure to higher temperatures; the higher temperatures facilitate a further breaking down of the carbon chains. In addition, at least some of the chemical bonds retaining sulfur atoms in the molecules are broken. This oil would be classified as a *young–deep* crude. Compared to other types of oils, we would expect that the young–deep crudes would be of moderate density and viscosity and moderate sulfur content. Sometimes in life, especially in business transactions, we encounter the expression "time is money." To a geologist, time is temperature. The same changes that were affected by a deep burial and exposure to high temperatures could occur with shallow burial (hence low temperatures) but over very long periods of time. Cooking again provides an analogy: If we need something to cook more quickly, we turn up the heat.* On the other hand, if not everything is yet ready, or if the meal needs to be delayed a while, we can let the cooking proceed for a longer period of time by turning the heat down. A young–deep crude developed its characteristic properties because it was "cooked" for relatively short times at high temperatures. We ought to be able to arrive at almost the same point by doing the geological cooking for a long time at low temperatures. Such oils would be called *old–shallow* crudes. Indeed these oils often show similarities with the young–deep crudes. Since these two kinds of oils have similar properties, and the properties are between the two extreme classifications, the oils are sometimes lumped together as *intermediate* oils.

The final case involves long periods of burial deep inside the Earth. An extensive breakdown of the molecules has occurred and produces an oil rich in low-boiling components, having low viscosity and low sulfur content (i.e., is sweet). The long-time exposure to high temperatures provides the impetus for breaking the carbon chains extensively to relatively small molecules; and, most of the sulfur has been driven out by this geological cooking process. The product, *old–deep* crudes, have a low density, low viscosity (i.e., they have little resistance to flow), and very low sulfur content. These low-sulfur oils are called *sweet crudes*. Old–deep crudes were found in Pennsylvania when the American oil industry was in its infancy. The term *Pennsylvania crude* or Penn-grade crude is sometimes used as a standard of comparison to denote oil of this high quality, regardless of its source. For example, Moroccan oil is often old–deep oil. Most refineries, if given a choice, would prefer to operate on Pennsylvania-crude-quality oil. It would provide a high yield of desirable products, would be easy to pump and handle, and would require little or no sulfur removal. Unfortunately, only about 2% of the world's remaining oil reserves is of Pennsylvania crude quality. That means that oil refiners have to work harder and harder to produce the products that we want, and that satisfy increasingly strict environmental regulations, from oils of lesser quality than old–deep crudes.

These four broad categories of crude oils range from lightly colored free-flowing liquids with a mild but not unpleasant odor to thick dark liquids with the consistency of road tar and that, frankly, stink. This is the stuff that goes into the refinery. At the other side of the refinery, we as customers expect and demand products that are of consistent quality. That demand arises because all of the items that we have built to use petroleum products—automobile engines, for example—are designed to handle a specific fuel of reasonably consistent properties and behavior. The first challenge that refiners face is how to take this highly variable mixture of "stuff" in these different kinds of oil and convert it into products of consistent quality.

---

* This gives rise to a delightful urban legend involving electric ovens that have a "self-cleaning" setting. On that setting, the oven might reach 500°C. It is claimed that putting the oven on "self clean" will roast an entire turkey in a half hour.

PRODUCTS FROM PETROLEUM: INTRODUCTION TO REFINING

If we were to analyze different samples of crude oils, we would find that they contain literally hundreds of individual chemical compounds. In principle, we could separate the crude oil into each of these components, which would certainly guarantee products of consistent quality. However, there are two serious problems with this idea. First, separating the crude oil into its individual molecular components would be so difficult that the products would be extremely expensive. Second, the percentages of all the components must, by definition, add up to 100 (since nothing can have more than 100% of ingredients). Since each sample of oil has several hundred components, most of them are, necessarily, present at less than 1% each. Therefore the yields of these very expensive components would be far too small to meet market demands. The compromise position separates the oils not into pure components, but into groups of components, of which each group has properties that are relatively consistent, and that vary only over narrow ranges. This compromise approach of separating crude oils of variable quality into affordable products of consistent properties is the essence of petroleum refining.

The technology for making this separation is distillation. This is a separation process that relies on differences in boiling behavior of the components of a mixture. In an oil refinery, distillation is done in a large, vertical cylindrical vessel called a distillation column or distillation tower (Figure 18.11). Distillation is the heart of an oil refinery and is the single most important process in refining petroleum into useful products. In the laboratory, we use distillation to separate individual components of a mixture at specific temperatures, their boiling points. In a refinery, a given distillation product is produced over a range of temperatures, the boiling range.

The first product from the crude oil in a distillation tower is gases that were dissolved in the oil. These are collected and liquefied by mild pressure and refrigeration. LPG can be used as a convenient fuel gas, especially in places where natural gas is not available, or can be converted to useful chemical products. Then, in increasing order of boiling range, the next products are gasoline or petrol; naphtha, which can be converted to gasoline, or used as a solvent; kerosene (sometimes called paraffin) and jet fuel; diesel fuel; and heating oil or fuel oil. Some materials will not distill at all. This is called the residuum, more commonly known as *resid*. Resids can be treated to make lubricating oils, asphalts, and paraffin wax.

Distillation is the most important step in petroleum refining.

Nothing happens in a refinery until the crude is first distilled. Many other processing steps are then applied to convert the distillation "cuts" into salable products. Some of these operations will be discussed in the coming chapters. The next few chapters focus on certain of these products and the refinery operations used to make them, particularly products relevant to transportation.

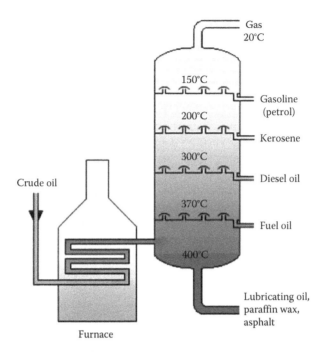

FIGURE 18.11   A schematic diagram of a petroleum distillation unit. This is the most important process in petroleum refining, separating families of petroleum components based on their boiling temperatures. (Reprinted from first edition, with permission from *Chemistry in Context*, by A. Truman Schwartz, Diane M. Bunce, Robert G. Silberman, Conrad L. Stanitski, Wilmer J. Stratton, and Arden P. Zipp, McGraw-Hill Book Company, New York, 1997.)

REFERENCE

1. Fénéon, F. *Novels in Three Lines*. New York Review Books: New York, 2007.

FURTHER READINGS

Anderson, R.O. *Fundamentals of the Petroleum Industry*. University of Oklahoma: Norman, OK, 1984. Though many of the statistics are now out of date, this well-illustrated book still provides a good overview of the industry.
Conaway, C.F. *The Petroleum Industry: A Nontechnical Guide*. PennWell: Tulsa, OK, 1999. A book designed for nonspecialists. It covers virtually every aspect of the petroleum industry, beginning with how the Earth was formed, through the marketing of petroleum products.
Schobert, H.H. *The Chemistry of Fossil Fuels and Biofuels*. Cambridge University Press: Cambridge, U.K., 2013. Chapters include the formation of petroleum, its properties and classification, distillation, and refinery operations for making various products. This book assumes that the reader has some background in chemistry.

19 Gasoline

As the twentieth century dawned, Henry Ford began to realize his dream of building a cheap, reliable automobile that everyone could afford. Ford is credited with a dominant contribution to making the automobile practical and affordable. He did not invent the automobile, just as Watt did not invent the steam engine, nor Stephenson the steam locomotive, nor the Wright brothers the airplane. The contributions of all of these individuals lay in building upon pioneering efforts made by others, and, in a sense, sliding the final piece into a complicated jigsaw puzzle. Not only did the coming of an affordable and practical automobile revolutionize transportation, it had significant effects on the petroleum industry. Automobiles were fueled by gasoline, at the time a low-value by-product of the oil industry. Kerosene remained the most important petroleum product until about 1920. Then two separate factors converged. First, the increasing availability of low-cost automobiles made it easy for more people to own cars. This led to more driving, which in turn increased the demand for gasoline. Second, the steady expansion of electric power networks made electric lighting increasingly available, and this in turn reduced the demand for kerosene for illumination.

Therefore the roles reversed. Gasoline became the dominant petroleum product with kerosene as a "throwaway." Kerosene has made a comeback since the 1960s, due to the increasing dominance of jet aircraft for civilian air transport (jet fuel is basically a refined form of kerosene), but in many countries, gasoline is still the main petroleum product by far. In recent decades, another factor has come into play, the increasing "dieselization" of the fleet of cars and light trucks throughout the world. This change, which has proceeded to different extents in different parts of the world, has caused an increase in demand for diesel fuel at the expense of gasoline. In the United States, the light vehicle fleet is still dominated by gasoline engines.

Two considerations dominate the production of gasoline in refineries: its performance in the engine and the yield (i.e., the amount produced) per barrel of crude oil. We will consider engine combustion performance first, since that issue has a bearing on strategies used for producing gasoline.

OTTO-CYCLE ENGINES AND THEIR PERFORMANCE

Virtually all automobile engines work on the four-stroke principle invented by Nikolaus Otto over a hundred years ago. Although engines come in various sizes, the combustion process is easiest to visualize if we consider a four-cylinder engine. It's also easiest if we assume two valves per cylinder. Sketches showing the operation of a simple engine of this kind are shown in Figure 19.1.

In the intake stroke, the piston moves downward, an intake valve opens, and a gasoline–air mixture enters the cylinder. This mixture is compressed by the upward

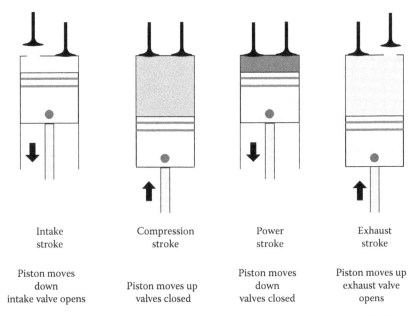

Intake stroke	Compression stroke	Power stroke	Exhaust stroke
Piston moves down intake valve opens	Piston moves up valves closed	Piston moves down valves closed	Piston moves up exhaust valve opens

FIGURE 19.1   A schematic diagram of the four-stroke Otto cycle, which is still the most common operation of a gasoline-fueled engine.

movement of the piston in the compression stroke. Then a spark plug ignites the gasoline–air mixture during the ignition, or power, stroke. Heat released from the burning gasoline raises the temperature and pressure in the cylinder. The high pressure pushes the piston down. This stroke converts the chemical potential ENERGY in the gasoline molecules into the mechanical WORK of propelling the car. Finally, in the exhaust stroke, the upward movement of the piston pushes the gaseous products of combustion out of the cylinder, making it ready for the next intake stroke. The engine produces WORK only in the ignition stroke; in the intake, compression, and exhaust strokes, the piston is being pushed up or pulled down by the operation of the engine, and not by ENERGY released from gasoline.

The ratio between the volume of the cylinder when the piston is at its lowest point and the volume when it's at its highest point is called the *compression ratio*. The compression ratio provides an approximate indication of the POWER of the engine, the acceleration available, and the maximum speed. It also indicates the maximum pressure in the cylinder, since the higher the compression ratio, the more the air/fuel mixture is compressed—raised to a high pressure—during operation. As examples, the engine in a Model T Ford from the 1920s had a compression ratio of about 4:1. Typical midsized cars on the market today have engine compression ratios around 8:1. High-performance cars may have compression ratios of 10:1 or higher. The compression ratio has an important effect on the combustion of the fuel.

During the power stroke, ignition of the gasoline/air mixture by the spark plug causes a flame to progress through the mixture as combustion proceeds. This causes an immediate rise in the temperature (and therefore the pressure too, thanks to Robert Boyle) of the mixture in the cylinder. As the hot mixture expands, it pushes

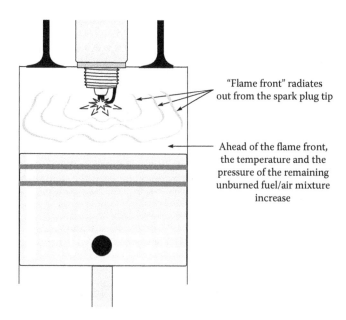

"Flame front" radiates
out from the spark plug tip

Ahead of the flame front,
the temperature and the
pressure of the remaining
unburned fuel/air mixture
increase

FIGURE 19.2   In normal operation, the gasoline/air mixture is ignited by the spark plug, with a flame progressing through the mixture until combustion is complete.

down on the piston, doing the WORK necessary to propel the car. In normal operation, the hot burning gasoline/air mixture expands smoothly and delivers its chemical ENERGY to the piston as WORK. Under certain conditions of engine operation, the pressure and temperature of the fuel become so high that the remaining unburned air/fuel mixture explodes, rather than continuing to burn smoothly (normal, smooth combustion is shown in Figure 19.2). When the explosion of the unburned air/fuel mixture occurs in the cylinder, it is so violent that that it can be heard inside the car. (On an equal-weight basis, a gasoline–air mixture is a more devastating explosive than dynamite.) This phenomenon is called *engine knock*.

Engine knock reduces fuel economy; that is, it wastes gasoline. It reduces the apparent POWER and acceleration of the car. It leads to considerable wear and tear on engine parts and increased maintenance. Engine knock became more and more of a problem beginning in the 1930s, as engine designs evolved to higher compression ratios. Two parameters affect engine knock: engine design and fuel quality. The key feature of engine design is compression ratio. The higher the compression ratio, the higher will be the pressure inside the cylinder at the moment of ignition. But the higher the initial pressure inside the cylinder at the instant the fuel ignites, the easier and more likely it will be for the pressure to get high enough to cause the engine to knock. In other words,

The higher the compression ratio, the more likely the engine is to knock (with a given type of fuel).

OCTANE NUMBER

To investigate the relationship between gasoline composition and engine knock, automotive engineers studied a variety of pure chemical compounds and their behavior in engines under standardized test conditions. The essence of their findings is summarized in Figure 19.3. To describe the knocking behavior quantitatively, two standards of comparison were established. The linear paraffin heptane was assigned a value of 0. The highly branched paraffin called, informally but quite incorrectly, "iso-octane" or even "octane"* was assigned a value of 100. The *octane number* of gasoline is equal to the percentage of "octane" in a blend of heptane + "octane" that has the same knocking characteristics as does the gasoline being rated, when both are compared in a standardized engine test. So, a gasoline of 90 octane number, for example, gives the same performance in a test engine as a blend of 90% "octane" with 10% heptane. The numbers 0 and 100 are purely arbitrary, and not fundamental properties of the compounds involved. They were selected to make a convenient numerical scale.

The higher the octane number of a gasoline, the lower is its tendency to knock. Therefore, increasing the octane number of the gasoline counteracts the effects of increasing the compression ratio of the engine.

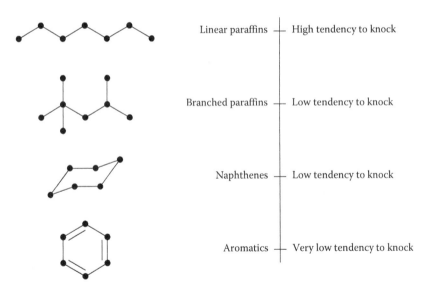

FIGURE 19.3   The different families of compounds in gasoline have characteristic knocking behavior. The molecules shown in this chart are representative examples of each family; only the carbon atoms are shown.

---

* In the formal nomenclature of organic chemistry, this compound is called 2,2,4-trimethylpentane. Iso-octane is 2-methylheptane. Octane itself is the linear paraffin containing eight carbon atoms. However, the term *octane number* is so heavily embedded in the terminology of fuels that it will surely be with us as long as gasoline is used. Ironically, the linear paraffin correctly named octane would be an even worse component of gasoline than is heptane. So, the octane number of true octane would be less than zero!

The higher the compression ratio of the ENGINE, the higher the octane number of the fuel needed to avoid knocking.

Nowadays there are three grades of gasoline sold at many service stations: 86–87 octane, intended for most small or midsized cars; 89 octane, used in some midsized or large cars; and 92–94 octane, used in cars with large engines or in high-performance cars.

Based on market demand, and on the need to meet engine performance requirements, there are two key points regarding gasoline: we want to have gasoline available in the 86–94 octane range, and we want to produce enough gasoline from each barrel of crude oil refined to meet the regional or national demand. The octane number of a gasoline can be increased either by increasing the percentage of branched-chain and aromatic hydrocarbons or by adding so-called octane enhancers (or a combination of both). Currently the United States probably represents the most formidable requirement for demand, equivalent to making 45–50 barrels of gasoline for every 100 barrels of petroleum refined.

PRODUCING GASOLINE IN THE REFINERY

STRAIGHT-RUN GASOLINE FROM DISTILLATION

One way to think about, and classify, refinery processes is to consider all the kinds of things we can do with molecules. We can separate them without any chemical change at all (i.e., these could be called physical processes, because they take advantage of physical properties of molecules, and not their chemical properties). We can take small molecules and put them together to make bigger ones, so-called buildup processes. We can take big molecules and break them apart to make smaller ones in breakdown processes. And, finally, we can change the structure of a molecule without changing its size (sometimes called change processes).

Distillation is an example of a physical process. Gasoline produced from the distillation of crude oil is called *straight-run gasoline*. Even with the best Pennsylvania-crude-quality oils, a refiner will likely get no more than twenty barrels of gasoline per hundred barrels of oil; with lesser quality crudes, the yield might be less than ten barrels of straight-run gasoline per hundred barrels of oil. Second, the octane number of straight-run gasoline is about 35. These factors result in a great need to enhance both the yield of gasoline and its octane number to meet present-day market demand and performance requirements.

The key to enhancing yield comes from the approximate relationship between boiling point and size of molecules. As long as we compare substances that are similar chemically, we find that the higher the boiling point, the larger the size of the molecules, and vice versa. Table 19.1 illustrates this, arranged in order of increasing boiling range *and* increasing molecular size (i.e., number of carbon atoms). To increase the yield of gasoline, two strategies can be used. We can combine molecules that have fewer than five carbon atoms to build up molecules in the gasoline-size range. Or, we

TABLE 19.1
Linear Paraffin Molecules in
Distillation Products Steadily Increase
in Size as Boiling Range Increases

Product	Number of Carbon Atoms	Boiling Range of Paraffins, °C
LPG	3–4	<–1
Gasoline	5–9	36–150
Kerosene	10–14	174–250
Diesel fuel	12–18	217–317
Fuel oils	12–20	217–350
Resids	(Very large)	>350

can also break apart molecules that have more than nine carbon atoms to make smaller molecules that are in the gasoline-size range. Normally, many refineries do both.

ALKYLATION

Alkylation builds up small molecules of three or four carbon atoms to products of six to eight carbon atoms. Buildup processes form larger molecules of high octane number, boiling in the gasoline range by combining smaller molecules. Cracking processes (discussed in the following text) produce by-product molecules too small to be in the gasoline range. Alkylation uses these small molecules to form compounds in the gasoline range with very high octane numbers. Usually the molecules combine in such a way that branched paraffins are formed. The product is called *motor fuel alkylate*. The octane number of motor fuel alkylate is about 100. Alkylates are important constituents of high-octane gasoline for automobiles or small piston-engine airplanes. Treating small hydrocarbon molecules with sulfuric acid, a process called sulfuric acid alkylation, provided most of the high-octane aviation fuel used during the Second World War and became, for a time, one of the most important processes in the refinery.

THERMAL CRACKING

Breakdown processes represent a third route to gasoline, in addition to physical processes and buildup processes. Breakdown processes encompass all types of "cracking" processes that use heat, pressure, and catalysts. The oldest of these operations is thermal cracking, now generally used only to make a solid coke and to increase the yield of light products.

Cracking breaks down molecules of more than nine carbon atoms to the range of five to nine carbon atoms. To do this, a refinery will select a material of high boiling range for which little demand or market exists. The simplest approach to cracking is to expose the material to high temperatures, a process called *thermal cracking.* The high temperatures break apart carbon–carbon bonds in molecules and form molecules of smaller size. Unfortunately, the products of thermal cracking tend to be in the 55–75 octane number range, not good enough for today's engines.

As the 1930s drew to a close, automobile engines were becoming more powerful, of higher compression ratio, and demanded gasoline of higher octane rating. The initial goal was to increase the yield of gasoline from a barrel of crude, which was no better than 20% for straight-run gasoline from a good-quality crude and had an octane rating below 50. By 1935, the octane rating had been boosted to 71 with corresponding increases in the gasoline-to-crude ratio. One of the thermal cracking processes important in the 1920s and 1930s was invented by a man with the extraordinary name of Carbon Petroleum Dubbs, who, understandably, preferred to be called by his initials "C.P."* Despite its oddity, his name was very apt, because one of his most important inventions was a technique to reduce the formation of undesirable solid carbon (or coke) when petroleum was processed to enhance the yield of gasoline.

CATALYTIC CRACKING

The most widely used and most important breakdown process today is catalytic cracking, which converts gas oils into high-octane gasoline and other lighter products. (Gas oils are petroleum fractions that boil in the range of about 230°C–425°C, such as diesel fuel, heating oils, and fuel oils.) Around the time of the Second World War, Eugene Houdry (Figure 19.4) discovered that if cracking was done in the presence of clay minerals, not only did the large molecules break apart, but also the products were converted to branched paraffins, naphthenes, or aromatics. That is, the yield and the octane number both increase. The clay does not enter into the cracking reaction; it serves only to speed up the reaction and to enhance the octane number of the product. Substances that affect the outcome of a chemical reaction without themselves participating as reactants or being consumed are called *catalysts*. Therefore, this kind of process is called *catalytic cracking*.

Houdry was the rich, successful heir to his father's French steel firm. He was a First World War hero, a good athlete, and interested in auto racing. The former Vacuum Oil Co. invited him to New Jersey and offered to purchase his laboratory. Their joint efforts succeeded, but a sudden drop in the price of crude oil caused a loss of interest. A third partner joined the effort, and in 1936, the first commercial plant

---

* Though he may have been with stuck with a bizarre name, Dubbs (1881–1962) was both an accomplished research scientist and a very fine businessman. Proof of the latter fact is that C.P. Dubbs managed to become a millionaire during the Great Depression. Not only did C.P. never murder his parents for giving him such an unusual name, he apparently continued the tradition. His two daughters are alleged to have been named Methyl and Ethyl. Other versions of the story claim that he named his son Carbon Petroleum Dubbs Jr.

FIGURE 19.4   Eugene Houdry (1892–1962), the inventor of the catalytic cracking process that is of critical importance in the production of high-octane gasoline. (From http://www. chemheritage.org/discover/online-resources/chemistry-in-history/themes/petrochemistry-and-synthetic-polymers/petrochemistry/houdry.aspx.)

went into operation using bentonite clay as the catalyst.* By the time the Second World War broke out, there were twelve plants in the United States, providing 132,000 barrels/day of high-octane gasoline, primarily for the war effort. By the end of the war, some 34 plants were in operation with a capacity of 500,000 barrels/day.

Catalytic cracking enables the production of 40–45 barrels of gasoline per 100 barrels of crude, with octane numbers over 90. Nowadays many of the catalysts used are zeolites.[†] The discovery of catalytic cracking revolutionized the refinery industry. Catalytic cracking (Figure 19.5) is the second most important process in a refinery, second only to distillation. Some historians of technology consider the development of catalytic cracking to be the greatest triumph of chemical engineering in the twentieth century.

---

* An enormous family of clay minerals exists throughout the world. The chemical structures of clays derive from the oxides of silicon and aluminum, usually with small amounts of other elements and some molecules of water. The chemical compositions and structures of the clays vary widely throughout the family. One major type of clay was first discovered in 1874 near the town of Montmorillon in France; these clays have since been called montmorillonites. Those montmorillonites that have significant commercial application are called bentonites. Large deposits of bentonites occur in, as examples, Wyoming and Alberta.

[†] Zeolites are another large family of naturally occurring minerals based on a framework of aluminum and silicon oxides along with atoms of such elements as sodium, potassium, calcium, or magnesium and some molecules of water. Zeolites were first characterized in the 1750s by the Swedish mineralogist Axel Cronstedt. One of the first commercial uses of zeolites was in water softeners. Today, many of the zeolite catalysts used in petroleum refining are synthetic, rather than natural materials. The synthesis of zeolites allows petroleum chemists to control precisely how rapidly the catalyst will facilitate reactions and control the size of the molecules that will react in its presence.

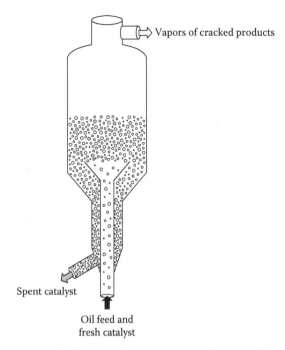

Vapors of cracked products

Spent catalyst

Oil feed and
fresh catalyst

FIGURE 19.5 A schematic diagram of a catalytic cracking unit. This process increases both the yield of gasoline (number of barrels of gasoline produced per hundred barrels of petroleum processed) and its octane number.

CATALYTIC REFORMING

Change processes represent a fourth route to gasoline. After the cracking processes have arranged the product distribution (i.e., the amounts of gasoline, jet fuel, diesel fuel, and heating oils) to match the market demand, some materials remain whose molecules are the right size to be in the boiling range of a particular product, but are of the wrong molecular configuration to have the properties desired in the product. For example, straight-run gasolines or naphthas have desirable boiling temperatures, but octane numbers that are too low.

In other words, the trouble with straight-run gasoline molecules is that they're the wrong shape—they're mostly low-octane-number linear paraffins. To enhance the octane number of straight-run gasoline, we don't need alkylation or cracking processes; all we need is to change the shape—to "reform," so to speak—of the molecules we already have. The process of "reforming" molecules also uses special catalysts (though different from the ones used for cracking). Thus this process is called *catalytic reforming*. In the presence of reforming catalysts, such as finely divided platinum dispersed onto aluminum oxide, straight-chain paraffins with low octane numbers can be reformed into their branched-chain isomers. This important gasoline-upgrading process produces gasoline of about 95 octane. Catalytic reforming can also be applied to straight-run naphtha.

These processes are summarized in the diagram of Figure 19.6.

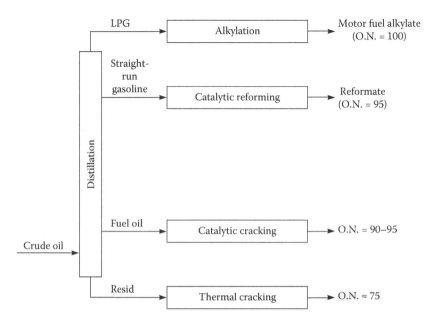

FIGURE 19.6   Gasoline is produced from many processes in a refinery: distillation, alkylation, reforming, catalytic cracking, and thermal cracking. O.N. denotes octane number.

OCTANE ENHANCERS

The octane number of gasoline can also be increased by adding special chemicals called antiknock agents or octane enhancers. In 1921, Thomas Midgley* discovered that the compound tetraethyllead was an outstanding antiknock agent for low-grade fuels. From 1925 to 1975, both regular and premium grades of gasoline contained tetraethyllead. This product was often called "leaded gasoline," although it most certainly did not contain metallic lead, as the name might imply. Sometimes it was also called "ethyl" gasoline (Figure 19.7). Eventually, toxicity of lead emitted from vehicles burning leaded gasoline encouraged fuel producers to remove tetraethyllead from gasolines; the driving force behind this decision was the worry that lead could

---

* Thomas Midgley (1889–1944) was a brilliant chemist who has gone down in history as the one individual who has personally done more to wreck the environment of the planet than anyone else. Tetraethyllead produces emissions of metallic lead when burned as a component of gasoline. Chronic exposure to low levels of lead results in a variety of adverse health effects; children can be particularly vulnerable. Lead emissions from vehicles accumulate in plants along highways, from plants to cows, from cows to milk, and finally from milk to humans. High levels of lead were also found in congested urban environments that were with dense traffic and high population density. Midgley also invented chloroflourocarbons (CFCs), or freons, once widely used in refrigeration and air conditioning equipment, as well as propellant in aerosol cans. Production and use of CFCs has been banned in many countries, because accumulation of these compounds in the atmosphere has a major role in creating the so-called ozone hole. The reduced concentration of ozone in the stratosphere allows higher levels of ultraviolet rays from the sun to reach the surface of Earth, causing increased sunburns and other skin effects, including cancer. During his career, Midgley was very highly regarded in his profession, being elected president of the American Chemical Society and receiving numerous prestigious awards. In those days, no one could foresee the possible environmental consequences of Midgley's inventions.

FIGURE 19.7 This sign on an antique gasoline pump (near Bismarck, North Dakota) advertises ethyl gasoline, containing tetraethyllead as an octane booster. (From http://en.wikipedia. org/wiki/File:EthylCorporationSign.jpg.)

"poison" the catalysts used in the catalytic converters installed on cars to reduce air pollution (Chapter 28). Without lead, less expensive gasoline blends with low octane ratings did not burn efficiently. This problem was partially alleviated by engine redesign and partially by blending higher octane components (such as benzene and toluene) into the fuel, but both approaches increased gasoline prices significantly.

The exhaust emissions of automobile engines contain CO, oxides of nitrogen, and unburned hydrocarbons, all of which contribute to air pollution. (We'll look at this problem in detail in Chapter 28, on vehicle emissions.) As urban air pollution worsened in the 1950s and 1960s, in the United States, the Congress passed the Clean Air Act of 1970, which, among many other provisions, required that 1975-model-year cars emit no more than 10% of the carbon monoxide and hydrocarbons emitted by 1970 models. The solution to reducing these emissions was a platinum-based catalytic converter. (In an interesting historical quirk, one of the first catalytic converters had been invented by Eugene Houdry.) The only problem was that it required lead-free gasolines, since lead deactivates the platinum catalyst by coating its surface.

Because tetraethyllead can no longer be used in the United States, the European Union, and other countries, other octane enhancers are now added to gasoline. Currently the most popular octane enhancer seems to be ethanol (also known as ethyl alcohol or grain alcohol). Ethanol has an octane number over 100. Production of ethanol from biomass resources that are, at least in principle, renewable and the potential use of ethanol itself as a vehicle fuel are discussed in Chapter 31.

BLENDING

Modern refineries can vary their mix of products to meet market demands, within rather wide limits, regardless of the composition of the crude coming into the refinery. The products illustrated in this chapter are called *blend stocks*. The refiner can blend various portions of these materials to obtain the 87, 89, and 93 octane gasolines available in service stations. This approach meets both the market demand and performance requirements.

The gasoline must also be stabilized. This involves a closely controlled distillation that removes enough lighter hydrocarbons to give the fuel the desired volatility. Too much volatility produces *vapor lock* in an engine; in cold weather, too little volatility leads to difficulty in starting an engine. Vapor lock happens when the gasoline turns to vapor in the fuel line before getting to the engine, instead of vaporizing when injected into the cylinder. This can occur, for example, on exceptionally hot days. Vapor lock will cause the car to stop because it interferes with the normal flow of liquid gasoline to the engine. Though often maddening and frustrating, there is usually little one can do other than coast to the side of the road and wait for the engine and fuel lines to cool. Refinery products must be blended to yield a final product of the desired viscosity, volatility, and octane number.

All gasolines are highly volatile. Thanks to this and the fact that the vapors are so easy to ignite, you can start your car even in the coldest of weather. However, this same high volatility means that some hydrocarbons get into the atmosphere as a result of accidental spills and evaporation during normal filling operations at the service station. Gasoline vapors also evaporate from storage tanks. Hydrocarbons in the atmosphere play an important role in a series of reactions that contribute to urban air pollution. Refineries need to adjust the volatility of gasoline to accommodate the different temperatures that accompany changing of the seasons.

Ethanol and other oxygenated compounds (i.e., chemicals that contain at least one atom of oxygen in their molecule) are now added to gasoline to promote cleaner air, smoother combustion, and higher octane numbers at low cost. A blend called "gasohol" introduced during the petroleum shortages of the 1970s contained about 10% ethanol. Ethanol's main drawback is that it absorbs water from the atmosphere, which causes corrosion to fuel system components and can also harm the combustion performance. Ethanol has a high octane number (108). Methanol (also called methyl alcohol or wood alcohol), often used as a gasoline line antifreeze, also has a high octane number (107), but it too absorbs water and causes severe corrosion. The less soluble butanol (butyl alcohol) avoids this problem, since it does not absorb water well; in fact, it reduces the tendency of gasoline blends to pick up water.

Although these oxygenated compounds have high octane numbers, a high octane number relates only to a tendency for engine knocking. An octane number does not itself relate directly to the ENERGY content—in MJ per kg, or MJ per liter—of a fuel. In general, all oxygenated additives contain less energy than gasoline because they are already partially oxidized. In a practical sense, this means that you get less knocking in the engine, but you also get lower fuel economy—the number of kilometers you can drive per liter of fuel. Despite that, because ethanol and methanol can be made from biomass or other renewable sources, they are considered to be good liquid fuel options by advocates of reducing our dependence on fossil fuels. Ethanol and methanol are not totally nonpolluting fuels. Both can produce toxic aldehydes* on incomplete oxidation, and like all carbon-containing fuels, they produce CO_2.

Refineries have always made gasoline, whether they wanted to or not. There was a time when gasoline was a nuisance, a dangerous by-product with little practical use. Once Ford dropped the price of automobiles into the range of affordability by the ordinary working person, gasoline became the premier petroleum product in many countries. It remains so today in the United States. A problem for refiners has been what to do with what is left over from the petroleum after the gasoline has been produced. There is, after all, a market for only so much kerosene, diesel fuel, heating oil, solvents, lubricating oil, and waxes. Increased use of diesel engines in automobiles and light trucks could likely alter this scenario, increasing the demand for diesel fuel at the expense of gasoline. Without "dieselization," satisfying the primary gasoline demand could potentially cause surpluses of other products. The solution has been and continues to be the development of new processes that allow the refiner to convert the residue from one process into the feed for another.

FURTHER READINGS

Berger, B.D.; Anderson, K.E. *Modern Petroleum*. PennWell: Tulsa, OK, 1992. This book is a useful primer on all aspects of the petroleum industry. Chapter 11 gives an overview of refinery technology.

McNeill, J.R. *Something New under the Sun*. Norton: New York, 2000. This book includes the story of Thomas Midgeley, inventor of leaded gasoline, an excellent scientist and fine person, and who, according to the author, did more to wreck the environment of our planet than any other person who has ever lived.

Queen, E. *The Roman Hat Mystery*. A murder mystery originally published in 1929 and available in numerous editions and bindings since then. Its relevance is that the murder victim, a nefarious lawyer, was poisoned by getting him to drink tetraethyllead, possibly the only such occurrence in all of crime fiction. It is remarkable in part because it was published not long after this compound began to be used as an octane booster. One of the characters remarks that "it is almost entirely new to us."

Schobert, H.H. *The Chemistry of Fossil Fuels and Biofuels*. Cambridge University Press: Cambridge, U.K., 2013. This book has individual chapters devoted to thermal and catalytic processes for making gasoline, which are discussed in some detail. A prior knowledge of chemistry is assumed.

Sperling, D.; Gordon, D. *Two Billion Cars*. Oxford University Press: New York, 2009. This interesting look at how we are going to cope when the world's population of automobiles reaches two billion, includes useful discussions on vehicle efficiency and technology development.

---

* Aldehydes are a type of organic compound containing, in addition to carbon and hydrogen, an oxygen atom. The oxygen atom is connected by a double chemical bond to the end of the chain of carbon atoms, as in acetaldehyde, $CH_3CH{=}O$, a material sometimes used as a flavoring agent in beverages and some foods. The simplest aldehyde is formaldehyde, $H_2C{=}O$, the water solution of which (formalin) is used as a preservative and embalming agent. Exposure to aldehydes can irritate the eyes, nose, and throat.

20 Impact of the Automobile

In 219 BCE, the Carthaginian general Hannibal assembled the greatest invasion army of the classical world. Hannibal first crossed the Pyrenees, and then the Alps, to invade Italy. Two thousand years later, in 1812, Napoleon assembled the greatest invasion army known prior to the twentieth century, for his invasion of Russia. *Both armies traveled at exactly the same speed.* In 45 BCE, Julius Caesar was the ruler of the greatest empire the world had yet known. In 1895, Queen Victoria was the ruler of the greatest empire in history. Nearly two thousand years after Caesar, Victoria moved around her capital, London, in the same way, and probably at the same speed, as Caesar had moved around his capital of Rome.

In perfecting Thomas Newcomen's engine in the late eighteenth century, James Watt developed the first practical steam engine. By 1829, George Stephenson had perfected Trevithick's early design of a steam locomotive. After several early European attempts, Robert Fulton developed the practical steamboat in the early years of the nineteenth century. So, in a relatively short time, Watt's engine was applied to railroads and ships. Yet it took almost another century before the practical development of the automobile. Without doubt, the automobile was the outstanding feature of land-transport technology in the twentieth century.

At the beginning of the twentieth century, the automobile offered a promise of personal freedom and affluence. Just like every other invention or technological development that was thought finally to lead humankind into the promised land of freedom, wealth, and ease, the automobile turned out not to be the economic and social panacea that many expected. Nonetheless, over the course of the twentieth century, the automobile probably changed the lives of most people, both in good ways and in bad ones, more than any other invention.* Figure 20.1 provides a time line of the developments in transportation.

THE CYCLING CRAZE

Clearly, in principle, a steam engine could be used to propel a "horseless carriage." A number of trials of this concept took place as early as the first attempts with locomotives and boats (Chapter 17). Eventually, improved engineering did lead to practical steam automobiles with good performance, the Scott–Newcomb (Figure 20.2, also known as the Standard) being a notable example. Other steam-propelled vehicles, such as steam tractors, were also developed. So why didn't equally ingenious individuals—the automotive counterparts of Trevithick, Stephenson,

---

* Television and the development of the Internet are perhaps the runners-up to this claim. However, television only began to penetrate most households in the 1950s and the Internet since the mid-1980s, whereas car ownership became common in the 1920s, years earlier.

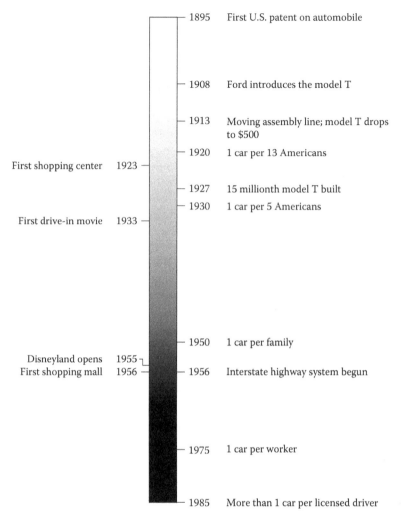

FIGURE 20.1   A time line in the history of the major developments of the automobile and its impact on society in the United States.

and Fulton—develop steam-powered automobiles early in the nineteenth century? A major stumbling block came from the lack of good roads at that time.

Before the coming of horseless vehicles, roads needed only to accommodate foot traffic or slow-moving horse-drawn vehicles. Roads were not paved, were full of potholes, were not lighted nor marked, and in many cases simply meandered through the countryside. The factor that caused the change to create roads and allowed the widespread use of the automobile was the bicycle.

The revival of interest in mechanical means of road transport came with a surprising reversion to the use of human ENERGY for transportation, in the form of the bicycle. They began with rather primitive devices. The Dandy Horse was just a framework with two wheels and a seat, with no pedals, steering, or brakes. The rider

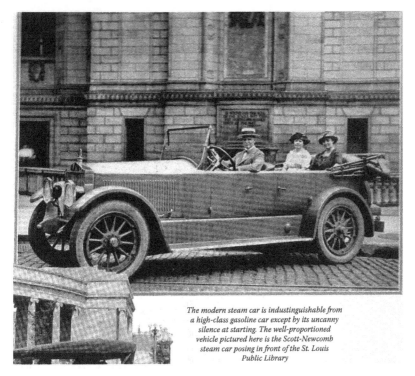

The modern steam car is industinguishable from a high-class gasoline car except by its uncanny silence at starting. The well-proportioned vehicle pictured here is the Scott-Newcomb steam car posing in front of the St. Louis Public Library

FIGURE 20.2   A Scott–Newcomb steam car from 1920. Externally, it differed very little from competing high-performance gasoline-fueled cars of the era. (From http://en.wikipedia.org/wiki/File:1920_scott.jpg.)

pushed himself along with his feet. The Pedestrian Circle, a variation of the Dandy Horse, somehow combined features of a hobbyhorse and bicycle. Pedestrian Circles were used for racing in some of London's busiest thoroughfares, likely drumming up plenty of business for doctors skilled at setting broken bones. Bicycles began to be popular in the 1870s and 1880s. In the 1890s cycling developed into an extremely popular sport (Figure 20.3). The bicycling craze of the 1890s brought with it numerous benefits. It demonstrated the possibility of long-distance travel over ordinary highways.* It led to the renovation of existing roads, including erection of signposts and the printing of road maps. It stimulated the construction of new roads. And, it led to the establishment of roadside repair facilities, which were the forerunners of the service station. Some of these benefits derived from pressure being brought on legislators by organized groups of cycling enthusiasts.

To establish a viable industry for a new form of transportation—the automobile—early innovators and entrepreneurs first had to perceive a market for personal transportation that did not rely on the horse. For most people of the nineteenth century,

---

* Long-distance cycling involved trips of many tens of kilometers; in some cases, more than 150. The first recorded bicycle race took place in 1868 and covered a distance of 130 km. Perhaps more remarkable than the distance ridden is the fact that the winning cyclist, James Moore, is said to have ridden a bicycle weighing 70 kg!

FIGURE 20.3 The popularity of bicycling—the so-called cycling craze—of the late nine-teenth century generated interest in personal transportation and public pressure for improvements to the road system, which later helped the automobile to gain popularity. (Reprinted from first edition with permission of Bygone Designs, Newport, Minnesota.)

personal transportation meant being pulled by, or riding on, a horse. The bicycle, and its extraordinary popularity in the 1890s, helped people, especially young people and women, to travel more widely than ever before. This in turn helped create a demand for even more freedom of movement in the future. The existence of the new market was established by the bicycle, which became an item of mass consumption in the 1880s and wildly popular in the 1890s. By 1915, there would be six million bicycles in the United States, five million in Britain, and four million in France. Not only did the bicycle help create a market of potential customers for the automobile, but also many of the techniques devised for bicycle construction were used later in car making. Light tubular frames, ball bearings, chain drive, pneumatic tires, gears, wire wheels, and brake cables were adapted from the bicycle by early car manufacturers. In some cases, the bicycle firms themselves switched from making bicycles to building cars.

HOW RAILROADS PUT PEOPLE IN CARS

A second factor that set the scene for the rise of the automobile was, curiously, the steady expansion of the railroads. This occurred in two ways. First, as railroads expanded and carried more and more passengers, there were a great number of highly publicized train wrecks. These caused considerable public indignation. Somehow, we seem to be willing to accept a steady diet of small crashes and disasters, whereas the large, spectacular ones attract our attention and outrage. Few people can escape a horrible feeling about a major airline crash (a total of 280 people were wiped out in two crashes alone—near Rawalpindi, Pakistan, and Lagos, Nigeria—in the first half of 2012, yet two days' driving in America manages to slaughter in automobile accidents the same number of people as died in those plane crashes). Except for the immediate families and friends of the victims of those car crashes, probably no one paid much attention.

Public concern about railway safety, because of the generally spectacular nature of passenger train wrecks, led to interest in alternative, and hopefully safer, modes of transportation.

It was cloudy in the west, looked like rain,
Round the bend come a passenger train.
North-bound train on a south-bound track,
He was all right a-leavin' but he won't be back.

—Pace[1]

Consider: 1864, Mont Saint Hilaire, Quebec, train fails to stop at open swing bridge, 99 dead. 1878, Ashtabula, Ohio, bridge collapses while train is crossing, 92 dead. 1879, Tay Bridge, Scotland, another bridge collapse, 90 dead.

So the train mov'd slowly along the Bridge of Tay,
Until it was about midway,
Then the central girders with a crash gave way,
And down went the rain and passengers into the Tay!

—McGonagall[2]

1882, Spuyten Duyvil, New York, train collides with the rear of another, passengers are forced to use snowballs to extinguish the ensuing fire. 1882, Tcherny, Russia, train derailment, 150 dead. 1885, St. Thomas, Ontario, train collides with a circus elephant. 1889, Armagh, Ireland, runaway passenger coaches collide with the following train, 88 dead. 1891, Müchenstein, Switzerland, yet another bridge collapse, 71 dead.

Second, the success of railroads meant that more and more people and goods were piling into the cities each day. This caused increased congestion on the streets (Figure 20.4), and especially problems with the horses used to haul buses, carriages, and freight wagons. One particular horse stable was a block long and seven stories high. New York City's horses produced 140 million kilograms of manure each year. In addition, horses can be very difficult to handle. When they are enraged or terrified, as in an accident, they can kick, bite, and bolt. In recent years the number of traffic deaths in the United States has been about 160 per million people. A century ago, a time that many people think was a kinder, gentler, slower paced era, the rate

FIGURE 20.4 The problem of congested city streets has been with us for a long time. The automobile is not the sole culprit. (Reprinted from first edition; image originally from General Electric Company, New York.)

was still 110 per million. So, it was clear for various reasons that the horse had to fade from the scene as the dominant source of energy for urban transportation. And then along came Ford.

AMERICAN DOMINANCE IN AUTOMOBILE MANUFACTURING

The major inventions that made the gasoline-powered automobile possible all came out of Germany, due to such workers as Nikolaus Otto, Wilhelm Maybach, Gottfried Daimler, and Karl Benz. After a time, the leadership in automobile technology passed to France. Such terms as *automobile*, *chauffeur*, *garage*, and *chassis* are all French in origin. The first patent on an automobile in the United States was issued to a lawyer, George Selden. The patent application was originally filed in 1879 and the patent finally issued in 1895.

In his original patent Selden claimed, 'The combination in a road vehicle equipped with an appropriate transmission, driving wheels, and steering, of an internal combustion engine of one or more cylinders, a fuel tank, a transmission shaft designed to run at a speed higher than that of the driving wheels, a clutch, and coachwork adapted for the transport of persons or goods.' Cars today still fall under that wonderfully broad claim.

—Reynolds[3]

By the second decade of the twentieth century, automotive leadership had passed to the United States. There are several reasons for this: a generally high standard of living, a large population living in cities separated by long distances, abundant supplies of petroleum, and a vigorous innovative and entrepreneurial spirit.

At first, automobiles were toys for the rich. Henry Ford recognized that for the automobile to be truly successful, it had to be priced within the reach of the ordinary working person of modest financial means. In his continuing effort to reduce the price of the Model T (first introduced in 1908), Ford developed in 1913 a revolutionary industrial idea—the moving assembly line (Figure 20.5). Some earlier assembly lines had been set up, as for example in the manufacture of military firearms, but normally the workers collected parts and brought them to one spot where the article was being assembled. On Ford's moving assembly line, the worker stayed in one place, and the car moved along the line. The success of the moving assembly line depends on two key factors: each worker repetitively carrying out one very simplified job and the parts being highly standardized and completely interchangeable. Though many people were involved in the development of the automobile industry in the United States early in the twentieth century, Henry Ford usually gets credit

FIGURE 20.5 Henry Ford's moving assembly line had a major role in reducing the price of a new automobile to a point that a family of modest financial means could afford to purchase one. (Reprinted from first edition, with permission from Bygone Designs, Newport, Minnesota.)

for putting the automobile within reach financially of persons of average means. Ford had little education and many character flaws—including racial bigotry and a domineering personality that drove away many talented associates. Indeed, Ford was about as ornery and reprehensible a person as most of us would care to know. Still, he is perhaps the most important industrialist of that century.

By 1913, due in large part to assembly-line production, Ford was able to sell the Model T for $500.* Before production of the Model T was curtailed in 1927, Ford had sold 15 million of them. Whereas there had been one car for every 265 Americans as late as 1910, there was one car for every five at the end of the 1920s. Though assembly-line production had a major role in this extraordinary growth, other factors contributed as well. For instance, a car purchase was made easier with installment plans; by 1923 over three-quarters of cars were purchased on credit.

Early in the 1920s, executives of car companies began to worry about growth of car sales and impending saturation of the market. They noticed, however, that many people did not want to possess exactly the same car as their neighbors. Instead, consumers seemed willing to accept some increase in cost to obtain variety in styles, colors, or other attributes that would make their car somehow "different." It began to be appreciated that car owners could get the idea of trading in their existing cars for a new one that would indeed be different from the one parked at the neighbor's house. This growing appreciation of consumer behavior led the car companies to produce a variety of different models and to change models on a regular basis, often annually, to encourage trade-ins.

A declining rate of technical improvements to the basic components of the car also made it worthwhile for car companies to call attention to changes in style and the diversity of models offered, rather than using small, incremental technical changes as a selling point. By the early 1920s, most important elements of the modern car had essentially been developed; brakes, tires, valves, air cleaners, oil purifiers, and shock absorbers are some of the examples. Although assembly-line production lowered the cost of cars, and made car ownership possible for most people, it brought with it a new problem. The capital investment required to build an assembly-line facility, or to make radical changes in an existing facility, was so high that it discouraged product innovation, because any major changes in the car would require expensive retooling of the assembly line.

By 1920, there was already one registered automobile for every 13 Americans. A decade later, the figure had risen to one for every five. Not long after the Second World War, there was one car per family in America and, in the 1970s, one car per

---

* In the early years of the twentieth century, automobiles were toys for the well-to-do. Many cost well over a thousand dollars. In fact, in 1908, the Model T was selling at $850. By 1916, further improvements in manufacturing and the use of the assembly line allowed Ford to drop the price to $400, making the Model T the lowest-priced car in America. To put these prices in context, this was an era in which an unskilled laborer might be paid $1.00, and a highly skilled craftsman $2.50 per day. (When Henry Ford offered to pay workers $5.00 per day, so many people showed up to apply for the jobs that a near-riot ensued.) Assuming a six-day workweek with no vacations, a skilled worker would bring home an annual income of $780. In 2011, the median household income in the United States was about $51,400 per year. Very roughly, an automobile purchased in 1920 might cost the equivalent of half a year's salary. Nearly a century later, with vast numbers of innovations in automobile technology, approximately the same ratio still holds. The average price of a new car was about $29,000.

worker. By 1970, there were more cars than there were households in the United States; in Los Angeles, there were more cars than there were people. By 1985, the number of cars in the United States actually exceeded the number of licensed drivers, and nowadays there are about 20% more cars than there are drivers. While it may seem bizarre that there could possibly be more cars than there are people to drive them, today many cars are owned, in large fleets, by corporations and rental companies. And, many families now own recreational vehicles or motor homes, which might be used only occasionally, in addition to their other vehicles that are used for everyday commuting to work or shopping.

EFFECTS OF GROWING AUTOMOBILE USE

What hath Ford wrought? Perhaps the most important single contribution is that the car gave the common person a feeling of independence, particularly independence from the rich and independence from transportation companies that provided travel only to fixed destinations on fixed schedules. This was true not just in America, of course, but virtually everywhere that cars were available. A person of average financial means could get around without being dependent on railroad companies or streetcar companies. Further, by being able to live outside of the center city, the average person was also no longer dependent on big-city landlords or rental companies.

CARS AND INDUSTRIAL GROWTH

The car benefited a large cross-section of industry. The obvious beneficiaries were the car companies and the producers and refiners of petroleum products. After the First World War, gasoline consumption by automobiles became the dominant market for petroleum (replacing lubricating oil and kerosene for illuminants). The enormous increase in gasoline consumption would not have been possible without significant advances in the technology of oil refining. From 1920 to 1940, tires and inner tubes accounted for 85% of the sales of the American rubber industry. The automobile also stimulated the application of continuous processing to making glass for windshields and windows. The Ford Company itself pioneered this method shortly after the end of the First World War. Glass output for automotive applications tripled in the 1920s. Besides stimulating the development of the oil, glass, and rubber industries, the automobile also impacted highway and bridge construction, and led to the development of "management science" (at first in the petroleum and construction industries) of how to run large engineering companies. In addition, all sorts of small, local businesses benefited: trucking companies, suburb developers, construction contractors, auto parts dealers and suppliers, auto mechanics, and service stations.

SUBURBS

The car made possible a major decentralization of cities. Actually, it was not the car that was first responsible for the flight to the suburbs. As early as 1814, people working in Manhattan began taking a steam-driven ferry to live in a quiet, rural,

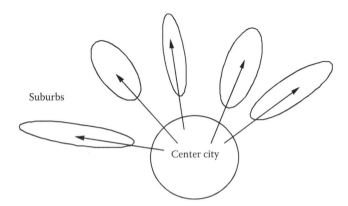

FIGURE 20.6 The development of suburbs initially proceeded in a pattern resembling the fingers on a hand, following the direction of railroad and trolley lines.

backwater town called Brooklyn. With the advent of railroads and streetcar lines, suburbs could continue to develop, but only wherever the tracks went. The pattern of development often looked like fingers on a hand (Figure 20.6). An example is Philadelphia's "Main Line" suburb. The streetcar companies had helped begin the exodus to the suburbs by extending their lines from the central cities into previously undeveloped districts; but in "streetcar suburbs," the only useful residential property was on or within walking distance of the streetcar line. The automobile would make it possible to develop suburban property that had previously been inaccessible; that is, to fill in the spaces between the spokes of the streetcar lines, as if filling the spaces between the fingers of the streetcar "hand." Furthermore, light trucks made it possible to transport raw materials and finished goods out to the suburbs, so that warehouses and factories could also be moved to the suburbs. These effects led to the phenomenon (or problem, depending on one's perspective) of suburban sprawl (Figure 20.7).

Probably the most important effect of the automobile was how it encouraged suburban sprawl, especially in the last half of the twentieth century. In the United States, the interstate highway system (begun in 1956) facilitated a rush to the suburbs. The suburb became accessible to less affluent people, thanks in large part to the car. For many people, the suburb offers the potential of owning a private home on one's own lot, instead of an apartment. As land and housing costs go up and up in cities, affordable housing may be restricted to apartments or condominiums. In the suburbs, there may still be the potential to be able to afford a private, single-family house.

Developers of suburbs assumed that families living in those new homes would have a car (or, in most cases, more than one car). Schools, hospitals, and shopping districts in the suburbs were dispersed across the landscape, rather than being clustered together near a streetcar or rail line, on the presumption that residents would be able to access these facilities with their cars.

A stereotyped view of commuting presumes that workers travel from the suburbs into the center cities and then back out when the workday is done. In fact, more than a third of metropolitan-area commuters go from one suburb to another, and about

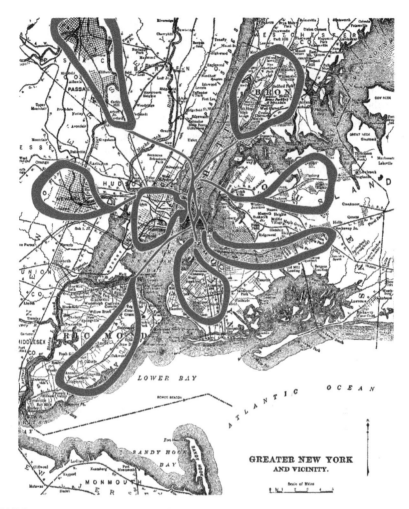

FIGURE 20.7 The phenomenon of suburban sprawl resulted in large part from the increased ownership and use of automobiles. (Reprinted from first edition, with permission from Bygone Designs, Newport, Minnesota.)

5% actually go from the center city out to the suburbs to work. Only about a fifth of metropolitan commuters travel from the suburbs to center city. Suburban living also gives rise to "two-commuter" families, in which two wage earners in a family commute to jobs in different locations—even in diametrically opposite directions—from the family home and gives rise to running multiple errands in cars, especially on weekends. With both adults away from the house on weekdays, home errands and recreation become compressed into the weekend days. As a result, Saturday afternoons are now a peak time for automobile use.

No doubt the automobile provided ordinary people with greater mobility than they had ever enjoyed previously. This gain had to be balanced against several negative aspects, particularly the ways in which interpersonal life changed. People with cars, and the mobility provided by cars, no longer depended only on the facilities and social life of their immediate neighborhoods. Most of the lifestyle associated with the city street neighborhood disappeared: street vendors, delivery boys from local stores, persons simply going for a walk, and, eventually, even the local businesses themselves. Those who chose to remain in the center city, or those who were economically trapped there, found the street radically changed. A place that had once been the location for social encounters with the neighbors degenerated into simply a route for transportation. A century ago, three-quarters of the streets, even in major cities, were either unpaved or gravel. The principal highways and thoroughfares in cities were surfaced with cobblestones, brick, concrete, or asphalt. In a few cases, even wooden blocks were used for paving. These kinds of road surfaces do a superb job of discouraging rapid driving. With slow-moving traffic, much of it horse-drawn, streets were places where children could play. Such traffic could make its way around or through other activities going on in the street, and with comparative safety to those using the streets for play or social activities. The rise of the automobile, along with the electric trolley, made travel, not neighborhood activities, the function of the street.

> *To leave our house meant that, once we had crossed our threshhold, we were in danger of being killed by the passing cars. I think back twenty years [to 1904], to my youth as a student: the road belonged to us then, we sang in it, we argued in it, while the horse-bus flowed softly by.*

> **—Le Corbusier**[4]

Improved paving provided streets with smoother surfaces that allowed traffic to move more quickly, and made it unsafe to conduct games or social functions in the street.

The internal combustion engine can drive trucks just as well as automobiles. Like automobiles, trucks can "fill in the blanks" between rail lines in the urban growth ring. Consequently, almost as soon as people started migrating to the suburbs, factories and warehouses started migrating with them. From the very beginnings of urban expansion, there were jobs available in the suburbs.

Road Building

The automobile revived the road transport systems that, in the nineteenth century, had languished in competition with the railways. By doing so, the automobile indirectly promoted a massive increase in facilities for personal transportation, such as gasoline stations and roadside restaurants. At the same time, the truck made it possible to move freight to the precise location required by the consumer, not to the local railroad depot, and often did so at a time suiting the convenience of the customer, not at the schedule established by the railroad. Cars and trucks required good roads that were surfaced for smooth, rapid travel.

Car and truck drivers quickly discovered, just as the bicycle riders of the late nineteenth century had done, that roads were not equipped to handle this new traffic. Local automobile clubs, usually voluntary associations of car owners, began forming early in the twentieth century. One of their objectives was to apply political pressure to governments at all levels—city; county, state, or province; and national—to improve the roads. Desired improvements included things taken for granted today: widening streets to accommodate more traffic; smoothing the original cobblestone or brick streets with paving materials such as asphalt; building wide, high-speed, limited-access highways; and paving dirt or gravel roads in the countryside. Many countries had a national network of highways in the 1920s, but often poorly constructed and maintained. In the United States, a major lobbying organization of that era was The Auto Club, which used the motto, "A Paved United States in Our Day." Ironically, not much progress was made in addressing the needs of an affluent, car-owning population until the Great Depression of the 1930s cut sharply into that affluence. Then, governments in some countries began massive public works projects, and began providing funding to states for such projects, to create jobs for people and to stimulate a devastated economy. Those projects included designing and constructing new roads, specifically to accommodate automobile traffic. The 1930s saw a second doubling in length of the paved road system in the United States. At about the same time, Germany began the construction of new *autobahn* highways.

The internal combustion engine made possible the cars and trucks that stimulated the demand for new and improved roads and wider city streets. At the same time, it also had a major role in reducing the costs of building the new roads. By about 1910, the horse-drawn road graders began to be replaced by power shovels and motorized graders. Similarly, horse-drawn wagons were replaced by trucks for the delivery of construction materials and for the removal of dirt and construction debris.

Before the Second World War, most Americans used their cars principally for local driving, that is, to get to work, to run errands, to make visits, all within a relatively short distance of home. High-speed driving over long distances on interstate highways or turnpikes did not occur until after the war. Countries that had been devastated by the war, or that had to invest almost all their resources into the war effort, could once again turn attention to highways. The decision by the Eisenhower administration to build a federally funded interstate highway system in the United States made long-distance travel by car much easier, but was a decisive blow to the health of urban communities. The interstate highway system contained a remarkable contradiction, which was perhaps not recognized, or even thought of, at the time the system was proposed. At first, it was thought that the interstate highway system simply would provide a means of alleviating downtown traffic congestion. But, as things turned out, the possibility of high-speed travel between cities meant that the interstate routes often bypassed downtown areas or cut straight through city neighborhoods that had limited access to the highway itself. Businesses began moving out of the city centers.

Transforming the Social Scene

The secret of success of Ford and of the other manufacturers who followed his example was cheapness. The only way that mass production of automobiles could be justified was to tap into the mass market. This mass market consisted of blue-collar workers and farmers. The automobile had to be made at a cost that was within the reach of the people who made the car in the first place. The successful mass production of automobiles had dramatic effects on workers' wages and on labor relationships. Mass marketing of cars led to the ways for providing credit to buyers. Mass marketing also led to ingenious ways of advertising by rival manufacturers. The automobile became a ubiquitous feature of twentieth-century society. Now, in the early years of the twenty-first century, there is still no alternative to the private automobile as a convenient means of personal transportation. A serious rise in the price of gasoline, a rise that most people would have to believe likely to be permanent, might encourage the development of alternatives to the internal combustion engine. Environmental concerns about air pollution from vehicles (Chapter 28) could give the electric car or hybrid vehicles significant market opportunities. In the 1950s, there was even a flicker of interest in the nuclear car, an idea that now seems extinct.* Some experimenters have tinkered with solar cars. But whatever the source of energy for the vehicle, it will still be recognizable as being a *car*. Furthermore, for these or other alternatives to achieve significant market share (at the expense of the gasoline-engine car), the rise of gasoline prices would have to be very dramatic.

Though the car revolutionized travel by liberating the individual from the timetables and routes of the streetcar and train, it also forced people to join the car culture, whether they wanted to or not. Mass ownership and use of automobiles led to the steep decline of public transportation, most notably in the United States. In the 1920s, car ownership forced working families to make choices about how best to use their discretionary income. For families of limited means, buying a car replaced the purchase of other "luxuries." A common rural saying was, "you can't go to town in a bathtub," implying that it was preferable to buy a car rather than indoor plumbing fixtures. By the 1930s, the car had often become a necessity for getting to work, thanks to the dispersion of people into suburbs and the decline of public mass transit. There is perhaps no story of greater power and impact than John Steinbeck's *The Grapes of Wrath* to illustrate the extraordinary dependence of impoverished people on their automobile.

The extension of the paved highway system and improvements in road surfaces made during the 1920s helped stimulate the new recreational activity of "auto touring" (Figure 20.8). Families could pile in the car and head off for the mountains or to the shore—from Massachusetts to Florida. Clever entrepreneurs lost little time in recognizing the potential of making money from these people on the move. As a result, this new recreation soon gave rise to the whole array of roadside tourist attractions, from fine museums and nature sites to tawdry reptile zoos and amusement arcades. Some of these new enterprises clearly embodied the old adage that you can never go broke by underestimating the taste of the American public. Auto touring also stimulated family trips to national parks. With the spread of the interstate highway system, beginning in the late 1950s, chains of motels clustered around the highway exits replaced the small-time "motor courts" that could be found on the older roads.

---

* It might be said of the nuclear car that not only is it an idea whose time has passed, it's an idea whose time should never have come in the first place.

FIGURE 20.8   Automobile touring became increasingly popular with the continued growth of ownership of automobiles. (Reprinted from first edition, with permission from Bygone Designs, Newport, Minnesota.)

The world's first drive-in theater appeared in 1933. By the early 1950s, there were 4000 of them all across America, the largest of which could accommodate some 2500 cars. Many cities claim the distinction of being the location of the first drive-in restaurant. In the 1940s, these drive-ins (often providing "car-hop" waitresses, who were far more popular than "tray boys") became the haunts of millions of adolescents and their cars (Figure 20.9). In 1955, the automobile helped make possible the opening of Disneyland, located along the Santa Monica freeway in Anaheim. Shopping centers, malls, or arcades date at least from the eighteenth century. They really flourished with the coming of the automobile. Nowadays the largest are in China, with several malls in the Philippines close behind in size. Now supermalls—combining shopping, office space, apartments, and amusement facilities in one—may represent the next wave. One of the first such examples is Turkey's Cevahir Mall.

All of these developments resulted in changing the environment through which the automobile moves, and the changes were done to fit the automobile. As much as half of total land area of a city is now dedicated to roads, driveways, parking lots, and service stations. Things not friendly or useful to the automobile, such as narrow streets and low clearances under bridges, were eliminated, along with most of the alternative forms of transportation. There were no bicycle lanes, nor racks for securely leaving a bicycle while shopping. Streetcar tracks were torn up; overhead electric wiring for trolleys came down.

The changes that the automobile caused in patterns of social life also came rapidly. The automobile, along with the refrigerator, totally redefined shopping. The car

FIGURE 20.9 The drive-in restaurant was another cultural phenomenon made popular by the growth of automobile ownership. (Reprinted from first edition, with permission from Bygone Designs, Newport, Minnesota.)

had a crucial effect in changing the role of housewives from women who produced the family's food from scratch and sewed their clothing to consumers who shopped for (and bought) national-brand canned goods, prepared foods, and ready-made clothes. As women drove to shop or to drive the children to school or after-school activities, the automobile became almost an extension of the home.

Automobiles provide a phenomenal amount of personal freedom. People are not restricted to going only where the railroad tracks run, not restricted to the schedule of public transportation of any form. Ownership of an automobile provides the means of going anywhere, whenever, and on any day of the year. Of course, cars have brought with them the problems of traffic accidents, suburban sprawl, and massive public investments in infrastructure such as roads and bridges. And, automobiles also contribute rather considerable pollution. The problems of air pollution by vehicle exhaust are discussed in more detail later (Chapter 28).

REFERENCES

1. Pace, K. Lyrics from "The Rock Island Line," traditional American blues song, ca. 1934.
2. McGonagall, W. from "The Tay Bridge Disaster," from *Poetic Gems Selected from the Works of William McGonagall*. The Folio Society: London, U.K., 1985, pp. 45–47. As this example demonstrates, McGonagall is considered to be possibly the worst poet in the history of the English language.
3. Reynolds, F.D. *Crackpot or Genius?* Chicago Review Press: Chicago, IL, 1993, p. 30.
4. Corbusier, L. *The City of Tomorrow and Its Planning*. Dover Publications: Mineola, NY, 1987.

FURTHER READINGS

Baritz, L. *The Good Life*. Knopf: New York, 1988. This book is a history of the middle class in America, from the founding up to the early 1980s. Several chapters include discussions on the impact of automobile ownership on the middle class, including such topics as Henry Ford and his cars in the early decades of the 20th century, and the rise of suburbia in the 1950s.

Berman, M. *All That Is Solid Melts into Air*. Penguin: New York, 1988. This book discusses the evolution of the concept, and the experience, of modernity, from the 19th century to the 1970s. Though not a history of technology per se, much of the discussion is devoted to the changing role of streets and the effect of building or rebuilding streets and highways, from the Haussmann's boulevards in Paris to the megalomania of Robert Moses, who boasted of driving parkways through New York City neighborhoods 'with a meat ax.'

Fletcher, S. *Bottled Lightning*. Hill and Wang: New York, 2011. The primary focus of this book is lithium batteries, and how they might impact technology in the future. Included is an interesting discussion on possibilities for electric cars.

Murray, V. *An Elegant Madness*. Viking: New York, 1998. A history of Regency England, with much useful information on transportation in the early 19th century.

21 Jet Engines and Jet Fuel

KEROSENE AS A REFINERY PRODUCT

Imagine a medicine so potent that it could cure almost any known disease or medical condition, yet so safe to use that it can be purchased without a doctor's prescription, or even without seeing a doctor. If this sounds too good to be true, that's because it *is* too good to be true. Yet humankind has sought such a nostrum probably since that day way back in prehistory when the first swindler met the first sucker. And many people still seek such a cure. In 1850s America, Samuel Kier* was peddling "Kier's Rock Oil," a patent-medicine cure-all that turned out to be just Pennsylvania crude in fancy, overpriced bottles (Figure 21.1). It may be surprising that Kier had far more oil than he had suckers to buy it, but there were no telemarketers or infomercials in his time. He decided to see if the oil could be processed into kerosene.

Kier sought the aid of a Philadelphia chemist, J.C. Booth, who converted an iron kettle into a primitive distillation unit. The vapors boiling off the five barrels of Pennsylvania crude yielded about twenty liters of kerosene product per day. Kier originally planned to market this product as medicine too, but his design of a new lamp, which did not use whale oil as the fuel, created a ready market for his product. Kier's first refinery for making what he called "carbon oil" was started up in the basement of a drugstore in Pittsburgh. Edison's development of a practical electric light put an end to the need for kerosene for lighting, although it took a long time to do so. Even in the United States, which emerged from the Second World War as the world's richest and most powerful nation, electricity was still not available in some rural areas until some years after the war.

From the origin of the modern petroleum industry† until the 1910s, kerosene was the most important petroleum product, because of its use for domestic lighting and for space heating. Gasoline was a nuisance by-product. After 1920, kerosene

---

* Samuel Kier (1813–1874) owned a number of salt mines in western Pennsylvania. Sometimes the salt would become contaminated by petroleum seeping into the mine. At first the workers just dumped the unwanted petroleum into a nearby creek, a practice with ample precedents in industry and which undoubtedly continues in many places of the world to this day. When there was so much petroleum contaminating the water that the creek actually caught on fire, Kier realized that it might be a good idea to find to find something else to do with the oil. As a result of his eventual development of facilities for making kerosene from crude oil, Kier is known as the father (or, sometimes, grandfather) of the petroleum refining industry.

† The birth of the American oil industry occurred on August 27, 1859. On that day, Colonel Edwin Drake struck oil in the first well drilled in America, near Titusville, Pennsylvania. Drake assembled an outfit based largely on the technology of drilling for salt and managed to hit oil at a depth of 21 meters. He sold the oil to producers of illuminating oils for lamps, at $20 per barrel, an astronomical price at the time. In 1859, the median family income in the United States was probably a few hundred dollars per year. If the ratio of the cost of a barrel of oil to median income had remained constant over the last 150 years, petroleum today would be selling at about $2000 per barrel!

FIGURE 21.1 A nineteenth-century advertisement for a patent medicine, made largely from petroleum. (From http://www.fda.gov/AboutFDA/WhatWeDo/History/Overviews/ucm 056044.htm.)

demand dropped, and gasoline became the dominant petroleum product, thanks to the enormous expansion of the private use of automobiles, which increased the demand for gasoline, and the expansion of electric power networks, which decreased the demand for kerosene in household applications. Gasoline still remains the dominant refinery product in some countries, especially the United States. However, the development of the jet engine and its subsequent adoption for most military and commercial applications has led to a modest comeback for kerosene as a petroleum product. (Jet fuel is essentially kerosene that has been carefully distilled and purified.) The increasing dieselization of the world's small vehicle fleet is also eroding demand for gasoline.

THE JET ENGINE

THE EARLY HISTORY OF THE JET ENGINE

The successful development of a jet-propelled aircraft emphasizes how few are the really new ideas and how long it takes to realize some of the old ones.

Hero of Alexandria described about AD 50 a machine demonstrating the principle of propulsion by the reaction of jets (of steam) and the Abbé Miolan attempted in 1784 to apply it to the navigation of a hot-air balloon. The project was dismissed as having 'no aviating merit,' but this did not deter J.W. Butler and E. Edwards from patenting in 1867 (36 years before the Wright brothers flew) a design for an aeroplane which envisaged propulsion by the reaction of jets of compressed air or gas, or by 'the explosion of a mixture of inflammable gas or air' emitted through jets.

—Davy[1]

The earliest powered aircraft, for example, the Wright brothers' plane of 1903, used a gasoline-fueled reciprocating piston engine. In 1929, Bofors, in Sweden, decided to develop a gas turbine engine for airplanes, to create, as they put it, an "airplane without a propeller." In such a plane the powerful air stream normally produced by the propeller would be replaced by a gas jet.* In 1934, two different versions of the "propeller-jet turbine" (the term turboprop, discussed later in this chapter, had not yet been coined) were developed. The Bofors project never got off the ground—both literally and figuratively speaking—because of deep cuts in the Royal Swedish Air Force budget in the mid-1930s.

At about the same time, the German aviation industry was trying to produce a gas turbine for use in airplanes. The aeronautical engineer Hans von Ohain patented a design for a jet engine in 1935, and began developing it in the following year for aircraft manufacturer Ernst Heinkel. On August 27, 1939, a Heinkel He 178 became the world's first jet-propelled aircraft to fly, at Warnemünde.

The hideous wail of the engine was music to our ears.

—Heinkel[2]

The Junkers firm started to develop another gas turbine engine in early 1940, and the first flights took place two years later with the Jumo 004 engine. This later became the first jet engine to be mass produced. For several years thereafter, the German authorities did not show an interest in the jet engine. For one thing, Hitler did not expect a long war, and, for another, rockets and guided missiles were accorded a higher priority. Eventually, a slightly modified version of the Jumo 004 engine was installed in the German twin-engine fighter-bomber aircraft, the Messerschmidt 262 "Stormbird" (Figure 21.2), which began to be produced in 1943.

The course of the development of the jet engine in Britain began with Frank Whittle. He entered the Royal Air Force as an apprentice and was, in due course, selected for training that would lead to a commission. During his training at the RAF College, he was required to write a number of essays; in his fourth term he chose, as a subject for an essay, the "Future Development in Aircraft Design." Whittle, who was 22 at the time, argued that high-speed long-distance flights would have to be at high altitudes, where air resistance was much less. Piston engines would not be suitable and he therefore considered the possibility of the turbine, which would perform far more satisfactorily at high altitudes. As we will see shortly, there was nothing new or novel about the idea of a gas turbine. At that time, Whittle thought of the turbine as driving a propeller (again leading to the "propeller-jet turbine" design— the turboprop—which we will discuss later). He imagined the first use for his turbine engine would be to propel a small, high-speed mail plane. Whittle patented a gas turbine in 1930. His superiors in the RAF showed little sympathy for his ideas at first, but in March 1936, he was given permission to devote part of his time to

---

* The word "jet" has several meanings. One, as used in this context, is a stream of liquid or gas that is shot out at high pressure from a small opening. The gases produced by combustion of the fuel produce the jet. A jet of water is produced by, for example, a fire hose or a water pistol. Another meaning is a fast and narrow current in the ocean or atmosphere, as in the "jet stream" we hear about in weather reports. "Jet" also refers to a jet engine or to an airplane equipped with jet engines.

FIGURE 21.2   The Messerschmidt Me-262 Stormbird was the world's first jet-engine military airplane to enter combat, in 1944. (From http://en.wikipedia.org/wiki/File:Bundesarchiv_Bild_141-2497,_Flugzeug_Me_262A_auf_Flugplatz.jpg.)

experimental work. He founded the company Power Jets Ltd. to develop his invention. The following year Whittle began to test his gas turbine. At the outbreak of the Second World War, the company received an order for exactly *one* engine while, at the same time, the Gloster Aircraft Co. was given the job of building an experimental airplane to fit it. This plane, the Gloster E28/39, made its first flight in May 1941.

> *'Frank, it flies!' 'That was what it was bloody well designed to do, wasn't it?'*
>
> **—Whittle**[3]

Further development was carried out mainly by British Thomson-Houston Co. and by Rolls-Royce.

By the end of the war, both Germany and Britain had put their first jet-engine aircraft into operational combat service. These planes came too late to have much effect on the war, but subsequently jet engines were adopted for most military uses because of the greatly improved performance that they gave in speed and power. From the point of view of the airplane, the great advantage of the jet engine is its high ratio of power to size, which more than compensates for its noise and high fuel consumption. Another advantage is that it enables airplanes to travel faster than the speed of sound, and this capability is now available in many military aircraft. All large passenger planes now use jet engines, even though the emphasis in passenger aircraft development has been on jumbo jets capable of carrying several hundred passengers on long-haul flights, rather than on the attainment of exotic speeds.

It is worth recalling the relationship of James Watt's steam engine and Charles Parsons' steam turbine. Watt's engine, which was a reciprocating design, was

---

TABLE 21.1

Dominant Forms of Engines Prior to Whittle's Development of the Aviation Gas Turbine (Jet) Engine

Type of Combustion	Reciprocating Engine	Rotary Engine
External	Watt's steam engine	Parsons' steam turbine
Internal	Otto four-stroke engine	????

The jet fills in the "blank" (as indicated by the question marks) for the internal combustion, rotary engine.

---

unquestionably successful. Nonetheless, these engines tended to be very large, very heavy, "rough" in operation due to vibration of the reciprocating motion, slow, and have a fairly low ratio of power output to weight of engine. The Parsons steam turbine could operate at much higher speeds, was smaller and lighter, and offered a higher power-to-weight ratio. In both, the steam needed for operation was generated in a separate boiler, so these engines are thought of as external combustion engines. Nikolaus Otto's four-stroke reciprocating internal combustion engine was also a successful development. The question remaining is whether an analogous internal combustion turbine engine could be developed. That is, what is the missing component of Table 21.1? In fact, the first suitable engine design was produced in 1791, by the English inventor John Barber. Demonstration of a working turbine engine that could be used in aviation was accomplished by the German engineer Hans-Joachim von Ohain in 1937.

In its design, the gas turbine is the simplest of the combustion engines. In 1897, Nils Dalén,* a Swedish engineer who later won a Nobel Prize in physics, tried to persuade a friend to become a partner in the development of a gas turbine engine. He described its principle as follows:

> *Imagine a stove! If you blow through the grate, the hot smoke goes out through the chimney. But the clever thing is, if you blow in one cubic meter, two or three will go out. A pump up there, in other words, could drive another pump down there to suck in the air. Now, if we substitute these pumps with turbines...*

—**Dalén**[4]

Many inventors had devoted themselves to finding a practical application for this simple principle.

---

* Nils Dalén (1869–1937), a highly talented engineer, held over one hundred patents. An area in which he contributed greatly was in the use of acetylene gas as the fuel for providing the brilliant light needed in lighthouses. Dalén invented the so-called sun valve, which would automatically shut off the acetylene when the sun was shining, since light from the lighthouse would be superfluous. He also invented what came to be called the Dalén flasher, which allows burning of acetylene in quick spurts, to create a bright, flashing light from the lighthouse. Acetylene can be tricky stuff to work with, and, under the wrong circumstances, can be violently explosive. Tragically, Dalén was blinded in an acetylene explosion in the same year that he won the Nobel Prize. He was too ill to attend the prize ceremony, so his brother, Albin, accepted it on his behalf. In an extraordinary irony, Albin Dalén was an ophthalmologist.

There are many problems in constructing a successful gas turbine, and they're not easy to solve. First, the speed of rotation must be very high to give adequate efficiency, a point that Watt noted when von Kempelen claimed, in 1784, that he had invented a steam turbine. (Von Kempelen's other bogus invention, the chess-playing Turk, is mentioned in Chapter 16.) Second, if this speed could be achieved, the temperature of the turbine blades would be extremely high, so that a heat-resistant material must be used. Many of these materials are expensive and difficult to shape or machine. Third, a gas turbine needs to operate at high efficiency, since two-thirds of the energy of the turbine is consumed by the compressor that forces air into the combustion chamber (described in the next section). The total efficiency must exceed 54% just to keep it going by itself. Not until the end of the nineteenth century did anyone achieve an efficiency exceeding this figure. One of the first was Ægidius Ellnig, a Norwegian engineer (Figure 21.3). He had patented his first gas turbine in 1884, which coincidentally was the year in which Charles Parsons was granted a patent for his steam turbine. Ellnig needed another twenty years to modify his original design sufficiently to make it efficient, but Ellnig's engine represents the first successful gas turbine, and the first to use the compressor in series with the turbine. Though almost completely forgotten nowadays, Ellnig contributed a very important step forward in jet-engine design.

When Germany's Messerschmitt Me-262 flew in 1942, it became the world's first jet fighter, with a top speed of 870 kilometers per hour. The somewhat slower British

FIGURE 21.3 Ægidius Ellnig (1861–1949), the brilliant Norwegian engineer who developed one of the first successful gas turbine engines. (From http://en.wikipedia.org/wiki/File:Aegidius_Elling.jpg.)

Gloster Meteor, with a top speed of 770 km/h, followed in 1943. The faster and superior German Me-262 first entered operational combat service in 1944 and was the first jet fighter to engage an enemy plane. More Me-262s were produced during the war than any other jet fighter, fortunately too late to help Hitler's cause. The Lockheed F-80 Shooting Star jet fighter in 1947 set the world speed record (at the time) of 1000 km/h. Civilian jet aircraft became operational in the mid-1950s. The British DeHaviland "Comet" was the first commercial jet airliner. As discussed in Chapter 17, it was not particularly successful, and the airplane that really established the jet age in passenger flying was the Boeing 707.

Compared to piston engines, all jet engines have the disadvantage of greater fuel consumption per unit of ENERGY output. However, jet engines have several advantages: They provide much greater speeds, especially at high altitudes; they are structurally and mechanically simpler; and they are of much lighter weight, about half the weight of a piston engine of comparable energy output.

THE TURBOJET ENGINE

Consider what happens when you blow up a balloon and let go of it. Contrary to what might be assumed, the balloon is moved by the pressure of the air inside, not by the air escaping. Air escaping from the balloon reduces the interior pressure on the "nozzle end" of the balloon. The high pressure continues to push out on the other end of the balloon. This unequal pressure propels the balloon in the direction of the greater pressure. In the same way, the jet engine moves in the direction of the greater pressure inside of it. The propulsion is possible because of Newton's third law of motion. According to this law, which is seldom known by its name, but is familiar when cited:

Every action has an equal and opposite reaction.

In the balloon, the *action* is the air rushing out of the neck. In the jet engine, the *action* is the hot gases rushing out the rear of the engine at a high pressure. The *reaction* in both cases is the unequal pressure inside.

Most jet engines nowadays are so-called turbojets. They are, in principle, very simple machines; simpler than the Watt or Otto engines, for example. A turbojet engine consists of a metal tube, open at both ends. At the leading (front) end is a rotary air compressor that draws in and compresses air. The importance of this part of the engine comes from the fact that fuel needs air (or, more specifically, oxygen) to burn, but at high altitudes air pressure is low, that is, the atmosphere is "thin." The compressor provides the air supply. This compressed air is then passed to combustion chambers, sometimes called burner cans (shown in the diagram in Figure 21.4). The combustion process involves injecting the fuel into the stream of compressed air. Injection takes place through nozzles arrayed around the burner can. The burning fuel maintains the air at a very high temperature and pressure. High-temperature, high-pressure combustion gases are directed against the blades of a turbine. A turbine is a device that converts the kinetic energy of a moving fluid—in

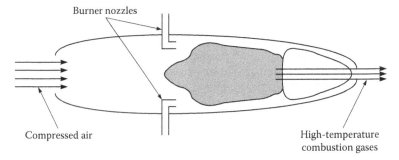

FIGURE 21.4 Schematic diagram of a burner can in a jet engine. The fuel is burned in a stream of compressed air, forming high-temperature, high-pressure gaseous combustion products.

this case, the hot, gaseous products of combustion—into rotary motion. Why do we need rotating motion in a jet engine? To run the air compressor. The combustion gases drive a turbine that is directly coupled to the compressor by a shaft. After leaving the turbine, the hot compressed air enters the tail pipe that ends in a nozzle. The function of this nozzle is to convert the high ENERGY of the air (at high pressure and temperature) into ENERGY of motion, or kinetic energy. The high temperature and pressure are created by release of chemical ENERGY in the fuel. The "reaction" of the mass of ejected exhaust gases pushes the plane forward. It is this basic scheme that gives rise to the turbojet engine, sketched in Figure 21.5.

During flight, air comes into the engine's inlet whatever the speed of the plane is, which could be in well excess of 800 kilometers per hour. Inside the inlet, the air slows down and its pressure increases, but its total ENERGY is unchanged. The air then passes through a series of fanlike compressor blades that push it forward, doing WORK on it and increasing both its pressure and its total ENERGY. By the time the air arrives at the combustion chamber, its pressure is far above atmospheric. The air will then be heated by release of ENERGY from the burning fuel.

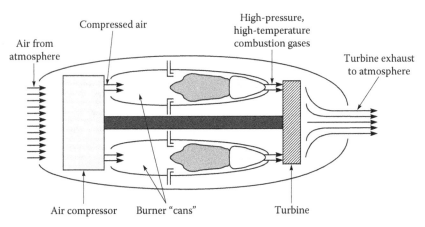

FIGURE 21.5 Schematic diagram of a turbojet engine. The turbine is operated by the gaseous combustion products; the air compressor is operated by the turbine.

Since a given mass of hot air occupies a greater volume than cold air (Charles' law), the hot exhaust gas takes up more volume than it did before combustion occurred. The hot exhaust gas exits the combustion chamber traveling faster than when it entered. Its pressure is still very high. It accelerates out of the back of the engine, exchanging its pressure for speed. On its route out of the engine, the exhaust gas does WORK on the turbine and gives up a little of its ENERGY in the process. The turbine drives the compressor for the incoming air. Overall, the engine has slowed the air down, then added ENERGY to it, and finally accelerated it back to high speed. Because the engine has added ENERGY to the air, air leaves the engine traveling faster than when it arrived. The jet engine has exerted a force on the air to accelerate it rearward; the air has pushed back and produced a thrust that propels the airplane forward.

In the turbojet engine, the working fluid in the turbine is the hot air and the products from burning the fuel. This hot gas mixture has two roles: It spins the turbine, but also provides some jet thrust. As in the steam turbine, the high-pressure, high-temperature gas stream expands going through the turbine and expands again as it exits the nozzle at the back of the engine. From the studies of Charles and Boyle, we should expect the temperature to drop, especially when the combustion gases expand as they exit the nozzle. Therefore, the jet engine is a type of heat engine. The thrust from the back end of the engine can be achieved only if the speed of the gases exiting the back of the engine is greater than the speed of the airplane itself. To ensure that this condition will prevail, the ENERGY supplied to the compressor by the turbine is used to raise the pressure, and of course the combustion of the fuel raises the temperature. Furthermore, the nozzle at the back end of the engine helps to accelerate the gases as they leave the engine.

Over the years, there has been a steady increase in POWER ratings of gas turbine engines. In the fifty years following the Wright's first flight, the maximum POWER of piston engines increased from 9 to 2,600 kW, but in only fifteen years the POWER rating of jet engines climbed to nearly 20,000 kW, more than seven times the maximum of a piston engine. The POWER of just one engine on the Boeing 777 passenger plane is 75,000 kW. Although jet engines are heat engines, as Watt's steam engine and Parsons' steam turbine, we should appreciate that the modern jet engine depends on materials that must be able to withstand temperatures and mechanical stresses that are far greater than the steam-operated devices experience. Additionally, the jet-engine components have to be machined to finer tolerances and greater accuracies than are required in a steam turbine and far beyond anything James Watt could ever have hoped to are see produced in the machine shops of his day.

FAN JET OR TURBOFAN ENGINES

Two more recent engine developments include the bypass engine, the so-called *fan jet*, and the "prop jet" or *turboprop*. Figure 21.6 is a sketch of the fan jet engine. If the highest possible speed is not needed (i.e., neglecting military aircraft), a fan jet engine is more efficient than a turbojet. Fan jet engines are now common on civilian aircraft. They're noticeable because of the conspicuously wider diameter of the engine (to accommodate the fan) at the front end.

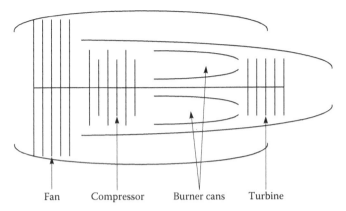

Fan   Compressor   Burner cans   Turbine

FIGURE 21.6   Schematic diagram of a fan jet or turbofan engine. The large fan at the front of the engine helps some air to bypass around the compressor and burner cans.

The thrust exerted by a jet engine must exactly equal the air resistance the plane encounters, assuming the plane is flying level and at a uniform speed. For a large commercial plane, flying below the speed of sound, the design goal is to achieve the optimum thrust from the engine for the minimum fuel consumption. Fuel consumption is an issue for two reasons. First, of course, the airlines must buy the fuel. Reduced fuel consumption translates directly into cost savings. Second, any given airplane is limited in the amount of fuel it can carry. The lower the fuel consumption, the greater the flying range of the plane for a fixed fuel load. In the fan jet engine, only a fraction of the air pulled into the engine is sent to the burner cans. The rest is blown out immediately, adding to the jet of air expelled from the engine. This means that the front section of the compressor must be of wider diameter than the rest, which is why the fan jet engines of commercial jet planes have such conspicuously large "bulbs" in front.

The turbojet engine moves too little air and gives the air too much kinetic energy. In doing so, it incurs an efficiency penalty. To make a jet engine more efficient, it should move lots of air. That way, the air will leave the engine traveling relatively slowly and will thus take away relatively little ENERGY. The thrust of an engine depends on the mass of air expelled per second and the increase in velocity of the air as it passes through the engine. If the total mass of air expelled per second can be increased at the same velocity, a greater thrust can be obtained for a given fuel consumption. As in the turbojet, the engine has slowed the air down, added ENERGY to it, and returned it to high speed. Again, the air leaves the engine traveling faster than when it arrived, but the fan jet engine moves more air than a simple turbojet engine does, giving that air less ENERGY and, therefore, consuming less fuel in adding ENERGY to the air.

This is why a fan jet is more economical than a turbojet, but why the turbojet gives higher speed. The fan jet, however, is not as economical as the turboprop. In the turboprop engine, little ENERGY is left for the jet. Most of it is used to drive the propeller so that as much air as possible "bypasses" the combustion chamber and turbine.

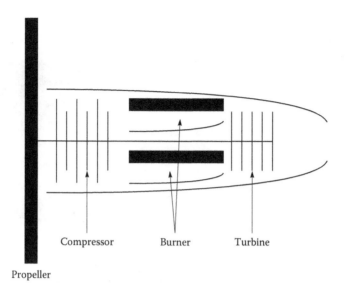

Propeller

FIGURE 21.7   Schematic diagram of a propjet or turboprop engine. The turbine operates a propeller that provides most of the propulsion.

Turboprop and Turboshaft Engines

The turboprop engine (Figure 21.7) represents, in a sense, the fruition of Frank Whittle's original idea. In a turboprop engine, about 80% of the total energy is used to spin the propeller, and only about 20% provides thrust via the nozzle at the rear. A related design is the turboshaft engine. In this engine, all of the ENERGY from the hot combustion gases is used to operate the turbine, and none is available for jet thrust. Turboshaft engines are used in helicopters. They are also used on some other types of aircraft as an auxiliary source of energy needed on board the aircraft.

Jet Cars?

The use of the jet engine for airplanes led observers to anticipate a similarly bright future for gas turbine engines in road vehicles (Figure 21.8). These hopes have never been realized. One reason is the high fuel consumption of the gas turbine engine. Another major concern of the production process for gas turbines is its required manufacturing accuracy, as mentioned above. The gas turbine must be machined to tolerances ten times finer than the usual automobile engine. Thus there is a high manufacturing cost combined with higher operating costs (because of the high fuel consumption).

In diesel engines (discussed in Chapter 22), the maximum gas temperature in the cylinder rises above 2000°C, but the cylinder must endure that temperature only periodically, at the peak of combustion. In the gas turbine engine, though, combustion is occurring continuously. As a result, the temperature of the combustion gases entering the turbine is limited to about 1000°C even when the turbine is made of special (which means expensive) heat-resisting alloys. If it were possible to develop

FIGURE 21.8 The gas turbine car has intrigued designers for decades, but has never managed to become a commercial success. (Reprinted from first edition, by permission from *The Romance of Engines* by Takashi Suzuki, Society of Automotive Engineers, Warrendale, Pennsylvania, 1997.)

and mass produce turbines produced from ceramic materials that would resist heat even better than specialty alloys, the turbine inlet temperature could be allowed to reach about 1370°C. The Carnot equation for heat engine efficiency (Chapter 11) shows that, even if no other design changes were made, this increased temperature alone can result in a 20%–30% reduction in fuel consumption.

ELECTRICITY FROM JET ENGINES

In Chapter 15, we saw that the essence of electricity generation is to find a way to provide WORK to the generator, to turn the generator as cheaply and reliably as possible. The solution is to connect the generator to a turbine. We've seen the use of water (Chapter 15) and steam (Chapter 16) as the working fluids in turbines used in electricity generation. Now we have met the gas turbine, and it would be reasonable to wonder if a gas turbine could also be used to operate a generator. The answer is yes.

Gas turbines can be used to provide electricity generation. These turbines are similar to those used as aircraft engines (indeed sometimes called aero-derivative turbines) but, in electricity applications, are usually fired with natural gas. In the past, these were used often in installations of smaller generating capacity than the hydro- or fossil-fuel plants. Stationary gas turbines for electricity generation offer many advantages. First, they can be purchased and installed relatively quickly, much faster than dams or large fossil-fuel plants. Therefore they offer a rapid way of expanding the total generating capacity of a utility system. Second, a gas turbine can be started and brought up to speed very rapidly, far faster than, say, a pulverized-coal-fired boiler. Gas turbines make excellent reserve units that can be quickly brought on line during periods of the day when demand for electricity peaks, or during unusual episodes of demand (e.g., on an extremely hot and humid summer day, when consumers are switching on air conditioners and causing electricity demand to skyrocket). Third, the fuel, natural gas, is very clean burning, so these plants do not need the emission-control equipment associated with a coal-fired plant.

In recent years some countries, including China, the United States, and South Africa, have verified the existence of immense deposits of natural gas tightly held underground in shale. This gas represents a potential bonanza of energy, the like of which has not been seen in many decades. If the estimated amounts of gas are reasonably accurate, and if it can be extracted without wrecking the environment, this shale gas will have a dramatic impact on patterns of energy use in those countries fortunate enough to have it. One likely application would be the use of shale gas in very large electric generating plants using gas turbines.

JET FUEL

A turbine engine can be designed and built to operate on a wide range of fuels, including natural gas, LPG, liquid products from petroleum, synthetic liquid fuels from coal, biofuels, and even specially treated solid coal.* This is not to imply that a given engine will run on any of these fuels interchangeably, but rather that, given the fuel of choice, it is possible to design a turbine engine that will use it. Today virtually all aviation gas turbine engines (i.e., jet engines) are designed to run on a refined grade of kerosene as the fuel. While jet-engine fuel has about the same boiling range as ordinary kerosene, it is usually more highly refined than the kerosene we would buy for space heating or kerosene lamps. The issues of concern in jet fuel performance include low-temperature behavior and smoke formation.

At high altitudes the atmosphere is very cold. At the very high altitudes encountered in transcontinental or intercontinental flights, the air temperature can be about −60°C. Any liquid is more difficult to pump or to get to flow at low temperatures. The resistance to flow, or the viscosity, of the fuel increases as the temperature decreases. At the low temperatures of high-altitude flight, the fuel still has to be able move through the lines from the fuel tank to the engine. Consequently the physical behavior of the fuel at very low temperatures is vitally important. As a rule, the larger the molecules, the more difficult it is for them to flow (i.e., the higher the viscosity) at any temperature. This relationship becomes especially pronounced at low temperatures. Thus the composition of jet fuel is adjusted to ensure that it will flow reliably at low temperatures. In addition, at very low temperatures some of the very large, wax-like

---

* All coals, as they come from the ground, contain a variety of mineral constituents that produce an ash residue when the coal is burned. The amount of ash produced varies from one coal to another, but can represent 5% to about 35% of the weight of the fuel. Firing coal directly into a turbine would be a sure-fire recipe for disaster, because the ash components would either erode and weaken the turbine blades, or would deposit unevenly on the blades. In either case this high-precision piece of machinery would become unbalanced. The eventual breaking off of one or more blades would lead to catastrophic failure. Some 15–20 years ago, several countries, with Japan in the lead, developed a process for making HyperCoal, a specially treated coal in which the ash residue was less than 0.1%, ideally about 0.02%. The intent was to use HyperCoal in turbines in electric generating plants. Although the development effort succeeded in reaching these low ash values, the process apparently has never been commercialized.

molecules that might be dissolved in the fuel will precipitate as a solid. "De-waxing" of kerosene helps to minimize this problem.

Properties used to measure the low-temperature behavior of fuels include the cloud point and pour point. The cloud point is the temperature at which the first crystals of wax (or other solid) begin to form as the fuel is cooled. This initial formation of small wax crystals gives the liquid a translucent, cloudy appearance, evident on visual inspection. The pour point, at even lower temperatures, is that temperature at which the liquid will not flow at all, even if one were to attempt to pour it out of an open container. While these may seem like "low tech" measurements, the determinations of the cloud point and pour point are carried out under very carefully standardized conditions, and the values of these properties are used to establish specifications for the fuel.

Smoke is a nuisance and a pollutant from civilian aircraft. In addition, it could possibly be used to detect military aircraft, so smoke formation is also a concern to the military. Aromatic compounds are mainly responsible for smoke formation, so smoke formation can be adjusted by reducing the amount of aromatics during refining. This property of the fuel is determined by the smoke point. Like the cloud and pour points, the smoke point also might appear to be a rather "low-tech" test, but it too is carried out under very rigidly standardized conditions. The smoke point determination involves adjusting the height of a wick used in burning the fuel being tested, to find the maximum height at which the flame can be sustained without producing visible smoke or soot.

Sulfur is also removed from jet fuel during refining, both to reduce SO_x emissions and because sulfur compounds in the fuel are mildly corrosive to the metal parts in fuel storage and handling systems. SO_x emissions are a concern because they can contribute to serious air pollution problems, notably acid rain (Chapter 27). As a secondary issue, many sulfur compounds have absolutely dreadful odors.

Jet fuels are also treated with a variety of specialized chemicals, added in very small quantities, to adjust or tailor the properties of the fuel. These so-called additives do a variety of jobs. They improve the lubricity of the fuel (its ability to lubricate) to help operation of fuel pumps. They prevent the buildup of static electricity, reducing the danger of an accidental fire caused by an electric spark. They help prevent the slow degradation of the fuel by oxygen during the time that it is in storage. They combat the formation of ice. After the dewaxing, desulfurization, and other refining steps, a fuel will be treated with a specified "additive package" before being sold and used.

Worldwide, the production of jet fuel is approaching six million barrels per day. This represents about 8%–9% of the amount of petroleum produced and processed in the world. However, not all of the products of an oil refinery are distillates (i.e., produced from the distillation tower). When expressed as a fraction of total distillate fuel production, the proportion of jet fuel is about 12%.

Production of jet fuel rose sharply in the 1950s and 1960s. In the mid-1950s, most operational jet airplanes were used by the military. Jet fuel accounted for a much smaller percentage of petroleum distillate fuels. By the mid-1960s, as jet airplanes steadily became more important in civilian air transportation (time line, Figure 21.9), production grew to be about 10% of total distillate fuels. The subsequent forty years have seen continued, but small, growth of the share of distillate fuel production

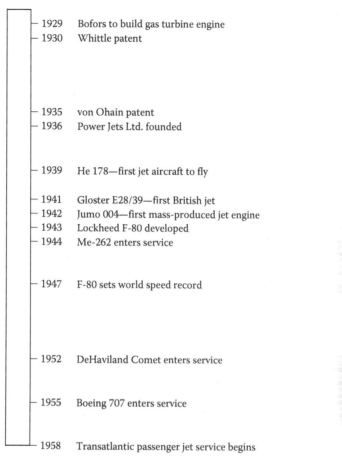

FIGURE 21.9 A time line for some of the major developments in the history of jet aircraft.

represented by jet fuel. It's important to remember that these figures are percentages, applied to the total amount of petroleum consumed. Even though the *percentage* of total distillate fuel production represented by jet fuel has held nearly constant over the past four decades, the actual *amount* of jet fuel produced has increased because of the increase of total petroleum consumption.

REFERENCES

1. Davy, M.J.B. Quoted in Gregory, K. *The Next to Last Cuckoo*. Akadine Press: Pleasantville, NY, 1997, pp. 67–68.
2. Heinkel, E. Quoted in Chase, A. *Technology in the 20th Century*. Bluewood Books: San Mateo, CA, 1997, p. 100.
3. Whittle, F. Quoted in Chase, A. *Technology in the 20th Century*. Bluewood Books: San Mateo, CA, 1997, p. 102.
4. Dalén, N. Quoted in Strandh, S. *The History of the Machine*. Dorset: New York, 1989, p. 150.

FURTHER READINGS

Guibert, J.C. *Fuels and Engines*. Editions Technip: Paris, France, 1999. Chapter 7 discusses engine operation, characteristics of jet fuel, and how fuels are formulated. Intended mainly for readers with technical background.

Gunston, B. *Jet and Turbine Aero Engines*. Patrick Stephens: Sparkford, U.K., 1997. A history of the development and applications of jet engines, written mainly for readers with little science or mathematics background.

Harvey, J.N. *Sharks of the Air*. Casemate: Philadelphia, PA, 2011. A recent addition to the literature on the Messerschmitt Me-262 "Stormbird" fighter jet. Told from the German perspective, and with biographical details on Willi Messerschmitt. Military aviation buffs will also enjoy the earlier volume by Hugh Morgan, *Me262: Stormbird Rising* (Osprey Aerospace: London, 1994). This book provides a history of the development and operation of the world's first successful turbojet military airplane.

Schobert, H.H. *The Chemistry of Fossil Fuels and Biofuels*. Cambridge University Press: Cambridge, U.K., 2013. Chapter 15 discusses the production and refining of kerosene and jet fuel. Intended for readers with some prior knowledge of chemistry.

Suzuki, T. *The Romance of Engines*. Society of Automotive Engineers: Warrendale, PA, 1997. Chapter 40 discusses many of the attempts to adapt the gas turbine engine to cars, trucks, and buses.

22 Diesel Engines and Diesel Fuel

LIFE WITHOUT MATCHES

The story of diesel engines actually begins with the question of how one actually lights a fire. The common, ordinary match (Chapter 7) is only about 150 years old. Practical lighters go back about a century.* Before the development of the match, there were various cumbersome ways of lighting fires, such as flint-and-steel sets to strike sparks or devices for rapidly rotating wooden rods and using the friction heat to ignite tinder. (Tinder is any readily combustible material such as dried cloth, wood shavings, or small twigs.) The easiest way to start a fire was to "borrow fire" from the neighbors, by obtaining an already glowing piece of coal or charcoal from them. But, there was still a need to have a portable and convenient way of starting a fire. One device for doing that, which might be considered an early forerunner of the lighter, was the fire piston.

The fire piston may have originated in primitive southeast Asia. In 1804, a French inventor, Joseph Mollet, produced a model that represents the European reinvention of the "fire piston" (Figure 22.1). Rapidly compressing air to one-fifth of its original volume produces a temperature high enough to ignite a little tinder on the underside of the piston. A sharp blow on the end of the piston compresses the air and heats it (remember Boyle) enough to ignite the tinder. The burning tinder could then be used to start a fireplace fire or whatever else it was desired to burn. Fire pistons were manufactured commercially in France, but were made obsolete by the development of the match in the mid-nineteenth century. By the late nineteenth century, fire pistons were museum pieces.† In Munich, about 1888, a lecturer on heat transfer lit his cigar by this method; in the audience was a young man named Rudolf Christian Karl Diesel (Figure 22.2), who later said that the demonstration was a major inspiration for his new engine, in which ignition is achieved by compression alone.

---

* Attempts to make a lighter that is both portable and practical actually go back many centuries. One of the earliest designs, apparently originating in Germany, was a modified version of a flintlock pistol. It's hard to imagine carrying such a thing in today's society.

† Though the fire piston is certainly two centuries old, and quite possibly much older, new ones can still be purchased today, sometimes made of first-rate materials such as titanium or cocobolo wood. Directions for making your own are also available. Fire pistons seem to appeal to survivalists, who anticipate the imminent collapse of civilization as we know it, due to a nuclear holocaust, running out of oil, bioterrorism, election of the opposite political party, or some other worldwide calamity. The fire piston is not such a bad idea, because matches have a way of getting lost, or getting too wet to work, or eventually just running out. As long as one has tinder and a fire piston, a fire can be started.

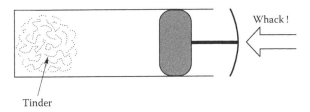

Tinder

FIGURE 22.1   The fire piston, a device that uses the high temperature generated by rapidly compressing air to ignite tinder.

FIGURE 22.2   Rudolf Diesel (1858–1913), one of the great engineers, and the only one so honored by having both an engine and its fuel bear his name. (From http://en.wikipedia.org/wiki/File:Diesel_1883.jpg.)

Diesel attended the Technische Hochschule (Polytechnic) in Munich. In 1878, he heard a lecture there on Carnot's theorem concerning the ideal conditions for expansion of gases in an engine's cylinder. In the lecture that inspired Diesel, his professor pointed out that steam engines transform only 6%–10% of the chemical ENERGY of their fuel into useful WORK (i.e., their efficiencies are in the range of 6%–10%). Diesel was convinced that he could develop an engine that would do better. Throughout the rest of his studies and into his early career, which involved research work on refrigerators, Diesel never let the idea drop. Eventually, in 1893, he came up with a scheme for an engine that would indeed operate much more closely to the ideal (Carnot) efficiency for a heat engine. The first diesel engine was built soon after.

THE DIESEL ENGINE AND HOW IT WORKS

EARLY DEVELOPMENT WORK

Diesel designed his engine on the general pattern of the gas engine perfected by Otto in 1876. He set about developing an engine that would be more efficient than existing steam or gasoline engines, and that could, in principle, be operated on a variety of fuels: gases, powered coal, coal tar (a liquid by-product of converting coal into the coke used in metallurgical furnaces), or higher-boiling petroleum products. (This is not to imply that all these fuels could be used in the same engine, but rather that variants of Diesel's fundamental design could be adapted to these different kinds of fuels.) The use of pulverized solid fuels quickly seemed impractical.* The vast majority of diesel engines nowadays operate on the petroleum product that we commonly call diesel fuel.

Diesel first intended to use finely powdered coal, a fuel that many designers were, at the time, thinking of using for various kinds of engines. Diesel soon abandoned that idea. Instead, he began to experiment with crude oil. Diesel's first engine, a single cylinder machine using oil as fuel, was built in 1893. In 1897, after a period of hard work on further development of the engine, which took a toll on his health, Diesel built an engine with efficiency clearly superior to that of any other combustion engines. The 1897 version of his engine (Figure 22.3) showed a mechanical efficiency of 34%, a great advance over all other engines.

In its original form, the diesel engine was designed to operate at low speeds. It was expected to serve as an alternative to reciprocating steam engines for use both in ships and in stationary installations. These engines usually ran at speeds from 150 to 250 rpm and weighed between 90 and 220 kilograms per kilowatt. By 1909, Diesel had developed a small four-cylinder engine of 22 kW rated at 600 rpm. It was intended for use as a heavy-duty automobile engine, but was not pursued at the time.

HOW THE DIESEL ENGINE FUNCTIONS

Diesel's engine can be considered as a variant of Otto's four-stroke cycle. The key components are the cylinders and pistons, intake and exhaust valves, and a fuel

---

* Several reasons contribute to making the idea of a solid-fuel diesel (or other internal combustion) engine impractical. Solids are harder to handle, to get to flow, and to meter or regulate than fluids. It is comparatively easy to inject a fluid into a chamber, such as an engine cylinder, at high pressure, whereas this can be difficult to do with solids. Solids can be abrasive, and steadily wear away the internal parts of the fuel-handling and injection system. Most solid fuels—particularly coal—leave an ash residue when they burn. This ash cannot be allowed to accumulate in the cylinder, but has to be swept out of the cylinder during each cycle of the engine. The ashes of many coals contain quartz, the principal mineral component of sand, a very hard and abrasive substance. Running the engine would result in, essentially, continuously sand-blasting the exhaust valves and other parts of the exhaust system. Also, the ash has to go someplace—either into the air or into some kind of special collection hopper on board the vehicle. Stimulated in part by the oil shortages of the 1970s, there has been occasional interest in operating a diesel engine on a suspension of finely pulverized coal in oil or in water. Doing so would eliminate the problem of handling solid fuel, and would probably reduce substantially the abrasion problem in fuel injection. However, it's hard to beat the ash problem.

FIGURE 22.3   Diesel's first engine. (From http://en.wikipedia.org/wiki/MAN_SE.)

injector. The intake and compression strokes deal only with air. The first, intake stroke (intake valve open) draws air into the cylinder (Figure 22.4). The second stroke (compression) compresses the air in the cylinder such that it is above the temperature needed to ignite the fuel (Figure 22.5). Note that both valves are closed: The compression ratio depends on the specific engine, but can easily be around 20:1, appreciably higher than in conventional gasoline engines.

Rapidly increasing pressure, as in the compression stroke, means that the temperature also has to go up.* We can see this phenomenon for ourselves in tire pumps or the small air pumps used to inflate various kinds of sports balls. The "barrel" of the pump can get quite hot after vigorous pumping (of course, some of this heat simply comes from the friction of the pump's piston against the cylinder walls).

In the diesel engine, the goal is to get the air hot enough to ignite the fuel. But increasing temperature (Figure 22.6) is a nonspontaneous change. Such a change can occur only if WORK is done on the SYSTEM (Figure 22.7). In this specific case, this WORK is sometimes referred to as *compression work*.

---

* At this point, it is worth noting the similarity of operation of the diesel engine and the fire piston. We need once again to recall Charles, who showed that the volume of a gas is proportional to its temperature; and to recall Boyle, who showed that the product of pressure and volume of a given quantity of gas is constant. If we combine the work of these two scientists into a single relationship, we can say that, for a given quantity of gas, the product of pressure and volume is somehow proportional to temperature. Hence a decrease in volume will increase pressure and increase temperature.

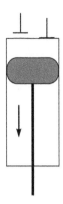

FIGURE 22.4   The intake stroke of a diesel engine draws air into the cylinder.

FIGURE 22.5   In the compression stroke of a diesel engine, only air is compressed as the piston rises. The compression ratio of a diesel engine is large enough that the temperature of the compressed air will be high enough to ignite the fuel.

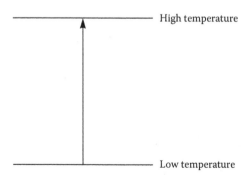

FIGURE 22.6   The compression stroke of a diesel engine raises the air from low to high temperature, an example of a nonspontaneous change.

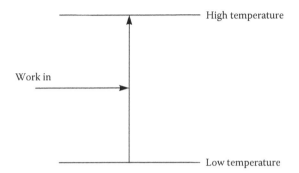

FIGURE 22.7   Because the temperature change in the compression stroke is nonspontaneous, we must supply ENERGY or WORK to make it happen. The necessary compression WORK is supplied by the engine in the course of its normal operation.

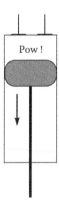

FIGURE 22.8   The diesel engine has no spark plugs. In the compression stroke, fuel injected into the cylinder ignites because of the high air temperature.

In the diesel engine, the first stroke of the piston draws air into the cylinder and the second stroke compresses the air to such an extent that the temperature rises far above the autoignition point of the fuel. In the third stroke (the power or combustion stroke), the fuel injector squirts fuel into the cylinder. The fuel is injected at high pressure as a fine spray into the compressed air, which is swirled to obtain turbulent mixing. Injection continues for a substantial part of the power stroke. Because the air is above the temperature at which the fuel will ignite, the fuel immediately begins to burn. The technical term for the diesel engine is the "compression ignition engine," called so because the compression of the charge is sufficient to cause the fuel to ignite. No spark plugs are needed. Both the intake and exhaust valves are closed (Figure 22.8). The piston begins moving downward. The consequent expansion of the gas in the cylinder should cause its temperature to drop, but, in this case, the temperature of the gas does not drop because heat is being released from the burning fuel. Initially, the heat ENERGY of the burning fuel is converted directly to useful WORK pushing down the piston in the cylinder.

The key difference between Diesel's engine and all other internal combustion engines is that heat from the burning fuel is not "wasted" to heat up the air in the cylinder.

The air is already hot from the compression work we supplied to it.

As the piston moves downward the air would normally be cooled, but the heat released by the burning fuel prevents this. Instead, the temperature of the air does not change significantly and the heat ENERGY of the burning fuel is converted into useful WORK driving the piston down the cylinder. This accords with Carnot's concept of maximizing efficiency that there must be no useless flow of heat (ENERGY) from the hot body (the burning fuel) to a cold body (the air). Compared to other engines, the operation of the diesel engine affords production of the highest temperature of the cycle not by combustion, but rather before the combustion has occurred, entirely by compression of the air in the cylinder.

Once the air and gaseous products of combustion have expanded without losing heat, the fuel injector shuts off, and the air can then expand further with the accompanying drop in temperature and pressure that we would expect. Temperature and pressure fall because the heat ENERGY in the air is now converted into useful mechanical WORK. When the temperature and pressure have fallen, the exhaust valve opens and the piston expels the air and burned fuel in the exhaust stroke. As in the Otto engine, the piston then returns, drawing in the next charge of air in the intake stroke. Although the ideal is to allow temperature and pressure to fall to the level of the atmosphere in order to maximize efficiency, this concept was not practical, because an enormously long cylinder would have been required.

The fourth stroke (exhaust) occurs as the piston sweeps the combustion gases out of the cylinder (Figure 22.9). In this case, the exhaust valve is open.

The best diesels have efficiencies of up to 42%, and performances above 30% are routinely achievable with any reasonably well-maintained engine. Although the ideal Carnot efficiency of the Otto cycle is 60%, the best practical performance of gasoline engines is only around 32%.

Usually, diesel engines have a weight per unit of POWER about 1.5 to 3 times greater than gasoline engines. The reason for this increased weight is the diesel's reliance on self-ignition, which in turn requires the high-temperature compressed air resulting from the high compression ratio. As a result, the pressure in the cylinders of a diesel engine is much higher than a gasoline engine. To withstand the high pressure, a sturdy, rugged cylinder is required. When the volume of the cylinders is the same (or, more specifically, the volume swept out by the piston through its stroke), the diesel engine produces only about two-thirds the POWER of the gasoline engine.

Owing to the high pressure generated in a compression ignition engine, the various parts have to be very strongly made. The necessary strength adds considerable weight, and although this may not be of great importance in stationary engines, it is a serious concern when the engine is intended for transport purposes. Another advantage of a diesel engine is that it has a higher durability, or a longer service life,

FIGURE 22.9 In the exhaust stroke of a diesel engine, the gaseous products of combustion are swept out of the cylinder as the piston rises.

as a result of the robust, heavy structure needed to sustain the high combustion pressures. Because of this long service life and excellent fuel economy, diesel engines are used extensively in commercial vehicles, such as long-haul trucks and many buses.

COMMERCIAL APPLICATIONS OF THE DIESEL ENGINE

Rudolf Diesel confidently predicted that his engine could have many possible applications. It could replace the steam locomotive for railway traction. It could be useful in road vehicles, such as trucks. It could be used for propulsion of ships and boats and even as a stationary power source in various kinds of devices. In 1910, Diesel's assertions were dismissed as ludicrous. The steam locomotive reigned supreme on the world's railways. The gasoline engine was steadily displacing electric motors and steam engines to become dominant in the automotive market. Small ships were powered by reciprocating steam engines and large ones by steam turbines. Steam engines of various sorts almost totally monopolized the applications for stationary power sources. It was quite obvious to everyone that there was no market for the diesel engine.

But sometimes the "obvious" can turn out to be dreadfully wrong. Diesels made their first major inroads in marine propulsion, both in ships and in submarines, during the First World War, and their dominance in this market (except for nuclear propulsion in the largest submarines) was complete by 1960. On land they began to be used first in heavy earth-moving and farming machines. As they became lighter, they displaced steam locomotives for railway use, and even lighter engines took over most of the truck and bus transport market virtually worldwide. Nowadays, even the small vehicle fleet (i.e., automobiles and small trucks) is being "dieselized" in many countries.

STATIONARY ENGINES

In the United States, the Diesel Motor Co. of America was founded in St. Louis in 1896, at the instigation of the brewer Adolphus Busch. The first diesel engine made in the United States was delivered in 1898 and generated 45 kW. It was used to operate an electricity generator. Nowadays the diesel engine totally dominates the market for stationary power sources, an example being its application to operate auxiliary or emergency electricity generators. Such generators may be either standbys in central electricity-generating stations or used in remote locations where a transmission grid from a central station would be too expensive.

MARINE APPLICATIONS

When the First World War broke out in August, 1914, the diesel engine found its first market niche—*das untersee Boot*—better known in the English-speaking world as the U-boat. Diesel engines offer numerous opportunities for submarine propulsion. One is their relatively compact size, much smaller than a steam engine of comparable power. A second is their good fuel economy, which extends the submarine's cruising range between refueling. A third is that the fuel itself is not nearly so flammable or explosive as gasoline, an important consideration in the confined space of a submarine.

The diesel engine has totally replaced the reciprocating steam engine in small- and medium-sized ships (though the steam turbine continues to dominate in large vessels). Virtually all large ships are now diesel powered. For such applications, the diesel engine has many advantages when compared with steam engines, especially those depending on coal-fired boilers. The diesel provides not only a substantial saving in the number of crew needed to operate the engine but also considerable saving in the space required for carrying fuel. In 1910 the first sea-going motor ship, the tanker *Vulcanus*, was built in Holland and fitted with a 485 kW engine. One of the largest diesel-engine vessels was the Cunard-White Star liner *Britannic*, which had two 7500 kW engines.

RAILWAYS

In several nations, including the United States, the diesel engine dominates as the "prime mover" in railway locomotives. In most so-called diesel locomotives, the diesel engine supplies the WORK to a generator. In turn, the generator supplies ENERGY to electric motors that actually propel the locomotive. Such locomotives should really be referred to as diesel-electric locomotives (Figure 22.10). The principal components in the diesel-electric locomotive are the diesel engine, the electric generator, and the traction motors.

The diesel-electric locomotive began to replace the steam locomotive for mainline operation in the 1930s. However, the enormous demands placed on transportation systems during the Second World War meant that mainline steam locomotives were not retired until the 1950s or, in some countries, even decades later. Like the automobile engine, the diesel locomotive engine requires some type of variable drive mechanism between it and the driving wheels to take care of varying speed and

FIGURE 22.10   An example of a relatively modern diesel locomotive. This locomotive is the Class 66 of the English, Welsh, and Scottish Railway. (From http://www.copyright-free-pictures.org.uk/trains/39-class-66-diesel-locomotive.htm.)

power demands in starting and in climbing hills, while the engine itself generally runs at constant speed. Neither mechanical gears nor a fluid (hydraulic) transmission, such as are used in automobiles, is adequate for the high power required in a locomotive, though some railways have tried diesel-hydraulic locomotives.

Compared with a steam locomotive of equal nominal POWER, the diesel locomotive has important advantages. The useful POWER of a diesel locomotive may be greater than that of a steam locomotive because the diesel POWER is available over a wider range of speeds. The engine terminal expenses are less with the diesel locomotive because routine maintenance and refueling require less work. There is no need to provide water for a boiler. The diesel-electric locomotive has an efficiency of about 30%, about five times that of the steam locomotive. Moreover, diesel locomotives can be coupled in multiple units under the control of one crew, and these multiple units may produce tractive effort far greater than a single steam locomotive. Finally, many diesel locomotives can operate readily in either direction, eliminating the need for a turntable for turning the locomotive around.

ROAD TRANSPORTATION

Almost all of today's large trucks and buses use a diesel engine as their power source. The first successful application (1924) of a diesel engine was in the Benz 38 kW, 5-tonne vehicle. The market position of the diesel was profoundly changed by the development of the high-speed engines designed to compete with gasoline motors. The speed of the engine was increased to 1000 rpm, and by the late 1930s maximum speeds of 3000 rpm were achieved in commercial types. The weight of the engine was greatly reduced; standard models were produced ranging in weight between 6 and 10 kg/kW. The diesel engine enjoys worldwide dominance in heavy trucks, is widely used in buses, and has steadily growing applications in cars and light trucks.

THE DIESEL AIRPLANE

The only internal-combustion-engine market in which diesels could not compete, let alone dominate, was in airplanes. The best-ever diesel airplane engine, the Jumo 207 developed in Germany for the Second World War, was still too heavy to compete with spark-ignition gasoline engines. But this was not for want of trying, nor for lack of some early, promising successes.

In September 1928, the Packard Company succeeded in the first flight with a diesel engine mounted on a Stinson-Detroiter plane (Figure 22.11). In May 1929, the plane crossed the United States in only 6.5 hours (a good time even by today's standards for a coast-to-coast flight), with a fuel cost of about one-fifth that for a gasoline engine. The next year, Packard began commercial production. In 1931, a Bellanca airplane with the Packard diesel engine established a new world endurance flight record, 84 hours and 33 minutes. A year later, this engine helped the Goodyear blimp (a type of airship) establish a world altitude record. These successes attracted the attention of the aviation community around the world. However, the Packard diesel engine cost about 35% more than the competing Wright Whirlwind engine. Also, the Packard engine was notorious for vibration, exhaust odor, smoke, and soot, characteristics that made it unpopular with pilots and passengers.

FIGURE 22.11 The Stinson Detroiter, a diesel-engine airplane. (Photograph courtesy of Kate Igoe, Smithsonian Institution Archives, http://www.sil.si.edu/smithsoniancontributions/AnnalsofFlight/pdf_hi/SAOF-0001.2.pdf.)

Currently aviation applications of diesel engines are seeing a modest revival. Modern engine designs and aircraft (usually small planes) are emerging that do away with some of the disadvantages experienced in earlier attempts. An advantage for small aircraft, such as bush planes, for flying into remote regions, is that diesel fuel is likely to be more available than the high-octane gasoline needed for spark-ignition aircraft engines. In some reasonably affluent, industrialized areas, including much of Europe, aviation gasoline is likely much more expensive than diesel fuel. One area of aviation in which the diesel engine clearly reigns supreme is in model aircraft. The engines used in most radio-controlled model airplanes are tiny diesels, running at constant speed.

DIESEL FUEL

Diesel fuel (occasionally also called diesel oil) boils in the range 175°C–345°C, higher than kerosene or jet fuel, and much higher than gasoline. The boiling range of diesel overlaps somewhat that of kerosene because commercial diesel fuel is often made as a blend of several refinery streams.

A diesel engine operates by self-ignition of the fuel in the cylinder, which, in a very crude sense, would let us say that the engine works by knocking. Thus the fuel characteristics we want are the inverse of gasoline's, for which self-ignition leads to knock. Because diesel fuel has a much higher boiling range than gasoline, its molecules are larger, on average, than those in gasoline. The octane scale for measuring fuel quality cannot be applied to diesel fuel. Instead, the standard of good combustion performance is cetane. The cetane molecule is a linear paraffin containing sixteen carbon atoms, which should autoignite (knock) very well. Cetane represents the standard of good diesel fuel performance, assigned a value of 100 (Figure 22.12). Recall that aromatic hydrocarbons are good antiknock components of gasolines, so would be undesirable in diesel fuel. For diesel fuel the aromatic compound 1-methylnaphthalene is given a value of 0. This makes it possible to assign a *cetane number* to diesel fuel as the percentage of cetane in a test blend of cetane and 1-methylnaphthalene that gives the same engine performance as the fuel being evaluated. Good-quality diesel fuels have cetane numbers around 50–55.

Other concerns of fuel quality are much the same as for jet fuel. Aromatics should be low, because of their role in smoke or soot production. Sulfur content should be low, to reduce air pollution from SO_x. Viscosity must be controlled to provide good flow properties at low winter temperatures. Cold-weather performance is even more of a concern than with jet fuel, in two regards. Cold-weather starts must overcome the problem of getting the fuel to vaporize when it is injected into the cylinder. And, flow of fuel in fuel lines is a problem at extremely low temperatures. In very cold weather, such as on Alaska's North Slope, it's not uncommon to allow diesel engines in trucks and machinery to run round-the-clock to keep them warm and avoid starting problems.

The diesel engine offers several advantages relative to other internal combustion engines. A prime advantage is its good fuel economy. Rudolf Diesel was right—he did develop a more efficient engine. The complicated electrical ignition

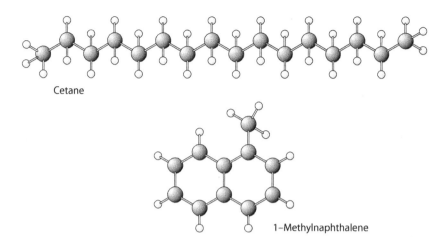

Cetane

1–Methylnaphthalene

FIGURE 22.12 The cetane number of a diesel fuel is determined by comparison with a blend of cetane (also called hexadecane), a linear paraffin of sixteen carbon atoms, and the aromatic compound 1-methylnaphthalene. Cetane itself is given a value of 100 and 1-methylnaphthalene a value of 0.

system is completely eliminated, there being no spark plugs in a diesel engine. The fuel is much less susceptible to accidental ignition and fires than is gasoline. Many diesel engines will run on less-refined fuel than is needed for spark ignition. Sometimes "pro-knocking" compounds can be added as ignition promoters. In those parts of the world where petroleum products are not readily available, diesel engines could be run using fuels of similar molecular weight and with abundant straight carbon chains, such as the oils from coconut, soy, and sunflower. Remarkably, one of Diesel's first major demonstrations of his engine, at the Paris Exposition of 1900, used peanut oil as the fuel. An emerging trend today is to use fuels derived from plant oils (usually not the oils themselves) as biodiesel. This is discussed further in Chapter 31.

DID HE JUMP OR WAS HE PUSHED?

Diesel foresaw that his engine would replace steam locomotives on railroads as well as be used for streetcars and other road vehicles; he supposed that it would be applied to propulsion of all sizes of ships and boats and that it would meet requirements for large and small stationary power sources on land. His engine came at a time when the electric motor was beginning to show great promise for the future, when the gas engine was proving itself economical, convenient, and reliable for a multitude of purposes, when oil and gasoline engines with flame or electric ignition were available, and when the marine steam engine was well established. The steam turbine, with steam now generated in oil-fired boilers, was a near-ideal power source for large, fast ships, while the simple, sturdy, and cheap reciprocating engine was well suited for smaller ships down to fishing boats and tugs. Yet his prophecies have proved, in retrospect, remarkably accurate. From its first commercial applications early in the twentieth century, the diesel engine slowly blossomed and virtually fulfilled all of Rudolf Diesel's predictions that were once dismissed as absurd. In fact, the diesel engine has become so successful that it is the only engine—and diesel fuel is the only petroleum product—that is commonly associated with the name of its inventor. Nobody ever talks about making an intercontinental flight on a Whittle, or about buying fifty liters of Otto fuel.

In 1913, Rudolf Diesel was taking an overnight ferry from The Hook in Holland to Harwich, England. Although he seemed to be in good health and in good spirits, after saying goodnight to the friend with whom he was traveling, he disappeared from the ship while it was en route across the North Sea. Some believe that Diesel committed suicide by throwing himself into the sea, despondent over the apparent lack of enthusiasm for his engine and, possibly, despondent over the likelihood that someday soon a major war would erupt in Europe. A more romantic, albeit less probable, version of the story suggests that the British secret service thought that Diesel was coming to England as a German spy, so they "helped him" throw himself overboard.

It is with great regret that we record the disappearance of Dr. Rudolf Diesel from the G.E.R. steamer Dresden *on the voyage from Antwerp to Harwich on the night of September 29; the circumstances are such as to leave no hope of his being alive. Dr. Diesel will be remembered as the inventor of the oil engine which bears his name. Born in Paris in 1858, of German parentage, his training included courses at the Augsburg technical schools and at the Munich Technical College. His first published description of the Diesel engine appeared in 1893; aided financially by Messrs. Krupp and others, the next few years were spent in arduous efforts to realize the principle of his engine in a commercially successful machine. The difficulties to be overcome were very great. In the earliest attempt, compression of the air was effected in the motor cylinder and the fuel injected direct. This engine exploded with its first charge and nearly killed the inventor. The modern Diesel engine compresses the air in the motor cylinder to a pressure above 400 lb. per square inch, during which process the air becomes hot enough to ignite the fuel. At the end of compression, the fuel is injected by means of a separate air supply at a pressure higher than that in the cylinder. Nothing of the nature of an explosion occurs in the cylinder; the oil burns as it is injected, and, as the piston is moving outwards at the same time, the pressure does not rise to any extent. The fuel consumption of these engines is remarkable, being roughly one-half of any other type of oil motor. Engines both of a two-stroke cycle and of a four-stroke cycle are now being developed by many firms both on the Continent and in Britain. In Dr. Diesel's opinion the two-stroke engine would probably be the standard type for marine purposes. Marine Diesel engines of very large power have not yet been constructed, but many important experiments in this direction are being made. Dr. Diesel's loss will be regretted by men of science on account of his efforts to interpret practically the Carnot ideal cycle, and by engineers on account of the immense strides which his untiring energy and indomitable pluck have made possible.*

—Nature[1]

REFERENCE

1. Gratzer, W. *A Bedside Nature*. Freeman: New York, 1998, p. 140 (*Nature* 1913, 92, 173).

FURTHER READINGS

Cummins, L. *Diesel's Engine*. Carnot Press: Wilsonville, OR, 1993. This book, the first volume of a projected multivolume series, treats the development of the diesel engine from Diesel's earliest ideas through 1918. An enormous amount of detail, including, e.g., original patent drawings, is presented.

Guibet, J.C. *Fuels and Engines*. Editions Technip: Paris, France, 1997. Chapter 4 provides a thorough discussion of diesel fuel, and an introduction to diesel engines. This book presumes that the reader has a background in science and mathematics.

Norman, A.; Corinchock, J.; Scharff, R. *Diesel Technology*. Goodheart-Willcox: Tinley Park, IL, 1998. Intended as a textbook for students in vocational-technical programs in diesel engines. This book provides a wealth of detail on engine components and operations.

Schobert, H.H. *The Chemistry of Fossil Fuels and Biofuels*. Cambridge University Press: Cambridge, U.K., 2013. The chapter on middle-distillate fuels covers the production and refining of diesel fuel. This book presumes that the reader has some background in chemistry.

Sperling, D.; Gordon, D. *Two Billion Cars*. Oxford University Press: New York, 2009. Among the topics covered are a history of automotive development, mainly in the United States, and discussion of possible innovations in efficiency and technology.

Suzuki, Tshi. *The Romance of Engines*. Society of Automotive Engineers: Warrendale, PA, 1997. This book treats the history of the development of internal combustion engines. It contains considerable information on diesel engines. In particular, Chapters 32 to 34 discuss the ill-fated Packard diesel intended for airplanes.

23 Atomic Energy

At the dawn of the twentieth century, most of the things we associate with modern-day life existed in some form. Edison had developed the incandescent light, along with all of the components of the electrical distribution system needed to get electricity into homes. Otto had developed the four-stroke gasoline engine, which Ford would soon use to make the automobile a reality for persons of average means. Steam locomotives—the evolution of the work of Watt, Trevithick, and Stephenson—ruled the rails and moved passengers at speeds up to 150 kilometers per hour. Charles Parsons' steam turbine allowed ship crossings of the Atlantic in a week or less, similar to times required today. The Wright brothers demonstrated the possibility of powered flight. Electrically powered subways and trolley cars were helping to move people in and between cities. Niagara Falls had been tapped as a source of hydroelectric power. The diesel engine was beginning to find applications in small ships and as a stationary power source. Many other developments not directly related to energy—photography, telephones, skyscrapers, phonographs, and radio, as examples—also came at about the same time.

Scientists and engineers of the time seemed to feel that they understood the basic underlying principles of these devices. Because of that, some people had come to believe that science was "over"; that everything that there was to discover had already been discovered. All that remained to be done was to improve the precision of measurements, and, by incremental, evolutionary design changes, to improve the speed and efficiency of devices already in existence. But, surprises were lurking.

THE DISCOVERY OF X-RAYS

People will say that Röntgen has probably gone crazy.

—**Röntgen**[1]

In 1895, the German physicist Wilhelm Röntgen* was studying the behavior of streams of electrons emitted from negatively charged electrodes, the so-called

---

* Wilhelm Röntgen (1845–1923, Nobel Prize, 1901) never sought to profit financially from his scientific work. Although his discovery of X-rays was considered so important as to be worth the first-ever Nobel Prize in physics, Röntgen donated all the prize money to the University of Würzburg, where he was serving as a professor at the time of his discovery. Also, he never patented any of his work. Sadly, he spent his final years in considerable financial distress. Unfortunately for science as a whole, Röntgen stipulated in his will that all of his personal scientific papers should be destroyed after his death.

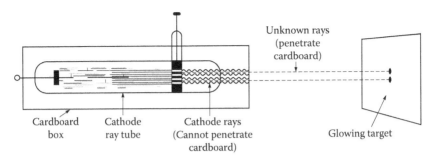

FIGURE 23.1   A schematic diagram of a cathode-ray tube experiment that was instrumental in Röntgen's discovery of X-rays.

cathode rays. Specifically, he wanted to investigate the luminescence* caused when cathode rays impinged on various chemicals. The apparatus Röntgen needed for his work was fairly simple. It consisted of a source of high-voltage electricity and a cathode-ray tube to discharge it through (Figure 23.1). The cathode-ray tube was a bulb of glass, evacuated to a fairly low pressure and equipped with metal electrodes to allow passage of the electric discharge.

To observe any faint luminescence, Röntgen darkened his laboratory room. In November 1895, he observed a flash of light coming some distance from the tube. A sheet of paper coated with a chemical compound known to be fluorescent was the source of the light. Because the cathode rays themselves were shielded such that they could not escape from the apparatus, the energy source causing the fluorescence had to be something other than cathode rays. Röntgen recognized that some kind of radiation was emerging from the cathode-ray tube, presumably produced by the impact of the cathode rays on the solid material with which they collided. That is, the kinetic energy lost by the cathode rays as they were stopped was converted into a different form of energy that could be observed as radiation. Besides finding out where the new radiation was produced, Röntgen also found that it traveled in straight lines, it would penetrate some materials but be stopped by others, and that it would "expose" a photographic plate, just as ordinary light would.† However, the most remarkable feature of this newly discovered radiation was that it could pass through considerable thicknesses of paper and even through thin layers of metals or other seemingly solid

---

* The phenomenon of luminescence is the production of light by means other than heat. Several kinds of luminescence are recognized. A fluorescent object, such as the screens on most early television sets, produces light while it is exposed to a source of radiation. A phosphorescent object continues to emit light even after the source of radiation is removed. A firefly is also luminescent because it produces light, but not heat, in this case, from chemical reactions in its body (chemiluminescence).

† What we recognize as ordinary, or visible, light is only a small portion of the entire electromagnetic spectrum. The electromagnetic spectrum comprises many kinds of radiation, all of which were shown to be fundamentally similar in the brilliant work of the English physicist James Clerk Maxwell. All electromagnetic radiation is propagated by waves; the distinction among the various forms rests in the wavelength—the distance from the crest of one wave to the crest of the next. The electromagnetic spectrum includes radio waves (with very long wavelengths, typically 1–2000 m); microwaves, used in the nearly ubiquitous kitchen appliance (0.1 m to about 0.5 cm); infrared radiation (0.5 to about 7.5×10^{-5} cm); visible light (4×10^{-5} to 7.5×10^{-5} cm); ultraviolet radiation, responsible for suntans (4×10^{-5} to 5×10^{-7} cm); and, at the shortest wavelengths, X-rays (5×10^{-7} to 1×10^{-7} cm) and γ-rays (less than 1×10^{-9} cm).

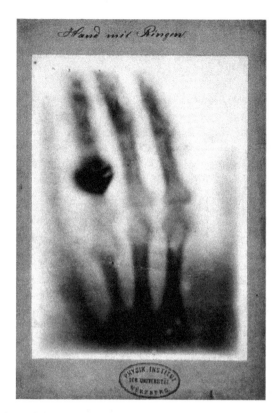

FIGURE 23.2 One of the first medical X-rays, that of Mrs. Anna Bertha Röntgen's left hand, complete with her rings. (http://commons.wikimedia.org/wiki/File:First_medical_ X-ray_by_Wilh.)

materials. At first, this new form of radiation was sometimes called "Röntgen rays" in honor of the discoverer, but Röntgen himself used the universal mathematical symbol of the unknown, calling the radiation "X-rays." (We still use the name even though the nature and origin of X-rays is now well understood.)

> *I have found rays that permit one to see the invisible, to see things inaccessible to your eyes.*

> —**Röntgen**[2]

Because metal and bone block the passage of X-rays, whereas thin wood and human flesh do not, Röntgen was able to produce pictures of a set of weights sealed in a small box and the bones of his wife's hand (Figure 23.2).

They were the first X-ray photographs. Röntgen's first scientific paper, presented in December 1895, began:

> *If the discharge of a fairly large Rühmkorff induction coil is allowed to pass through a vacuum tube, and if one covers the tube with a fairly close-fitting mantle of thin black cardboard, one observes in the completely darkened room that a paper screen painted*

with barium platinocyanide placed near the apparatus glows brightly, regardless of whether the coated surface or the other side is turned toward the discharge tube. The fluorescence is still visible at a distance of two meters from the apparatus.

—Röntgen[1]

On New Year's Day, 1896, Röntgen mailed copies of this first publication on X-rays to the leading physicists of Europe, along with the X-ray pictures he had taken. X-rays immediately caught the fancy of both the scientific community and the general public. Much of the fascination lay in the ability of X-rays to penetrate some kinds of matter, but not others. Particularly amazing was the effect of X-rays on humans. As most of us have seen for ourselves, the so-called hard tissue, such as bones and teeth, show up easily on an X-ray photograph, because they stop the X-rays, while soft tissue, such as flesh and blood, does not.

Prof. Röntgen, of Wurzburg, at the end of last year published an account of a discovery which has excited an interest unparalleled in the history of physical science. In his paper read before the Wurzburg Physical Society, he announced the existence of an agent which is able to affect a photographic plate placed behind substances, such as wood or aluminum, which are opaque to ordinary light. This agent, though able to pass with considerable freedom through light substances, such as wood or flesh, is stopped to a much greater extent by heavy ones, such as the heavy metals and the bones; hence, if the hand, or a wooden box containing metal objects, is placed between the source of the Röntgen rays and a photographic plate, photographs... are obtained. This discovery appeals strongly to one of the most powerful passions of human nature, curiosity, and it is not surprising that it attracted an amount of attention quite disproportionate to that usually given to questions of physical science. Though appearing at a time of great political excitement, the account of it occupied the most prominent parts of the newspapers.... The interest this discovery aroused in men of science was equal to that shown by the general public. Reports of experiments on the Röntgen rays have poured in from almost every country in the world, and quite a voluminous literature on the subject has already sprung up.

—Thomson[3]

By early 1896, experimenters in many countries began to publish pictures of bones. The usefulness of this work to medicine was obvious, and X-rays were immediately applied in medicine and in dentistry. Röntgen became one of the most celebrated scientists of the time. Unfortunately, the danger lurking in X-rays—that extensive exposure of a person to X-rays can cause cancer—was not appreciated for a number of years.

HENRI BECQUEREL'S EXPERIMENT

In January 1896, members of the French *Académie des Sciences* were shown the first French X-rays of the bones of a hand. One of the attendees, Henri Becquerel* (Figure 23.3), had devoted himself to the study of luminescence. What most

---

* Henri Becquerel (1852–1908, Nobel Prize, 1903) succeeded his father and grandfather as the professor of physics at the Muséum National d'Histoire Naturelle, in Paris. In an odd twist of fate, Niepce de Saint-Victor, a colleague of Becquerel's father, had noticed as early as 1861 that uranium emits some kind of radiation, but apparently never followed up this observation.

FIGURE 23.3 Henri Becquerel (1852–1908) in his laboratory. His experiments led to the discovery of radioactivity. (http://en.wikipedia.org/wiki/File:Becquerel_in_the_lab.jpg.)

interested him about X-rays was a report that they came from the fluorescent spot on the wall of the cathode-ray tube. Röntgen himself had specifically stated that where the cathode-ray tube fluoresced the strongest was the center from which X-rays radiated. Becquerel wondered whether there could be a connection between X-rays and fluorescence. That is, if the fluorescence of the cathode rays contained X-rays, then it seemed reasonable to suspect that X-rays could be produced in other forms of the fluorescence. In other words, does a fluorescent material also produce X-rays?

> *I thought immediately of investigating whether the X-rays could not be a manifestation of the vibratory movements which gave rise to the fluorescence and whether all phosphorescent bodies could not emit similar rays. The very next day I began a series of experiments along this line of thought.*

> —Becquerel[4]

Seeing fluorescence in the visible region is easy, because our eyes respond to light in that region of the electromagnetic spectrum. We can't see X-rays. Becquerel needed some way, other than vision, to test for the production of X-rays during fluorescence. X-rays do "expose" some kinds of photographic films—enabling physicians or dentists to make permanent records of our X-ray examinations. Becquerel's experiments involved wrapping a photographic plate in dark paper to prevent visible light from exposing the plate. He put the wrapped plate onto a window sill, where it would be in the sunlight, but, because of the wrapping, the sunlight itself could not affect the plate. Then he put specimens of the fluorescent minerals he was testing on top of the plate. Because of the wrapping, any visible light that was emitted by the mineral as it fluoresced would not affect the plate any more than would the sunlight. X-rays, though, could easily penetrate the paper wrapping. Any blackening of the photographic plate would mean that X-rays had been given off by the mineral specimen.

Samples of a fluorescent compound that Becquerel's father had worked with, potassium uranyl sulfate, were still on hand in the laboratory. Becquerel quickly discovered that after exposure to the sun, the fluorescent radiation from the compound would penetrate black paper (which is opaque to ordinary light) and darken a photographic plate on the other side. This test gave Becquerel a positive result— the photographic plate was blackened, suggesting (but not proving!) that the uranium compound was giving off X-rays when it fluoresced. Becquerel had designed and performed this experiment to test his hypothesis that fluorescent objects also emit X-rays, and it had given the results as predicted. So far, so good—except that, had Becquerel stopped his work at that point, the first, and apparently positive, results would have been quite misleading.

To prove that the rays causing the plate to blacken really were X-rays, Becquerel had to demonstrate that the rays behaved in all respects like X-rays, not just in their ability to penetrate paper. (More generally, the fact that some material or phenomenon *A* behaves in one respect, or has one characteristic, similar to *B* does not itself prove that *A is B*; rather, that proof requires that they be alike in *all* possible respects or characteristics.) Röntgen had already shown that a characteristic property of X-rays was their tendency to travel in straight lines, and that the penetrating power of X-rays is different for different materials. Continuing with his work, Becquerel designed an experiment in which he would put a small copper cross between the uranium sample and the photographic plate. Figure 23.4 presents a schematic diagram of Becquerel's apparatus.

If the rays from the uranium sample traveled in a straight line, and penetrated materials to different extents (specifically, passing through paper easily, but not through copper), just as X-rays were already known to do, then an outline of the copper cross should appear on the photographic plate. All that remained actually to perform the experiment was to put the cross between the mineral and the carefully wrapped plate, set this assembly in the sunlight so that the uranium mineral would fluoresce, and then develop the plate.

Henri Becquerel (1852–1908), Professor of Physics at the Ecole Polytechnique *in Paris, read about X-rays soon after their discovery. He thought that similar penetrating rays might be emitted by phosphorescent substances when exposed to sunlight. So he took some phosphorescent crystals of a uranium compound, in the form of a thin crust, and placed them on a photographic plate which he had previously wrapped in thick black paper to keep the light out. Then he exposed the whole thing to the sun for a few hours. When he developed the photographic plate he found a silhouette of*

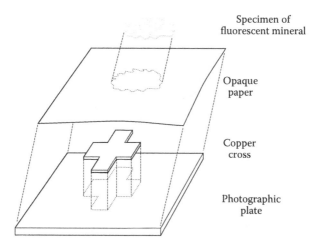

Specimen of
fluorescent mineral

Opaque
paper

Copper
cross

Photographic
plate

FIGURE 23.4 A conceptual diagram of Becquerel's experimental setup. It consisted of a fluorescent mineral specimen; a piece of copper cut in the shape of a cross, opaque both to light and to X-rays; thick paper, opaque only to light but not X-rays; and a photographic plate.

the crystals in black on the negative—which seemed to confirm his idea that sunlight made the crystals emit radiation.

His discovery that they emitted radiation even in the dark was a matter of chance. The sun did not shine in Paris for several days, but, as he had set up his apparatus, he decided to develop the plate nevertheless...

—Carey[5]

Then an unexpected complication came up—bad weather. On the day that Becquerel wanted to try his experiment, the sky was completely overcast. Rather than take the whole uranium mineral/copper cross/wrapped photographic plate assembly apart, Becquerel decided to store it, intact, inside a drawer in one of the laboratory cabinets. The inside of the drawer provided a dark place that would ensure that the mineral couldn't accidentally be exposed to light and fluoresce until he was ready to try the experiment again. Five days went by without his being able to try exposing the assembly to sunlight. Finally, he took the apparatus out of the drawer and developed the photographic plate, even though the experiment had not been completed as planned.

...experiments had been made ready on Wednesday the 26th and Thursday the 27th of February and as on those days the sun only showed itself intermittently I kept my arrangements all prepared and put back the holders in the dark in the drawer of the case, and left in place the crusts of uranium salt. Since the sun did not show itself again for several days I developed the photographic plates on the 1st of March, expecting to find the images very feeble. The silhouettes appeared on the contrary with great intensity. I at once thought that the action might be able to go on in the dark...

It is important to notice that this phenomenon seems not to be attributable to luminous radiation emitted by phosphorescence. The radiations of uranium salts are emitted not only when the substances are exposed to light but when they are kept in the dark, and for more than two months the same pieces of different salts, kept protected from all known exciting radiations, continued to emit, almost without perceptible enfeeblement, the new radiations.

All the salts of uranium that I have studied, whether they become phosphorescent or not in the light, whether crystallized, cast, or in solution, have given me similar results. I have thus been led to think that the effect is a consequence of the presence of the element uranium in those salts...

—Magie (quoting Becquerel)[6]

This is the curious aspect of Becquerel's work. There really was no reason for him to have developed the plate, because he certainly knew that it had not been exposed to sunlight. That is, he had not actually done the experiment the way he had planned it. But, when he did develop the plate anyway, he discovered that it had the image of the copper cross. Since the assembly—uranium mineral, copper cross, and photographic plate—had not ever been placed in sunlight, Becquerel's original idea, that exposure to the sunlight caused the mineral to fluoresce, and the light given off as a result of the fluorescence exposed the photographic plate, could not be correct. The only explanation for the source of the image of the cross on the photographic plate was that some kind of radiation had to be coming out of the uranium mineral itself. Since the mineral had not been exposed to sunlight, the radiation was *not* the result of fluorescence. Becquerel hadn't heated the mineral, exposed it to other chemicals, or indeed done anything else other than simply store it in a drawer in his laboratory for those five days. So, the radiation that produced the image of the copper cross on the photographic plate could not have been caused by anything that Becquerel did to the mineral specimen. Something in the mineral was giving off radiation.

In fact, Becquerel found that the uranium compound constantly and ceaselessly emitted a strong, penetrating radiation. This property of constantly emitting penetrating radiation was termed *radioactivity* by the Polish–French physicist Marie Curie* in 1898. She went on to show that all uranium compounds that she tested were radioactive, and that the intensity of radioactivity was directly proportional to the uranium contents of the various compounds tested. Uranium seemed to be the source of the radioactivity.

What was puzzling about all this was the ENERGY involved. It took ENERGY to expose the photographic plates. This ENERGY was apparently some that the crystals had somehow stored away. We can think of this in terms of the energy balance:

$$E_{in} - E_{out} = E_{stored}$$

In the SYSTEM Becquerel was studying, E_{out} is the ENERGY emitted by the uranium-containing compounds that darkened the photographic plates. Because the uranium

---

* Very few scientists ever manage to become international celebrities; Marie Curie is one of those very few. Marie Curie (1867–1934, Nobel Prizes 1903, 1911) can claim such distinctions as being the first woman to win a Nobel Prize, and the only scientist of either gender to win the Nobel in two different scientific fields. Her long struggles, working alongside her husband Pierre, to isolate polonium and radium have become the stuff of scientific legend. She was the subject of a 1943 biographical film, in which she was portrayed by the British actress Greer Garson. Tragically, Pierre was killed in 1906 by being run over by a horse-drawn wagon. Marie Curie's death in 1934 was due to a form of anemia believed to have been caused by her exposure to radiation over many decades of scientific work.

compounds had not been exposed to light, nor heated, nor treated with other chemicals, E_{in} is zero. Therefore, there can only be ENERGY coming out of the crystals (E_{out}) if there is a substantial amount of ENERGY stored in them somehow (E_{stored}). Becquerel observed another strange aspect of this radiation. It is common experience that a hot object, for example, a hot cup of coffee, gradually cools until it is the same temperature as its surroundings. The amount of heat given off steadily becomes less and less until no more is available to exchange with the immediate surroundings. The same is true of fluorescence—the light being emitted weakens and "goes out." But this wasn't the case for the radiation coming from the mineral. It went on steadily, at apparently the same intensity, for months.

The constant factor in these experiments was uranium. So long as a test material contained uranium it did not matter whether it was fluorescent or not, whether it was exposed to light or not, whether it had been heated or not, or even whether it was solid or in solution. At that time, elemental uranium had not yet been isolated, but, coincidentally, Henri Moissan, at the School of Pharmacy in Paris, was working on a process for producing pure uranium.* When Becquerel had the opportunity to test a sample of pure uranium metal, it proved to be more intensely radioactive than any other substance yet tested. This substantiated Marie Curie's observation that the intensity of radioactivity was proportional to the amount of uranium in a sample; the most intensely radioactive specimen of all was, at least nominally, 100% uranium.

RADIATION

Early in the study of radioactivity, it appeared that more than one kind of radiation was being given off from such radioactive substances as uranium and thorium. For example, in a magnetic field, part of the radiation was deflected in one direction by a very slight amount, part was deflected in the opposite direction by a considerable amount, and some wasn't affected at all. Since the nature of the radiation was, at that time, not understood, the various kinds of radiation were simply named using the first three letters of the Greek alphabet: α-rays, β-rays, and γ-rays, respectively. The γ-rays displayed an ability to penetrate through materials, much like X-rays. In contrast, β-rays were much less penetrating, and α-rays were scarcely penetrating at all. From the direction and extent of the β-ray deflection, Becquerel recognized that it must contain negatively charged particles like those in cathode rays. β-rays proved to be streams of rapidly moving electrons. Figure 23.5 contrasts these kinds of radiation.

---

* Ferdinand Frédéric Henri Moissan had quite a career, in which his work on isolating uranium was almost a sidelight. His most enduring accomplishment was isolating fluorine in its elemental form. Fluorine is perhaps the most voraciously reactive of the elements. Moissan's work on fluorine brought him the Nobel Prize in chemistry (1906), but also ruined his health. Moissan invented the electric arc furnace, which enabled scientists to achieve extremely high temperatures (well over 2000°C). The arc furnace led to Moissan's most enduring embarrassment. He seemed to have produced diamonds in his furnace and claimed to be the discoverer of synthetic diamond. Sadly, that work is now widely believed to be false. Moissan was apparently the unwitting victim of overzealous assistants, who "salted" the experiments with real diamonds in a genuine, but grossly misguided, effort to be helpful.

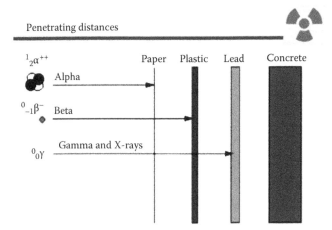

FIGURE 23.5 A chart of the penetrating distances of the major types of radiation. (Illustration courtesy of Arnold Edelman, Argonne National Laboratory, Lemont, Illinois, http://www.gtcceis.anl.gov/includes/dsp_photozoom.cfm?imgname=PenetratingRad. gif&caption=Ionizing%20radiation%20penetration%20distances&callingpage=%2Fg uide%2Frad%2Findex.cfm&callingttl=Radiation%20Basics&source=Penetrating%20 distances%20for%20alpha%2C%20beta%2C%20gamma%20and%20x-ray%20radiation.)

THE STRUCTURES OF ATOMS

All atoms consist of a nucleus surrounded by one or more electrons that orbit around the nucleus. Electrons have a negative charge and are very light weight. The nucleus consists of two kinds of particles: *protons*, which have a positive charge, and electrically neutral *neutrons*. (Actually, many other kinds of subatomic or nuclear species have been shown or postulated to exist, but for our purposes we need to only concern ourselves with protons and neutrons.) Since the mass of a proton or neutron is about 1800 times greater than the mass of an electron, the nucleus contains almost the entire mass of the atom. Even the lightest nucleus—that of the hydrogen atom—is 1836 times the mass of an electron.

Ordinary chemical reactions depend on the donation, transfer, or receipt of electrons. The chemical behavior and identity of an element is determined by the number of its electrons; since in a neutral atom the number of negatively charged electrons and positively charged protons must be equal, the chemical identity of an element is also determined by the number of protons in its nucleus. But, since the number of electrons might be changed by the atom's gaining or losing some in a chemical reaction, the number of protons in the nucleus is a more reliable indicator of the chemical identity of an element. The number of protons in the nucleus is called the *atomic number* of the element.

The atomic number (which also measures the nuclear charge) is a fundamental property of atoms. Electrons can be removed from atoms, or added to them by heat, by light, or by any chemical processes. On the other hand, the nuclear charge (atomic number) most certainly could not be altered by any method known to the chemists and physicists of 1900. Furthermore, no alteration of the number of electrons associated

---

TABLE 23.1

Nuclei of the Three Isotopes of Hydrogen Have the Same Atomic Number but Different Mass Numbers

Number of Protons	Number of Neutrons	Atomic Number	Mass Number
1	0	1	1
1	1	1	2
1	2	1	3

---

with an atom, either increasing or decreasing their number, would alter the atomic number of the nucleus. Therefore, the atomic number (or the size of the nuclear charge) best characterizes the different varieties of atoms and best discriminates among the chemical elements.

For nuclear reactions—that is, ones involving the nucleus—we disregard the electrons, and instead must focus on the behavior of the nucleus. A second characteristic property, the *mass number*,* represents the sum of the numbers of protons plus neutrons in the nucleus. (Note that the number of neutrons in the nucleus is the difference between the mass number and the atomic number.) In cases in which it is convenient to lump the protons and neutrons together, as in determining the mass number, we refer to them by the generic term *nucleons*. Thus the mass number is equal to the number of *nucleons*.

In an electrically neutral atom, the number of protons must equal the number of electrons. Since neutrons are electrically neutral themselves, it is not necessary to restrict the number of neutrons to achieve electrical neutrality. Thus it is entirely possible that nuclei of the same chemical element could have different numbers of neutrons. Consider the nuclei listed in Table 23.1. The atomic number (the number of protons) is the same in all three cases. Therefore, all of these nuclei are atomic nuclei of the same chemical element, which, in this example, is hydrogen, H. Atoms or nuclei having the same atomic number by different mass numbers are called *isotopes*.

When we write nuclear reactions, it will be helpful to be able to keep track of the fate of the protons and neutrons. We will use a notation scheme that shows the chemical symbol of the element, the atomic number as a subscript, and the mass number as a superscript. The three isotopes of hydrogen shown earlier are therefore written as $_1H^1$, $_1H^2$, and $_1H^3$, respectively.

---

* The mass number is a convenient whole number that allows us to characterize a particular nucleus. It is close to, but not identical with, atomic weight. The accepted convention for atomic weights now in use was established fifty years ago. Atomic weights are measured in atomic mass units (amu), defined to be 1/12 the mass of a single atom of a specific isotope of carbon, C^{12}. The value of the atomic mass unit is known to an accuracy of better than one part per million: 1 amu $= 1.660540 \times 110^{27}$ kg. The atomic weight of $_6C^{12}$, in this convention, is exactly 12.00000. However, the atomic weight of $_1H^1$ is not 1.00000 but 1.007825. In terms of amu, the atomic weight of any pure isotope is approximately equal to its mass number.

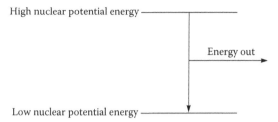

FIGURE 23.6 Radioactivity is a spontaneous change from high to low nuclear potential energy, which means that ENERGY is released from the SYSTEM.

RADIOACTIVITY IS A NUCLEAR PROCESS

Continued scientific investigation made it clear that radioactivity was not like any known chemical process. Chemists had learned a number of ways to influence the rate or course of a chemical process: as examples, by changing the temperature or pressure, by adding a catalyst, or by exposing the system to light. None of these has the slightest effect on the rate of radioactivity. Therefore, if radioactivity is not a chemical process, it must not be occurring in the swarm of electrons, because chemical processes involve gaining, losing, or sharing of electrons. If radioactivity is not occurring among the electrons, the only other place it can be happening is in the nucleus.

Marie Curie defined radioactivity as the *spontaneous* emission of radiation. She also showed that radioactive materials produce a prodigious amount of ENERGY— millions of times more than any chemical process. (On an equal mass basis, nuclear reactions release about ten million times as much energy as chemical reactions.) This ENERGY keeps pouring out in apparently endless quantity. We have seen that a spontaneous change is one that involves a transition from a state of high potential energy to one of low potential energy. Thus we can immediately draw an energy diagram (Figure 23.6). To begin to understand how to tap into this enormous and seemingly limitless supply of ENERGY, we must begin by inquiring into the nature of the nuclear potential energy and how to measure it.

NUCLEAR BINDING ENERGY

Since the nuclei of all atoms (other than hydrogen) have more than one proton and since electrical charges of the same sign repel each other, we might ask why then do nuclei stay together. In other words, why do any atoms larger than hydrogen even exist? Certainly if we consider an atom the size of uranium, with 92 protons, there would be a tremendous release of energy (from repulsion of particles of the same electric charge) if the nucleus could just fly apart. Some force stronger than the tendency of like charges to repel each other must be holding the nucleus together—as if we had somehow wrapped a tiny, nucleus-sized rubber band around the nucleus to hold all the pieces in place.

Inside the nucleus, nucleons experience two competing forces. In addition to the repulsion between protons, a second force, one of attraction instead of repulsion, holds the nucleus together. This second force is called the nuclear binding energy and is strong enough to overwhelm the weaker force of electrical repulsion.* However, the nuclear binding energy acts only over very small distances, indeed, only when the nucleons are in contact.

The nuclear binding energy is, in effect, the "rubber band" that prevents the nucleons from flying apart. For most nuclei, an external force that adds ENERGY to the nucleus to help overcome the nuclear binding energy is necessary to break the nucleus apart; otherwise, it will stay together forever. While the nucleons normally stay in contact with one another for an extremely long time, a very large nucleus always has a small chance that nucleons will find themselves separated by a distance beyond the reach of the binding energy, but still within the realm of the repulsive force between two objects of the same charge. If that happens, the nucleons will suddenly be free of one another; the electric repulsion of similar charges "wins" and pushes the nucleons apart, overcoming the binding energy. This process we recognize as radioactivity.

The more protons in a nucleus (i.e., the higher the atomic number), the more the repulsion between one another and the more likely the nucleus is to experience radioactive decay. Having additional neutrons in the nucleus reduces the proton-to-proton repulsion by increasing the size of the nucleus without adding to its positive charge. Adding too many neutrons can also destabilize the nucleus, by making it so big that the short-range binding energy becomes less effective. So, a stable nucleus results from a finely tuned balancing act among the neutrons and protons.

Many elements with atomic numbers up to 20 have equal numbers of protons and neutrons. The situation is different for elements with larger nuclei, because protons repel each other electrically and neutrons do not. A nucleus with 30 protons and 30 neutrons, for example, could be less stable than a nucleus of 28 protons and 32 neutrons, the number of protons and neutrons no longer being equal. This inequality of numbers of neutrons and protons that begins to develop above atomic numbers of about 20 becomes more pronounced for even heavier elements, as indicated in Figure 23.7. All nuclei having more than 82 protons are unstable. For example, in $_{92}U^{238}$, which has 92 protons, there are 146 neutrons. If the uranium nucleus were to have equal numbers of protons and neutrons, it would fly apart at once because of the electrical repulsion forces. The extra 54 neutrons (extra, i.e., in the sense of making up the difference between the 92 to equal the number of protons and the 146 that actually exist in this nucleus) are needed for relative stability. Even so, the $_{92}U^{238}$ nucleus is still unstable. That is, because there is an electrical repulsion between

---

* Atomic physicists have proposed the concept of the "strong nuclear force" to explain how nuclei are held together. To break nuclei apart, the strong nuclear force must be overcome, that is, ENERGY or WORK must be supplied to the nucleus to overcome this force. The amount of WORK needed to separate a nucleus into its individual neutrons and protons is the binding energy of that nucleus. The greater the binding energy of a nucleus, the more stable it is, because of the more WORK that would have to be done to break the nucleus apart. The origin of the strong nuclear force lies in the fact that the protons and neutrons themselves are made up of smaller and more fundamental particles called quarks. The force that links quarks together to make protons and neutrons also links the protons and neutrons together to make nuclei, and is responsible for the strong nuclear force.

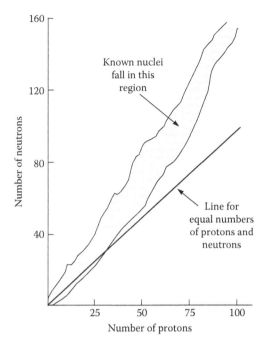

FIGURE 23.7   In small nuclei, the numbers of protons and neutrons are about equal, but in larger nuclei, neutrons dominate.

every pair of protons in the nucleus, but there is not a substantial nuclear binding force between every pair. Every proton in the uranium nucleus exerts a repulsion force on each of the other 91 protons. However, each proton (and neutron) exerts an appreciable nuclear attraction only on those nucleons that happen to be near it.

The relationship between the number of nucleons and the average binding energy per nucleon is called the *curve of binding energy* (Figure 23.8). The highest binding energy occurs at a mass number of 56, an isotope of iron. This is the most stable nucleus. At low mass numbers (<56), the curve rises very steeply. At high mass numbers (>56), the curve drops, but slowly.

There is one point that we need to interject in regard to the curve of binding energy. *Relative to all of the energy diagrams we've seen previously, the curve of binding energy is upside down.* In the curve of binding energy, stability increases as we go up the curve. Therefore,

High nuclear potential energy represents low binding energy; and low nuclear potential energy is a high binding energy.

The binding energy per nucleon indicates the stability of a nucleus. The higher the binding energy, the more stable the nucleus. Nuclear processes release far

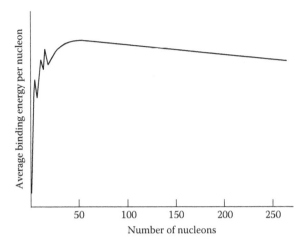

FIGURE 23.8 The curve of nuclear binding energy, which plots the average binding energy per nucleon as a function of the number of nucleons. The higher a given nucleus falls on this curve, the more stable it is.

more ENERGY per reaction than do chemical processes because the "glue" holding nuclei together is much more powerful than is the electric force that binds electrons to their atoms. The binding energy of a nucleus is the ENERGY needed to take the nucleus apart into its constituent nucleons.

In small nuclei with low atomic numbers, the nuclear binding energy overwhelms the repulsive force and the nucleons are tightly bound together. The rising slope of the binding energy curve for small nuclei tells us that the average binding energy of the nucleons would increase even if these nuclei had more protons and neutrons than they already do. At the exact crest of the curve of nuclear binding energy, for nuclei with 26 protons, the nuclear binding energy and the electrical repulsive force are well balanced. Such nuclei are extremely stable. For these nuclei, there is no increase in the average binding energy of their nucleons by adding or subtracting nucleons.

In contrast to the situation for small nuclei, the situation is reversed for nuclei with many protons: electrical repulsion now overwhelms binding energy. For these large nuclei, the nuclear binding energy cannot hold the nucleus together. Rather, these large nuclei break apart in the process of radioactivity (also called radioactive decay). The decreasing slope of the binding energy curve in the region of large nuclei signals that average binding energy of these large nuclei would increase—they would become more stable if they had *fewer* nucleons, not more. The greater stability afforded by a smaller number of nucleons is achieved by breaking apart. Nuclei that contain more than about 210 nucleons are so large that the short-range forces holding them together are barely able to counterbalance the long-range electrostatic repulsive forces of their protons. Such a nucleus can reduce its bulk and, in doing so, achieve greater stability by emitting an α-particle.

TRANSMUTATION OF ELEMENTS

The isotope $_{92}U^{238}$ emits α radiation. This type of radiation consists of the nuclei of helium atoms; it contains two neutrons and two protons. In nuclear equations, it can be written as $_2\alpha^4$ or as $_2He^4$. We begin to write the process of uranium "decay" as

$$_{92}U^{238} \rightarrow {}_2He^4 + \cdots$$

This partially completed equation shows the loss of two protons (because $_2He^4$ has an atomic number of two) and two neutrons (the difference between the mass number and atomic number of $_2He^4$) from the nucleus of $_{92}U^{238}$. Therefore, there must be another product that has 90 (92 − 2) protons and 144 (146 − 2) neutrons. Since this product has a different atomic number (90), it must be a different chemical element and is in fact thorium, Th. Thus the complete nuclear reaction can be written as

$$_{92}U^{238} \rightarrow {}_2He^4 + {}_{90}Th^{234}$$

The essential feature of writing nuclear reactions such as this is that the sums of the atomic numbers and the sums of the mass numbers must be equal on each side of the equation. In this example, a quick check can be made: $92 = 90 + 2$ and $238 = 4 + 234$.

In this radioactive process, one chemical element, uranium, has turned into a different one, thorium. The change of one element into another is called *transmutation*. Transmutation was a goal of early scientists, the alchemists, for centuries. Their objective was to turn inexpensive metals such as lead or iron into precious metals such as gold. With the hindsight of knowledge of atomic structure, not developed until centuries after the heyday of alchemy, we know that transmutation of elements through chemical reactions is impossible, because such reactions involve only electrons, and not nuclei. To transmute one element to another requires, we now know, a change in atomic number—that is, a change in the number of protons in the nucleus. No chemical reaction can do this, because chemical reactions occur only among the electrons.

The alchemists of centuries ago had no way of knowing it, but processes of transmutation were constantly going on all the time, all around them. They are going on all the time, all around us, too. Radioactive decay of minerals in rocks has been occurring since the Earth was formed. But alchemists had no knowledge of radioactivity nor of how to measure or detect it.

As a consequence of changing uranium into thorium, we have moved slightly along the curve of binding energy, as shown in Figure 23.9. Because a nucleus of mass number 234 has a slightly higher binding energy than one of mass number 238, it is slightly more stable. When a SYSTEM spontaneously changes from being less stable to being more stable, some ENERGY is given off, just like in the change of gravitational potential energy in a waterfall. Therefore, the standard energy diagram (Figure 23.10) applies to this nuclear process.

Where does the ENERGY come from? Making a very scrupulous account of all the mass on both sides of the equation would show that some very tiny amount of mass appears to be "missing" after the reaction. This very tiny amount of "missing mass" has been converted to energy. The relationship between missing mass and energy is given by Albert Einstein's (Figure 23.11) equation, $E = mc^2$ probably the

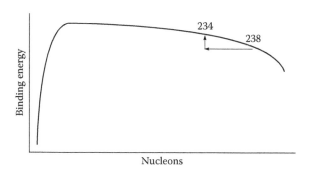

FIGURE 23.9 The transformation of uranium to thorium is illustrated on the curve of binding energy. For illustration purposes, the curve has been "smoothed" and the horizontal scale is exaggerated.

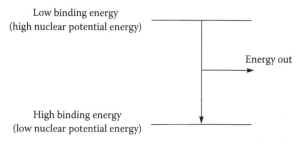

FIGURE 23.10 Because a high binding energy represents a stable nucleus, a spontaneous change in a nuclear SYSTEM proceeds from low to high nuclear binding energy.

most famous equation in all of science. Here E is the ENERGY liberated in the nuclear process, m is the amount of mass that is lost (the missing mass), and c is the speed of light. Although m may be a very tiny number, c is very large (and consequently c^2 is extremely large); thus the amount of ENERGY liberated, E, can be enormous.

> *...where I and my assistant engineer have racked our brains for days, this young fellow comes along and solves the whole business in a mere quarter-hour. He'll go far one day.*
>
> **—Einstein**[7]

When it became appreciated that natural radioactivity involved transmutation, nuclear physicists tried to see if transmutation could be induced in the laboratory, by deliberately causing nuclear reactions to occur.

> *From 1900 onwards the Curies had been in correspondence with scientists from all over the world, responding to requests for information. Research workers from other countries joined the search for unknown radioactive elements. In 1903, two English scientists, Ramsay and Soddy, demonstrated that radium continually disengaged a small quantity of gas, helium. This was the first known example of the transformation of atoms. A little later Rutherford and Soddy, taking up a hypothesis considered by Marie Curie as early as 1900, published their Theory of Radioactive Transformation,*

FIGURE 23.11 Albert Einstein (1879–1955), one of the great scientific geniuses in history. His equation relating energy to mass explains the enormous output of energy from nuclear processes. (Reprinted from first edition with permission from Photo Researchers, Inc., New York, New York.)

affirming that radioactive elements, even when they seemed unchangeable, were in a state of spontaneous evolution. Of this Pierre Curie wrote: 'Here we have a veritable theory of the transformation of elements, but not as the alchemists understood it. Inorganic matter must have evolved through the ages, following immutable laws'.

—**Carey**[8]

The first artificial transmutation was performed by Ernest Rutherford (Figure 23.12)* in Cambridge, England. He bombarded nitrogen with α-particles and observed the reaction

$$_7\mathrm{N}^{14} + {}_2\mathrm{He}^4 \rightarrow {}_8\mathrm{O}^{17} + {}_1\mathrm{H}^1$$

---

* Ernest Rutherford (1871–1937) ranks as one of the greatest scientists of the early twentieth century and was a pioneer in the study of nuclear processes. Probably his most famous discovery lay in determining that atoms consisted of a very tiny, heavy nucleus with a swarm of electrons outside the nucleus. An even more enduring legacy to science lies in the fact that four of Rutherford's students and one of his young colleagues went on to win Nobel Prizes themselves. Rutherford came from very humble beginnings on a farm in New Zealand. Supposedly Rutherford was in a field digging potatoes when a family member brought him the news that he had won a fellowship for study at Cambridge University. According to the legend, he set his pitchfork down and said, "That's the last potato I've ever dug."

FIGURE 23.12   Ernest Rutherford and Hans Geiger in the Physics Laboratory, Manchester University, England. (From http://natlib.govt.nz/records/22608563.)

(Note that the sums of mass numbers are equal on both sides of the equation $[14 + 4 = 17 + 1]$, as are the sums of the atomic numbers $[7 + 2 = 8 + 1]$.)

The Curies had shown that radium emits various kinds of 'radiation', and one of these was now known to consist of a stream of electrically charged particles. These 'alpha particles' were identical to helium atoms with their electrons removed; but they originated not from helium gas but sprang spontaneously from the radium atoms as they disintegrated.

Even though atomic disintegration was still little understood, Rutherford saw these high-speed alpha particles as useful projectiles. He intercepted them with a thin sheet of gold foil, to see what happened as they passed through. If atoms were diffuse spheres of electrical charge, as Thomson had imagined, then most of the alpha particles should have gone straight through; a few should have been deflected slightly. But some of the alpha particles bounced straight back again. It was like firing artillery shells at a piece of tissue paper, and getting some of them returning in the direction of the gun.

Rutherford could only explain this by postulating that these alpha particles were hitting small, massive concentrations within the atoms. He thus concluded that most of an atom's mass resided in a minute, positively charged nucleus at the centre, while the electrons went around the outside—very much like the planets orbiting the massive sun. Most of the atom was just empty space. If an atom were expanded to the size of the dome of St. Paul's Cathedral, virtually all its mass would lie within a central nucleus no larger than an orange.

In 1919, he started firing alpha particles at nitrogen atoms. Nothing much should have happened. A great deal did.

... the alpha particles came from radium. This time he was directing them down a tube filled with nitrogen gas. At the far end, he found he was detecting not just alpha particles, but also particles with all the properties of hydrogen nuclei. There was, however, no hydrogen in the tube. With his high-speed alpha-particle projectiles, Rutherford had actually broken them off the nuclei of the nitrogen atoms.

The discovery of radioactivity had earlier shown that certain, rare types of atom could spontaneously disintegrate. Now Rutherford had shown that ordinary atoms were not indestructible. By knocking out a hydrogen nucleus (later called a proton) from the nucleus of nitrogen he had converted it into another element, oxygen. Rutherford had, to a limited extent, achieved the dream of alchemists and changed one element to another.

—**Snow**[9]

THE DISCOVERY OF NUCLEAR FISSION

Rutherford's successful artificial transmutation of nitrogen to oxygen was an important step forward in nuclear physics, but had little practical application. The alchemists wanted to turn inexpensive lead into gold. No one is likely to get rich making oxygen out of nitrogen. Oxygen constitutes about 21% of the atmosphere, 88% of the oceans, and is the most abundant element in the Earth's crust.

In 1934, the brilliant Italian physicist Enrico Fermi* (Figure 23.13) and his colleagues were studying a process called β decay, the emission of β radiation during radioactivity. They were adding neutrons to the nuclei of many different atoms to see what nuclear processes would stimulate β decay. Nuclear physicists had recognized that bombarding a nucleus with neutrons can stimulate the emission of β-rays. These rays, which are high-energy electrons created in the nucleus, have an "atomic number" of −1. Therefore, the atomic number of the product after β emission will be one unit *higher* than the nucleus that emitted the β. (Why? Because the atomic number changes by −[−1], which is mathematically equivalent to +1.)

In the mid-1930s, Fermi and his research team showed that dozens of new radioactive materials are produced as a result of neutron bombardment. In most cases, neutron absorption is followed by β emission, yielding an element one higher in the periodic table. When they added neutrons to uranium, Fermi and his colleagues observed the production of some very short-lived radioactive systems. They thought that they had formed ultra-heavy nuclei of atomic numbers greater than 92. Fermi felt that $_{92}U^{239}$ might be formed from $_{92}U^{238}$. By emitting β particles, this would become an isotope of the element of atomic number 93, and then possibly, by a second β emission, of element 94, neither of which were known at the time.

Fermi indeed discovered the presence of several new radioactive species in neutron-activated uranium. Their chemical properties did not match those of uranium, nor of any of the elements in the range of atomic numbers from 82 to 92, which would be the logical products of known (at the time) nuclear processes involving uranium. Fermi thought at first that he had indeed demonstrated the transmutation to an element of atomic number 93, which he referred to as "uranium x." However, the German chemist Ida Noddack found a flaw in Fermi's reasoning. In 1934, she wrote,

---

* Ernico Fermi (1901–1954, Nobel Prize 1938) is possibly the last-ever physicist to be recognized as a master of both theoretical and experimental physics. As a very young man, Fermi's interest in physics was stimulated by a lengthy physics textbook, written in Latin, found in a used-book stall. In Italy, Fermi's most productive years were at the University of Rome. Fermi and his family emigrated to the United States because newly enacted racial laws in fascist Italy endangered his wife, who was of Jewish ancestry, and caused a number of his research staff to lose their positions. Fermi soon became involved in the Manhattan Project, which developed the first atomic bombs. He was especially instrumental in building the first working nuclear reactor. Fermi and two of his student assistants died of cancer, possibly as a result of exposure to the radiation.

FIGURE 23.13 Enrico Fermi (1901–1954) perhaps the last physicist to master both theoretical and experimental physics. He led the design and construction of the first-ever nuclear reactor. (Photograph courtesy of Diane Minton, Xeno Media, Inc., http://vms-db-srv.fnal.gov/fmi/xsl/VMS_Site_2/000Return/photography/r_online_mrdetail.xsl?-db=VMS_Frames&-lay=WWWBrowse&-recid=94064&-find=.)

It is conceivable that the nucleus breaks up into several large fragments which would, of course, be isotopes of known elements but would not be neighbors of uranium.

—Noddack[10]

Though she did not use the term in any of her formal scientific publications, she was the first scientist to call attention to the possible occurrence of nuclear fission.

The radioactive by-products produced from neutron bombardment of uranium have a variety of chemical properties. Otto Hahn, Lise Meitner, and Fritz Strassman, working in Germany, attempted to identify these materials. All nuclear reactions known up to that time (the mid-1930s), whether natural or artificially produced, had involved the emission of small particles, no more massive than an α emission. It was not unreasonable, then, that they and other nuclear scientists tried to associate the various types of radioactivity in the bombarded uranium with atoms only slightly smaller than uranium. In 1938, Hahn and his coworkers found that when barium compounds were added to the bombarded uranium, one form of the radioactivity they were measuring followed the barium through all of the subsequent chemical manipulations they carried out. Since barium is chemically similar to radium, and since radium (atomic number 88) has a nucleus only slightly smaller than that of uranium, Hahn and his colleagues supposed, not unreasonably, that they were dealing with a radium isotope. However, no chemical process known would separate the added barium carrier from the radium that they supposed was accompanying it. When separation after separation failed, and it became quite clear that the presumed "radium" could not be separated from the barium, the only remaining conclusion was that it was not a radium isotope formed in the process, but rather a radioactive barium isotope.

It is an old maxim of mine that when you have excluded the impossible, whatever remains, however improbable, must be the truth.

—Holmes[11]

Barium isotopes have an atomic number of 56, that is, 36 less than that of uranium. To form a barium nucleus, a uranium atom would have to lose 36 protons, which, by the processes known to nuclear scientists at the time, could be done only by emitting 18 α-particles. Such a vast production of radiation would be a veritable flood of α-particles; nothing like this was ever detected in the neutron bombardment of uranium.

Lise Meitner (Figure 23.14), Hahn's long-term collaborator on the research program that had led to the discovery, was Austrian. She also happened to be Jewish. She had worked with Hahn in Berlin at the Kaiser Wilhelm Institute for Chemistry until the intensifying persecution of Jews by the Nazis, coupled with some political vulnerability as an Austrian citizen after the annexation of Austria, forced her to flee (with Hahn's help) in July 1938. With hindsight, it seems likely that her departure may have saved her from the Holocaust. A few months later, Hahn wrote to Meitner of the remarkable conclusion to their research.

Perhaps you can suggest some fantastic explanation. We understand that it really can't break up into barium. So try to think of some other explanation.

—Hahn[12]

Like Hahn, Meitner at first found the production of barium puzzling and difficult to explain. Over the Christmas holidays, she discussed it with her nephew Otto Frisch, also a German refugee, visiting her from Denmark where he worked with Niels Bohr.* Together they came up with the explanation.

But how can one get a nucleus of barium from one of uranium? We walked up and down in the snow trying to think of some explanation. Could it be that the nucleus got cleaved right across with a chisel? It seemed impossible that a neutron could act like a chisel, and anyhow, the idea of a nucleus as a solid object that could be cleaved was all wrong; a nucleus was much more like a liquid drop. Here we stopped and looked at each other.

—Frisch[13]

---

* By any reckoning, Niels Bohr (1885–1962, Nobel Prize 1922) was one of the greatest of twentieth-century physicists, perhaps second only to Einstein for his impact on the science. Bohr's active career lasted nearly half a century. Among his numerous contributions to science, perhaps the two best known are the liquid drop model of the atomic nucleus and his model of atomic structure. Bohr's model of atomic structure, proposed in 1913, explained almost perfectly the patterns of light emitted from hydrogen atoms and introduced the concept of specific "energy levels" of the electrons orbiting the nucleus. Though nearly a century old, Bohr's model remains an excellent conceptual tool for thinking about atomic structure. The liquid drop model suggests that the nucleus behaves like a drop of liquid—if it is hit very hard by a small object (i.e., a neutron) it can split into two smaller droplets of roughly equal size.

FIGURE 23.14 Otto Hahn (1879–1968) and Lise Meitner (1878–1968) in their laboratory in Dahlem, Germany, around 1913. Their analyses of the products of bombarding uranium with neutrons led to the discovery of nuclear fission. (From http://commons.wikimedia.org/wiki/File:Otto_Hahn_und_Lise_Meitner.jpg.)

The uranium nucleus, activated by neutron bombardment, had split roughly in half. To Frisch, this splitting process was so similar to the division of a biological cell that he suggested the name *fission* for the new phenomenon.

A few years earlier, Bohr and his former student, the American physicist John Wheeler,* had argued that a large nucleus should behave like a droplet of liquid. When struck by a particle, it could become lopsided. If it were struck hard enough, it

---

* John Archibald Wheeler (1911–2008), an American physicist who had studied with Bohr and collaborated in the development of the liquid drop model, has made numerous contributions to science on his own, including such fields as cosmology and nuclear weapons control. Wheeler's memoir of his life in science, *Geons, Black Holes, and Quantum Foam* (Norton, 1998), is an engaging and readable discussion of aspects of his career and developments in physics and cosmology over the last half-century. Even as a highly productive researcher and world-class physicist, Wheeler continued to teach freshman and sophomore physics.

could split up into two smaller droplets. Frisch and Meitner pointed out to Bohr that his theory could explain nuclear fission.

> *I had hardly begun to tell him about Hahn's experiments and the conclusions that Lise Meitner and I had come to when he struck his forehead with his hand and exclaimed, 'Oh, what idiots we have been. We could have foreseen it all! This is just as it must be!' And yet even Bohr, perhaps the greatest physicist of his time, had not foreseen it.*

—**Frisch**[13]

Bohr and Wheeler postulated that fission was more likely to occur in the light isotope of uranium ($_{92}U^{235}$) than in the regular or natural uranium ($_{92}U^{238}$):

$$_{92}U^{235} + _0n^1 \rightarrow _{56}Ba^{144} + _{36}Kr^{89}$$

(Notice that this equation is not balanced. We will see below how to complete the equation to balance it.) The products of this reaction are, however, roughly half the size of the original $_{92}U^{235}$ nucleus. Although the atomic numbers sum on each side of the equation, the right-hand side is "short" by three mass numbers. The missing product must have an atomic number of 0. There is no nuclear particle with atomic number of 0 and mass number of 3 that would complete this equation. However, the neutron is the species with atomic number 0, and since it has a mass number of 1, the equation can be balanced by having three neutrons (giving a total mass number of 3) as products. Thus

$$_{92}U^{235} + _0n^1 \rightarrow _{56}Ba^{144} + _{36}Kr^{89} + _0n^1 + _0n^1 + _0n^1$$

When a uranium nucleus breaks in two in this fashion, it does not always divide in exactly the same way. The nucleus may well break at one point in one case and at a slightly different point in another. For this reason, a great variety of radioisotopes is produced, depending on just how the division takes place. They are lumped together as *fission products*. The possibilities are highest that the division will be slightly unequal, with a more massive half in the mass number region of 135–145 and a less massive half in the region of 90–100.

From Figure 23.7, the ratio of neutrons to protons decreases as the atomic number decreases, so that, when fission into two pieces occurs, there are extra neutrons in the product nuclei. These are rapidly emitted to allow the product nuclei to become stable. Most fission reactions end up producing two or three "extra" neutrons. Usually the fission products are still somewhat unstable, and may undergo β decays or γ decays to achieve the appropriate neutron-to-proton ratios for stability. Because of this, many fission products are themselves radioactive, a problem to be discussed in Chapter 25.

We saw how the transmutation of uranium to thorium resulted in a move to a small distance along the curve of binding energy, and released a small amount of energy. Now, however, we have a different situation, indicated in Figure 23.15.

Uranium fission will yield about four times as much ENERGY, gram for gram, as ordinary radioactivity. Why doesn't uranium spontaneously undergo fission rather than radioactive decay? Before fission can occur, the nucleus must absorb the small quantity of ENERGY needed to distort it sufficiently to overcome the short-range

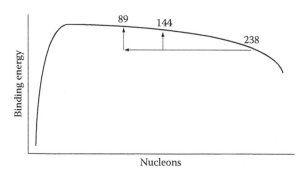

FIGURE 23.15 Nuclear fission is illustrated on the curve of binding energy. As in Figure 23.9, the curve has been "smoothed" and the horizontal scale is exaggerated.

nuclear binding energy. Essentially, the neutron absorbed into the nucleus is analogous to the small input of ENERGY from a match that initiates the much larger release of ENERGY from a fire. Many substances—gasoline, for example—are very flammable yet never ignite spontaneously; some small amount of ENERGY must be provided first from a match or spark plug. Similarly, though a nucleus can release large amounts of ENERGY from fission, it cannot do so spontaneously.

The ENERGY release from fission is prodigious. On an equal weight basis, the fission of $_{92}U^{235}$ releases two million times as much ENERGY as is obtained from burning coal. For example, fission of a kilogram of $_{92}U^{235}$ would release the energy equivalent to about 2700 tons of coal. A 600 MW electricity-generating station could run for a year on the $_{92}U^{235}$ contained in 90 tonnes of naturally occurring uranium (which, as we will see in Chapter 24, contains only a small fraction $_{92}U^{235}$). The ENERGY equivalent of a full 80 liter tank of gasoline is a tiny sphere of $_{92}U^{235}$ about 0.25 millimeter diameter.

What is the source of all this ENERGY? In the chemical process of combustion (Chapter 5), ENERGY is needed to break the chemical bonds in a molecule of fuel, such as methane. At the nuclear scale, ENERGY would be needed to break apart a nucleus completely into its constituent nucleons. ENERGY is released in a chemical combustion process when the constituent oxygen, carbon, and hydrogen atoms reassemble into the product carbon dioxide and water molecules. Similarly, ENERGY, the nuclear binding energy, would be released when a product nucleus is assembled from its constituent nucleons. Since the binding energy of the original uranium nucleus is smaller than the combined binding energies of the product nuclei, there will be ENERGY left over—and released from the SYSTEM—when fission occurs. In chemical or nuclear reactions, it takes ENERGY to undo binding—whether of chemical bonds or nuclear binding energy. Conversely, ENERGY is released whenever a SYSTEM becomes more bound, in the formation of product molecules or product nuclei. The energy balance equation describes this situation.

$$E_{in} - E_{out} = E_{stored}$$

E_{in} is the ENERGY supplied to the SYSTEM to, conceptually, decompose the uranium nucleus into its constituent nucleons. E_{out} would be the ENERGY released when the nuclei of the fission products are formed from those nucleons. Since the fission products are more stable than the original uranium, E_{out} is greater than E_{in}. In that case, E_{stored}, the ENERGY stored in the SYSTEM, is negative, which tells us that some ENERGY must be released from (or withdrawn from) the SYSTEM. That ENERGY is liberated to us.

CHAIN REACTIONS

The practical utilization of the abundant store of energy locked up in every atom of matter is a problem which only the future can answer. Remember, at the dawn of electricity, it was looked on as a mere toy.

—Becquerel[14]

Because each fission event liberates two or three neutrons, while only one neutron is required to initiate fission, a rapidly multiplying sequence of fission events can occur. In principle, it should be possible to take one of the newly produced neutrons to start a second reaction that will generate three more neutrons. One of the product neutrons from that second reaction can start a third reaction, and so on. Any process in which the product of one step is the reactant in, or initiator of, a subsequent step is called a *chain reaction*.

An uncontrolled chain reaction releases an immense amount of energy in a short time. If the two neutrons emitted in each fission can induce further fissions, and if a microsecond (i.e., one millionth of a second) elapses between the emission of a neutron and its subsequent absorption by another nucleus, a chain reaction starting with a single fission will release nearly 21 *gigajoules* of energy in less than a thousandth of a second. This is the basis of the atomic bomb. (For comparison, if we took a coal that would release, say, 23 MJ/kg, that energy release from the atomic bomb is the same as would be achieved by burning one tonne of that coal in one millionth of a second.) But, a properly controlled chain reaction (in which only one neutron per fission causes another fission) can be made to release ENERGY at a constant output. If we could figure out how to control a chain reaction of nuclear fission, then we should be able to obtain a sustained release of ENERGY that we can use for some useful purpose, that is, WORK.

Whenever an energy transformation involves a spontaneous change from high potential to low potential, if we are clever we can insert some kind of device into the SYSTEM to capture some of that ENERGY as useful WORK for us. By 1939, physicists recognized that although the nucleus represented tremendous quantities of potential energy, there was no practical way of converting it into useful work. Some eminent scientists believed that no practical method ever could be found. For example, in 1933, Ernest Rutherford declared that the development of a practical source of large-scale nuclear energy was an idle dream.

These transformations of the atom are of extraordinary interest to scientists but we cannot control atomic energy to an extent that would be of any value commercially, and I believe we are not likely ever to be able to do so. A lot of nonsense has been talked about transmutation.

—Rutherford[15]

Leo Szilard* was the sort of person of the "never say never" philosophy. In a flash of inspiration, said to have happened while waiting for a traffic light to change as he was walking to work, he realized that if an atom requires one neutron for fission but releases two or three, then it is possible to establish a chain reaction by capturing one of the product neutrons to initiate the next step. He patented the idea, in a 1934 patent application that contains the term chain reaction. Szilard wanted to keep the patent secret to prevent the fundamental scientific ideas from being misused. He assigned the patent to the British Admiralty, so that it was not published until after the Second World War. Szilard's basic idea for the chain reaction process leads to some important questions about putting it into practice. First, of course, is the question of whether it is actually possible to create and control the chain reaction process in a practical device. Assuming that were possible, then it becomes important to know how much energy is liberated when the fission chain reaction takes place. It is also crucial to know how to control the process while it is going on, and how to stop it when desired. The translation of Szilard's fundamental scientific idea into engineering practice is the topic of Chapter 24.

The key to release of useful energy from nuclear fission is to establish a sustained, controlled, chain reaction.

REFERENCES

1. Röntgen, W.; quoted in Glashow; Sheldon, L. *From Alchemy to Quarks*. Brooks-Cole: Pacific Grove, CA, 1994, pp. 417–423.
2. Röntgen, W.; quoted in Miller; Arthur. *Einstein, Picasso*. Basic Books: New York, 2001, p. 26.
3. Thomson, J.J.; quoted in Gratzer; Walter. *A Bedside Nature*. Freeman: New York, 1998, p. 100.
4. Becquerel, H.; quoted in Glashow; Sheldon, L. *From Alchemy to Quarks*. Brooks-Cole: Pacific Grove, CA, 1994, pp. 417–423.
5. Carey, J. *The Faber Book of Science*. Faber and Faber: London, U.K., 1995, pp. 188–192.
6. Magie, W.F. *A Source Book in Physics*. McGraw-Hill: New York, 1935.
7. Einstein, J. (Albert's uncle); quoted in Miller; Arthur. *Einstein, Picasso*. Basic Books: New York, 2001, p. 46.
8. Carey, J. *The Faber Book of Science*. Faber and Faber: London, U.K., 1995, pp. 260–261.
9. Snow, C.P. *The Physicists*. Macmillan: London, U.K., 1981.
10. Noddack, I.; quoted in Glashow; Sheldon, L. *From Alchemy to Quarks*. Brooks-Cole: Pacific Grove, CA, 1994, p. 570.

---

* Leo Szilard (1898–1964) was born in Hungary and began a university career in Germany. As a young man serving in the Austro-Hungarian army in the First World War, Szilard became sick and was sent to hospital just at the time his unit was sent into the front lines. Szilard recovered; his unit was wiped out. In 1929, he had published a seminal paper in the science that is nowadays called information theory. By the early 1930s, Szilard became certain that Hitler would eventually come to power, and that another European war was inevitable. He kept two bags packed in his room, to be ready for precipitous flight, and by 1933 he managed to move to England. Szilard and Albert Einstein collaborated on the invention of a refrigerator that functioned with no moving parts.

11. Doyle, A.C. *The Adventure of the Beryl Coronet.* Available in any of the numerous editions of the cases of Sherlock Holmes that are always in print.
12. Hahn, O. Letter to Lise Meitner, December 1938. www.mphpa.org/classic/HISTORY/H-02e.htm
13. Frisch, O.; quoted in Glashow; Sheldon, L. *From Alchemy to Quarks.* Brooks-Cole: Pacific Grove, CA, 1994, p. 570.
14. Becquerel, H.; quoted in Glashow; Sheldon, L. *From Alchemy to Quarks.* Brooks-Cole: Pacific Grove, CA, 1994, p. 535.
15. Rutherford, E.; quoted in Boorstin; Daniel, J. *Cleopatra's Nose.* Random House: New York, 1994, p. 149.

FURTHER READING

Blom, P. *The Vertigo Years.* Basic Books: New York, 2008. A fascinating history of the early years of the twentieth century, up to the outbreak of the First World War. Chapter 4 discusses the discoveries of the early atomic scientists.

Bodanis, D. $E = mc^2$. Walker: New York, 2000. Subtitled *"A Biography of the World's Most Famous Equation,"* this book provides an excellent survey of the scientific discoveries that led up to the development of this equation, from Faraday through Rutherford, Fermi, Meitner, and Einstein himself. Requires little previous scientific background.

Curie, M. *Radioactive Substances.* Dover: Mineola, NY, 2002. This is a reprint of the original 1904 work that provides an eyewitness account of the early days of developing the understanding of radioactivity.

Glashow, S. *From Alchemy to Quarks.* Brooks/Cole: Pacific Grove, CA, 1994. An excellent introductory physics text primarily for liberal arts students, though requiring some mathematics background; or, as the author says, "physics for poets who can count." Chapter 10 discusses the discoveries of x-rays and radioactivity, as well as the discovery of the nucleus itself. Radioactivity, the structure of nuclei, and the discovery of fission are treated in Chapter 14.

Hewitt, P. G. *Conceptual Physics.* Addison-Wesley: Reading, MA, 1998. Another splendid introductory physics text, with very informative illustrations. Chapter 32 discusses the atomic nucleus and the phenomenon of radioactivity; Chapter 33, nuclear fission.

Keller, A. *The Infancy of Atomic Physics.* Dover: Mineola, NY, 2006. A reprint of a book first published in 1983, providing a history of the early work in this field, culminating with the discovery of fission.

Mahaffey, J. *Atomic Awakening.* Pegasus Books: New York, 2009. This book, intended for persons of little background in science, provides an interesting history of the development of nuclear science, as well as a discussion of nuclear power plants.

Malley, M. *Radioactivity.* Oxford: New York, 2011. This book provides a history of the development of our understanding of radioactivity and its health effects.

Mann, T. *The Magic Mountain.* (Numerous editions of this classic are available, including relatively inexpensive paperbacks.) This novel is frequently cited as one of the greatest works of literature of the twentieth century. For our purposes, it is noteworthy in containing perhaps the first account in fiction of the use of X-rays in medicine—in this case, in a tuberculosis sanatorium.

Norton, T. *Smoking Ears and Screaming Teeth.* Pegasus Books: New York, 2011. This engaging book ranges over many fields of science, involving those intrepid, or crazy, individuals willing to perform experiments on themselves, family, or friends. The chapter entitled "That Unhealthy Glow" discusses the uses of X-rays or other forms of radiation.

Pullman, B. *The Atom in the History of Human Thought.* Oxford University Press: New York, 1998. A survey of concepts of atoms and our knowledge about atoms through almost the entire span of human history, from the ancient Greeks to the 1990s.

24 Nuclear Power Plants

For any SYSTEM undergoing a transition that liberates ENERGY, it must be possible, at least in principle, to develop a device or piece of hardware that captures a portion of the ENERGY and transforms it into useful WORK. Nuclear fission produces enormous amounts of ENERGY, far more than from any chemical reaction. Much of the ENERGY resulting from fission occurs as heat. When a source of heat is readily available, using that heat to make steam is often a good idea. The steam can be used in a turbine to operate an electrical generator. The ENERGY release from nuclear fission can be transformed to electrical ENERGY. Once that has happened, the electricity is just as useful as electricity made in any other way. Steam turbines can also be used to supply propulsion ENERGY used in some types of submarines and surface ships. The challenge lies in the "piece of hardware," designing and operating it to capture the ENERGY released from fission in a safe, controlled way.

COMPONENTS OF A NUCLEAR REACTOR

Niels Bohr and John Wheeler determined from theory that $_{92}U^{235}$ was much more likely to undergo fission than was $_{92}U^{238}$. Their prediction was soon confirmed by experiments. Naturally occurring uranium is about 99.3% $_{92}U^{238}$, and only 0.7% fissionable $_{92}U^{235}$. Consequently, natural uranium, even if produced in very high purity, offers essentially no chance for a sustained nuclear chain reaction, because 993 atoms out of every 1000 would be the $_{92}U^{238}$ isotope, which absorbs neutrons without undergoing fission. To sustain a nuclear chain reaction based on uranium fission, the uranium fuel has to be prepared such that the fissionable isotope is present in greater concentration than in natural uranium. The process for doing this is called *uranium enrichment* (shown conceptually in Figure 24.1). It requires effecting a separation of isotopes, which is not an easy task. This is because isotopes are, after all, the same chemical element, and therefore can't be separated by processes that rely on differences in chemical behavior. The only distinction is the slight difference in mass number—a difference of about 1.3% in the case of the two uranium isotopes. Therefore, separation processes must rely upon small differences in those physical properties that depend on differences in atomic or molecular weights.

One approach to uranium enrichment involves the formation of uranium hexafluoride (UF_6). This compound, despite its very high molecular weight, boils at about 57°C, a temperature well below the 100°C boiling point of water. UF_6 molecules containing U^{235} will be slightly lighter than those containing U^{238}, and, in the gaseous state, will diffuse very slightly faster. As the molecules diffuse over very long distances, a portion of the gas will contain a higher proportion of

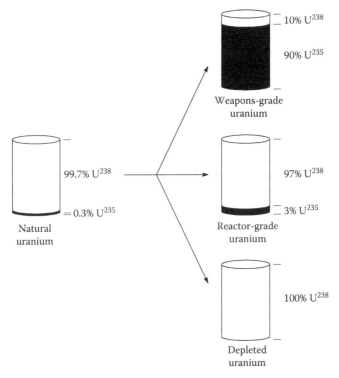

FIGURE 24.1 Uranium enrichment separates the isotopes in natural uranium ore to produce weapons-grade uranium, about 90% U^{235}; reactor-grade uranium, about 3% U^{235}; and depleted uranium, 100% U^{238}.

the lighter $U^{235}F_6$ molecules, while the gas lagging a bit behind will have a higher proportion of the heavier $U^{238}F_6$.

To make a suitable fuel, the uranium must be enriched to at least 3%–4% U^{235}, depending on the design of the reactor. This form of enriched uranium is sometimes called "reactor-grade" or "power-grade" uranium. A second form of enriched uranium, containing about 90% U^{235}, has been used for the construction of nuclear weapons, and is sometimes referred to as "weapons-grade" uranium. The difference in U^{235} concentration between reactor-grade and weapons-grade uranium removes the possibility of a nuclear reactor exploding like an atomic bomb, because the concentration of the fissionable U^{235} isotope in the reactor is too low for a nuclear explosion.

Enriched UF_6 is converted to solid uranium dioxide (UO_2), then shaped into pellets typically 2.5 cm diameter and about 2 cm tall, each pellet being encased in a ceramic shell. These are packed into long, thin zirconium alloy tubes (nominally about 2.5 cm diameter and 4 meters long) to make *fuel rods*. A group of fuel rods packaged together makes up a *fuel assembly* (Figure 24.2).

As the reactor operates, the concentration of fissionable U^{235} drops steadily, since it is being consumed to release ENERGY. At the same time, fission products,

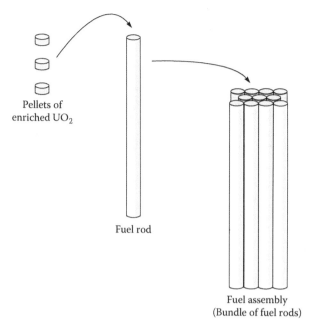

Pellets of
enriched UO$_2$

Fuel rod

Fuel assembly
(Bundle of fuel rods)

FIGURE 24.2 Individual pellets of enriched, reactor-grade uranium are loaded into metal tubes to make fuel rods. A group of fuel rods together makes up a fuel assembly.

some of which are good neutron absorbers, increase. Eventually a time comes when the combined effects of losing U^{235} and building up neutron-absorbing fission products leave a fuel rod unable to contribute further to the ENERGY production in the reactor. At that point the fuel rod is said to be *spent*, but it still contains some 1%–2% U^{235}.

To control the nuclear chain reaction, something is required that will readily absorb the neutrons without reemitting them. The chain reaction is controlled with neutron-absorbing rods, *control rods*, usually containing cadmium or boron. Control rods can speed up, slow down, or shut down the fission process altogether. The rate of the chain reaction depends on the number of control rods inserted into the reactor, and the extent of their insertion into the reactor (Figure 24.3). Control rods can be moved into and out of the reactor to control the flux of neutrons. They are somewhat analogous to the accelerator in a car, in providing the operator a way of speeding up or slowing down fission. Control rods fully inserted into the reactor stop the chain reaction. A typical reactor design might employ 48 fuel rods and 12 control rods in a single fuel assembly. Several hundred such fuel assemblies would be used in a single reactor.

When fission occurs, disintegrating $_{92}$U^{235} nuclei emit fast-moving neutrons having high kinetic energy. Atoms of $_{92}$U^{238}, which still amounts to about 97% of the uranium atoms in the fuel, readily capture the fast neutrons released in fission. $_{92}$U^{238} does not undergo fission. Slower-moving neutrons are far more likely to be captured by $_{92}$U^{235} than by $_{92}$U^{238}. The slower neutrons would be able to contribute to sustaining the chain reaction when absorbed by the fissionable U^{235}.

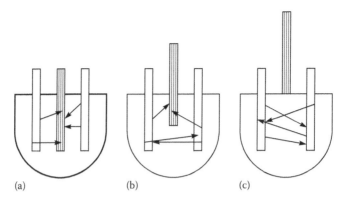

(a)   (b)   (c)

FIGURE 24.3 This diagram of a hypothetical reactor with two fuel rods and one control rod illustrates the role of the control rod. In (a), the control rod is inserted into the reactor and absorbs neutrons, shutting down the reactor. In (b), the control rod is partially withdrawn, allowing some neutrons to enter fuel rods and cause fission. In (c), a fully withdrawn control rod allows the maximum amount of fission, so that the reactor can operate at full output.

A *moderator* slows the neutrons so that U^{235} nuclei will absorb them and continue the fission process. The ideal moderator should reduce neutron energies as efficiently as possible, quickly and with minimum neutron loss. The best moderators are nuclei of low mass that rarely or never absorb neutrons, and that don't undergo fission themselves. Hydrogen, deuterium, helium, and carbon are all good moderators.

When a fast-moving neutron hits the nucleus of an atom in the moderator, the collision resembles that between a golf ball and a baseball. If a golf ball smacks into a baseball, it will bounce back, but at much lower velocity than it had before the collision. Similarly, when the neutron rebounds from a relatively light nucleus, it loses considerable velocity, and, consequently, loses kinetic energy. In comparison, when a golf ball whacks into a solid wall, it flies back with considerable velocity, as empirically demonstrated by anyone who has ever tried to practice golf indoors. In the same way, if a neutron bounces back from a heavy nucleus, it loses hardly any of its velocity or kinetic energy. As an additional requirement for a good moderator, it should not contain nuclei that would absorb the neutrons upon collision, because that would cause their loss from the chain reaction.

Hydrogen exists as three isotopes, $_1H^1$, the most common form and called ordinary or light hydrogen; $_1H^2$, called deuterium or heavy hydrogen; and the radioactive $_1H^3$, tritium. Water in which the light isotope $_1H^1$ is the dominant form of hydrogen is called light water. In principle, hydrogen should be best as a moderator because it has the lightest atoms, which means that a neutron gives up its energy in the fewest collisions. Every molecule of ordinary water contains two atoms of hydrogen, is cheap, and has the further advantage that it can also act as a coolant (described later). But, light hydrogen is also a good absorber of neutrons.

Heavy water slows neutrons quickly without absorbing them at all. However, heavy water is expensive because only 0.015% (or 1 out of 7000) of hydrogen atoms is deuterium, and separating heavy water from light water—as is the case with any isotope separation—is difficult. Because the neutrons need more collisions with deuterium atoms than with light hydrogen, heavy water reactors tend to use large quantities of heavy water—roughly 500 tonnes in a large reactor. Some Canadian reactors have used heavy water as moderator.

Graphite (the preferred form of carbon for nuclear applications), though relatively inexpensive and easy to work with, is a less efficient moderator than heavy water, so graphite-moderated reactors had to be big. Furthermore, graphite, when exposed to air at high temperatures, can ignite and burn. Graphite fires were partly responsible for two of the world's four major reactor accidents (Chapter 25). Though the carbon atoms are relatively heavy, compared to light hydrogen or deuterium, neutron loss by absorption is only a quarter of that in light water. Because graphite is a solid, it cannot be pumped or made to circulate through the reactor. Consequently, unlike light or heavy water, graphite cannot do double duty as both moderator and coolant (discussed below). Some reactors in Britain and in countries of the former Soviet Union, including the notorious Chernobyl-type reactors, are graphite moderated.

As fission occurs, large amounts of heat will be released. The heat provided by a fission reactor comes from the kinetic energy of the fragments of the broken-up nuclei. About 85% of the liberated energy goes into the nuclear fragments, which rapidly collide with other atoms within the fuel rod. As these nuclear fragments collide within the fuel rod, they give up their kinetic energy in the form of heat, thereby heating the rod. The temperature in a light-water reactor (the most common type of reactor) typically reaches 320°C, while in the fuel rods it may reach 2200°C. A *coolant* serves two purposes: It keeps the core's temperature under control, and it moves the heat generated in the core to some device that is able to make use of it. Light or heavy water can be used as the coolant as well as being the moderator. In a graphite-moderated reactor, gases, such as carbon dioxide, can be used as coolants.

The heart of any nuclear plant is the reactor *core*, which typically contains the fuel assemblies (including control rods), moderator, and coolant.

The essential components of a nuclear reactor core are the fuel, control rods, moderator, and coolant.

The core is enclosed in a very thick, very strong steel shell called a *pressure vessel*. In most reactors, the pressure vessel is enclosed in a *containment structure* (sometimes called containment building), a steel-reinforced concrete structure typically at least one meter thick that serves as the outermost barrier between the plant and the environment, preventing most radiation from escaping the plant. As its name implies, the function of the containment building is to contain the reactor contents in the event of a leak or an accident. This structure is used in virtually all reactors in modern industrialized nations. As we will see (Chapter 25), one reason why the 1986 Chernobyl reactor accident was so severe is that power reactors of that type built in the former Soviet Union do not have containment structures.

PROCESSES IN THE REACTOR'S CORE

A sustained nuclear chain reaction depends on using a neutron produced in a fission event to start the next fission. Chain reactions would not be achievable if, on average, fewer than one neutron was produced per fission. Not enough neutrons would be available to allow one new fission to take place for every one that has just occurred. Even if *exactly* one neutron was produced per fission event, that would not be sufficient to sustain a chain reaction. Inevitably, some neutrons would escape from the core, or would be absorbed by non-fissioning nuclei, and would not be available to interact with further U^{235} nuclei. If, on average, one of the fission neutrons produces a further fission the reaction is self-sustaining. In this case, we say that the reaction is *critical*. The fission rate does not change over a period of time. If more than one fission neutron produces a further fission, the reaction is said to be *supercritical*, in which case the number of fissions increases with time. If less than one fission neutron produces a further fission, the reaction is *subcritical*, and the number of fissions gradually decreases with time.

To produce a working, controllable nuclear reactor, it must be possible to start the chain reaction, allow the number of nuclei involved to increase gradually to a fission rate sufficient to generate the required energy, and then to maintain this rate. Of course, it must also be possible to decrease the reaction rate in a controlled manner to shut down the reactor. Control of the chain reaction is achieved by controlling the number of neutrons present in the core. Whether the reactor is supercritical or subcritical can be established by manipulating the control rods that are inserted into the core. Pulling control rods out of the core increases the chance that each neutron will induce a new fission, rather than being absorbed in a control rod, and so moves the core toward supercriticality. Inserting control rods into the core increases the chance that each neutron will be absorbed before it can induce a new fission, and moves the core toward subcriticality.

The three possibilities are defined by the reaction *multiplication factor*. If the multiplication factor is less than 1, at each succeeding step in the chain reaction, fewer atoms will undergo fission and, hence, fewer neutrons will be produced. The nuclear chain reaction rapidly comes to a stop. On the other hand, if the multiplication factor is greater than 1, at each step in the chain reaction a larger number of uranium atoms undergo fission, and a greater number of neutrons are produced. Right at critical mass, where the average fission induces only one subsequent fission, the fission rate remains essentially constant.

U^{235} nuclei undergoing fission release an average of 2.47 neutrons, which induce other fissions within a thousandth of a second. Some of the neutrons released in fission emerge immediately, called *prompt neutrons*. Some of the fission fragments are unstable nuclei that decay and release neutrons, *delayed neutrons*, long after the original fission. On average, each U^{235} fission eventually produces 0.0075 of these delayed neutrons (or 0.75%), which can still go on to induce other fissions. Their time delays range from about one second to about one minute, averaging about 14 seconds. Delayed neutrons allow greater operational safety in reactors by giving operators some time to start shutting the reactor down if any indication of serious problems or danger is detected.

To summarize the processes that take place in the reactor's core: Fission reactions in the fuel rods result in the highly energetic neutrons and fission fragments moving rapidly away from the site of the reaction. These fission fragments quickly collide with other nuclei in the same fuel rod and, in doing so, give up their kinetic energy as heat. Most of the high-energy neutrons escape from the fuel rod and enter the moderator. In the moderator, these neutrons experience multiple collisions with hydrogen or carbon nuclei, each time losing a little of their kinetic energy to the nuclei of the moderator. The moderator slows the neutrons so that their capture by another fissionable U^{235} nucleus is more likely. Some neutrons may encounter a control rod and be absorbed and are no longer able to initiate another fission reaction. The coolant flowing among the rods carries away the heat generated.

THE FIRST REACTOR

Enrico Fermi led the construction of the first nuclear reactor in a squash court underneath the stands of the University of Chicago's football stadium. Fermi had emigrated from Italy to the United States in 1938, but had not yet become an American citizen. Since the United States was at war with Italy, Fermi was technically—and rather ironically considering his contributions to the development of the atomic bomb—an "enemy alien."

In 1942, pure uranium was available as both the metal and its oxide. It was not enriched, so the amount needed for a reactor to become critical was extremely high. A very large atomic pile had to be built. (It was called a "pile" because it was, literally, a pile of bricks of uranium, uranium oxide, and graphite.) Three fates are possible for a neutron in a fission reactor. It can cause fission of another U^{235} atom, can escape into nonfissionable materials, or can be absorbed by U^{238} without causing fission. To make the first fate more probable, the uranium was divided into small pieces and buried at regular intervals in about 400 tonnes of graphite (Figure 24.4). When this first nuclear reactor was completed, it was 9 meters wide, 10 meters long, and 6.6 meters high. It weighed 1270 tonnes, of which 47 tonnes were uranium. The uranium, uranium oxide, and graphite were arranged in alternating layers with holes into which cadmium control rods could be inserted.

On December 2, 1942, at 3:45 p.m., the control rods were pulled out just enough to produce a self-sustaining chain reaction. On average, slightly more than one neutron was produced per fission that was not absorbed by either U^{238}, or by some impurity, or lost to the pile. This meant that, for the first time, a controlled, man-made nuclear chain reaction was in process. That day and minute are taken to mark the beginning of the atomic age. Because of wartime security considerations, the news of this success was relayed to Washington by a telegram reading, "The Italian navigator has entered the new world." There came a questioning message in return, "How were the natives?" The immediate reply was: "Very friendly." By December 12, the pile was releasing enough heat to be the equivalent of the energy needed to operate two light bulbs (i.e., 200 watts). Fermi's success showed that both nuclear energy and the atomic bomb were practical possibilities.

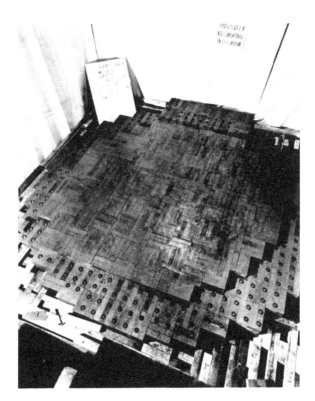

FIGURE 24.4 The first reactor to achieve sustained, controlled nuclear fission, constructed at the University of Chicago in 1942. (From http://photoarchive.lib.uchicago.edu/db.xqy?keywords=apf2-00502.)

FIRST APPLICATIONS OF NUCLEAR FISSION

In 1954, the first nuclear submarine, the USS *Nautilus*,* was launched in the United States (Figure 24.5). Its energy was obtained entirely from a nuclear reactor, and, unlike conventional submarines, it did not have to rise to the surface at short intervals to recharge a set of batteries. Since it could remain underwater for extended periods of time, it was that much safer from enemy detection and attack. The first American atomic surface ship was the NS *Savannah*, launched in 1959.

In the mid-1950s, nuclear power stations were designed for the production of electricity for civilian use. The Soviet Union built a small station of this sort in 1954, with a capacity of 50 MW. The British soon followed with one of 92 MW capacity at

---

* The first nuclear submarine was named for An earlier U.S. Navy submarine of the same name, which served with distinction in World War II, and was very likely inspired by the fictional *Nautilus* the electrically powered submarine in Jules Verne's classic science fiction novel of 1870, *Twenty Thousand Leagues under the Sea*. Though parts of this book are rather fanciful in light of today's knowledge, in places it is remarkably prophetic. This great book is almost always in print. Verne is said to have selected the name for his submarine from Robert Fulton's early (about 1800) attempt at making a submarine, also called the *Nautilus*.

FIGURE 24.5 The *Nautilus*, the first nuclear submarine, undergoing its sea trials. (From http://en.wikipedia.org/wiki/File:SS-571-Nautilus-trials.gif.)

Calder Hall. The first American nuclear reactor for civilian purposes began operations at Shippingport, Pennsylvania, in 1958. These early commercial reactors were designed for what we would nowadays consider a modest generating capacity of 100 MW. Later reactors were much larger; by the late 1970s, plants in the range of 1000 MW were being built. (A 1000 MW generating capacity provides approximately enough electricity for about a half-million homes, based on an average consumption.)

THE BOILING WATER REACTOR

More than three-quarters of the commercial nuclear electric generating plants worldwide are light-water-moderated reactors of two types: boiling water reactors (BWRs) and pressurized water reactors (PWRs). Both types use ordinary, or light, water as both coolant and moderator and require enriched uranium fuel.

Nuclear plants differ from fossil-fuel plants only by using the heat from a nuclear reactor to raise steam for their steam turbines. The steam turbine–electricity generator sets are virtually identical with those of conventional fossil-fuel-fired stations. Temperatures and pressures of the steam fed to turbines in nuclear plants are not so high as those of fossil-fuel-fired systems. This is due partly to the more severe conditions that the materials used in a nuclear plant must withstand. Because the steam temperatures (T_{high} in Carnot's efficiency equation) and pressures are lower, the resulting thermal efficiencies are also lower (a 34% turbine efficiency is considered good). Since the conversion of ENERGY to WORK is lower, the waste energy—in this case, waste heat output—is necessarily higher. Heat losses should be the least if the coolant itself becomes the steam, and this is the method adopted in *boiling water reactors*. Since water is used as the coolant,

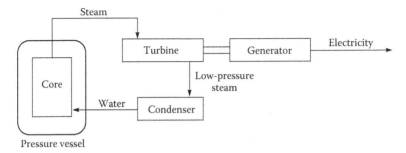

FIGURE 24.6 In a boiling water reactor, water in the reactor's core is allowed to boil, to produce the steam needed to operate the turbine.

steam can be generated directly from the water. This leads to the simplest design of a power reactor system, a BWR (Figure 24.6).

Fuel assemblies in BWRs are surrounded by the water coolant. As fission occurs, fission fragments move through the fuel elements and the water; their kinetic energy becomes transformed, via many collisions with other atoms, to thermal energy. This thermal energy—heat—boils the water in the core. The steam produced is taken from the core to the turbines. The steam drives the turbine, after which it is condensed to water and returned to the reactor, very much like in a fossil-fuel plant (Chapter 16).

The BWR works perfectly well for electric power generation. However, some of the radioactive fission products can diffuse through the metal cladding of the fuel rods. In the United States, reactors are allowed to stay in operation with up to 1% of the fuel rods actually leaking. As water circulates through the core, it dissolves some of the fission products, some of which are highly radioactive. Thus the steam leaving the containment building and the water returning to it are radioactive. If an accident occurred that ruptured the steam or water lines outside the containment building, then radioactivity could be released to the environment. (This does *not* mean that radioactive material is released during normal operation of the reactor.)

PRESSURIZED WATER REACTOR

The more commonly used PWR addresses this potential emission of radiation into the environment by sealing the cooling water in a closed loop and adding a heat exchanger inside the containment building (Figure 24.7). Water in the reactor core is kept under high pressure (10–15 MPa, much higher than in a BWR) so, even though the water gets very hot, above 300°C, it does not boil to steam.*

---

* Though we accept 100°C as the boiling point of water, in actuality the boiling point of any liquid depends on the prevailing pressure. At pressures higher than our normal ambient pressure, the boiling point increases; conversely, at reduced pressures the boiling point is lower. Strictly speaking, the boiling point of water will change very slightly with swings in atmospheric pressure caused by changing weather. Those changes are so miniscule that they can be disregarded in all but the most precise of laboratory work. Sometimes, as in the case of the PWR, we can exploit the pressure dependence of boiling point in practical devices.

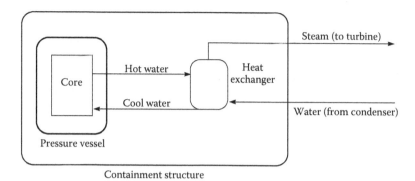

FIGURE 24.7   In a pressurized water reactor, hot, high-pressure water from the core is circulated through a heat exchanger to generate the steam used in the turbine. The steam fed to the turbine is not radioactive.

The hot water flows through a heat exchanger that serves as the steam generator, where this extremely hot, high-pressure water converts a secondary water supply into steam to operate the turbine. The water in the heat exchanger vessel boils as a result of the heat provided by the reactor (via the hot, high-pressure water), and this steam runs through a turbine just as in a conventional fossil-fuel plant. The ideal Carnot efficiency for this process is nearly 50%, but the actual efficiency is about 30%.

With a PWR, there is no release of radiation if pipes outside the containment building are ruptured. Because heat from the high-pressure water inside the core is used to raise steam in the heat exchanger, no radioactive material is ever outside the containment building during normal operation. There is no radioactive steam going to the turbine. This of course does not mean that there can never be a problem with operating such a reactor, nor that there are no potential environmental problems. (These issues will be discussed in Chapter 25.) However, the likelihood of an accidental release of radiation to the environment is much, much less from a PWR than from a BWR. Most nuclear power stations worldwide use the PWR. A typical PWR design uses uranium oxide fuel, enriched to around 3% U^{235}, encased in zirconium alloy fuel rods. Water is used as a moderator and a coolant.

Operators start the reactor by retracting the control rods. As rods are removed, the flux of neutrons grows. Soon, the chain reaction begins. As more control rods are retracted, the reactor passes from subcritical through the critical stage, and temporarily into the supercritical range to sustain growth. As the number of fission reactions increases, the thermal power rate grows as well. This growth continues until the full power level is attained, at which time control rods are adjusted to keep the reaction rate steady in time. Under normal conditions, the reactor is operated in the critical range. Thermal energy is carried by the coolant. The water travels in a loop from the reactor to the steam generator and back. The process is self-contained and discharges no effluent to the atmosphere other than steam exhaust and waste heat. There are no piles of fuel or waste products outside the plant, nor any fuel deliveries via railroad cars or barges around the plant.

PRESSURE TUBE REACTORS

The graphite–water reactors used in the former Soviet Union are known as *pressure-tube reactors* (known by the initials RBMK). They operate on very slightly enriched uranium fuel and differ from BWRs and PWRs in that it is possible to refuel the reactor as it operates, without shutting down. RBMK reactors use graphite moderators. A 100 MW reactor design incorporates about 2500 columns of graphite blocks, over 200 boron control rods, and over 1600 zirconium alloy fuel elements. The water system for this reactor has been characterized as a "plumber's nightmare." This type of reactor, the RBMK1000, was involved in the 1986 disaster at Chernobyl in Ukraine.

THE STEAM CYCLE

The steam cycle in a nuclear fission plant is very similar to that discussed in Chapter 16 for fossil-fuel-fired plants. Thus a very similar set of energy diagrams (Figure 24.8) apply. A high-temperature, high-pressure head of steam is fed to the turbines. There, the thermal ENERGY of the steam is converted to mechanical WORK, which is used to turn the generators. The steam leaving the turbines is at lower pressure and temperature. That steam is led to a condenser, where it is condensed back to liquid water, which is returned to the steam generator. The condensers require a source of

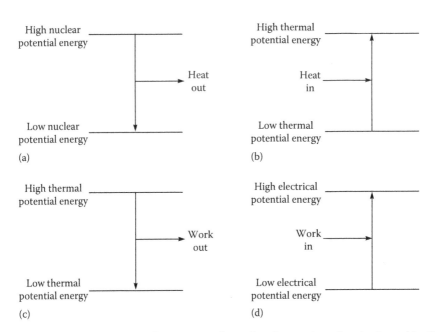

FIGURE 24.8 The sequence of energy transformations in a nuclear plant is almost identical to that of a fossil-fuel-fired electrical plant, except that the heat necessary to generate the steam comes from a nuclear transformation, rather than from the chemical process of combustion. (a) Reactor core, (b) heat exchanger (steam generator), (c) turbine, and (d) generator.

FIGURE 24.9 Cooling towers help to reduce the temperature of the cooling water leaving the condenser before it is returned to the environment. Cooling towers have come to symbolize nuclear power, especially in some news media, even though no nuclear processes occur there.

cool water; often this is obtained from a natural source such as a river. The cooling water that leaves the condenser is considerably warmer, because it has absorbed heat from the steam as the steam condenses. To avoid upsetting the local ecosystem by discharging large quantities of warm water from the condensers, this water is usually passed through a cooling tower (Figure 24.9) before its ultimate return to the environment. Because most cooling towers have a distinctive shape, and often dwarf the reactor building, cooling towers have sometimes been used in the news media as a symbol of nuclear power. This unfortunate choice has led some to believe that the cooling tower *is* the reactor, whereas in fact it is probably the most innocuous part of the plant. The cooling tower plume is sometimes alleged to be a cloud of deadly radiation escaping from the "reactor" when in fact it is just a cloud—condensed water vapor just like any other cloud.

Fossil-fuel–fired boilers, such as the pulverized-coal–fired water-tube boiler discussed in Chapter 16, and the nuclear reactors discussed here, are simply different ways of boiling water to feed a head of steam to the turbines. All of the public concern about the environmental consequences of burning fossil fuels and all of the concern, or occasional uproar, over nuclear reactor safety and nuclear waste are both ultimately focused on the question of how to boil water.

REACTOR SAFETY SYSTEMS

Of course, it is always appropriate—indeed, important—to ask what happens if something does go haywire in the reactor. The first line of defense, so to speak, is the *s*elf-*c*ontrolled *r*emote *a*utomatic insertion of control rods, known from the first letters of its name as *scramming*. This procedure is the emergency insertion of *all* the control rods into the core. With all control rods inserted, this should shut down the reactor. The second safety feature is the emergency core cooling system—the *ECCS*—that floods the core with large volumes of cold water. This flooding should reduce the temperature in the core substantially. In addition, the water acts as a moderator to help slow down neutrons.

RADIATION FROM NUCLEAR PLANTS

In ordinary operation, the radiation from a nuclear power plant is virtually nil. By one scale of radiation dosage (the millirem, discussed in Chapter 25), the background radiation that we are all exposed to in our environment, from cosmic rays and naturally occurring radioactive materials, amounts to about 100 millirem per year. The *additional* radiation experienced by people working directly in the vicinity of a reactor is about one millirem per year. The additional radiation to which the general public is exposed amounts only to 0.05 millirem per year. Virtually all radioactive materials are contained within the reactor system, and small amounts only of short-lived radioactive elements are allowed to vent into the atmosphere. Shielding of the reactor is required by various laws and regulations to protect the health and safety of workers inside the plant. This shielding, along with the limitations on radioactive discharges, is also designed to meet public health standards for the population outside the plant.

THE END OF THE STORY?

The history of civil nuclear energy has been a checkered one indeed. After the Second World War, when the general public first became aware of the potential of releasing enormous quantities of energy via nuclear fission, there were widespread hopes of extremely cheap energy. The expression "too cheap to meter" came into currency, meaning that electricity could be produced in such vast quantities, and so incredibly cheaply, in fission reactors, that it would not be cost-effective to install electricity meters at consumers' sites, nor to bother mailing out bills for the electricity. We could use all the electricity we wanted, free. In 1956, the first nuclear power station at Calder Hall, part of the Windscale complex in Britain, began to feed electricity into the public electricity supply. At the time, Britain was dependent on increasingly expensive, locally mined coal.

Unfortunately, it soon became clear that electricity from nuclear plants was not, and would never be, "too cheap to meter." This realization was followed by the first of the four major, well-publicized reactor accidents. This one, in October 1957, involved the Windscale reactor and resulted in a fire that polluted an area of farmland in the northwest of England. In 1979, the Three Mile Island accident, followed in 1986 by the major disaster of Chernobyl, resulted in significant public disillusionment with, indeed even outright hostility for, nuclear energy. Chernobyl was followed in turn by the 2011 disaster at Fukushima. (A time line of nuclear events is given in Figure 24.10.) In addition, there remain lingering concerns, even serious feelings of guilt on the part of some individuals, over the atomic bombing

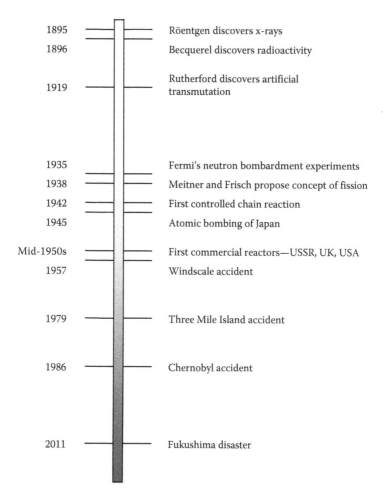

1895	Röentgen discovers x-rays
1896	Becquerel discovers radioactivity
1919	Rutherford discovers artificial transmutation
1935	Fermi's neutron bombardment experiments
1938	Meitner and Frisch propose concept of fission
1942	First controlled chain reaction
1945	Atomic bombing of Japan
Mid-1950s	First commercial reactors—USSR, UK, USA
1957	Windscale accident
1979	Three Mile Island accident
1986	Chernobyl accident
2011	Fukushima disaster

FIGURE 24.10   A time line of the major events in the history of nuclear energy.

of Hiroshima and Nagasaki. Nuclear energy is now regarded as, at best, a dubious blessing. The inability of engineers to give a cast-iron guarantee that there can be no future accidents, and the inability of most of the public to understand that *no* technology can expect to have such a guarantee, increases the negative attitude toward nuclear energy. The situation is complicated even further by the unsolved problem of the disposal of nuclear waste—where are we going to put the stuff to store it safely?

The collective impact of concerns about economics (i.e., it's not "too cheap to meter"), nuclear accidents, and what to do with the radioactive waste have, not surprisingly, led to enormous problems for the nuclear power industry in many countries around the world. Another potential problem is nuclear weapons proliferation, the possibility of subverting an ostensibly peaceful civilian nuclear energy program to make weapons-grade uranium instead of reactor-grade material. This is quite possibly happening in Iran today. The uncertainties of governmental regulations—combined with what seems at times to be a never-ending, but seemingly always-escalating, sequence of lawsuits, appeals, counter-suits, and counter-appeals—have ended the construction of new nuclear reactors in some countries, including the United States. One example of the impact of delays caused by regulation and litigation is the dramatic increase in construction costs. In 1973, the Shoreham, Long Island nuclear power station was projected to cost $300 million to build. By the time it was completed (in 1984), its cost had escalated to $5.5 billion. And, it has never been put into operation because of public opposition.

Currently, nuclear generation accounts for about 15% of electric energy generated in the United States and nearly 20% worldwide. As we begin the new century, the world has about 450 nuclear power stations in thirty different countries. They represent about one-sixth of the world's total electric generating capacity and produce a little over one-ninth of the total annual output of electricity.

While many observers expect, hope for, or actively work to bring about, the demise of the nuclear industry, it may be premature to consign nuclear energy to the proverbial dustbin of history.* Concerns are increasing about the impact of human sources of carbon dioxide—mainly from the burning of fossil fuels—on changing the global climate (Chapter 29). Proponents of the large-scale use of nuclear energy have sometimes argued that nuclear energy is the only "emission-free" source of electricity. Though one might argue about this choice of words, certainly in the event that major reductions in CO_2 emissions are mandated, nuclear fission is the only existing, demonstrated technology for large-scale electricity generation that does not release CO_2.

As evidence for global climate change continues to accumulate, and that human activities contribute to climate change, there seemed the possibility of a modest revival of nuclear energy for electricity generation. Despite the potential problems, which will be discussed in Chapter 25, right now nuclear energy is the only large-scale carbon-free route to electricity generation.† New reactor designs have been developed to address such legitimate concerns as reactor safety and radioactive waste. Some of these designs will be discussed in the next chapter. Attitudes toward nuclear energy were beginning to shift. And then came the fourth nuclear accident, when a tsunami hit the Fukushima reactors in Japan. Once again, it remains to be seen what the future will be for nuclear energy.

---

* The phrase is thought to have been first used by Leon Trotsky, a Bolshevik, in castigating the Mensheviks in a squabble over which faction should govern the Soviet Union. The Mensheviks have indeed been relegated to the dustbin of history. So have the Bolsheviks. So has Trotsky. So has the Soviet Union.

† Of course it can be argued that hydroelectricity falls into this same category. But, as already discussed in Chapter 15, hydro has its own set of issues. In the terrible event that the Three Gorges Dam were to experience catastrophic failure, the ensuing death toll would be orders of magnitude higher than all the nuclear accidents put together. Putting aside calamities that can be sincerely hoped never to happen, not all places in the world possess adequate sites for hydroelectric plants, even combined with dams. A nuclear plant can be built, at least in principle, just about any place, so long as fuel can be brought to it, and there is a way to run the condensers.

FURTHER READINGS

Bryce, R. *Power Hungry*. Public Affairs: New York, 2010. This book lays out the case that various sources of "green" energy will not be adequate to meet all energy needs, and that both natural gas and nuclear energy will have a significant role in the future.

Cassedy, E.S.; Grossman, P.Z. *Introduction to Energy*. Cambridge University: Cambridge, U.K., 1998. An excellent introductory book on the benefits and problems of many forms of energy technology. Chapter 7 is a good overview of nuclear reactors.

Cohn, S.M. *Too Cheap to Meter*. State University of New York: Albany, NY, 1997. The title derives from the dream of the early days of the nuclear industry: that electricity could be produced so cheaply there would be no reason to keep track of its use nor to send out bills. The book treats the development of the nuclear industry in a societal context.

Josephson, P.R. *Red Atom*. Freeman: New York, 2000. A history of the nuclear energy program in the former Soviet Union. This book traces the evolution of the early hope of the Communist party that nuclear energy would provide massive amounts of cheap energy, to the tragic reactor accident at Chernobyl. As the dust jacket indicates, it is a story of "big science run amok."

Koerth-Baker, M. *Before the Lights Go Out*. Wiley: Hoboken, NJ, 2012. A reasonable and rational discussion of possible energy technologies and options for the future, including nuclear energy.

Ngô, C.; Natowitz, J. *Our Energy Future*. Wiley: Hoboken, NJ, 2009. A survey of many kinds of energy sources likely to be important in the future. Chapter 10 discusses nuclear energy.

Silverstein, K. *The Radioactive Boy Scout*. Random House: New York, 2004. This is the remarkable story of the high-school student who designed and built a nuclear reactor in a garden shed, using openly available information and blueprints, and materials scrounged from a variety of reasonably accessible sources. The book also has much good discussion about nuclear reactors in general.

Smil, V. *Energy at the Crossroads*. MIT Press: Cambridge, U.K., 2003. This book, subtitled "global perspectives and uncertainties," is dense with facts and figures relating to many energy sources. Chapter 5 includes a discussion of the possibilities for nuclear energy.

Suppes, G.; Storvick, T. *Sustainable Nuclear Power*. Elsevier Academic Press: Burlington, MA, 2007. Chapters 10 through 12 of this book relate to the material presented here.

Zoellner, T. *Uranium*. Viking Penguin: New York, 2009. This book provides a history of humankind's relationships with uranium, from the days when it was simply a nuisance to miners looking for other ores (recall the early history of gasoline!) to the present, when uranium just might have a role in the fight against global climate change.

25 The Nuclear Controversy

In the years following the Second World War, the prospects for ENERGY from nuclear fission seemed to be so extensive that humankind might have entered a promised land of energy. Electricity would be so cheap that there would be no reason to install meters in homes or businesses, or to waste the effort of mailing out monthly electric bills.

> It is not too much to expect that our children will enjoy in their homes electrical energy too cheap to meter.

—**Strauss**[1]

Serious plans were being drawn up for nuclear airplanes, nuclear ships, nuclear cars, and miniature nuclear electric or heating plants in the basements of homes.

> Even more intriguing than the atomic submarine, but also more complicated, is the idea of an airplane with a nuclear-driven power plant. It could fly nonstop around the world at supersonic speeds, ride out any bad-weather traffic stacked over an airport and would never be subject to a power failure on take-off or landing.

—**Benford**[2]

Food would be irradiated for safe, long-term storage so that refrigerators would be obsolete.

> In the more distant future, say 1975, is an atomic wrist watch, a time piece operated by a midget nuclear power plant.

—***New York Times* (1956)**[3]

Instead of that promise coming to fruition, today ENERGY from nuclear fission is likely the most contentious, and the most feared, of all the ENERGY sources available to us. In the past half-century, something went badly wrong. There are many reasons why nuclear fission is of great concern to the public. Among them are four issues to be discussed in this chapter: the fear of exposure to radiation, the possibility of a catastrophic accident involving a nuclear reactor, the ongoing debate about what to do with the waste products from a reactor, and the possibility that a rogue nation or a terrorist group could acquire technology and components to build nuclear weapons.

No one any longer seriously believes that the day will come when electricity from nuclear plants will really be "too cheap to meter." No one expects, or wants, to see nuclear cars and nuclear airplanes, let alone a nuclear furnace in the basement. Yet many people believe seriously that nuclear fission must be considered as an ENERGY source for the future, especially as part of a shift away from fuels that produce carbon dioxide emissions. New reactor designs, intended to allay many of the concerns about fission energy, are in various stages of development. Examples are discussed at the end of the chapter.

HEALTH EFFECTS OF RADIATION

A major concern about radiation is that we can't see it, detect it with our other senses, or even feel it affecting us. Most other energy sources and forms of fuels are materials we can see (such as coal or gasoline) or at least feel (such as heat or sunlight). Not long after X-rays were discovered, the scientists working with them learned that overexposure to X-rays caused skin inflammations and burns that healed very slowly. The same proved to be true of the radiations from radioactive substances. The acute effects of exposure to massive doses of radiation are known from studies of the victims of the bombs dropped on Hiroshima and Nagasaki in the Second World War, and from accidents in the early days of the atomic weapons and nuclear electricity programs. Most of this knowledge has come the "hard way," from studying the effects of radiation on those who have, whether inadvertently or deliberately, been exposed to very large doses. The main factors determining the health effects of radiation are the type of radiation and its intensity of the radiation (i.e., the rate at which it is being produced).

The three major types of radiation were discussed in Chapter 23. α-Particles travel only a few centimeters in air and can be stopped by a thin shield of material like cloth, cardboard, or even tissue paper. Although α-particles are easily stopped, if they do penetrate tissue they are an order of magnitude more dangerous to health as γ-rays, but they can cause such harm only if the radioactive source is eaten or inhaled into the body. β-Particles have intermediate penetrating ability and health effects intermediate between the other two types of radiation. β-Particles can travel up to a few meters in air and can be stopped by sheets of such metals as aluminum, iron, or lead. γ-Rays, essentially identical to high-energy X-rays, can penetrate thick layers of materials and cause serious health effects. γ-Rays are stopped only by substantial thicknesses of lead or thick concrete walls. These relationships are shown schematically in Figure 25.1.

The second issue is the rate at which the radiation is being emitted. In other words, how radioactive is the substance that is giving off radiation? One answer comes from the rate at which a radioactive material is decomposing, given by its *half-life*. The half-life of a radioactive isotope is the time required for half the amount on hand to decay. For example, suppose we have 100 grams of an isotope that has a half-life of 400 days. If we measured the amount of isotope on hand, we would find the relationship summarized in Table 25.1. Though the half-lives of radioactive substances vary enormously, from fractions of a second to billions of years, a plot of the amount remaining as a function of time (Figure 25.2) would always have the same shape. In a sense, the half-life issue is a "good news–bad news" situation. A material with a short half-life will be intensely radioactive, but won't be around long. A material with a long half-life has a low level of radioactivity, but is going to be around for a long time. The amount of radioactive nuclei present in a material can be characterized by the number of nuclear

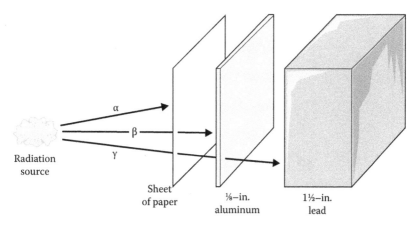

FIGURE 25.1 The different kinds of radiation have markedly different abilities to penetrate matter. γ-radiation is the most dangerous and can only be stopped by thick concrete or metals such as lead.

TABLE 25.1

Example of the Decay of 100 g of a Hypothetical Radioactive Substance with a Half-Life of 400 Days

Time Elapsed (Days)	Number of Half-Lives Elapsed	Amount Remaining (g)
0	0	100
400	1	50
800	2	25
1200	3	12.5
1600	4	6.25

---

disintegrations per unit time. The unit of radioactivity, the *becquerel*, is defined as one disintegration per second.

The half-life and the type of radiation emitted combine to determine the prospective danger of a material. An α-emitter with a very long half-life (e.g., the common isotope of uranium, $_{92}U^{238}$) is relatively safe. A γ-emitter with a short half-life is very dangerous. Radiation affects us through its action on cells. The energies of X-rays, γ-rays, or high-speed α- or β-particles can be high enough to break chemical bonds in molecules. When that happens, very reactive molecular fragments, radicals or ions,*

---

* Radicals are created when a chemical bond breaks in such a way that one of the electrons goes with each of the fragments. Using the fictitious X:Y molecule as an example, the breaking of the bond between the X and Y atoms can be represented as X:Y → X· + ·Y. Ions form when such a bond breaks so that both electrons wind up with one of the fragments, such as X:Y → X⁺ + :Y⁻. Unlike ions, radicals have no electric charge. Both ions and radicals are, in most instances, extremely reactive species and readily interact with other nearby molecules.

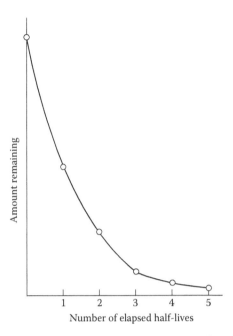

FIGURE 25.2   This generic graph illustrates the decay of any radioactive substance through five half-lives.

are formed. These very reactive species often quickly react with other molecules. Also, an α- or β-particle absorbed by an atom will alter its nature (i.e., the reaction of an atomic nucleus with either an α or β will change the atomic number, and hence chemical identity, of that nucleus, transmuting one element into another); so it will alter the nature of the molecule of which that atom is a part. If the newly formed atom is radioactive and decays, that process will also break apart the molecule even if it had survived intact to that point. The biological damage caused depends not only on the type of radiation and its energy, but also on the chemical composition and types of atoms in the body tissue absorbing the particles.

Many of the pioneers in the study of radioactivity suffered health effects. This is not because they were stupid or careless; it is because in that era no one, including the scientists themselves, truly understood the effects and dangers of this remarkable new phenomenon. Pierre Curie tried an experiment of strapping a glass vial of uranium compounds to his arm, and wearing it for a time. The resulting radiation burn took more than six weeks to heal.

Chemical changes caused by radiation disrupt the many complex biochemical processes inside a cell. When that happens, these changes upset the sequences of reactions that control how a particular cell functions, and upset the various interactions among cells. In some cases, biological changes may be induced that allow the unrestrained growth of certain kinds of cells at the expense of their neighbors. This is the pathological condition that we recognize as cancer. Most vulnerable to the effects of radiation are the skin, which of course is always the first part of the body experiencing radiation exposure, and those parts of the body that produce blood

FIGURE 25.3   Marie Curie (1867–1934) at work in her laboratory. Scientists of her era did not understand the grave dangers of working with radioactive materials. Notice the complete absence of safety glasses, a laboratory coat, gloves, or any kind of shield. (From http://commons.wikimedia.org/wiki/File:Madame_curie_3334194920_e4014f35a4_o.jpg.)

cells, such as the lymphoid tissue* and bone marrow. Leukemia, the unrestrained production of white blood cells (a condition that is slowly, but invariably, fatal) is one of the more likely results of excessive exposure to radiation. Both Marie Curie (Figure 25.3) and her daughter, Irène Joliot-Curie,[†] died of leukemia, which may have resulted from the long exposure to radiation that both of them experienced during

---

* The lymphoid tissues constitute the body's immune system. The lymph nodes, spleen, bone marrow, and thymus are all lymphoid tissue. The lymph nodes help to overwhelm bacteria. The spleen filters bacteria and other microorganisms from the blood. The thymus shrinks and apparently loses its functioning around the onset of puberty, but if it is damaged in infants, severe problems with the infant's immune system can occur. Injury or damage to the bone marrow sharply reduces the production of blood cells.

† Irène Curie (1897–1956, Nobel Prize 1935), daughter of Marie Curie, married the physicist Frédéric Joliot, with whom she collaborated in numerous scientific endeavors relating to radioactivity, fission, and the development of nuclear energy. Irène and Frédéric published their work using their hyphenated surnames, Joliot-Curie. Among the many contributions of this brilliant couple were the discovery of artificial radioactivity and studies contributing to the eventual discovery of the neutron by James Chadwick. The Joliot-Curies had observed neutrons first, but did not want to suggest the existence of another subatomic particle to explain their observations, so discovery of the neutron is credited to Chadwick. The Joliot-Curies shared the 1935 Nobel Prize in Chemistry for their splendid contributions to science. Becoming increasingly aware of the dangers to world peace cause by the rise of fascism in the 1930s, the couple voluntarily stopped publishing their work on nuclear fission in 1939 and had that information locked up for ten years by the French Academy of Sciences.

their scientific careers. These, and many other deaths of pioneering nuclear scientists, happened not because the persons involved were stupid—they rank among the most brilliant scientists ever—but because the effects of exposure to radiation were unknown or poorly understood. When radiation exposure is particularly severe, very extensive destruction of tissues can completely break down the functioning of cells. This extreme condition can cause death in a period from months to as little as a few days. Such radiation sickness was observed on a large scale among the survivors of the atomic bombings of Hiroshima and Nagasaki.

The effects of radiation on living tissue can be considered in separate categories, called somatic and genetic effects. Somatic effects occur in the person who was exposed to the radiation; genetic effects, which might be even more frightening, appear in children of such a person. Though the death of an individual from radiation exposure is a sad and unfortunate event, there is also danger—perhaps even worse than the death of the person affected—in that long-term genetic effects may continue over many generations. Sometimes damaging or destroying a molecule by radiation may not affect the individual very seriously, especially if the damage is confined to only a few cells. But, the molecules damaged by radiation may be in the genes and chromosomes of the individual who had been exposed. Then the radiation damage could be transmitted to a child of that person, and, if that happens, the child might have the damaged molecule in every cell of its body. The child might suffer far more than the parent, who was actually the person exposed to the radiation.

During the natural progression of generations, mutations in genetic material occur as a result of random chemical imperfections that become incorporated in key molecules as they are reproduced. Mutations can also arise as the result of natural background radiation (discussed later), such as from radioactive substances in the soil, or cosmic rays from outer space. All of us have imperfect genetic material to some extent, and it is passed on to the next generation. However, the rate of genetic mutation in a population increases as the total amount of radiation in the environment—from natural background sources plus that introduced from human activities—increases. Commonly, genetic mutations are generally for the worse, in that the affected individual may be less strong, less healthy, or deformed as a result of the mutation. (We must recognize, though, that mutated genetic material that is injurious to some people could be advantageous to others in different circumstances. This helps give humans—and other organisms—the genetic ability to adjust to change and to evolve.) Some mutations can be fatal, condemning the afflicted individual to an early death. The nonlethal mutations are the greatest long-term worry for society, however, because the person having a genetic mutation may have children of his or her own. If those offspring survive, they in turn can pass on the damaged genetic material to yet another generation, spreading the mutation further and further into the population.

Units of Radiation

There are several ways of measuring the radiation dose received by an individual. The *rad* (the name being short for "radiation") is a direct measure of energy. One rad is equivalent to the absorption of enough radiation in any form to liberate

0.01 millijoules of ENERGY per gram of absorbing material. The rad is approximately equal to the Röntgen, a unit that is discussed below.

Doses larger than 10,000 rad damage the central nervous system. Death occurs relatively quickly, often within one or two days. A dose of 1,000–10,000 rad has a major effect on the gastrointestinal tract, causing vomiting, fever, and dehydration. Death likely follows within one to two weeks. In the dose range 300–1000 rad, blood cells are the worst affected. Hemorrhaging and diarrhea can lead to death in three to six weeks. Doses below 300 rad, and certainly below 50 rad, seldom lead to death, even within thirty days. There may be loss of hair, loss of appetite, and diarrhea. There is also an increased risk of leukemia and other cancers for the survivors, often developing years after the actual radiation exposure. The biological effects of radiation that cause, or are capable of causing, mutations are proportional to the radiation dose rate down to about one hundred rad. A rule of thumb suggests that about a hundred additional cancers per million people will result from a 1-rad dose of radiation. It is not certain whether this rule-of-thumb relationship holds for even lower doses.

The röntgen is formally defined as the quantity of X-rays or γ-rays required to produce a number of ions equivalent to one electrostatic unit of charge* in a cubic centimeter of dry air at 0°C and one atmosphere pressure. A given quantity of radiation does not produce the same effect on all living species. To take this difference among species into account, the "röntgen equivalent man," or *rem* for short, is defined to be that quantity of radiation which, on absorption by the tissues of a living human, produces the same effect as the absorption of 1 R of X-rays or γ-rays. As a rule of thumb, exposure to 0.5 rem, or 500 millirems (mrem) per year, is tolerable.

α-Particles are particularly dangerous. One rad of X-rays, γ-rays, or β-particles is nearly equivalent to one rem. However, 1 R of α-particles is equivalent to a dose of 10–20 rem. Absorption of α-particles is at least ten times as dangerous as the absorption of the same amount of β-radiation. In that respect, it's fortunate that α-particles are the least penetrating form of radiation and the easiest to protect ourselves against.

Many fission products are radioactive. Although previous chapters have illustrated the fission of $_{92}U^{235}$ as its transformation to barium and krypton, actually a variety of possible fission products form, whose mass numbers are roughly half that of uranium. Not only are some of the original fission products themselves radioactive, they can decay into other, secondary products that are also radioactive. Among the more dangerous fission fragments are radioactive isotopes of strontium and cesium. The isotope $_{38}Sr^{90}$, a β-emitter, has a half-life of 28 years. Even after a century, $_{38}Sr^{90}$ will have undergone less than four half-lives, so a material containing this isotope is still dangerously radioactive. Strontium is in the same chemical family as calcium. Therefore, $_{38}Sr^{90}$ in the environment can become concentrated in the milk

---

* Electrostatic units derive from the gram–centimeter–second system of measurement. When two equal electrical charges are separated by a distance of 1 cm and exert a force of 1 g·cm/s^2 (i.e., 1 dyne) on each other, the magnitude charges is one electrostatic unit (abbreviated *esu*). The electric charge represented by 1 R is equivalent to the formation of about two billion ions. Two billion ions may seem to be an enormous quantity, but, to put this into perspective, 1 cm^3 of air at these conditions will contain about 25 quintillion—that is, 25 billion billion—molecules.

of cows that graze on $_{38}Sr^{90}$-contaminated grass. Children who drink milk contaminated with radioactive strontium will concentrate the $_{38}Sr^{90}$ in their bones. Calcium compounds are the major inorganic constituents of bone tissue, and strontium atoms, in the same chemical family as calcium, are able to replace calcium atoms in bones. Once bone tissue has formed, the atoms in the bones are replaced only very slowly. Consequently $_{38}Sr^{90}$, once assimilated, will remain in the bones for a long time and stay in dangerously close contact with the blood-cell forming tissues. $_{55}Cs^{137}$, with a half-life of thirty years, remains in the soft tissues. It is exchanged with other atoms more quickly than is $_{38}Sr^{90}$ in the bones, so stays in the body for a shorter time. However, $_{55}Cs^{137}$ emits energetic γ-rays and, for the time it remains in the body, can do significant damage.

BACKGROUND RADIATION

While public attention usually focuses on radiation from nuclear waste, atomic weapons testing, or reactor accidents, we are all constantly bathed in natural radiation. Anyone, anywhere in the world, will be exposed to radiation. This unavoidable background radiation comes in the form of cosmic rays from outer space, from naturally occurring radioactive substances in the environment, and from medical procedures. On average, we receive 50 mrem per year from outer space and another 50 mrem per year the natural radioactivity of the soil. Radioactive $_{90}Th^{232}$ and $_{92}U^{238}$ occur in various kinds of rocks and soil. When these elements happen to be present in clays used to make bricks, the bricks then have a low level of natural radioactivity. The estimate of radiation exposure for medical uses (such as diagnostic X-rays) is about 75 mrem per year. In fact, we irradiate ourselves to a small extent (about 25 mrem per year) from radioactive potassium and carbon isotopes, $_{19}K^{40}$ and $_6C^{14}$, which become incorporated in our body tissues.* The total background radiation for most people is in the range of some 125–350 mrem per year, well below the 500 mrem US background radiation standard for the general public. Despite what many people may believe, *less than 1%* of our background radiation comes from sources such as nuclear electricity-generating plants and their wastes and nuclear fuel reprocessing.

The isotope of radon $_{86}Rn^{222}$, an α-emitter, occurs naturally in the environment as a result of decay of $_{92}U^{238}$. Radon is an inert gas in the same chemical family as helium and neon. Because radon comes from natural uranium deposits, the amount present in the environment depends highly on the local geology, that is, on the amount of uranium present in the rocks and soil of a particular locality. In some extreme cases, radon can amount to about 55% of natural background radiation (Figure 25.4). Because radon is inert chemically, when it is produced it will not be trapped by chemical processes in the surrounding soil or water. That trapping in soil or water, if it could have taken place, would have immobilized the radon. Instead, radon seeps

---

* We are "carbon-based life forms." All of our soft tissues are made of compounds of carbon; carbon atoms are the stuff we are made of. The isotope $_6C^{14}$ occurs in trace quantities in our bodies. Potassium is essential to maintaining the proper balance of fluids inside and outside the cells of our bodies. (The fluid balance is determined by the Na^+/K^+ ratio.) About twelve potassium atoms out of every 100,000 (0.012%) are the radioactive isotope $_{19}K^{40}$.

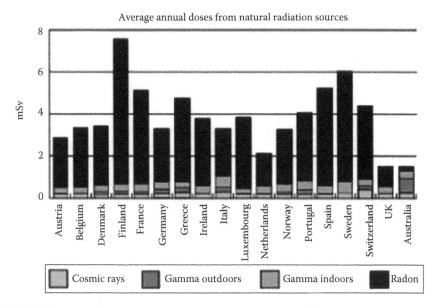

FIGURE 25.4 Radiation is always present in the environment. This chart illustrates average annual radiation doses from natural sources. (From http://www.world-nuclear.org/info/inf05.html.)

from the ground into basements or lower levels of buildings through cracks or pores in concrete walls or floors.

When air containing radon is breathed, radon decays inside the lungs to give polonium, a radioactive element that is neither a gas (so it is not likely to be removed in normal exhalation) nor chemically inert (so it is, unfortunately, likely to interact with the surrounding tissue). The isotope $_{84}Po^{218}$ is also an α-emitter and on doing so produces a lead isotope, $_{82}Pb^{214}$. That lead isotope is a β-emitter. So, as the radon initially inhaled decays, it produces other radioactive materials that are more likely to stay in the body. Because the range of an α-particle in air is quite small, it is relatively easy to shield against external α-particles. But in this case, the α-emitters are already inside the body, where α-radiation can cause serious biological damage. In the lungs, the range of penetration of the α-particles is about equal to the thickness of the cells that line the respiratory tract passages. Radiation damage to these cells can induce lung cancer.

Is There a Threshold Radiation Exposure?

Over the years, attempts have been made to determine the amount of radiation that can reasonably be tolerated by individuals (and by humankind in general) without representing a serious danger. Radiation effects can be divided into acute effects, due to large amounts of radiation received in a single dose, and chronic effects, due to small amounts of radiation but repeated exposures over some period of time. Acute effects, though they can be horrific in the worst cases, are at least straightforward, and there seems to be consensus about what those effects will be. In contrast, the effects of long-term exposure to low levels of radiation are not well known. In fact, it is not even clear whether there *are* such effects, let alone what they might be if they do exist, and whether they can be distinguished statistically among the potentially large number of similar effects that happen to be due to other causes.

It is often possible to establish a relationship between health effects and radiation exposure in one person, and then, from that, to *estimate* the incidence of that same effect in the population as a whole. Doing so makes it possible to predict that, if the population were exposed to a particular radiation dose, there would be a corresponding incidence of that related health effect. There are two problems with this approach. First, individuals vary widely in how their bodies function. We all see, for instance, among our family and friends that some individuals gain or lose weight, or catch a cold, much more easily than others. Second, many kinds of observable health effects likely have multiple causes. Because of these problems, it becomes very difficult to detect effects specifically due to the one event of radiation exposure within a background of similar effects that originated from other causes. For example, in the United States, about 550,000 people die each year of cancer. If the population happened to be exposed to a radiation dose double the background level, there would, in principle, be an additional 2000 cancer deaths. However, it would be extremely difficult to determine unequivocally that a cancer death rate of 552,000 instead of 550,000 was due to radiation exposure and not to other fluctuations in the population, such as an increased use of tobacco products.

Some experiments have compared the effect of delivering the same total dose of radiation either in a single large dose (that is, equivalent to an acute exposure) or by the accumulation of many separate, smaller doses (simulating chronic exposure). A single larger dose can be fatal in cases where the several smaller doses have little apparent effect. These observations suggest a "threshold" dose below which there is no radiation damage. They also suggest that a repair mechanism allows the body to recover from small amounts of radiation damage. However, there also seems to be a reduced life expectancy for those receiving the several smaller doses. That is, the body's repair mechanism may not provide permanent repairs. Though there are vigorous proponents on both sides, it seems on balance that it is not yet possible to show unequivocally that a safe threshold exists for chronic exposure. Also, it's not yet possible to sort out clearly the effects of radiation exposure in a population from other natural causes or just from random, year-to-year statistical fluctuations.

NUCLEAR REACTOR SAFETY

Several concerns focus on the safety of nuclear reactors. One is the issue of whether a nuclear reactor can blow up just like an atomic bomb. That event is impossible. The amount of fissionable $_{92}U^{235}$ in a reactor is about 3% of the total amount of uranium. In an atomic weapon, the proportion of $_{92}U^{235}$ would be about 90%.

A second issue concerns whether a nuclear reactor emits radiation to the environment during normal, routine operation. The PWR design (Chapter 24) now accounts for about two-thirds of the "power reactors" now in operation. The heat exchanger system ensures that all radioactive water is confined inside the containment building. The pressure vessel around the core acts as a radiation shield. The containment structure around the pressure vessel and heat exchanger also serves as a radiation shield. During normal operation, the radiation level outside the plant is so low that it poses no concern to human health or safety. The containment structure is expected to withstand more than the most severe weather or other natural conditions, such as floods or earthquakes, ever anticipated in the region. Indeed, some people have suggested quite seriously that in the event of a weather emergency, people should go to the reactor, since it is probably by far the safest building around. The Fukushima disaster, discussed below, proved them wrong.

Some low-level radioactive material is periodically released to the environment, for example, through vents or in cooling water, during routine operation. Those countries that have electric plants based on nuclear reactors will have some sort of government agency or oversight board that is charged with monitoring radiation releases and setting limits both on worker exposure inside the plants and on releases to the environment. From the design stage onward, nuclear plants must prove that they are operating within these boundaries.

Of greatest concern for most people is the prospect that a nuclear reactor could release substantial amounts of radiation to the environment in the event of a serious accident. This is amply demonstrated by the four most famous (or notorious) nuclear reactor accidents: Windscale in England, Three Mile Island (TMI) in the United States, Chernobyl in the Ukraine, and Fukushima in Japan. We'll examine these accidents in some detail later in this chapter.

The most feared nuclear accident is a *meltdown*, in which the core overheats, most likely because of a loss of the coolant, causing the fuel to melt. Molten fuel could conceivably burn its way through the steel pressure vessel and then through the containment structure, contaminating the surrounding environment. In the early days of the nuclear industry, it was thought that the molten fuel could continue to burn its way deep into the Earth, possibly even coming out the other side. That quite erroneous belief led to the term *China syndrome* to describe this fearsome event. (An old, and quite incorrect, notion is that a hole dug straight through the ground in the United States would eventually come out in China.) Currently it is thought that the molten fuel would only (!) penetrate about three meters into the ground below the plant. In the event of a meltdown that breached both the pressure vessel and the containment structure, the molten core would continue to melt its way downward until the rate at which heat released by the core became equal to the rate at which the heat can be conducted away from the molten rock around the core. It turns out that these rates become equal at a depth of about three meters. By then, a volume of soil would have melted and congealed into a glass that might, it would be hoped, contain the remains of the fuel. The radiation that escaped during this event could certainly contaminate a huge area around the plant, especially if it happened to get into the surrounding groundwater supplies.

WINDSCALE

The Windscale plant (now called Sellafield, north of Liverpool, England) had been built to produce plutonium for use in nuclear weapons. The Windscale reactors used graphite as the moderator, and air as the coolant. Air was drawn into the reactors and then swept out through enormous chimneys (Figure 25.5).

FIGURE 25.5   The Windscale reactor. (From Garland, J.A. and Wakeford, R., *Atmospheric Environ*, 41(18), 3904, 2007.)

The purpose of any moderator is to absorb collisions of high-energy neutrons, to adjust their kinetic energy to a point at which they are more likely to cause fission when they collide with a uranium nucleus. With graphite, collisions between carbon atoms and high-energy neutrons can be so energetic as to knock some of the carbon atoms out of the positions they occupy in the atomic structure of graphite. The displaced carbon atoms are in a higher state of energy. When these atoms are able to move back to their normal positions in the graphite structure, this "extra" energy is released.* If too much energy is released at once, the graphite could catch fire. To forestall this, the graphite can be heated under very carefully controlled conditions to allow the displaced atoms to move back to their normal positions, a process called annealing.

---

* The energy absorbed by carbon atoms during collisions with energetic neutrons is called Wigner energy, in honor of Eugene Wigner (1902–1995, Nobel Prize, 1963), a native of Hungary who was trained in chemical engineering and who excelled at theoretical physics, mathematics, and nuclear engineering. Wigner was the leader of the group that designed the nuclear reactors used at the Hanford, Washington plant during the Manhattan Project. Wigner impressed his numerous friends and colleagues by his remarkable sense of politeness. There is a story that Wigner had been cheated by an unscrupulous used-car salesman, and returned the car to the salesman saying, "Go to hell, please."

In October 1957, plant operators were conducting a controlled annealing of the graphite moderator in one of the reactors. Poorly distributed, and possibly malfunctioning, temperature sensors led operators to believe that the moderator was cooling properly, as it should when annealing is finished. Then it became apparent that there was a "hot spot" in the core. Visual inspection showed that at least four sets of fuel containers were red hot. Unbeknownst to anyone, the reactor core had been on fire for about two days. Radiation monitors at the top of the chimneys were at maximum reading. Increasing the flow of air coolant only made things worse. At perhaps the worst point, some ten tonnes of uranium were on fire. Water failed to douse the fire. Eventually, the situation was brought under control—after four days—by shutting off every possible source of air, including the coolant and air used for normal ventilation.

A large amount of radiation escaped, contaminating the surrounding countryside, and spreading across the rest of Britain into Europe. (Nonetheless, the Windscale accident released far less radiation than Chernobyl, discussed below.) Nearly two million liters of milk contaminated with radioactive iodine were thrown away (into nearby rivers or the Atlantic Ocean) in the weeks following the accident. Fortunately, a survey of the workers who had been directly involved in dealing with the fire and its aftermath, conducted in 2010, showed no long-term health effects. The unit that had actually caught on fire was left totally unusable, and the other reactor was shut down for safety reasons.

Three Mile Island

The TMI, near Harrisburg, Pennsylvania, incident began on March 28, 1979. It remains the worst commercial nuclear reactor accident in the United States. Clean-up, waste storage, entombment, and decommissioning was expected to cost at least two billion dollars. It was not until 1990, eleven years after the accident, that most of this work had been completed.

The TMI reactors were water-moderated PWRs. The problem began in Unit 2 when a cooling system water pump failed. Good engineering design always involves redundancy—extra or backup units of critical components of a system. There were three backup pumps to feed water to the reactor if the main pump stopped working. Normally they would swing into operation and easily remedy the problem. But, two weeks before the accident, two valves were mistakenly left closed after a routine maintenance check, shutting off water to these pumps. With no cooling water flowing, control systems automatically stopped the turbine, which halted the generation of electricity. However, the reactor itself remained at its full level of energy production. Temperatures in the cooling water circuit quickly rose, increasing the pressure inside the reactor vessel. An automatic reactor shutdown, scramming, occurred because of the high pressure. To this point, the various control systems were doing their jobs appropriately, trying to effect an orderly shutdown of the turbine and reactor.

But, the operators misread the pressure gauges and were not aware that the pressure was excessively high. As the pressure continued to increase, eventually something had to "give." When the pressure rose, unnoticed by the operators, past 15.5 MPa, an automatic pressure-relief valve opened—exactly as it was designed to do—to reduce the pressure. The pressure fell back to the value designed for normal operation, but then the valve failed to close as the pressure dropped past the normal operating pressure. Pressure inside the reactor continued to drop, well below design level. When the pressure reached 11 MPa, the high-pressure coolant injection system, part of the emergency core cooling system (ECCS), automatically turned on.

The plant operators were relying on a backup cooling system to keep the reactor core covered with water. Assuming that the backup system would be adequate for this job, they turned off most of the emergency coolant pumps, in part because the water level gauges incorrectly indicated an acceptable water level in the reactor. What the operators did not know was that their backup system was turned off as well. This act, based on erroneous information, guaranteed that the core would be uncovered. At this point, no water was being supplied to the core from any source. What little coolant was left in the system turned into steam, partially exposing the reactor core. The now-exposed core began to heat up. As temperature continued to rise, the zirconium metal cladding of the fuel rods began to rupture. So much heat was generated that the zirconium, a metal deliberately chosen to withstand extremely high temperatures, began to burn. Molten, burning zirconium, molten zirconium oxide, ZrO_2, and molten UO_2 from the fuel pellets began flowing to the bottom of the reactor.

At the worst point, 15%–30% of the core was uncovered and remained so for about 14 hours. An estimated 45% of the core melted and as much as 70% was damaged (Figure 25.6). The fuel assemblies had burst open, and, along with a few hundred tons of radioactive rubble, the vessel also contained twenty tonnes—more than half the core—of uranium that had melted, then cooled, and hardened. For uranium to melt, temperatures of at least 2800°C must be reached.

Remarkably, the pressure vessel managed to contain the molten material, although theoretically it should not have been able to do so. It was found later that the molten fuel had almost completely burned through the twenty-centimeter-thick steel reactor vessel. With erroneous information displayed on the control system and reactor safety systems unwittingly disabled, the operators had no idea of what was really happening. Six hours after the accident started, a representative of the US Nuclear Regulatory Commission became convinced that the core had begun meltdown. At about the same time, plant engineers also deduced that the core was uncovered. The engineers convinced the operators to increase coolant flow. A total meltdown was prevented only by a last-minute rush of cooling water. The sudden injection of cold water caused some of the extremely hot fuel assemblies to fragment, just as we sometimes accidentally shatter a drinking glass that's hot by putting ice or a cold beverage into it. Also, extremely hot zirconium will react chemically with water to form hydrogen and zirconium oxide. This reaction produced a bubble of hydrogen inside the reactor. But when mixed with air (specifically, the oxygen in air), hydrogen can, under the "wrong" circumstances, blow up. As workers strove to bring the situation under control, there were fears that the hydrogen bubble could wreak havoc by exploding. (It didn't.)

Because the pressure-relief valve failed to close properly after reducing the too-high pressure in the reactor, water flooding into the reactor was able to exit via this

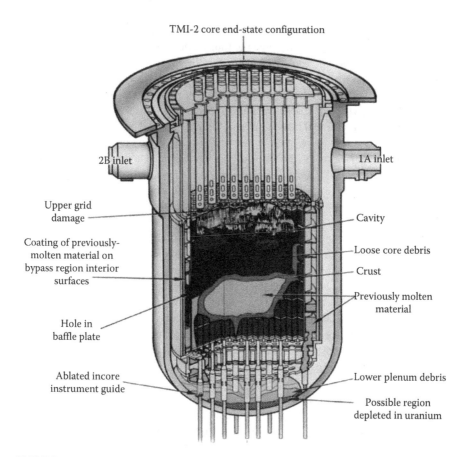

FIGURE 25.6 A schematic diagram of the core of the Three Mile Island reactor after the accident. (Image courtesy of Douglas Akers, Idaho National Laboratory, Idaho Falls, Idaho, http://www.inl.gov/threemileisland/inl.shtml.)

valve. So much water came pouring out that it ruptured a special holding tank that had been designed to contain it. At the end of the incident, the core had been totally ruined and there were 1.5 million liters of radioactive water sitting on the floor of the containment structure. Radioactive gas had also flooded a water supply tank. Some of it was released through venting to the air and was carried toward nearby towns by the wind. The governor of Pennsylvania at the time, Richard Thornburgh, decided to evacuate young children and pregnant women from an eight kilometer radius surrounding the plant. It is estimated that more than 150,000 residents voluntarily fled the area.

CHERNOBYL

Sometime in late 1957 or early 1958, an accident occurred in the former Soviet Union, at the Chelyabinsk-40 facility in the Ural Mountains. A tank holding radioactive gases exploded, contaminating thousands of square miles around the facility.

Little is known in the West about this accident, which is believed to have taken place at a plutonium processing plant. Since the thugs who ruled the Soviet Union did not exactly encourage the open exchange of information, it was not until some thirty years later, in 1988, that Soviet officials finally acknowledged that an accident had even occurred. The region around Chelyabinsk is now sealed off, and the names of about thirty towns in the area simply disappeared from Soviet-era maps. In the bizarre fantasyland of Soviet officialdom, nothing could have happened to the people in those towns, because there never were such towns. Rumors persist that even more horrific nuclear accidents occurred in the former Soviet Union or in China in the 1950s and 1960s, but, if such accidents really occurred, the totalitarian regimes in those countries suppressed dissemination of information about them. Details will not be forthcoming until (or if) their archives are opened to the public. We may never learn the truth.

All electric generating plants—nuclear, fossil fuel, or hydro—require electricity themselves, for their operation and safety. Pumps, control valves, and other equipment usually operate with electricity. Electricity is needed for the various computer-based control and data acquisition systems. To ensure that the plant will run safely is to bring the electricity needed for its operation from some other plant. That way, if there is a problem at Plant 1, say, the electricity needed to operate its control systems, pumps, and other equipment is still available, because it's coming from Plant 2. What if the supply from Plant 2 should fail? To be prepared for that situation, most power plants have backup generators, usually operated by diesel engines, to supply emergency power.

Diesel-engine generators are reliable, rugged machines. Their only fault is that it takes 15–60 seconds to reach full output. The question then becomes how to obtain the necessary backup electricity in the tiny 15–60 seconds window of time, between the instant of the power failure and the diesel coming up to full output. Watch what happens when an electric fan is turned off. When the switch is turned "off," the fan does not stop dead; rather it will continue to turn for a short time as it coasts to a stop. In the same way, when an electric plant shuts down, the turbine and generator do not stop instantly, but, like the fan, coast to a stop. This suggested that, for those critical seconds before the diesel generator is up to full capacity, it might be possible to continue to draw some electricity from the turbine and generator as they "wind down." The engineers at the Chernobyl plant designed a test to see how far the plant could be turned down while retaining the ability to restart the reactor.

The Chernobyl reactor complex is located on the Pripet River about 130 kilometers from Kiev, at the time, a city in the Soviet Union and now the capital of Ukraine. The accident began late on the night of April 25, 1986; a time line is given in Figure 25.7. The operators undertook a test of one of the turbines connected to Unit 4, a boiling-water, graphite-moderated reactor. Ironically, this reactor, now infamous for being the site of a nuclear disaster, had performed flawlessly since going online in 1983. Furthermore, the method of using the residual kinetic energy of the turbines to supply electricity to the vital pumps, computers, and other components had already been successfully tested at other generating stations, including earlier tests at Chernobyl.

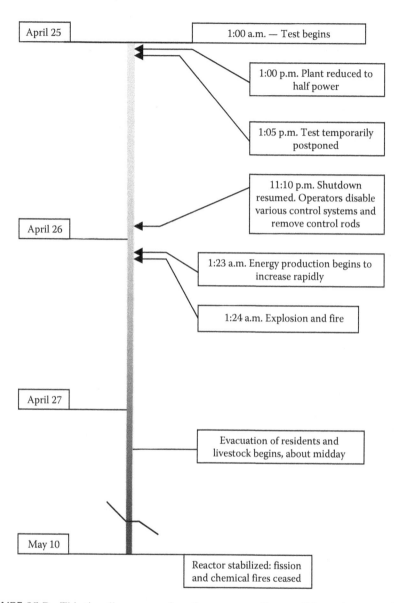

FIGURE 25.7 This time line summarizes the sequence of some of the major events in the Chernobyl reactor accident.

Initially, the plant was brought down to half power, generating only half the normal electrical output, over a twelve-hour period. This took until 1 o'clock the following afternoon. At 1:05, one of the turbines was switched off. At this point, the shutdown was stopped because the electricity being produced from the reactor, even at half capacity, was needed in the Soviet distribution system. Stopping a shutdown in this manner was a violation of experimental and operating protocol. At 11:10 p.m., the shutdown resumed. When the test was resumed, the operators shut off the local automatic control system, because the reactor, at such low output, would be close to the level at which the automatic control system would shut the reactor down. Such an automatic shutdown would abort the test. A year's delay would occur before the test could be tried again.

This temporary stoppage of the test to continue to produce electricity for the grid allowed xenon (a fission product) to build up inside the reactor. Xenon is an excellent absorber of neutrons, so its presence in the core had the same effect as if there were extra control rods in place. Normally the Chernobyl plant operated at a capacity of 3200 MW. At the start of the test, the operators had brought the plant down to 1600 MW, half power. Plans called for an attempt to restart the reactor when the power had been brought down to 700–1000 MW. But, with the excellent neutron absorber xenon abundantly present in the core, the actual power output dropped to 30 MW, when the operators shut off the local automatic control system.

Being desperate to restart the reactor, the operators pulled out almost all of the manually operated control rods. In fact, they were pulled out beyond the physical limits that would allow for their easy reinsertion. Only six of the reactor's eight water pumps were running to feed water into the core, which caused problems with the stable flow of water into the reactor. The operators turned off the system that would automatically shut down the reactor in case of an upset or interruption in water flow. Then, the operators started up the remaining two pumps. The two additional pumps that were started caused the flow rate to jump and the reactor steam pressure to drop toward the emergency trip level. This was a violation of operating instructions.

Eight pumps were now supplying water to the reactor. In normal operation, the number is four. The additional water meant that the core was getting cold, and steam generation had almost stopped. The reactor's control system recognized the abnormally low steam generation, and responded by withdrawing the automatic control rods. At 1:23 a.m. the operators turned off the system that closes the control valves in an emergency, yet another violation of all operating procedures. They also ignored the computer command for total shutdown of the reactor. All of these deliberate disablings of control systems were intended to prevent an automatic shutdown of the reactor that would abort the test, forcing a year's delay before it could be tried again.

To let the core temperature rise, water flow, which had been twice the normal amount (i.e., eight pumps running instead of four), was reduced to 75% of normal (three pumps running). As it should, the temperature started rising, boiling began again in the core, and the power output rose to 530 MW. The All-Soviet regulations for operation of a nuclear reactor required a minimum of thirty control rods to be in the core under any and all circumstances. In principle, even if Mikhail Gorbachev, leader of the Soviet Union at the time, had been standing in the control room himself, he did not have the authority to order or allow removal of more control rods. By now all control rods were out of the core, and all automatic circuits that would "trip" and shut down the reactor if some design parameter were exceeded had been shut off. In another 2.5 seconds, the power output surged to 3840 MW, more than the designed maximum production of 3200 MW. In only 1.5 seconds more, the power skyrocketed to an incredible 384,000 MW. The rate at which heat was being generated was much,

much faster than the rate at which it could be transferred out of the core. Fuel rods disintegrated and evaporated the cooling water. All of the liquid water in the reactor was converted to gaseous steam virtually instantaneously, a phenomenon called *flashing*. Extremely hot zirconium reacted with water to produce hydrogen, as at TMI. The Chernobyl reactor used graphite as the moderator; graphite, which is pure carbon, can also react with steam to make hydrogen and gaseous carbon monoxide. The combined result of these factors was that the pressure inside the reactor became extremely high. The high pressure broke the seals around some of the pipes that penetrated the pressure vessel, allowing oxygen inside. This led to immediate ignition of the hydrogen at the high temperature inside the core. The lid of the pressure vessel, which weighed 2200 tonnes, was blown off. Some accounts suggest that it was hurled into the air like a gigantic coin that had been tossed. The hydrogen ignited and exploded. The remaining graphite ignited to cause a chemical (rather than nuclear) fire and caused flames to carry radioactive reactor contents up and out of the pressure vessel (Figure 25.8). Since there was no containment building, they were immediately released to the atmosphere. Flaming debris from the reactor showered over the complex's buildings, setting them on fire. As cooling water reacted with the graphite core, hydrogen gas built up, causing a second explosion that shot radioactive material nearly two kilometers into the sky.

FIGURE 25.8 An aerial view of the explosion and fire at Chernobyl. A view of smoke from the graphite fire and core melt down. The photo was taken from a helicopter on May 3, 1986, of the destroyed Unit 4. (From http://en.wikipedia.org/wiki/File:Chernobyl_burning-aerial_view_of_core.jpg.)

FIGURE 25.9 The aftermath of Chernobyl—lumps of graphite moderator ejected from the reactor core during the explosion. A channel for a control rod can be seen in the largest piece. (From http://en.wikipedia.org/wiki/File:Ejected_graphite_from_Chernobyl_core.jpg.)

At 1:24 a.m. there was a loud explosion, followed seconds later by a fireball and two more explosions. The explosions pulverized fuel rods, and the rising plume from the blasts carried the debris upward at least 450 meters (Figure 25.9). This tremendous plume was probably responsible for protecting the region right around the reactor from immediate contamination. Evidence of severe irradiation of firefighters and workers did not appear until the next day in most cases. Flames rose at least thirty meters into the air when the graphite moderator caught fire. The resulting fires raged throughout the whole reactor complex. High flames carried most radiation upward so that relatively few deaths occurred (all 31 of the immediate fatalities were plant workers or firefighters). The fire itself was put out by about 5 o'clock that morning, although it took ten days to control the smoldering core. Several thousand tonnes of radiation-absorbing boron and lead were dropped into the exposed reactor from helicopters. Nitrogen was pumped under the exposed reactor vessel to cool it, and a huge concrete slab was placed below the reactor to keep molten fuel from burning through. By the end of 1986, the plant was completely entombed in a steel-and-concrete reinforced radiation containment structure.

Local residents were not evacuated until over 36 hours after the start of the accident. Within two days of the accident, 50,000 people had been evacuated from the town of Pripet. Eventually more than 100,000 people and nearly as many cattle were removed from nearby villages and towns and permanently resettled elsewhere.

So great was the release of radiation from Chernobyl that the first inkling Western nations had that something was wrong came when radiation detectors in

Sweden began responding, three days after the initial accident. The level of radiation detected at the Forsmark nuclear reactor outside Stockholm was so high as to cause worry that there must have been a release of radiation from a Swedish reactor. Other Scandinavian countries showed similarly high readings. Winds were coming from the direction of the Soviet Union, leading to some suspicion that a disaster may have occurred there, but no information was available from the Soviet government. Later that day, an official announcement of the accident was broadcast on Soviet television.

Debris from the accident was carried so far because it was transported upward by the hot gas plume from the graphite fire. The accident released about 2.5% of the radioactivity in the core into the environment. That makes it the largest nonweapon release of radiation publicly known. For at least a month after the accident, all the buildings in Kiev were hosed down daily to remove radioactive particles. Approximately 2500 square kilometer of land around the reactor complex were contaminated, possibly permanently. Financial losses have been estimated at ten billion dollars, but may eventually prove to be much higher. Half of the radioactive fallout dropped within fifty kilometers of the reactor complex. The remainder traveled in a radioactive cloud carried by prevailing winds over parts of the Soviet Union, Scandinavia, continental Europe, Great Britain, and Ireland. Smaller amounts were measured as far away as the United States and the Middle East. The Chernobyl accident is the most terrible nuclear event, other than the atomic bombs dropped on Hiroshima and Nagasaki in August 1945, of which we have detailed records. The human suffering caused by radiation exposure continues to this day.

FUKUSHIMA

By 2010, memories of Chernobyl had faded. Concerns about global climate change, and the likely impact of emissions of carbon dioxide from human activities, began to make nuclear energy look perhaps not quite so bad. Certainly nuclear energy would never fulfill the wild dreams of the 1950s. Certainly nuclear energy had issues of concern, as discussed here. But maybe, given global climate issues, nuclear wasn't so bad after all. And then the tsunami hit.

On March 11, 2011, a colossal earthquake in the Pacific Ocean, off Tohoku, Japan, created an immense tsunami. The earthquake itself was one of the most powerful natural events ever known. It shifted the island of Honshu (the main island of Japan) about 2.5 meters eastward and shifted the Earth's axis by some 10–25 cm. It also created a tsunami with waves about 40 meters high. Standing in the path of the onrushing tsunami was the Fukushima nuclear plant.

The Fukushima plant consisted of six BWRs. It is probably very fortunate that the fuel had been removed from one of the six reactors, and two others were already shut down for maintenance. When the earthquake was sensed by the control systems, the automatic response was to shut down the remaining three reactors, which had been operating up to that point. As designed, emergency generators began operating, to supply backup electricity to keep the coolant circulating and to keep the control systems functioning. Had there only been an earthquake, these design features, which were operating as intended, might well have salvaged the situation without further consequences. But probably no one had anticipated a forty-meter-high tsunami.

When the wave hit, the plant's connections to the electrical grid were severed. Water flooded the emergency generators, knocking them out of action. Their output probably would not have mattered anyway, because the control rooms were flooded, and no longer functioning. This sequence of events was compounded by the effects of the tsunami inland, which were unimaginable outside of a horror film. The tsunami moved many kilometers inland, destroying buildings or sweeping them off their foundations, and tossing vehicles about as if they were toys. Because of the extensive flooding and destruction inland from the nuclear plant, it was near impossible for rescue and relief workers to reach the plant.

The effects of loss of cooling water and lack of any sort of control system led to consequences that should be familiar from the previous discussions: meltdown of the cores of all three reactors, compounded by hydrogen explosions. With so much civil infrastructure damaged, the Japanese government authorized using seawater to cool the reactors. However, compared to ordinary, fresh water, seawater is corrosive.* The consequence of this desperate, and probably necessary, act is that the three units that had experienced meltdown are now permanently ruined. The population was evacuated from a twenty-kilometer radius around the plant.

The release of radiation into the air is a small fraction, probably no more than 10%, of that from the Chernobyl accident. Radioactive materials also contaminated water and soil around the plant. The government banned the sale of food grown anywhere within fifty-kilometers of the plant. For a time, residents of Tokyo were warned not to use tap water to prepare food for children. Although some people died as a result of the destruction caused by the earthquake and tsunami, no one died as a result of radiation released during the accident. Several hundred people received large doses of radiation, and a small number of workers, probably fewer than ten, have exceeded the total radiation dose to which they can legally be exposed during a lifetime of work.

Nine months after the tsunami hit, the Japanese government declared the situation to be "stable." No doubt it will take many years, even many decades, to rebuild the area and make sure that it is fully decontaminated of radiation (Figure 25.10). Whether long-term health effects will emerge remains to be seen.

RADIOACTIVE WASTE

One has a queasy feeling about something that has to stay underground and be pretty well sealed off for 25,000 years before it is harmless.

—David[5]

---

* This phenomenon is familiar to anyone who has experienced the rapid rusting of tools or sporting equipment in a saltwater environment. Corrosion of metals in the presence of water or aqueous solutions is an electrochemical effect. Salt dissolved in water makes the water a much better electrical conductor than is pure or fresh water. A drop of saltwater on a metal surface contains a lower amount of oxygen than does the nearby atmosphere. The differential concentration of oxygen, abetted by the good electrical conductivity of saltwater, essentially sets up a tiny battery that facilitates the corrosion of many metals.

FIGURE 25.10 An example of the damage and destruction following the disaster at Fukushima. (From http://en.wikipedia.org/wiki/Fukushima_Daiichi_nuclear_disaster.)

Fission products fall into three general categories. The first category is fission products that themselves are radioactive. Most of these unstable products emit β- or γ-radiation; collectively they account for about three-fourths of the radioactivity of the spent fuel. Fortunately, the half-lives of most of these products are relatively short, usually no more than a few years. A second category is the so-called heavy isotopes, meaning elements of atomic numbers above that of uranium (i.e., >92). These products belong to the family of elements called *actinides*. They are produced from nuclear processes, but not fission reactions. Most of these heavy isotopes emit α- or γ-radiation. Their half-lives tend to be longer than those of the fission products, ranging from some tens of years to hundreds of thousands of years. They account for the remainder of the radioactivity (roughly one-fourth) in the spent fuel rods. These two categories of products remain inside the fuel rods. Finally, as the reactor operates, there is an intense flux of neutrons in the core. Not all of these neutrons interact with the fuel. Neutrons also bombard the materials in the fuel rod cladding and the internal structures in the pressure vessel. Those nuclear reactions also produce new isotopes, some of which are radioactive.

During the course of operation, the amount of fissionable $_{92}U^{235}$ in a fuel rod gradually decreases, because it is being consumed by the fission reactions to liberate ENERGY. In BWRs and PWRs used in commercial power plants, fuel rods have a life expectancy of three to four years. Eventually the amount of fissionable material remaining will not be sufficient to sustain a chain reaction. When reactor fuel is depleted to this point, it must be removed and replaced. Routine plant maintenance includes shutting down the reactor, removing old fuel assemblies, and replacing them with fresh fuel. Each year, one-third to one-fourth of the fuel rods are removed and replaced.

Some fissionable $_{92}U^{235}$ still remains in the rods, and it could be recovered and recycled to the reactor. Many of the other products in the categories mentioned above cause significant concern. One of the actinides, plutonium ($_{94}Pu^{239}$), is fissionable and can be used to make atomic weapons.* The Windscale reactor was originally intended to make plutonium for the British nuclear weapons program. Various other radioactive isotopes are useful in technology or in medicine, an example being the radioactive cobalt that is used in radiation therapy for cancer. And, some radioactive material is produced for which there is no use.

The spent fuel rod is intensely radioactive. When removed from a reactor core, the spent rod emits millions of rems of radiation. Normally these rods will be stored at the reactor site until the most intense radiation has decayed. Nuclear plants generally store spent fuel rods on-site in lead-lined concrete pools of water. These pools keep the rods relatively cool and contain the γ-radiation. An average commercial nuclear electric plant would put about sixty spent fuel assemblies into temporary storage each year. Until all the issues surrounding permanent disposal of radioactive waste are resolved, the wastes are being temporarily stored until some decision is eventually reached on permanent disposal.

Nuclear waste is categorized as being high-level or low-level material. Both kinds must eventually be dealt with. High-level waste consists of spent fuel rods from both commercial power plants and military weapons production. High-level waste derives its name from the fact that it can emit large amounts of radiation for hundreds of thousands of years. Low-level wastes come from a much wider range of sources. This waste includes the radioactive materials produced in a plant other than the spent fuel rods themselves; as an example, the radioactive water is classified as a low-level waste. Low-level wastes also include radioactive isotopes used in a range of medical, industrial, and scientific research processes. Mill tailings (discussed below), a waste from the processing of uranium ore, are also classified as low-level waste. Wastes in this category usually release smaller amounts of radiation for a shorter amount of time than do the high-level wastes. However, the term "low-level" does not mean that these wastes are not potentially dangerous. Though the radioactivity is less than that of high-level waste, low-level waste can still be radioactive—and dangerous—for tens of thousands of years.

Slightly more than half of low-level wastes come from nuclear reactors. This low-level waste can be further divided into fuel and nonfuel categories. The fuel wastes are products of fission reactions that leak out of fuel rods and into the surrounding cooling water. Nonfuel wastes are produced as a result of interactions of neutrons with anything in the core other than the fuel itself. The rest of the material comprising low-level wastes comes from medical, industrial, or research facilities. These wastes are often kept on-site for the short time (usually a few days or weeks) needed for them to decay to safe levels.

---

* Like $_{92}U^{235}$, the isotope $_{94}Pu^{239}$ is fissionable. Therefore, it can be used as an ingredient in atomic weapons. It could, of course, also be used as the fuel in a nuclear fission reactor, but nowadays no serious consideration seems to be given to that use, due to the potential of the plutonium falling into the hands of criminal or terrorist organizations. Plutonium was the explosive in the atomic bomb dropped on Nagasaki, Japan, in August, 1945. Plutonium is so dangerous to human health that the estimated amount that can be tolerated in the body without harm is about 0.1 mg.

Mill tailings are the material that is left over when uranium ore is processed. Only about 1% of uranium ore is actually uranium. After processing, the remaining 99% is left on-site as a residue, the so-called tailings. These tailings are left outdoors in piles. The action of wind and rainwater blows radioactive thorium, radium, and radon into the surrounding air and leaches the thorium and radium into water. Twenty years ago, there were over 125 million tonnes of mill tailings that had accumulated in the United States alone. More is added each year. When measured by volume, mill tailings constitute the largest source of any form of radioactive waste. Mill tailings were sometimes used as foundation and building materials until the time came when someone realized that this material is radioactive, and therefore that exposure to it brings a risk.

In the early years of the nuclear industry, it was believed that spent fuel could be reprocessed to re-enrich the uranium content and extract plutonium. The highly radioactive spent fuel presents a serious disposal problem. The process of spent fuel reprocessing extracts the plutonium, the fissionable U^{235}, and useful isotopes (such as radioactive cobalt). Several serious concerns affect the use of nuclear fuel reprocessing. An accident during transportation of spent fuel to a reprocessing plant could release radioactivity to the environment. Criminals or terrorists could steal plutonium or U^{235}. These stolen materials could be used to fabricate crude nuclear weapons. Consequently, reprocessing requires extreme security measures to prevent the theft of weapons-grade plutonium. Some residual, unusable radioactive waste still requires disposal. Nevertheless, the remaining waste would be both smaller in volume and dangerously radioactive for a much shorter time period—10,000 rather than 240,000 years—relative to the untreated spent fuel rods.

In the early days of the nuclear industry, the difficulty of waste disposal was not considered a deterrent to commercial electricity generation. It was assumed that somehow, someone would come up with a good idea to allow waste to be reprocessed or buried. From the beginning of the nuclear era in 1940s, and continuing through the 1960s, barrels of radioactive waste were frequently dumped into the oceans.

The practice was halted in the United States by 1970; by 1980 at least one-fourth of these barrels were leaking, and therefore slowly discharging radioactive material into the oceans. The disposal of waste on land is extremely controversial, and many concerns have been raised about this idea. Transportation of waste from the reprocessing plant to any burial site is an issue, because of the potential for accidents that would release radioactivity, or theft of radioactive materials. Unfortunately, the deceptively simple problem of finding safe ways to store high-level radioactive wastes has proved more difficult than anticipated. In 1983, the US National Academy of Sciences estimated that it will take three million years for radioactive waste to decay to background levels.

We nuclear people have made a Faustian bargain with society. On the one hand, we offer... an inexhaustible source of energy.... But the price we demand of society for this magical energy source is both a vigilance and a longevity of social institutions that we are quite unaccustomed to.

—**Weinberg**[6]

Currently the basic concept for handling the radioactive waste involves burial deeply inside the earth, in geological formations that should be stable for at least 10,000 years. One technique for packaging high-level wastes involves melting them with glass and pouring the molten material into impermeable containers. For safe, secure, long-term storage of high-level waste these containers must be waterproof and leak proof. The kind of packaging used has to be tailored to the volume of the waste, the isotopes it contains, how radioactive it is, the half-lives of the isotopes it contains, and how much heat it generates. These containers then must be put into a repository that is geologically stable; an earthquake could split open the cavern in which the waste is stored and release radioactivity to the environment. Groundwater naturally seeping through the earth could dissolve some radioactive substances, if the containers split open or if the groundwater eventually corrodes its way into the interior from the outside. Dissolved radioactive substances could potentially get into public water supplies. In that case, the radioactive species could then be absorbed by vegetation or ingested by marine and animal life. Eventually humans could ingest these radioactive materials with drinking water and food. What horrors might have ensued if the 2011 Pacific earthquake and tsunami had hit a nuclear waste repository?

In the United States, the 1982 Nuclear Waste Policy Act established a plan for the first permanent high-level commercial nuclear waste storage repository. Five years later, a site was chosen at Yucca Mountain, Nevada, about 160 kilometers northwest of Las Vegas. It is not yet in use, as a result of numerous political, regulatory, and legal hassles. While many people involved with the project believe the site to be dry, so that groundwater poses no risk, others disagree. Nonetheless, funding for this site was terminated by the US Congress in 2011, primarily on political arguments. The waste disposal issue remains controversial, due in part to the NIMBY syndrome. (NIMBY stands for *not in my back yard!*). Because of the political, social, and technical problems associated with radioactive waste disposal, no one wants it stored near them.*

At the end of its useful working life, a nuclear plant must be decommissioned, or taken out of service. The plant contains some radioactive material in the fuel rods and spent rods. Some radioactive material has likely contaminated the pressure vessel, piping, pumps, or other components because it has leaked out of fuel rods into the reactor during operation. In addition, the high flux of neutrons inside the reactor during the regular operation may have made some parts of the pressure vessel or other internal components radioactive via nuclear reactions that the neutrons induced. As a result, the decommissioning of a nuclear reactor, along with the allied process of decontamination, can be a difficult, expensive, and possibly hazardous operation.

---

* As one example of the knotty problems associated with long-term storage of nuclear waste, consider the issue of how to create warning signs that people will still be able to read several thousand years from now. Think of the problem this way: Readers will find some unfamiliar words or passages in the novels of Dickens, written about 150 years ago. Many words or phrases in Shakespeare's plays, about 400 years old, sound strange to us now. Chaucer's works, written some 700 years ago, are extremely difficult to read in his original English. *Beowulf*, about 1200 years old, is unreadable in the original by anyone other than specialists knowledgeable in medieval English. How much will our early twenty-first-century version of English have evolved by the year 3200?

NUCLEAR WEAPONS PROLIFERATION

Radium could become very dangerous in criminal hands.

—P. Curie[7]

The question of nuclear weapons proliferation centers around the production of $_{94}Pu^{239}$ and the handling of enriched uranium. Both of these isotopes are fissionable. Therefore, they are, in principle, capable of being used in atomic weapons. *Any* nation that has a nuclear energy program based on uranium-fueled fission reactors inevitably produces these isotopes. So far, outside of the former Communist nations, four countries have nuclear fuel reprocessing capabilities—France, Britain, India, and Japan. (Although the United States has the technology, it does not presently reprocess nuclear fuel.) An argument raised by other countries is that their energy and economic development is restricted by the near-monopoly on nuclear fuel reprocessing held by these four nations. This is perceived to be unfair, especially by developing countries that seek to enhance their gross domestic product and overall standard of living by relying on energy from nuclear reactors. The counterargument is that the more reprocessing plants there are—especially if they are built in countries that are politically unstable—the greater is the risk of despotic governments or terrorists obtaining weapons-grade material.

Further, *any* nation that is in the business of producing enriched uranium has no particular requirement to stop the enrichment process at reactor-grade material. Why not go all the way to weapons-grade stuff and build a bomb?

The newspapers have published numerous diagrams, not very helpful to the average man, of protons and neutrons doing their stuff. But curiously little has been said, at any rate in print, about the question that is of most urgent interest to all of us, namely, "How difficult are these things to manufacture?"

—Orwell[8]

This issue currently affects relations with Iran. The leader of this oil-rich nation argues that a peaceful, civilian nuclear energy program is vitally needed for domestic ENERGY production and to raise standards of living. He also argues that Israel must be obliterated. Other countries now seem to be having a hard time developing a strategy to come to grips with this situation.

The concern about terrorists relates in part to the fact that plutonium is, on a weight basis, perhaps the most poisonous and deadly substance known. Plutonium is sometimes called "the element from hell."

Plutonium will be the fuel of the future [and its value] may someday make it a logical contender to replace gold as the standard of our monetary system.

—Seaborg[9]

A nuclear bomb operates by having a charge of conventional explosives blast two portions of the fissionable material together to create a supercritical mass. In a crude device made by terrorists, even if the plutonium does not become supercritical and set off a nuclear explosion, the deadly, poisonous plutonium would be blown over a very large area by the conventional explosive, causing health effects from the radiation and making buildings and land uninhabitable. This concept is the basis of *dirty bombs*, which may not cause a nuclear explosion, but would create massive death toll, long-term health effects, and destruction by scattering radioactive material over an area. Further, it takes only a small amount of plutonium to make an atomic bomb, much less than the amount of U^{235} needed for a bomb based on uranium. An effective nuclear bomb might be the size of a suitcase, making it easy to transport and smuggle into a country.

Perhaps it is this concern, in a politically unstable world that never seems truly to settle down, that may ultimately have a greater effect on the nuclear industry than issues of reactor safety and waste disposal.

IS THERE A FUTURE FOR NUCLEAR ENERGY?

Despite some notorious accidents, it does appear that nuclear power plants—at least PWRs with containment structures—can be operated with good safety records. The TMI accident, in which no one was killed or seriously injured, effectively halted the development of the nuclear power industry in the United States. The Chernobyl accident, with dreadful human suffering and contamination of prime agricultural land, nearly killed any likelihood of a comeback of nuclear power in the United States and other countries. Just when the comatose body was showing a few signs of life, the Fukushima disaster may have driven a stake through its heart. These accidents have certainly fueled mounting public resistance to nuclear energy. Will nuclear energy ever come back? Of course, no one knows for sure. The one factor that is likely to lead to the resurrection of the nuclear power industry is the greenhouse effect (Chapter 29), global climate change that may be due, in part, to emissions of carbon dioxide from the combustion of fossil fuels.

Nuclear energy, for all its faults—perceived and real—has the undeniable advantage of producing no carbon dioxide. Therefore, it represents a real option if there is concern for significant cuts in carbon dioxide emissions while maintaining a high level of energy production and consumption. For that reason, new reactor designs are being developed that are "inherently safe." Some designs are also based on using thorium as the fuel, to avoid making and handling weapons-grade uranium and plutonium.

WESTINGHOUSE AP1000 REACTOR

The Westinghouse AP1000 is a PWR designed in part to address issues of mechanical failure and human error that contributed to some of the previous nuclear accidents. In the event of a failure, the reactor is designed to achieve a safe shutdown even without intervention of the operators, and without need for AC electricity coming from outside the plant or generated on site. Should something go wrong, cooling of the pressure vessel would be achieved by air circulation and water cooling. The water would drain by gravity onto the pressure vessel, eliminating the need for pumps. The plant is designed for a 72-hour supply of water for cooling the pressure vessel; more water could be added if required. The reactor building is designed to withstand a direct hit by an airplane.

The AP1000 system has been designed to have fewer valves and pumps, and less piping, than earlier PWR designs. Fewer components mean lower investment costs to build such a plant. This also translates into reduced costs for maintenance, testing, and inspection. Fewer components mean fewer things to go wrong. The overall plant layout, including turbine, generator, and other buildings, requires less area than other reactor designs. Spent fuel rods would be stored on-site in water.

At present, twelve AP1000 reactors are said to be under construction in China, to be operable by the middle of this decade. China apparently has very ambitious plans to deploy a large fleet of AP1000 units by the 2020s. Seven sites in the United States have applied for construction and operating licenses for AP1000s, with two reactors at each site. Other possible deployments include India and the Czech Republic.

Pebble Bed Reactor

The pebble bed reactor (or PBR, Figure 25.11) represents a significant design change from the more conventional BWR and PWR designs discussed previously. The PBR is a graphite-moderated, gas-cooled reactor. The preferred coolant is helium, though nitrogen and carbon dioxide have also been suggested. Helium offers the advantage of being a poor neutron absorber, so has little likelihood of becoming radioactive. The hot coolant gas could be sent directly to a turbine, just as hot gases are in a jet engine; alternatively, the hot gas could be sent to a heat exchanger to generate steam for a steam turbine, as in a PWR.

Unlike other reactor designs, the PBR, as the name implies, uses fuel in the form of "pebbles," though these pebbles are actually about 60 mm in diameter, which is roughly the size of a tennis ball. The pebble contains tiny particles of the fuel, for example, enriched uranium (plutonium or thorium are also candidate fuels for a PBR) coated with silicon carbide (Figure 25.12). This material is hard, strong, and resists high temperatures. The carbide-coated fuel particles are then dispersed in graphite to make the pebble. Every pebble contains fuel, a barrier to prevent escape of fission products (the silicon carbide), and the moderator (graphite). Criticality is achieved by accumulating enough pebbles—more than 300,000—in the core. The pebbles recycle through the core some ten to fifteen times. Several hundred spent pebbles would be withdrawn every day, so that the PBR is continuously refueled, and does not require a shutdown to replace fuel assemblies.

The PBR core operates above the annealing temperature of graphite, avoiding potential problems of the kind that triggered the Windscale accident. Also, U^{238} becomes a better neutron absorber the hotter it gets. In the event of some accident or malfunction that would raise the core temperature, more and more neutrons would be absorbed by the nonfissionable U^{238}, causing reactor output to drop. Rather than experiencing a core meltdown, the PBR would shut itself down automatically as it got hotter, reaching a "hot idle" point. Since the core has no water present, hydrogen cannot form, eliminating the potential of a hydrogen explosion.

Unlike spent fuel rods from more conventional BWR or PWR designs, the spent fuel pebbles are less hazardous and comparatively easier to handle, though the volume of spent pebbles would be much greater than that of spent fuel rods from a

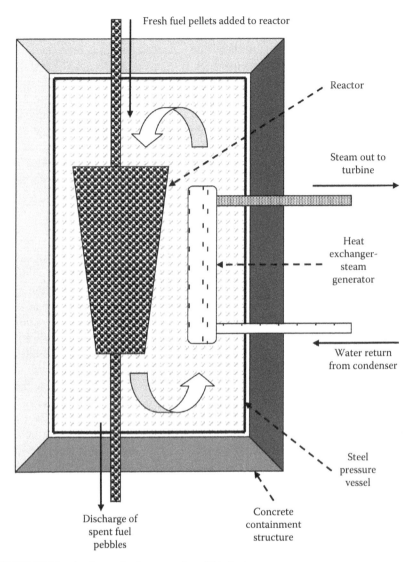

FIGURE 25.11   A schematic diagram of a pebble bed reactor.

reactor of similar output. The fission products are permanently encased in the pebbles, which should make permanent disposal easier and less contentious, at least in principle.

Currently a PBR is under development in China for electricity generation. South Africa had an active PBR program up to about 2010. The national electric utility, Eskom, postponed plans to build a nuclear electric plant in 2009, apparent due to public opposition. The South African government then canceled further funding for the PBR in 2010. A small PBR that could be used for ship propulsion is under development in the Netherlands, though apparently none has actually been built.

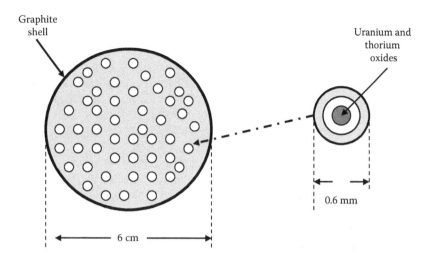

FIGURE 25.12   A schematic diagram of a fuel "pebble" for the pebble bed reactor.

THORIUM REACTORS

Naturally occurring thorium occurs almost entirely as a single isotope, $_{90}Th^{232}$, with only small traces of other isotopes. $_{90}Th^{232}$ is radioactive, being an α-emitter with a half-life of 14 billion years. This isotope cannot undergo fission, but is "fertile," meaning that it can be converted readily into fissionable material. In particular, $_{90}Th^{232}$ absorbs a neutron, emits two β-particles, and becomes the fissionable material $_{92}U^{233}$. Thorium is approximately four times as abundant as uranium and costs about one-third the price of uranium. In principle, thorium could be an attractive fuel for nuclear reactors. This has been recognized since the 1940s, but little has been done about it, compared to the massive efforts in developing uranium reactors.

Thorium-fueled reactors offer numerous potential advantages relative to uranium reactors. The entire fuel enrichment issue goes away, because the desired fertile $_{90}Th^{232}$ isotope is the naturally occurring one. Thorium cannot sustain a chain reaction itself, so in the event of a problem with the reactor, an automatic, natural shutdown will occur, without operator intervention. Thorium absorbs neutrons very well, so it acts as its own neutron shield. While it cannot be claimed that there is no chance of nuclear weapons proliferation with a thorium reactor, the possibility is much less than for a uranium reactor.

The liquid fluoride thorium reactor (LFTR, pronounced "lifter") is one of numerous proposed designs for a thorium-fueled reactor. A mixture of molten salts containing dissolved thorium fluoride provides both the fuel and the coolant. The hot molten salt mixture can be used in a heat exchanger to generate hot gas for operating a turbine; alternatively it can be used to heat a second, nonradioactive molten salt that generates steam in a second heat exchanger. No water in the core means that no hydrogen can be produced, eliminating the risk of hydrogen explosions. Gaseous fission products, such as the xenon that exacerbated the problems with the Chernobyl reactor, do not stay in the molten salt, so they do not accumulate in the core.

The reactor operates at low pressure, reducing the likelihood of a buildup of extreme pressures that would burst the reactor vessel. The reactor vessel has a "freeze plug" of solidified salt at its bottom. This plug must be kept cool to keep it solid, using, for example, an electric fan. If there is a failure of electricity supply to the plant, the freeze plug cannot be kept cool; it melts, draining the molten salt out of the reactor.

India appears to be the world leader in thorium reactor technology; perhaps not coincidentally, India also has about one-third of the world's thorium supply. Development and demonstration efforts are underway in several other countries, including China and Australia.

The AP1000, PBR, and LFTR represent new-generation reactor technology designed to reduce or eliminate many of the issues that affected the earlier reactors. Time will tell how a balance will be achieved among public skepticism, or even fear, of nuclear energy, and the increasing perception of the need for safe, large-scale, carbon dioxide-free electricity sources.

REFERENCES

1. Strauss, L. (speaking in 1954!). Quoted in Flatow, I. *Present at the Future*. Collins: New York, 2007, p. 116.
2. Benford, G. *The Wonderful Future That Never Was*. Hearst Books: New York, p. 159.
3. From the *New York Times* in 1956. Quoted in Schlesinger, H. *The Battery*. Smithsonian Books: Washington, D.C., 2010, p. 241.
4. David, E. (science adviser to President Nixon). Quoted in Schumacher, E.F. *Small Is Beautiful*. Harper and Row: New York, 1973.
5. Weinberg, A. Social institutions and nuclear energy. *Science* 1972, 175, 27–34.
6. Curie, P. (1904). Quoted in Blom, P. *The Vertigo Years*. Basic Books: New York, 2008, p. 78.
7. Orwell, G. (1945). Quoted in Silverstein, K. *The Radioactive Boy Scout*. Random House: New York, 2004, p. 185.
8. Seaborg, G. (1968). Quoted in Silverstein, K. *The Radioactive Boy Scout*. Random House: New York, 2004, p. 115.

FURTHER READINGS

Blackwell, A. *Visit Sunny Chernobyl*. Rodale: New York, 2012. Virtually unique in the genre of travel literature, this book describes visits to seven of the most dreadfully polluted places on our planet. Chapter 1 is devoted to the Chernobyl area as it is today.

Cassedy, E.S.; Grossman, P.Z. *Introduction to Energy*. Cambridge University: Cambridge, U.K., 1998. Chapter 8, on the nuclear fuel cycle, discusses issues relating to nuclear waste and nuclear weapons proliferation.

Flatow, I. *Present at the Future*. Collins: New York, 2007. This book is a collection of essays about various aspects of science and technology, for persons with little previous science background. Chapter 13 discusses nuclear energy, including plant safety, problems of radioactive waste storage, and the PBR.

Hersey, J. *Hiroshima*. Vintage: New York, 1989. Originally published in 1946, this book describes the events of August 6, 1945, the day the atomic bomb was dropped on Hiroshima. It has been in print almost continuously since then. Recent editions contain an additional chapter, on the author's visit to the city forty years later. The effects of acute exposure to radiation are discussed.

Josephson, P.R. *Red Atom*. Freeman: New York, 2000. This book provides a history of the nuclear energy and nuclear weapons programs in the former Soviet Union and present-day Russia. It blends discussions of nuclear physics, science policy, and political decisions, all of which culminated in the tragedy at Chernobyl.

Kalfus, K. *Pu-239*. Milkweed: Minneapolis, MN, 1999. The title story in this collection of short stories about modern Russia is a harrowing account of a worker from a Russian nuclear plant attempting to sell stolen plutonium on the black market.

Mahaffey, J. *Atomic Awakening*. Pegasus Books: New York, 2009. A historical account of the development of nuclear science and nuclear plants. Intended for persons with little or no prior background in science.

Martin, R. *Superfuel*. Palgrave Macmillan: New York, 2012. This book discusses development of thorium-fuelled reactors. The book provides an interesting blend of the history of efforts to develop thorium as a nuclear energy source, the science and technology of thorium-based energy, and issues relating to the business development of thorium reactors.

Silverstein, K. *The Radioactive Boy Scout*. Random House: New York, 2004. The remarkable story of how an ingenious high-school student managed to acquire plans and materials to build his own nuclear reactor.

Zoellner, T. *Uranium*. Viking Penguin: New York, 2009. A history of humankind's interaction with this element. The book spans the eras from when uranium was simply a nuisance to its use in nuclear reactors, to today's interest in nuclear energy as a to possible defense against global warming.

26 Energy and the Environment

All energy sources, regardless of what kind and where they come from, share three characteristics: Each one has certain technological advantages and technological disadvantages. Each has economic benefits and economic disincentives. Each impacts the environment in certain ways and may have potential benefits for the environment. Working toward a sensible energy policy for a particular region or nation requires selecting a mixture of energy sources that best reach a balance or compromise among the issues of technology, economics, and environmental impact. That problem would be vexatious enough, but is further complicated by the fact that no unique solution works in all parts of the world at all times. Some places enjoy excellent resources for hydroelectric development; in other places, hydro would be a "nonstarter." Some areas are ripe for the massive development of solar energy; elsewhere, solar might not be such an attractive option. These same things could be said about coal, nuclear, wind, or biomass, as examples.

This chapter considers some of the implications of energy use for the environment. Because fossil fuels continue to dominate the energy economy in many parts of the world, the focus of this chapter lies heavily, but not exclusively, on the environmental consequences of fossil-fuel utilization. Virtually all of our transportation energy, about 70% of our electricity, and most of our industrial and domestic heating derive either directly from burning of fossil fuels or indirectly via electricity produced in fossil-fuel power plants.

Broadly, the use of energy sources involves four main steps. Extraction, harvesting, or collecting gathers the energy source so that we can use it. Mining of coal or uranium ore provides an example. Beneficiation, upgrading, or refining encompasses the various steps in improving the quality of a fuel before it's used. Utilization involves obtaining ENERGY from the source, quite often by burning it, to produce useful WORK. Finally, any responsible country or society will take steps to prevent degradation of the environment, or work to rectify any environmental damage that had occurred.

EXTRACTION OF FUELS: MINES AND WELLS

Those who live in coal-producing regions of the world may have witnessed the tremendous environmental devastation caused by surface mining (also known as strip mining or open-cut mining). Surface mining involves removal of the topsoil and any plants that happen to be growing in it, and then removal of subsoil to expose the coal seam (Figure 26.1). Topsoil and subsoil are heaped into "spoil" piles that continue

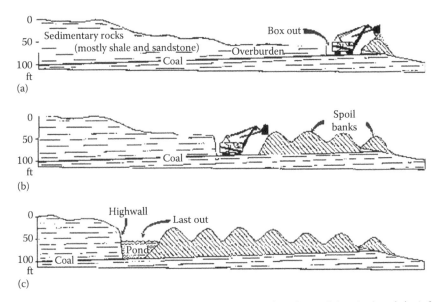

FIGURE 26.1 A schematic diagram of the progress of surface mining (strip mining) for coal. (a) Mining begins, (b) mining continues, and (c) mining ends. (Image courtesy of Connie Smith, Oklahoma Geological Survey, Norman, Oklahoma.)

to grow as mining progresses. Rain can wash some of the spoil away, polluting whatever bodies of water it runs into.

Inevitably a strip mine site is an eyesore during mining. The temporary disruption of the environment is something that must be tolerated while the mine is in active operation. The crucial issue, affecting permanent disruption of the environment, is what happens after the coal has been removed, or the mine ceases to operate. Countries, states, or provinces have passed *mined land reclamation* laws that require the mining company to restore the mining area to the condition in which it was before mining began.

Mined land reclamation laws have two major components, re-contouring and fertilizing and seeding (or replanting). The objective is to restore the land to approximately the same topography it had before mining, and to restore the plant community that existed before mining. Mined land reclamation laws sensibly applied do provide excellent restoration of the environment. A superb example is in the region near Cologne, Germany, which is home to one of the world's largest coal strip mines (Figure 26.2). A visitor could scarcely guess that some of the land is in fact reclaimed mines. Some have argued that the requirement of restoring the land to *exactly* the same condition it was prior to mining is too limiting. For example, if the mined land had no particular use prior to mining, wouldn't it be a benefit to society to restore the land not to the way it was, but, say, into productive agricultural land?

The real problems occur if there are no mined land reclamation laws or if the mining company somehow avoids compliance. (This can be done via the so-called grandfather clauses, so that the mine is exempt from reclamation if it was in operation prior to passage of the law, or by the simple expedient of the mining company's

(a)

(b)

FIGURE 26.2 An illustration of the impact of good, and well enforced, mined land recla-mation policies, this example from Germany. Part (a) shows the area with mining in progress and (b) shows the same area after reclamation. (Images courtesy of Michelle Nottebrock, RWE Power, Essen, Germany.)

declaring bankruptcy and vanishing.) Unreclaimed strip mines are serious problems for several reasons. The obvious one is that they are ugly eyesores. They continue to pollute the local environment by the washing away of the spoil banks. They are also dangerous. Some partially fill with water and are used as unsupervised swimming holes, occasionally resulting in accidental drownings. Some fools who erroneously believe that a four-wheel-drive vehicle can be driven literally anywhere enjoy the challenge of off-road driving in abandoned strip mines, only to discover that "4×4s" or sport utility vehicles can tip over or go out of control and result in injury or death.

Usually, underground coal mines do not disrupt so wide an area as surface mines. Today, underground mining is a remarkably safe occupation in many countries (for which China seems to be the exception), with an annual death rate lower than that for general manufacturing industries. Such was not always the case, and still is not the case in some parts of the world. The early days of coal mining are history written in blood. In the anthracite-mining region of Pennsylvania, the annual death toll of miners in the years at the turn of the twentieth century was about 3%—that is, three out of every hundred miners were killed every year. Even today there are potential long-term consequences for miners, most notably black lung disease, which is a result of chronic inhalation of coal dust.*

Underground mining has other hazards as well. Water percolating through mines can react with sulfur compounds in the coal to produce, through a sequence of reactions, what is essentially a solution of sulfuric acid. This so-called acid mine drainage can be so acidic as to kill all aquatic life forms in a stream. Some streams are so acidic that they have been used by nearby residents as a folk medicine to remove warts! Abandoned mines can cave in, causing subsidence to the ground above and seriously damaging buildings.

In contrast, the extraction of petroleum and conventional natural gas usually does not do so much harm to the environment, except for localized disturbances around the well sites. However, dreadful conditions can arise in those cases when companies are not committed—or required—to protect the environment. Consider this from the late days of the Russian Empire (Figure 26.3):

> *The oil fields left in my memory a brilliant picture of dark hell. I do not joke. The impression was stunning.... Amidst the chaos of derricks, long low workers' barracks, built of rust-colored and gray stones and resembling very much the dwellings of prehistoric peoples, pressed against the ground. I never saw so much dirt and trash around human habitation, so much broken glass in windows and such miserable poverty in the rooms, which looked like caves. Not a single light in the windows, and around not a piece of earth covered in grass, not a tree, not a shrub.*

> —Gorky[1]

Fortunately, such conditions would not be tolerated in most industrialized nations nowadays.

Technology for extracting oil from offshore wells, especially in deep water locations, improves steadily. A new Korean rig has the capability of drilling in water twelve kilometers deep. But, drilling for oil in offshore locations has created some

---

* Black lung disease, also called anthracosis, is a specific form of the general disease pneumoconiosis, caused by the inhalation of irritants; the severity depends on what kind of irritant is inhaled, and how much. Coal miners, particularly those working in underground mines, inhale fine particles of coal dust. The very small particles penetrate into the air sacs in the lung. There, the surrounding lung tissue is converted into scar tissue in the body's effort to encase these foreign, irritating particles. Over time, the victim's breathing becomes more difficult, and the ability of the lungs to exchange waste carbon dioxide for fresh oxygen is reduced. The victim also has a much higher likelihood of developing chronic bronchitis or emphysema. Those who have spent much of their working lives in underground coal mines seem particularly at risk. The last few, unpleasant years of the victim's life become a struggle to breathe.

FIGURE 26.3 The oil fields around Baku, Azerbaijan, circa 1891. (From http://pages. uoregon.edu/kimball/mfgR.htm#nrgCOAL.)

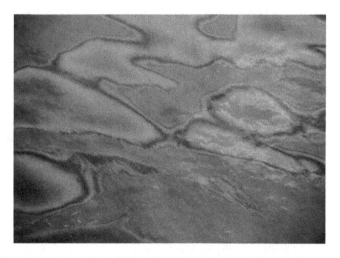

FIGURE 26.4 A portion of the oil slick from the Deepwater Horizon disaster in the Gulf of Mexico, 2010. (From http://publicdomainclip-art.blogspot.com/search?q=Deepwater+Hor izon+oil+slick.)

notable oil spills. The most recent was the BP Deepwater Horizon incident in 2010 (Figure 26.4), in the Gulf of Mexico. Eleven workers died in what became the worst oil spill of any sort in US history. An even worse disaster occurred in 1979 off the coast of Mexico, and took a year to stop.

Many of the spectacular environmental problems usually blamed on the oil industry are associated with transportation of crude oil to refineries. Empty tankers returning to their home port for another load often carry water as ballast. Some tankers, when switching their cargo from one petroleum product to another, used to flush their tanks with water between loads. In either case, oily water used to be discharged into the ocean before the ship entered port to be loaded. Fortunately, this practice has largely been discontinued. When an oil tanker or supertanker breaks up or starts leaking, its oil is spilled and can cause severe environmental damage. As much as a century can be required for the local area to recover from such an ecological disaster. The worst disaster of the past half-century was the break-up of the *Atlantic Empress* in 1979, off the coast of Tobago. This spilled 287,000 tonnes of oil. In contrast, the 1989 *Exxon Valdez*, running aground in Prince William Sound, Alaska, was considered to be a particularly heinous example, yet spilled "only" 37,000 tonnes.

Oil spills have numerous negative consequences. Ingested crude oil can kill or injure fish, other aquatic animals, and aquatic plants. They contaminate local beaches. Perhaps worse, oil carried long distances by ocean currents can contribute to pollution in areas far away from the original spill. Fortunately, some experience suggests that the damage, however unpleasant it may be for a short time in a local area, can be healed over the long term. Submarine warfare in the Second World War resulted in the sinking of many oil tankers, with the spillage of millions of tonnes of oil into the oceans around the globe. There seems to be no permanent environmental damage attributable to those dreadful times. (However, the typical tanker launched in the 1930s or 1940s carried substantially less oil per ship than today's supertankers.) Indeed, the oceans are far more polluted by plastic packaging wastes or discarded plastic items tossed from ships than from oil spills.

Oil spills sometimes bring with them attempts at clean-up that verge on the bizarre. In 1967, the supertanker *Torrey Canyon* broke up off the coast of England. In an attempt to prevent the oil spill from coming ashore, the British bombed it with napalm, hoping to set it afire and have it burn away. When that didn't work, chemicals were added to the oil to try to disperse it into tiny particles that would mix with the seawater and sink. Unfortunately, these chemicals were lethal to some of the aquatic organisms that had managed, up till then, to survive the oil spill itself. A different course was taken in response to the *Exxon Valdez* disaster. Very hot water was squirted onto the contaminated shore through high-velocity pumps (Figure 26.5). This approach, which cost about two billion dollars, again killed organisms that had survived the spill itself, and the high-velocity water blasted barnacles and limpets loose from their abodes.*

On land, oil is usually transported in pipelines. Concerns with pipelines include the fact that they will, necessarily, disrupt some bit of the environment as they are built through agricultural land, forests, or wilderness. Once the pipeline is in operation, the potential exists for oil spills from corrosion of the pipes, leaking joints or valves, or catastrophic accidents that rupture the pipe. The most controversial pipeline project has been the Alaskan pipeline, which transports oil from the North

---

* Barnacles are crustaceans, the class of aquatic animals that includes lobsters, crabs, and shrimps. Barnacles cement themselves to all sorts of things in the water—rocks, ships' hulls, driftwood, and even whales. Barnacles on a ship's hull can be a particular nuisance. The natural cement secreted by barnacles is so strong and resists water so well that it has been studied to learn how to make superior dental cements. Limpets are types of marine-living snails; most cling to rocks near the shore.

FIGURE 26.5 The clean-up from an oil spill is a hard and difficult job, and sometimes unavoidably causes its own disruption to the environment. (Reprinted from first edition, with permission of Photo Researchers, Inc., New York, New York.)

Slope of Alaska to a tanker loading facility in Valdez. It went into service in 1977; proponents argue that the environmental impact has been minor.

In recent years, enormous deposits of natural gas associated with shale have been discovered in several parts of the world. Unlike the reservoir rocks of so-called conventional natural gas, shale is not porous. Gas cannot diffuse through it easily. To make it easier to extract this gas, hydraulic fracturing ("fracking," Figure 26.6) is used to break up the shale. The procedure uses high-pressure water, containing various additives. The controversy about fracking pits the gas industry and, in some parts of the United States, state agencies against ordinary citizens. People are concerned about possible contamination of drinking water by the chemicals used in the frack water, or by components of the shale that are liberated to dissolve in underground water, or by the gas itself. Gas companies and regulatory agencies regularly contend that fracking has no serious environmental or health consequences. Arrayed on the other side are numerous anecdotal reports of drinking water developing unpleasant odors or taste, and occasional tales of ordinary tap water containing so much dissolved gas that it can be lighted with a match.

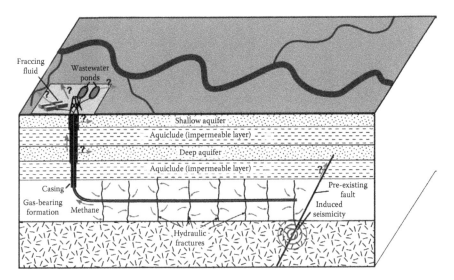

Fraccing fluid

Wastewater ponds

Shallow aquifer

Aquiclude (impermeable layer)

Deep aquifer

Aquiclude (impermeable layer)

Casing

Pre-existing fault

Gas-bearing formation

Methane

Induced seismicity

Hydraulic fractures

FIGURE 26.6 A schematic diagram of hydraulic fracturing ("fracking") of shale to help release the natural gas it contains. The question marks indicate the potential escape of components of fracking fluid into the environment. (From http://en.wikipedia.org/wiki/File:HydroFrac.png.)

REFINING AND BENEFICIATION

In most industrialized nations, oil refineries operate under very stringent environmental regulations. These regulations are intended to control potential environmental problems associated with leaks from process equipment or pipes, production of waste materials (such as sulfur compounds removed from sour crudes), and spills or leaks of some of the chemicals used in the refining process. Waste materials that can be produced in refineries include such noxious gases as ammonia and hydrogen sulfide, as well as poisonous materials such as cyanides. Usually these materials are produced in small quantities that are contaminants of other process streams. The contaminated streams are collected and purified; the contaminants are removed and destroyed.

Coal is not ordinarily "refined" in the same sense that oil is.* However, many coals are "cleaned" prior to use. All coals contain some amount of incombustible material that got incorporated in the coal as it formed in nature. As coals are mined, they may also become adulterated with some of the surrounding rock material that is unavoidably removed with the coal. These incombustible minerals and rocks transform to ash when the coal is burned. Ash has no calorific value, so its presence in the coal is useless in terms of the production of energy. Worse, the ash needs to be collected and disposed of.

---

* Technology is well known for converting (or "refining") coal into clean liquid transportation fuels or into a substitute natural gas. South Africa leads the world in the former area. However, in most instances these processes have never been economically competitive with petroleum. Despite vast amounts of research, development, and demonstration in many countries, the production of synthetic liquid or gaseous fuels from coal has never penetrated deeply into the commercial marketplace.

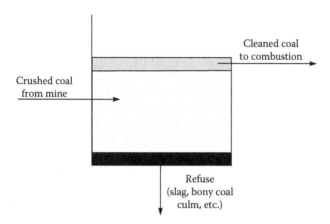

FIGURE 26.7 Processes for cleaning coal take advantages of differences in physical properties between the coal itself and the rock or mineral impurities. In this example, coal is separated in a liquid medium that is dense enough to allow the coal to float but the minerals to sink.

Coal cleaning, or coal beneficiation, reduces the amount of ash-forming material. Because much of the sulfur in many coals is associated with the mineral pyrite, coal cleaning also reduces the sulfur content. A very simple strategy for coal cleaning relies on the difference in density between the coal itself and the rocks or minerals. If crushed coal is placed into a liquid that has a density intermediate between that of the coal and that of the minerals, the coal will float and the minerals will sink (Figure 26.7). The separation is never perfect. The cleaned coal still contains some minerals, and the refuse contains some coal.

The refuse from coal cleaning has many names that often vary from one coal mining region to another: culm, slag, and bony coal are examples. For many years the refuse was dumped on the ground, creating ugly miniature mountains that supported very little in the way of plant life. In 1966 a slag heap outside the town of Aberfan, Wales, gave way and slid downhill. Tragically, the village school stood directly in the path of this avalanche of muck. One hundred sixteen children and 28 adults died that day. In recent years these culm banks or slag heaps have become regarded as potential energy sources, since about 30% of the material is actually coal. Combustion equipment is now coming into use to utilize this unusual form of fuel, generating useful energy and cleaning up the environment at the same time.*

---

* The principal technology is fluidized-bed combustion. Air is blown upward through a bed of limestone particles, at a velocity sufficient to suspend the particles in the air stream. This process is called fluidization, because the bed of suspended particles looks and acts much like a fluid. The behavior of the particles in the air stream is roughly similar to demonstrations of vacuum cleaners in stores, in which a table-tennis ball is suspended in mid-air by a stream of air. Coal, culm, slag, and bony coal (or most any other solid fuel) can be burned in a fluidized-bed combustor. The use of limestone as a bed material captures SO_x as it forms. Ash is retained in the bed. The temperature of combustion is much lower than that of a pc-fired boiler, substantially reducing thermal NO_x formation.

FUEL UTILIZATION

All fossil fuels and fuels from biomass contain carbon and hydrogen. Petroleum and coal contain some amounts of sulfur, nitrogen, and oxygen. Biofuels usually contain no sulfur, but may contain oxygen and nitrogen. Coal and wood contain some incombustible materials that produce ash. For our purposes, we need not worry about the specific chemical compounds present in the fuels. Instead, fuel combustion can be treated as if the fuel were simply a mixture of the individual chemical elements.

Depending on the combustion conditions, particularly the amount of oxygen available, the carbon in fuel can produce carbon dioxide, carbon monoxide, or soot (Chapter 7). Sulfur is capable of forming more than one oxide (Chapter 16), conveniently lumped together with the symbol SO_x. Similarly, nitrogen can produce more than one oxide; they are lumped together as NO_x. The situation with NO_x is further complicated by there being two sources of nitrogen in a combustion system, fuel NO_x and thermal NO_x (Chapter 16). Sometimes the generic symbol R is used to represent the metallic elements (such as silicon, aluminum, and iron) present in the various constituents of ash. Then the notation RO_x can be taken as a symbol for ash. It's then possible to refer to the potential pollutants as socks, knocks, and rocks.

Ultimately, *all* processes of fuel utilization involve combustion of the fuel to release chemical potential energy in the form of heat, as shown in the energy diagram of Figure 26.8. The heat can be used directly for warmth, cooking, or carrying out various manufacturing processes. It can be used in a heat engine, or to raise steam, which we then use in a different kind of heat engine. In all these cases, the objective is to capture some of the ENERGY of the fuel to do useful WORK.

Every combustion process is necessarily accompanied by formation of products of combustion, which may include CO_2, CO, soot or unburned hydrocarbons, H_2O, SO_x, NO_x, and ash (Figure 26.9). For many centuries, the combustion products were either ignored or considered simply to be an inevitable nuisance consequence of fuel utilization. (There are a few notable examples to the contrary, such as complaints about the pollution from coal fires in England that date at least from the fourteenth century. King Edward I issued a proclamation in 1306 banning the burning of coal under certain circumstances.)

No process that produces, consumes, or transforms ENERGY from one form to another does so with 100% efficiency. Some amount of ENERGY is wasted, and that wasted ENERGY is dumped into the environment, commonly in the form of heat.

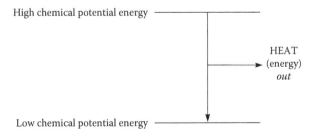

FIGURE 26.8 Fuel utilization takes advantage of the spontaneous change from high to low chemical potential energy to liberate useful ENERGY, usually in the form of heat.

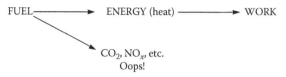

FIGURE 26.9   When fuel is burned to produce heat, by-products are produced as well. These include carbon dioxide and oxides of nitrogen and sulfur.

For example, in a coal-fired electricity-generating plant, only about one-third of the chemical potential energy in the coal is ultimately transformed to electrical energy leaving the plant to the consumers. The other two-thirds is wasted, much of it in the form of heat that leaves the condenser or the cooling tower.

Transferring waste heat directly to aquatic systems—for example, if the heated water leaving a condenser goes directly back into a nearby lake or river—can cause several serious problems. The ability of gases to dissolve into liquids decreases as the temperature of the liquid increases. (This is why, e.g., a warm carbonated beverage foams much more readily when it's opened than does a cold one; the carbon dioxide responsible for the carbonation is less soluble in the warm beverage and comes out of solution when the can or bottle is opened, generating the foam.) Warm water holds less dissolved oxygen than colder water. Oxygen dissolved in water is essential for reacting with, and helping to destroy or neutralize, any biological or chemical pollutants in the water. Water warmed by waste heat contains less dissolved oxygen and is less able to cleanse itself of other pollutants. Warm water can also sustain higher rates of aquatic plant growth. The warming of water by waste heat could lead to unwanted amounts of aquatic plants, such as algae. When these algae eventually die, their decomposition uses up even more dissolved oxygen. Many kinds of fish are sensitive to the surrounding water temperature; some prefer relatively cool water, while others are tolerant of warmer water. Heat stress on fishes can affect their resistance to disease and their reproductive cycles. Even cooling towers, intended to dissipate waste heat before it can enter a river or lake, have impacts on the local environment. Evaporation of water from cooling towers can increase the incidence of fog in the area. It can also change local patterns and frequency of rain or snow.

It was only in the last half of the twentieth century—indeed, in about the last quarter of that century—that many countries seriously considered environmental issues associated with combustion products, most notably acid rain (Chapter 27), vehicular smog (Chapter 28), and global climate change (Chapter 29). We now clearly recognize that the production of ENERGY brings with it the unavoidable production of substances that are harmful to the environment, and, because we are inescapably part of the environment, that are also harmful to us. But, we cannot lose sight of the fact that environmental problems associated with energy use are *not* inevitable, and, should they occur, *are* able to be corrected.

We wring our hands over the miscarriages of technology and take its benefactions for granted. We are dismayed by air pollution but not proportionately cheered up by, say, the virtual abolition of poliomyelitis. (Nearly 5,000 cases of poliomyelitis were recorded in England and Wales in 1957. In 1967 there were less than thirty.) There is a tendency, even a perverse willingness, to suppose that the despoliation sometimes produced by technology is an inevitable and irremediable process, a trampling down of nature by the big machine. Of course it is nothing of the kind. The deterioration of the environment produced by technology is a technological problem for which technology has found, is finding, and will continue to find solutions.

—Medawar[2]

ADDRESSING THE ENVIRONMENTAL CHALLENGE

Broadly, two strategies can be adopted. The first is to reduce substantially our ENERGY consumption. But since there appears to be some connection between ENERGY consumption and standard of living (Chapter 1), at least up to a point, this strategy does not seem likely to be willingly adopted. The second strategy is to develop technological "fixes" to reduce or remove potential pollutants before release to the environment. The operation of this second strategy is governed by one of life's most important rules:

There's no such thing as a free lunch.

In other words, any action whatsoever has some consequence for the environment, and those consequences have costs associated with them. The costs may be those for the installation and operation of control equipment to prevent pollutants from escaping into the environment; costs for the refining or beneficiation of fuel to prevent pollutants from forming in the first place; or the costs of clean-up and remediation of damage caused by pollutants in the environment. Sometimes those costs are fairly easy to calculate. For example, an expert in coal beneficiation can calculate the additional cost per tonne of coal resulting from reducing the amounts of sulfur and ash-forming minerals in the coal. Sometimes, though, costs are much more difficult to assess. Many health-related issues fall into this category, because it is often hard to attribute a health problem unequivocally to a single factor in the environment. There is, in a sense, a third strategy as well—simply to do nothing. For a long time in human history, this was the operant strategy, but it appears now, at least in many of the industrialized nations, that this third strategy is no longer socially or politically acceptable.

All energy technologies produce some amount of pollution. Regardless of what specific device or system is considered—for example, whether a nuclear electric plant, an automobile, or an oil refinery—it serves as a source of pollution. That pollution can be measured in terms of its emissions, as for example, a number of kilograms of SO_x emitted per megajoule of ENERGY generated. Those emissions can ultimately be quantified as a burden on the environment, which might be expressed as parts per million of SO_x in the local air. In turn, the burden has identifiable impacts. Increased concentrations of SO_x in air may lead to human health effects, corrosion of marble and limestone buildings, or crop damage. It should, in principle, be possible to put a cost on each of those impacts. From that information, one could assess the cost of environmental impacts of pollution from a particular source.

Unfortunately, we are a long way from being able to do that with much degree of certainty, except in the most egregious cases of emissions causing immediate, local damage. In working backward, from impacts to burdens to emissions to pollution, every link of this chain has uncertainties. First, it is not always clear that a particular impact can be the unequivocal result of one specific environmental burden. Suppose, for example, a person has respiratory tract problems and has been living for a time in a region of relatively high airborne SO_x concentration. Suppose further that that individual is also a heavy smoker. Can it be said with certainty that his or her problems are due to the SO_x? The situation is further complicated by the fact that some impacts might not be immediate; that is, it might be years before an organism, or a whole ecosystem, shows the effects of pollution. A second uncertainty lies in relating the burden to the emissions. Weather patterns may transport pollutants long distances from their sources. Pollutants first emitted to the environment, called primary pollutants, might undergo chemical reactions in the atmosphere, water, or soil to produce new materials, secondary pollutants. Formation of secondary pollutants in the environment complicates even further the job of tracing the path of pollution back to its origin. Issues like these make it very difficult both to draw a direct line from environmental impact to a specific pollution source and to quantify and place a value on the costs of that environmental impact.

Continuing the example of reducing SO_x emissions, three strategies could come into play: switching to low-sulfur coal (unfortunately, there is no such thing as a truly sulfur-free coal); removing or, more realistically, reducing the sulfur content of the fuel before combustion; or capturing the SO_x after the fuel has burned but before it can be released to the environment. All of these options have real financial costs associated with them. Companies that own deposits of low-sulfur coals are not fools; they know they have a premium commodity and can charge accordingly. Coal cleaning to reduce the sulfur content could add, say, three to four dollars per tonne to the cost of the coal. For a coal-fired generating plant that burns 10,000 tonnes of coal per day, coal cleaning would represent an extra cost of, say, 30,000 dollars per *day* to the utility, and, ultimately, to us in our electric bills. Scrubbers (Chapter 16) now do an excellent job of SO_x capture. However, they can represent a capital investment roughly equivalent to a third of the cost of the entire power plant. Once purchased and installed, scrubbers represent a continuous cost for lime or limestone, maintenance, and the workers who would operate or maintain them. These costs have to be recouped somehow, often through the medium of the inevitable monthly electric bill.

While no coals are sulfur free, other fuels either are completely sulfur-free or have sulfur content so low that SO_x emissions from their use would be negligible. Examples include natural gas, wood, and ethanol. All are more expensive than coal, particularly when on a dollars-per-MJ basis. Furthermore, since natural gas is, of course, a gas, and ethanol a liquid, significant modifications to the combustion equipment itself might be required to handle these fuels instead of coal. This retrofitting would also represent a cost to be considered.

Options for sulfur control also have "energy costs" associated with them. Running a coal cleaning plant requires ENERGY that must be purchased from some place. Most conveniently, this ENERGY would come from the generating plant itself. But, this decreases the net efficiency of converting the chemical potential energy in the coal to electrical potential energy leaving the plant to the consumers. Roughly, the energy cost of a scrubber system reduces the so-called coal pile to busbar efficiency, meaning the conversion of chemical potential energy in the fuel to electrical potential energy leaving the generator, from about 34% to about 30%.

The problem of dealing with NO_x is compounded by the fact that much of the NO_x produced is actually thermal NO_x, not fuel NO_x. In stationary sources, there are three approaches to NO_x control. Post-combustion NO_x capture devices represent an initial capital cost and a continuing cost for operation. A second approach, low NO_x burners, could also represent an incremental capital cost for a new boiler. The third, and perhaps most straightforward approach, is simply to reduce the combustion temperature. Unfortunately, this approach can seriously impact the efficiency of a process. NO_x emissions from mobile sources are primarily addressed by the catalytic converter on the vehicle (Chapter 28). This NO_x is almost entirely thermal NO_x. Nitrogen incorporated in the fuel is reduced during refining and, similarly to sulfur removal, by treatment with hydrogen. A catalytic converter adds several hundred dollars to the cost of a car.

Carbon monoxide and unburned hydrocarbons are generally not problems in stationary sources. Emissions of these substances from mobile sources are generally handled via the catalytic converter. Soot emissions from diesel-powered vehicles are addressed by reducing the concentration of aromatics in the fuel during refining. This, like sulfur and nitrogen removal, is achieved by reacting the fuel with hydrogen, at a cost of about several cents per liter. Because aromatics are undesirable for good cetane rating of diesel fuel (Chapter 22), the reduction of the aromatic content to avoid sooting gives, at the same time, the bonus of improved cetane number.

A direct connection exists between air pollution and energy utilization. Very significant strides have been made in the industrialized nations at reducing this pollution over the past forty to fifty years. Progress continues to be made in those countries. The problem for the future may be in the developing nations. These countries represent nearly two-thirds of the total population of the world. Many are in the stage of economic development in which there is a strong relationship between per capita energy consumption and gross domestic product (Chapter 1). For these nations, the key to increasing their standard of living is to increase energy use. In many such countries, the financial resources or the legislative mandates (or both) are lacking to install and use pollution control devices such as scrubbers. Likely, the world could be in for a continued period of increasing air pollution as these developing nations continue to industrialize, and continue to increase their energy use. The argument has even been raised that we might induce the developing nations to take on the burden of operating highly polluting processes so that the rest of us needn't be bothered.

I've always thought that under-populated countries in Africa are vastly underpolluted. Shouldn't the World Bank be encouraging more migration of the dirty industries [to such countries]? I think the economic logic behind dumping toxic waste in the lowest-wage countries is impeccable.

—Summers[3]

Some air pollution comes from natural, rather than human, sources. Gaseous sulfur compounds are emitted into the environment from volcanoes. Natural forest fires produce smoke and soot. Lightning flashes generate NO_x. Natural oil seeps contribute about 10% of the oil in the oceans. This is by no means a complete catalog of natural processes that result in materials classified as pollutants. Rather, these examples should remind us that the most comprehensive, indeed draconian, worldwide clean air act imaginable will never reduce air pollution to zero. Mother Nature wins every time.

How do we handle this? Carbon dioxide emissions are an inevitable consequence of the combustion of fossil fuels and biomass. *All* of the other emissions we have been discussing are not inevitable since they can, at least in principle, be addressed by pre-combustion treatment of the fuel, some modification of the combustion process, post-combustion capture of the product, or some combination of these. But not CO_2! Dealing with CO_2 emissions will likely prove the most difficult problem of all. A significant part of the problem is the immense amount of CO_2 to be dealt with. Burning a tonne of coal that is about 70% carbon will produce 2.5 tonnes of CO_2. A "typical" 10,000 tonnes of coal per day electric plant produces 25,000 tonnes of CO_2 per day.

Pre-combustion removal of carbon won't work. Depending on the specific fuel, fossil and biofuels contain 50%–90% carbon. In essence, more than half the fuel would be thrown away. Many ideas, some brilliant, some sound, some seriously weird, have been proposed for post-combustion CO_2 capture. The problem lies in the sheer volume of CO_2 produced around the world every day.

The most realistic short-term solution to the problem of CO_2 emissions would seem to be simply to produce less. There are two ways to do that: reduce the amount of useful energy being produced at the present levels of efficiency, or maintain about the same amount of energy consumption, but produce it at much greater efficiency. Since it is unlikely that many people would voluntarily select the first option, the key seems to lie in enhanced energy efficiency. Using the same amount of energy, but produced with greater efficiency, means burning less fuel. For example, doubling the fuel economy of a vehicle allows driving exactly the same amount of distance, but consuming only half the amount of gasoline (and hence producing half the amount of CO_2 as before).

There now exists an overwhelming consensus in the scientific community that CO_2 emissions and global climate change are directly linked and constitute a serious problem for humanity. This will likely be the major issue that moves us as a society away from our present heavy reliance on fossil fuels. At present, the only large-scale technology available for generation of electricity without CO_2 emissions is nuclear. Though nuclear energy unquestionably has its own set of problems (Chapter 25), it can produce electricity on a scale of tens or hundreds of gigawatts with no CO_2, no SO_x, no NO_x, and no ash emissions. Other energy sources, notably wind (Chapter 32) and solar (Chapter 33), can also make the claim of being emission free, but have yet to be developed to the scale of nuclear energy. Hydroelectricity (Chapter 15) also offers emission-free electricity generation, but may already be at a maximum scale of development in parts of the world.

The following several chapters examine in greater detail some of the environmental issues associated with energy use.

The environment is not an 'other' to us. It is not a collection of things that we encounter. Rather, it is part of our being. It is the locus of our existence and identity. We cannot and do not exist apart from it.

—**Lakoff and Johnson**[4]

REFERENCES

1. Gorky, M. Quoted in Knight, A. *Who Killed Kirov?* Hill and Wang: New York, 1999, p. 96.
2. Medawar, P. *The Strange Case of the Spotted Mice*. Oxford University Press: Oxford, U.K., 1996, p. 117.
3. Summers, L. Quoted in from Arnold, G. *Africa: A Modern History*. Atlantic Books: London, U.K., 2005, p. 776.
4. Lakoff, G.; Johnson, M. *Philosophy in the Flesh*. Basic Books: New York, 1999, p. 566.

FURTHER READINGS

Blackwell, A. *Visit Sunny Chernobyl*. Rodale: New York, 2012. A remarkable tour of seven of the most polluted places in the world. Chapter 2 discusses oil sand extraction in Alberta; Chapter 3, oil refining in Texas; and Chapter 6, coal mining in China.

Caudill, H. *Night Comes to the Cumberlands*; Llewellyn, R. *How Green Was My Valley*; Zola, É. *Germinal*. This trio of books depict, in different ways, the hard lives of coal miners. They are set in respectively, 1950s Appalachia in the U.S., 1920s and 1930s Wales, and 1860s France. They are almost always in print, usually as trade- or mass-market paperbacks from various publishing houses.

Crane, H.; Kinderman, E.; Ripudaman, M. *A Cubic Mile of Oil*. Oxford University Press: New York, 2010. The title comes from the fact that the total use of oil worldwide, per year, is equivalent to a volume of one cubic mile (4.2 cubic kilometers) of oil. The importance of energy efficiency and energy conservation is discussed.

Freese, B. *Coal: A Human History*. Perseus Publishing: Cambridge, MA, 2003. This book provides a useful discussion of the environmental history of coal use.

Gebelein, C.G. *Chemistry and Our World*. Brown: Dubuque, IA, 1997. Chapter 18 of this introductory chemistry text discusses air pollution associated with energy production.

Greene, M.F. *Last Man Out*. Harcourt: Orlando, FL, 2003. A remarkable account of the coal mine disaster at Springhill, Nova Scotia, in 1958. Nineteen miners were trapped for over a week in pitch black conditions, without food or water.

House, S.; Howard, J. *Something's Rising*. University Press of Kentucky: Lexington, KY, 2009. One of the most environmentally devastating techniques for coal mining is called mountain-topping. The top of a mountain is simply cut off to expose the coal below; the rock, soil, and trees are pushed down into the nearby valleys and streams. This book presents an oral history of the fight against mountain-topping in the Appalachian region of the United States.

Margolis, J. *A Brief History of Tomorrow*. Bloomsbury: New York, 2000. Chapter 3, 'Global Warning,' focuses on predictions of the future as they apply to the environment. The entire book is an interesting look at the discipline of 'futurology.'

McNeill, J.R. *Something New under the Sun*. Norton: New York, 2000. Subtitled 'An Environmental History of the Twentieth-Century World,' this book reports on the global-scale, and essentially uncontrolled, experiments that we are performing on ourselves in altering the environment. Though not all of the discussion relates to energy use, this is an important and thought-provoking book.

Quigley, J. *The Day the Earth Caved In*. Random House: New York, 2007. This book describes the remarkable fire in an underground coal seam that has been burning since 1962, and has basically depopulated the little town of Centralia, Pennsylvania. Not the least of the problems was a fairly consistent bungling and chicanery by government officials.

Thorsheim, P. *Inventing Pollution*. Ohio University Press: Athens, OH, 2006. Centuries ago, society's idea of pollution was that it represented the sometimes awful odors associated with the decay of plants and animals. During the nineteenth century, a new concept arose, linking pollution to coal smoke and emissions from coal combustion. This is the story of how that change of perspective came about.

27 Acid Rain

Combustion of *any* fuel converts its constituent chemical elements into their respective oxides. Some biofuels and petroleum products, and all coals, contain nitrogen and sulfur. These two elements form combustion products—SO_x and NO_x—of great concern because of their effects on the environment, specifically through formation of *acid rain*. Before discussing the origins and environmental effects of this problem, we will first digress briefly to the subject of what an acid is and how to quantify acidity.

THE pH SCALE

In aqueous solutions (i.e., solutions in water), acids are conveniently defined as substances that release hydrogen ions. Since an atom of hydrogen has only one electron, and its nucleus is a single proton, the H^+ ion is the same thing as a proton. Protons will not exist by themselves in solutions. Rather, when an acidic substance such as hydrogen chloride dissolves in water the H^+ ions that would be released associate with one or more water molecules, forming the H_3O^+ ion (called the hydronium ion). The resulting solution has the characteristics that are normally associated with an acid, such as a sour taste and the ability to corrode metals, because of the presence of H_3O^+ ions. Often reactions of aqueous acids are written as if the active species were H^+, but this is done for convenience or simplicity and the actual acidic species is H_3O^+.*

The acidity (or basicity) of any solution in water can be described in terms of the pH scale (Figure 27.1). Pure water is chemically neutral, with a pH of 7.00. pH values less than 7 represent acidic solutions; the smaller the number, the greater the acidity and the higher the concentration of H^+ in solution. pH values greater than 7 represent basic (alkaline) solutions; the larger the number, the greater the basicity. The pH scale is logarithmic; pH is the negative of the logarithm of the H^+ ion concentration in units of moles per liter. Because of the logarithmic nature of pH, every unit in the pH scale represents a change by a factor of ten from the previous unit. For example, a solution of pH 2 is not two times as acidic as one of pH 4 (i.e., a change of $4 - 2 = 2$), it is a hundred times (i.e., 10^2) more acidic.

Substances encountered in everyday life have a wide range of pH values. Bases include household ammonia, 11; baking soda, 9; and blood, 7.5 (nearly, but not quite, neutral). Acids include milk, 6.5; black coffee, 5; tomatoes, 4; lemon juice, 2; stomach acid, 1.5; and the acid in automobile batteries, 0. These relationships are illustrated

---

* It's ironic that water, a simple three-atom molecule and one of the most abundant substances on the planet, is also an extraordinarily complex material. The H_3O^+ ion, which is the H^+ ion hydrated by a water molecule, itself can associate with more molecules of water. Chemists have characterized such ions as $H_5O_2^+$, $H_7O_3^+$, and $H_9O_4^+$. That's why it's convenient simply to write H^+ when considering the reactions of acids in solution, even though the H^+ ion never exists by itself in solids or liquids.

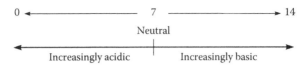

FIGURE 27.1 The key relationships in the pH scale.

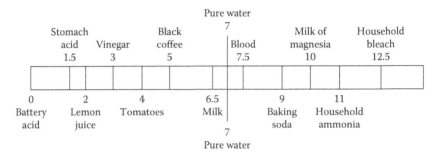

FIGURE 27.2 The pH values of a variety of common substances.

in Figure 27.2. Because the pH scale is logarithmic, our stomach fluid is not roughly six times more acidic (7.5 − 1.3 = 6.2) than blood, it's 1.5 million ($10^{6.2}$) times more acidic! The tastes of some foods come in part from the acids that occur naturally in them, or that are added to produce them. Lemons (2.2) contain citric acid, vinegar is a solution of acetic acid (2.5), and apples contain malic acid (3.0). Carbonated water, such as club soda, has a pH of about 4.8 because it is made by dissolving carbon dioxide (forming carbonic acid in solution) into water in the carbonation process. Many colas have pH values close to 3, because they contain phosphoric acid in addition to the carbonic acid from the carbonation process.

"NATURAL" ACID RAIN

Rain falling in a pristine, unpolluted environment (assuming such a place still exists somewhere) might be expected to have a pH of 7, since pure water is chemically neutral. However, carbon dioxide, a naturally occurring constituent of the atmosphere, is slightly soluble in water. As rain falls through the air, some carbon dioxide dissolves to form a weakly acidic solution of carbonic acid. This mildly acidic solution of carbon dioxide in rainwater has a pH of 5.6. In other words, even rain falling in a completely nonpolluted environment will have a pH of 5.6 and be mildly acidic. Only when the pH of rain is below this value can we suspect the presence of pollutants.

Acid rain is rainfall having pH less than 5.6.

Small amounts of other natural acids, such as formic and acetic acids, are almost always present in rain and contribute slightly to its acidity. Though it is common to speak of acid "rain," in principle, acidic pollution can be deposited not just as rain but as snow, hail, sleet, fog, mist, or dew as well. Sometimes acids can become attached to dry particles that simply drop out of the sky without any form of water present. A more encompassing term, therefore, is acid precipitation, or acid deposition.

Even in the absence of pollution sources, precipitation can temporarily become even more acidic (i.e., of pH less than 5.6) because of natural events such as volcanic eruptions, which send large amounts of sulfur oxides into the atmosphere, and forest fires and lightning flashes, the high temperatures of which promote formation of thermal NO_x. Decaying organic matter also contributes nitrogen and sulfur compounds to the atmosphere. These natural phenomena all contribute to acidic precipitation. Therefore, an acidic rainfall or other precipitation is not necessarily evidence of pollution caused by human activities.

The worst examples of acid precipitation can be very acidic indeed. In 1982, fog in Pasadena, California, had a pH of 2.5. Being outdoors in this fog would be roughly equivalent to breathing a fine mist of vinegar, roughly a *thousand times* more acidic than normal, unpolluted rain (pH 5.6) and ten times more acidic than would kill all fish in most lakes. In a small city on the Tibetan Plateau of China, which one might naively assume to be as far away from human influence as one can get, rain of pH 2.25 has been observed—like strolling in a shower of lemon juice. Fogs at Corona del Mar, south of Los Angeles, have been observed that are ten times more acidic than even the one at Pasadena, with a pH of 1.5.

SO_x AND NO_x AS CAUSES OF ACID RAIN

In 1852, a British chemist, Angus Smith,* noticed the problems of the rusting of metals and fading of dyed fabrics in the heavily industrialized city of Manchester. He recognized sulfuric acid in the air as the likely source of these problems. About twenty years after his original insight, he published a book, *Air and Rain: The Beginnings of a Chemical Climatology*, which qualifies Smith to be considered the "father" of acid rain chemistry. Unfortunately, the book rather promptly vanished into obscurity, taking along with it the concept of acid rain.

> ... *truth prematurely uttered is of scarcely greater value than error.*

—Kennan[1]

Then, in the 1950s, the effects of acid rain were rediscovered. Scientists working in the northeastern United States, in Scandinavia, and in the Lake District of England noticed the acidification of lakes and its resulting environmental effects. Their investigations began a new era in the studies of acid rain.

---

* Angus Smith (1817–1884), born in Glasgow, spent most of his career in Manchester. For the last twenty years of his life, he headed up the Alkali Inspectorate, the original role of which was to regulate the emissions of hydrogen chloride (which dissolves in water to make hydrochloric acid) from a factory complex producing sodium carbonate or soda ash.

FIGURE 27.3   Professor Eville Gorham of the University of Minnesota, sometimes known as the "grandfather of acid rain." (From http://www.cedarcreek.umn.edu/about/people/person.php?input=Gorham,%20Evile.)

During the 1950s, the Canadian scientist Eville Gorham (Figure 27.3)* recognized that acid rain could overwhelm the neutralizing capacity of lakes, soils, and bedrock. Many natural soils or rocks contain basic compounds—for example, limestone, which is calcium carbonate, $CaCO_3$. Basic substances in rocks and soils, or dissolved in natural water, react with acids—as in acid rain—to neutralize them. However, continual exposure to acids over a long period of time could eventually exhaust this natural neutralizing ability. Once that point had been reached, then the effect of further acidic precipitation would be to make soils and water increasingly acidic. Gorham discovered a number of acidified lakes in England's Lake District. He suggested that the acidification of these small mountain lakes (called tarns) was caused by air pollution.

A decade later, Svante Odén,† a Swedish scientist, suggested that the increase in acidity in the lakes of his home country was a result of pollution blowing into Sweden

---

* Born in Canada and receiving his PhD from the University of London, Eville Gorham (b. 1926) spent most of his career at the University of Minnesota. One of his first projects in environmental science showed that radioactive debris had been deposited on vegetation near the Windscale plant (Chapter 25). Gorham's studies on the chemistry of acid rain and other environmental problems have been recognized with numerous prestigious awards.

† Svante Odén (1924–1986) spent his career at the Swedish University of Agricultural Sciences in Uppsala. He did much to establish the link between air pollution in Germany and Britain with the acidification of natural waters in Sweden. In turn, he linked the acidification to destruction of fish populations, reduction of crop yields, and human health issues. His work put Sweden in the forefront of acid rain research. Later, Odén became interested in using tiny changes in wave patterns to spot incursions of foreign submarines (likely Soviet) into the Baltic Sea. In December 1986, Odén put to sea in his research vessel to try some experiments. The boat was found two days later, apparently undamaged, but Odén and all his equipment were missing.

from other countries. Odén found that both rain and surface waters throughout Europe were becoming more acidic, and that winds were responsible for spreading airborne sulfur and nitrogen compounds across the continent. His work also demonstrated how acid rain could cause declines in fish populations, reduction of forest growth, and corrosive disintegration of materials.

The dominant contributors to acid rain are SO_x and NO_x. In virtually all parts of the world, increases in acid deposition correlate very well with local increases in SO_x emissions, and the most acidic rain is found where SO_x levels are highest. SO_x and NO_x are not, strictly speaking, acids themselves in the way that we have defined the term.* Rather, they would be called acid anhydrides, or "acids without water." They dissolve in water—including rain—to form sulfurous and sulfuric acids and nitrous and nitric acids, respectively. The two major acids in acid rain are sulfuric and nitric acids. In their concentrated forms, sulfuric and nitric acids are very strong acids.

Combustion of fuels is *not* the only human source of SO_x and NO_x. Production of such metals as copper and nickel generates huge quantities of SO_x. The common ores of these metals are sulfides, compounds of the metal with sulfur. Conversion of sulfide ores involves heating them with oxygen (a process called "roasting") and is responsible for the SO_x emissions. The world's largest smelter, in Sudbury, Ontario, produces nickel from nickel sulfide. The environment in the immediate vicinity of the plant has been ruined by the uncontrolled release of SO_x in years past. Some who have visited the region suggest that it is about as close to a barren lunar landscape as we might see on Earth. The contribution of the metal industry was dramatically highlighted by the effect of a labor strike at copper smelters in the southwestern United States in 1980. During the time the workers were out on strike, SO_x emissions in Arizona and New Mexico fell by about 90%.

Worldwide, the atmosphere receives about as much sulfur from human activities as it does from natural sources, 75–100 million tonnes per year from each. Sulfur is released naturally by plankton in the oceans, decaying vegetation in swamps, and from volcanoes. The eruption of Mount Pinatubo in the Philippines, in June 1991, injected between 15 and 30 million tonnes of SO_x into the stratosphere. The emitted SO_x reacted relatively quickly to form small droplets of sulfuric acid, which remained suspended in the atmosphere for more than two years, where they reflected and absorbed sunlight. In late 1991, after the eruption, and continuing through 1992, there was a decrease in the average global temperature. This temperature drop was contrary to an otherwise steady increase in average global temperature (Chapter 29) and has been attributed to the effects of the tiny droplets or frozen crystals of sulfuric acid absorbing or reflecting sunlight.

Air provides nitrogen for production of thermal NO_x. The energy necessary for the reaction of the relatively unreactive nitrogen molecules in the air with oxygen

---

* Chemists recognize more than one classification of acids (and bases). Those substances that behave as acids in water by forming H^+ ions are called Brønsted, or Brønsted–Lowry acids. For acid rain and its effects, this is only type of acid with which we need be concerned. Another important category of acids is the Lewis acids. In this classification, an acid is a substance capable of accepting a pair of electrons from another species. Lewis's classification is much broader than Brønsted's, because it includes species that don't even contain hydrogen. For example, SO_2 cannot act as a Brønsted acid until it interacts with water to produce H^+ from water molecules, but it can function as a Lewis acid in some reactions.

comes from the high temperatures in combustion systems, such as in the burning of the gasoline in automobile engines. There is also concern over high-altitude NO_x emissions from jet engines and NO_x production from diesel engines. The same reaction of nitrogen molecules in the air occurs when air is heated to a very high temperature in the furnace section of a coal-fired electricity-generating plant. Such plants can potentially contribute both sulfur and nitrogen oxide emissions to the formation of acid precipitation. Nitrogen oxides are fairly reactive compounds and are of concern not just because of their contribution to acid precipitation. At high altitudes, NO_x can react with upper-atmosphere ozone, contributing to the destruction of the ozone layer.* Closer to the surface of the Earth, NO_x reacts with other components of vehicle exhaust to create the pollution problem known as photochemical smog (Chapter 28).

On a worldwide scale, human activities release almost twice as much SO_x as NO_x. If we were to restrict the comparison only to SO_x and NO_x from fossil fuel combustion, then the annual NO_x emission is less than 40% of the SO_x emissions. The effect masked by a simple global comparison is the major decrease in SO_x emissions from the industrialized nations. (Scrubbers work!) Unfortunately, the drop in SO_x emissions from industrialized nations was more than offset by a large increase in SO_x emissions from the developing countries. Many developing nations do not have the financial ability to afford emission-control technologies or low-sulfur fuels that have been adopted by the industrialized nations. Some of the developing nations have not enacted environmental control legislation or regulations. NO_x emissions might be the more serious long-range problem, because they are more difficult to control than SO_x.

As with SO_x emissions, human activities are not the only sources of NO_x. Plants, soil, and lightning are major natural sources of NO_x. A lightning bolt, for example, provides a splendid way to generate thermal NO_x. In a year not marked by some dramatic event like the Mount Pinatubo eruption, natural emissions would account for about 20% of SO_x and about 40% of the NO_x released to the atmosphere.

ENVIRONMENTAL CONSEQUENCES OF ACID RAIN

Acid precipitation is an environmental problem for many reasons. It corrodes the limestone, marble, or metals used in buildings. Some ancient monuments and statues that have survived unscathed for millennia, since the time of ancient Egypt, for example, are now being rapidly destroyed by acid rain. It accumulates in and acidifies

---

* Some 10–30 kilometers above Earth's surface, the atmosphere contains increased concentrations of ozone, the form of oxygen with three oxygen atoms per molecule (O_3). The thickness of the ozone layer varies from one season to another, and also varies with latitude. The ozone layer is *not* some sort of spherical shell of pure ozone that surrounds our planet. Under the best of circumstances the ozone concentration in the "ozone layer" is relatively small. However, the ozone layer is crucial for human health. High-altitude ozone intercepts some of the ultraviolet light that comes to Earth as part of the incoming solar radiation. Problems attributed to ultraviolet exposure include an increased susceptibility to skin cancer and cataracts, and, possibly, reduced ability of the immune system to ward off disease. For every 1% reduction in ozone concentration, the amount of ultraviolet reaching the surface increases by about 2%. NO_x reacts with ozone; an increase in high-altitude NO_x concentrations could therefore contribute to reduced ozone concentrations, which in turn lead to more exposure to ultraviolet for us.

lakes and streams and contributes to the destruction of aquatic life. It destroys terrestrial plant life, probably the most dramatic example being in the Black Forest of Germany, where more than two-thirds of the trees are now dead. In the United States, acid rain appears to be affecting forests in northern New England and in the southern Appalachians. Inhalation of acid rain mist irritates or exacerbates human respiratory problems.

EFFECTS ON BUILDINGS AND STATUES

Priceless, unique marble or limestone statues and buildings are being attacked by acid precipitation (Figure 27.4). In many cases, these buildings or monuments are irreplaceable artifacts of human civilization. The beautiful sculptures adorning the exteriors of cathedrals, churches, or other historic buildings throughout Europe are eroding. Buildings famous around the world, such as the Parthenon in Greece, the Taj Mahal in India, and United States Capitol, show signs of corrosion that is likely due to acid precipitation. Old gravestones, which may be sources of genealogical or historical data, are no longer legible. The ancient monuments of classical Greek civilization, including the Parthenon, have deteriorated more in the past three or four decades than they had in the previous two millennia.

ACIDIFICATION OF NATURAL WATERS

Whenever acidified water enters a lake, the pH of the lake will necessarily drop unless the lake itself, or something in the surrounding soils, contains basic compounds

FIGURE 27.4 An example of acid rain damage to an outdoor monument. The type of damage shown here is relatively minor (so far); in worse cases, many of the details of the figures would have been obliterated. (From http://www.flickr.com/photos/widnr/6588126917/in/photostream/.)

that can neutralize the acid. The capacity of a lake to resist change in pH when acids are added is called its *acid neutralizing capacity*. In those parts of the world where the surface rock is mostly limestone, the limestone itself can neutralize acids. Unfortunately, many lakes in, e.g., Scandinavia and in the northeastern United States are surrounded by granite rather than limestone. Granite does not react with acids nearly so readily as limestone. There will be very little acid neutralizing capacity in such lakes, and they will necessarily show a steady, though usually gradual, acidification.

Two other mechanisms also affect the ability of a body of water to withstand acidification. First, some lakes or streams may possess a buffering capacity.* The natural weathering of surrounding rocks may add bicarbonate ions to the water; they act as natural buffers that prevent a substantial drop in pH when acid precipitation occurs. Second, the H^+ ions added to the water by acid precipitation can exchange with other positively charged ions (such as Ca^{2+} or K^+) that are naturally present in soils or in sediments on the lake bottom. This process is called ion exchange.† It helps counteract the effects of acid addition because the calcium or potassium ions that exchanged for the H^+ ions are not acidic. But, just as in the case of the limestone lakes, the continual addition of acid can eventually get to a point at which the natural buffering capacity of the water, or the ion exchange capacity of soils and sediments, can be overwhelmed.

The decreasing pH is often accompanied by a decline in the population of fish and other aquatic species. By the time the pH is reduced to 4, a lake is essentially dead. Ecosystems that are especially at risk are lakes in areas with high levels of SO_x and NO_x emissions coupled with thin soils, low buffering capacity, or granite rather than limestone. In southeastern Ontario, the average pH of lakes is now about 5.0. In many of these places, the water contains increased concentrations of sulfate ions, consistent with sulfuric acid being the acidifying agent.

Effects on Aquatic Ecosystems

In many cases, pollution has a visual impact, in addition to its chemical or biological effects. As examples, we see sooty clouds of smoke, the discoloration of water by something dumped into it, or accumulated dust and dirt. The acidification of

---

* Buffers are substances that resist changes in pH, even when strong acids or bases are added to the system. Most common buffers are aqueous solutions of a weak acid (i.e., one that does not produce much H^+ in solution) and one of its salts. An example would be a solution of acetic acid and sodium acetate. When a strong acid is added, the H^+ ions that it provides are consumed in converting acetate back to acetic acid. A strong base added to the system converts acetic acid to acetate. In either case, little pH change occurs. An application in daily life is the use of buffers as food additives, to prevent pH changes during storage. Potassium hydrogen tartrate is an example of a buffer used in this way.

† Some substances, such as the zeolites, contain ions that are relatively loosely held to their molecular structure. When such substances are exposed to a solution that contains ions of a different kind, some of the loosely held ions can enter the solution, and some of the ions originally in solution can be taken up instead. An everyday example is the home water softener. Most such units use a synthetic polymer containing sodium ions. In the presence of the so-called hard water, which contains calcium and magnesium ions, an exchange occurs such that sodium ions from the polymer enter the water, and calcium and magnesium are taken up by the polymer. This ion exchange does away with the behavior—forming scums with soap and leaving residues in teakettles, for example—associated with hard water.

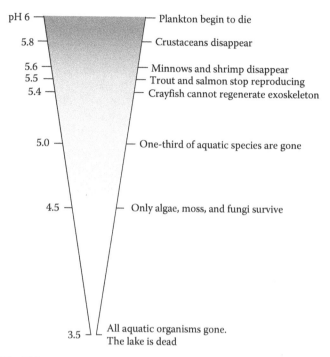

pH 6 — Plankton begin to die
5.8 — Crustaceans disappear
5.6 — Minnows and shrimp disappear
5.5 — Trout and salmon stop reproducing
5.4 — Crayfish cannot regenerate exoskeleton
5.0 — One-third of aquatic species are gone
4.5 — Only algae, moss, and fungi survive
3.5 — All aquatic organisms gone. The lake is dead

FIGURE 27.5 This section of the pH scale summarizes some of the effects experienced in aquatic ecosystems as they become increasingly acidic.

natural waters is unusual in this respect. A lake or stream damaged by acid precipitation does not usually have the appearance of being polluted. The water seems to be pristine and sparkling clear. It would be easy to misinterpret this appearance as signifying clean healthy water, whereas the truth is just the opposite—the water is biologically dead. Figure 27.5 illustrates the effect of continued acidification and decreasing pH.

Any organism has a particular range—sometimes a rather narrow range—of pH at which it thrives best. One of the first warning signs that acid rain was a serious problem, noticed in the late 1960s, was a decline in fish populations in Scandinavian lakes. At about the same time, similar observations were made in New Hampshire, which correlated with an increase in the acidity of lakes and streams in New England. The effect on fish is often the first noticeable impact, because jobs and revenue are lost when commercial fisheries become less productive and recreational opportunities are lost for sport fishing. However, long before fish are lost, other, more humble life forms that have critical positions in the aquatic food chain of the ecosystem are lost.

When the pH of a lake drops below 6, many kinds of small organisms begin to die off. For example, phytoplankton, the tiny, free-floating aquatic plants that serve as the base of the aquatic food chain (Figure 27.6), have low reproduction and high mortality at pHs below 6. Similarly, many species of free-floating

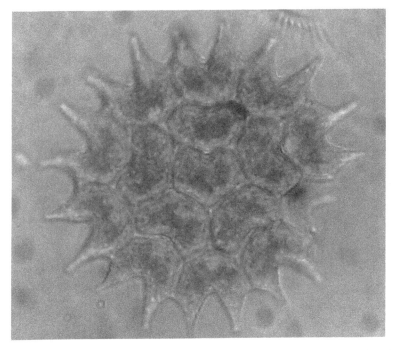

FIGURE 27.6   An example of one of the great many kinds of phytoplankton. (From http://
commons.wikimedia.org/wiki/File.Pediastrum minus 1.jpg.)

aquatic animals (zooplankton) are adversely affected. The loss of these organisms
begins the disruption of the entire food chain, thus threatening the viability of
many other species.

Irreversible changes begin to occur in lakes when the pH drops to about 5.8. At
this pH, the tiny crustaceans that are another important link in the aquatic food chain
(copepods), disappear.* As the pH drops further, to 5.6, the populations of minnows,
sculpins (Figure 27.7), and at least one type of shrimp begin to decline.† The shrimp
are a major food source for lake trout.

Acidity then begins to affect the reproduction of fish. Many fish, particularly
those of the family that includes trout and salmon, will not reproduce successfully
at pHs below 5.5. As the pH drops below this value, most species of fish are endangered. Numerous problems beset the fish at this acidity: growth is retarded, the salt
balance in the blood is upset, calcium is depleted from the skeleton (leading to mal-
formation), aluminum clogs their gills, and mercury may be absorbed. Fish are not

---

* Copepods, though perhaps known only to specialists in aquatic ecology, have enormous ecological
  importance because they provide a food source for many species of fish. Most copepods are very tiny,
  usually no more than three millimeters long. They eat phyto- and zooplankton, and each other.
† Sculpins are small fish characterized by wide and heavy heads but long, tapered bodies. They occur
  in both fresh and salt waters of the northern hemisphere. Sculpins stay close to the bottom of the
  water and usually are not very active. They are of no direct use to humans (in contrast to, e.g., trout or
  salmon) but of course are still a vital link in the total ecosystem.

FIGURE 27.7 The sculpin, an important component of the aquatic food chain, and a potential victim of acidification of water due to acid rain. (From http://www.arctic.noaa.gov/aro/russian-american/photos-fish.htm.)

the only organisms to suffer under these conditions. Amphibians are also hard hit by the effects of an acidic environment.

Leaching of such elements as cadmium, aluminum, and mercury from surrounding soils or rock can poison fish. (Toxic effects on fish can be detected at aluminum concentrations as low as 100 parts per *billion*.) Fish contaminated with potentially toxic elements, such as mercury, pose a threat to humans as well. Such elements leached into the water supply can affect humans if these elements make their way into aquifers and from there into drinking water.

At pH 5.4, additional species of copepods vanish. Crayfish suffer extensively, by being unable to regenerate their hard and protective exoskeletons after molting.* Fish are unable to reproduce at all at pH values below 5.4. By the time the pH has dropped to 5, about a third of all aquatic species have been eliminated from the ecosystem. Most adult salmon or trout die because of the environmental stress resulting from life in water with pH below 5. As an example, many of the rivers in Nova Scotia that provide locations for salmon fishing have become so acidic that they no longer can support fish life. At least eight such rivers have pHs below 4.7, with another twenty having pHs in the range 4.7–5.4. A century's worth of sport fishing records show that these rivers provided regular good catches until the 1950s, after which they began steadily to decline. Twenty years later, essentially no fish were being caught there, and the salmon spawning runs have ceased.

Below a pH of 4.5, most life is gone except for some very hardy aquatic plants, such as algae, moss, and fungi. At pH 3.5, virtually all aquatic organisms die off. At that point the stream or lake is "dead." Many lakes in Norway and Sweden are effectively dead in this sense.

---

* Crayfish are crustaceans, the class of animals that includes lobsters, shrimp, and crabs. Unlike mammals, birds, and fish, whose skeletons are inside the body, crustaceans have an exoskeleton, that is, their body parts are supported by a skeleton or shell on the outside of the body. Such organisms can grow only by the periodic shedding of the exoskeleton—molting. Molting occurs at longer and longer frequencies as the animal reaches adulthood. Molting is usually quick, but the animal is almost totally helpless, especially vulnerable both to enemies and to pollution, until a new exoskeleton has grown.

Remediation of Aquatic Systems

Although the acidification of natural waters can have these devastating effects on ecosystems, there is hope that, with proper treatment, many acidified lakes can be reclaimed. Since the problem is caused by acids, short-term remediation relies on neutralizing the acid. This requires adding a basic compound to the water. The base selected for this application must be inexpensive, because large quantities will be needed, and should not be poisonous or otherwise harmful in itself. The best candidate for this application is lime (calcium hydroxide). Large-scale addition of lime to acidified waters has been tried in various locations in the United States and Europe. "Liming" can raise the pH of the water to levels that allow aquatic species to live successfully. The affected lake can be restored within a few years. However, over a period of time the added lime could itself be overwhelmed by further acid precipitation events. The only long-term solution is to reduce acid precipitation by reducing the amounts of SO_x and NO_x emitted to the environment.

Destruction of Terrestrial Plants

Acid precipitation by itself—that is, in the absence of any other factors—does not appear to kill plants. However, acid precipitation in conjunction with other environmental stresses can have significant impacts on plants (Figure 27.8). Acid precipitation

FIGURE 27.8   This photograph illustrates the effects of acid rain damage on a forest. (Reprinted from first edition, with permission from Photo Researchers, Inc., New York, New York.)

may weaken the plant to the point where other environmental factors, which might not kill a healthier plant, would kill or seriously harm the acid-weakened plant.

Weather-related stresses likely have significant responsibility for the damage. Acid precipitation slowly weakens trees, reducing their ability to withstand extremes of cold or heat. Acid rain makes conifers more susceptible to cold. In the fall, conifers normally prepare for the freezing temperatures of the coming winter by withdrawing water from their needles. This "cold hardening" constitutes part of the natural life cycle of the tree. The tree roots slow down the release of nitrogen-bearing nutrients into the soil, storing them up in the tree instead. But, nitrogen-containing acids soaking into the needles might override the signal from the roots, because the tree could mistakenly sense an abundance of nitrogen compounds. That "wrong signal" caused by acid rain soaking the needles could delay the cold-hardening process and leave the tree vulnerable to freezing. Maple trees can also be damaged by weather-related stress that has been exacerbated by acid precipitation.

Acid precipitation weakens the ability of trees to resist attacks by insects or by viruses. Gypsy moth damage becomes more severe if the tree has also been exposed to acid rain. Similarly, blight-causing viruses cause more damage to trees weakened by acid rain. The destruction of fir trees in the southern Appalachians of the United States demonstrates this type of problem.

The effects of ozone or other air pollutants can also be made worse if the trees have also been exposed to acid precipitation. Ozone and nitrogen oxides attack the waxy coating on leaves. Once this coating has been breached, the H^+ ions can deplete vital nutrients, such as magnesium, calcium, and potassium. The waxy coating also retards water loss from the tree. In addition, ozone can damage the chlorophyll that facilitates photosynthesis. If that happens, photosynthesis, and, consequently, tree growth, is slowed.

Acidification of soil beneath the trees mobilizes toxic metals, such as aluminum, which attack the tree roots and prevent the absorption of nutrients and water. At the same time, elements such as potassium, calcium, and magnesium, essential for plant growth, are leached out of the soil. Potentially toxic elements are mobilized and attack the tree, while at the same time nutrient elements are mobilized and lost. If the toxic metals were originally chemically combined in soil particles, they may have been biologically unavailable to the plant, and essentially harmless. Only when acidification of the soil makes them soluble do they become accessible for uptake by the plant. The essential elements were originally present in the soil in chemical forms from which their uptake could proceed consistent with the normal biological needs of the plant. When these nutrients become soluble as a result of leaching by acid precipitation, they could flow (in the water) away from the root zone and no longer be available to meet the requirements of the plant. Some soils are better able to withstand acidic precipitation because they contain more alkaline substances, such as limestone, which can neutralize or buffer acids. As in the case of lakes and streams, the neutralizing or buffering capacity declines with continual, long-term exposure to acid precipitation.

Acid precipitation also interferes with the mechanisms by which trees obtain essential nutrients such as nitrogen and phosphorus. Many plants obtain essential nitrogen by capturing nitrogen from the air and converting it into a chemically useful form. This so-called nitrogen fixation process is facilitated by microorganisms that live in soil and are often symbiotically associated with the root structure of the plants. Acid rain can inhibit nitrogen fixation by the microorganisms, by killing them.

In summary, there is strong circumstantial evidence that acidic precipitation is at least a contributing factor to the declining health of trees. These effects leave the trees susceptible to destruction by such other factors as disease, insects, or weather extremes. The damage is enormous on a worldwide scale.

In North America, forest damage has been most dramatic in portions of the Adirondack and Appalachian Mountains and southeastern Canada. Acid precipitation has contributed to the decline of red spruce in the northern Appalachians by reducing the cold tolerance of those trees. About 70% of all British beech trees and 80% of all yew trees show signs of injury associated with acid precipitation. In western Europe, tree loss spread during the 1980s. Damaged forests were reported in Italy, France, the Netherlands, Sweden, Norway, and Germany. Almost one in four European trees has lost more than a quarter of its leaves or needles. Trees in northern Germany have been stripped of many of their leaves. About fifty million hectares of European forests have been damaged by acid precipitation, about one-third of Europe's total forested land. Over 80% of the forests in eastern Germany, Poland, and the Czech Republic may be damaged. In some regions of these countries, forests have been totally destroyed by air pollution. In northern Siberia, reindeer have changed their habitat because lichen, their principal food, has suffered severely from acid precipitation.* In this heavily polluted region, lichen growth has dropped to just 1%–2% of normal levels, and the number of lichen species has dropped from about fifty to three or four.

Trees on mountainsides or tops of mountains face a more serious risk of damage from acid precipitation. Because they are at higher elevations, they are the first trees to be struck by the acid precipitation. In some regions, these trees can often be shrouded for long periods in clouds and mist, which can be just as acidic as the precipitation itself. The problem is compounded because these are the trees more likely to be stressed by cold.

HUMAN HEALTH EFFECTS

Acid precipitation affects human life in many ways. The damage or destruction caused to ecosystems can deprive us of various kinds of outdoor recreation or of the aesthetic pleasure of enjoying the natural landscape. The damage to ecosystems more directly affects the benefits that we derive from them, such as clean drinking water and food in the form of fish or agricultural crops. But, perhaps most importantly, acid precipitation can directly lead to severe effects on human health.

When a person is outdoors during a rainstorm, or even in an acidic fog or mist, tiny droplets of acid precipitation are deposited directly in the sensitive tissues of the nose,

---

* Lichens are not one plant, but two, growing together in symbiosis—a relationship that benefits each. Lichens consist of algae growing together with fungi. The alga performs photosynthesis and manufactures food for both itself and the fungus. For its part, the fungus shelters the alga and keeps it moist, allowing the alga to grow in harsh environments where it could not survive on its own. Lichens are very abundant in the treeless regions of the Arctic. The plant (or, strictly speaking, plant community) known as reindeer moss, which is a valuable food item for reindeer, is a lichen.

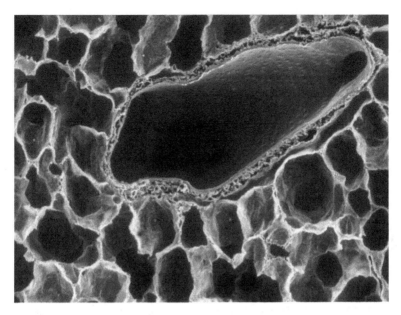

FIGURE 27.9 A magnified view of human lung tissue. Each of the tiny passageways is lined with sensitive tissues that can be harmed by chronic exposure to acid rain. (Reprinted from first edition with permission from Phototake, New York, New York.)

bronchial tract, and lungs (Figure 27.9). Three groups in the population are especially at risk. One group is the elderly, those whose respiratory systems are starting to wear out or shut down. A second group is the very young, whose respiratory systems may not yet be fully developed. The third group is people of any age who already have other respiratory problems such as asthma, chronic bronchitis, or emphysema.

One of the worst recorded instances of pollution-related respiratory illness occurred in London in December 1952 (Chapter 16). This particular pollution incident lasted for five days and claimed about four thousand lives. A similar incident in Donora, Pennsylvania, in 1948 resulted in twenty deaths and caused illness in about 40% of the residents. Figure 27.10 is a photograph taken in Donora at noon during this incident.

Many potentially toxic elements, such as lead, cadmium, and mercury, are normally chemically combined with minerals present in rocks and soil. When continually soaked by acid precipitation, some of these elements can be leached into the acidic water. Once that happens, the elements can eventually wind up in the public water supply, where they can lead to chronic toxicity or serious health threats. Aluminum, one of the most abundant elements in the Earth's crust, appears to be harmless to most animals as long as acceptable levels of calcium are maintained in the body. If calcium becomes in short supply, aluminum may replace it in tissues or organs. Aluminum can be toxic in humans and does not appear to be an essential nutrient. No disease or condition resulting from aluminum deficiency has ever been identified in living organisms. An excess of aluminum has been found in damaged brain tissue of victims of Alzheimer's disease and of amyotrophic lateral sclerosis

FIGURE 27.10 Smog in Donora, Pennsylvania, around the time of the 1948 air-pollution disaster. This photograph was taken in mid-day! (From http://www.erh.noaa.gov/ctp/features/AirQuality/CTP AirQuality.php.)

(often known as Lou Gehrig's disease). However, it is *not* proven that an excess of aluminum in the body *causes* Alzheimer's or Lou Gehrig's disease; and, even if such connection were to be established unequivocally, it is *not* proven that the aluminum was mobilized as a result of acid precipitation. Much remains to be learned about how acid precipitation or other forms of pollution result in the liberation of potentially harmful elements and the impact of those elements on human health.

ACID RAIN ON THE MOVE

Consider these two statements: (1) What goes up must sooner or later come down. (2) The solution to pollution is dilution. The first statement implies that acid rain is often regional in character, in that what goes up—SO_x and NO_x—comes down as acid precipitation in the same general locality. The second statement has been put into practice by the building of very tall stacks on emission sources such as electricity-generating plants and smelters. Tall stacks were built to eject pollutants high into the atmosphere where they could become diluted, thus improving local air quality, and ultimately come down somewhere else, sometimes hundreds of kilometers away. As a result, local pollution problems are converted into regional ones.

Geographic regions with the most acidic precipitation might be expected to be the regions of highest emissions of SO_x and NO_x. This relationship is generally confirmed. SO_x emissions tend to be highest in regions where there are many coal-fired electric power plants, steel mills, and other metallurgical smelters. Although electric generating plants also generate nitrogen oxides, the highest NO_x emissions are usually found in large urban areas with high population density, factors that translate directly into much automobile traffic. However, acid precipitation does not recognize state, province, or national boundaries, thanks in part to the use of tall stacks. This leads to political controversies and accusations. SO_x and NO_x emissions generated in Germany, Poland, and the United Kingdom are carried into Norway and Sweden. SO_x and NO_x emissions from the midwestern United States, carried to the northeast by prevailing winds, contribute to acid precipitation in the northeastern states and in eastern Canada.

Over a century ago, the Norwegian playwright Henrik Ibsen* complained about airborne pollution from Britain crossing the North Sea to Scandinavia. Not until the 1960s did Scandinavian scientists show that acid precipitation, derived in part from burning of high-sulfur coals in Britain, had severely affected rivers and lakes of southern Sweden and Norway. (Norway and Sweden receive 90%–95% of their SO_x-related pollution from other countries.) Subsequent meteorological studies showed enormous international flows of air pollution, generally moving from west to east with the prevailing winds across Europe. By 1994, satellite monitoring had shown a region of regular, highly acidic precipitation virtually circling the globe, from Britain to all the way to Central Asia.

DEALING WITH SO_x AND NO_x EMISSIONS

Most industrialized nations have made, and continue to make, significant strides in controlling SO_x and NO_x emissions. In this century, the most significant threats from acid precipitation will be in the developing or underdeveloped nations, which become significant emitters of SO_x and NO_x themselves. Many such countries lack the legislative or regulatory structures needed to control emissions. Impoverished countries lack the financial resources to install modern, but expensive, emission control equipment. These countries have rapidly growing populations; more people account for more energy consumption, and increased energy consumption brings with it increased emissions to the environment, especially in the absence of regulatory structures and the ability to fund pollution-control equipment.

SULFUR OXIDES

The fact that most of the anthropogenic SO_x emissions come from a limited number of stationary sources makes the problem of reducing or controlling these emissions easier to attack. Chapter 16 introduced some of the approaches to doing this. They include precombustion strategies of coal cleaning, switching to a fuel of lower sulfur content, or switching to a different kind of fuel, as well as the postcombustion strategy of flue gas desulfurization (FGD).

---

* The Norwegian writer Henrik Ibsen (1828–1906) is generally regarded as one of the great playwrights of the late nineteenth and early twentieth centuries, perhaps one of the best ever. Many of his plays are noted for a realistic portrayal of the tragic aspects of society. Some of his better-known works include *Peer Gynt*, *Hedda Gabler*, and *A Doll's House*. Ibsen's major works are always in print, including inexpensive paperback editions. At the end of his life, Ibsen suffered a series of strokes. His very last words were uttered when a visitor asked Ibsen's nurse how he was doing. The nurse assured the visitor that Ibsen was doing better; overhearing this, Ibsen managed to respond, "On the contrary!" They were his last words. He died very early the next morning.

Coal cleaning is reasonably easy. Most coal cleaning processes take advantage of some difference, often density, between the carbonaceous portion of the coal and minerals associated with it, as described in Chapter 26. The process can add several dollars per tonne to the cost of coal.

Regions having large numbers of electric generating plants and heavy industries that burn coal can approach reducing the SO_x emissions by switching these facilities to coals of lower sulfur content. Shipment of low-sulfur coal over very long distances (e.g., from Wyoming to the midwest or northeastern United States) would roughly double the cost of fuel to these facilities. Not only would the fuel cost increase, but there also would be significant social and economic impacts on those places that produce high-sulfur coal, in the form of job loss and reduced tax revenues. Nevertheless, switching to different coal supply, of lower sulfur content, may be the least expensive (least expensive, i.e., to the consumer of the coal) of the possible strategies for reducing SO_x emissions.

FGD, carried out in scrubbers, relies on the facts that SO_x solutions in water are acidic and react chemically with bases. The cheapest bases are lime (calcium hydroxide, $Ca(OH)_2$) and limestone (calcium carbonate, $CaCO_3$). The calcium sulfate product appears as an insoluble material, gypsum. This material can be removed from the system as "scrubber sludge." There is some interest in possible commercial applications of sludge, such as using it to manufacture gypsum wallboard. A diagram of the scrubber operation is shown in Figure 27.11. Scrubbers work. They can remove more than 90% of the SO_x from flue gas. It is important to notice that a scrubber does *not* destroy pollution. What the scrubber does is to convert the pollution from a rather dilute, difficult-to-handle form—a small concentration of SO_x in flue gas—into a far more concentrated form (scrubber sludge) that is much easier to collect, handle, and dispose of.

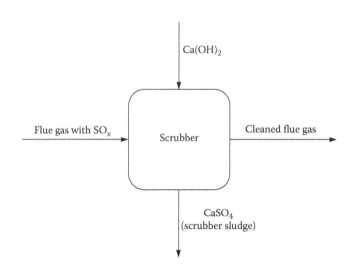

FIGURE 27.11 A schematic flow diagram of the operation of a scrubber, a device for removing SO_x from the flue gas of a boiler. The SO_x is captured as insoluble calcium sulfate (scrubber sludge).

A combination of coal cleaning and FGD allows a plant to meet stringent SO_x emission regulations. However, economic issues always come into play. First, scrubbers can represent about a third of the capital cost of a plant, a cost that must be recovered somehow (usually in the form of the rate paid for electricity by the utility's customers). Scrubbers themselves require energy to operate. This energy has to come from somewhere, the obvious source being the plant itself. The use of some energy internally in the plant for scrubber operation necessarily reduces the net amount of energy available for sale to customers. In a plant without a scrubber, about 34% of the chemical energy in the coal is converted to electricity leaving the plant; in an otherwise identical plant with a scrubber, the comparable figure might be about 30%. If the sludge cannot be marketed as a useful by-product, then there will be a disposal cost associated with handling it.

Another approach to dealing with SO_x emissions is the concept of "emissions trading." In this system, each company or utility that produces SO_x would operate under a permit that specifies the maximum level of SO_x that can be legally released. Exceeding the maximum could result in legal penalties such as fines. A goal of the system is simply that each company achieve, and not exceed, its individual allowance level. Those companies that actually perform better, and have SO_x emissions below the legally permitted level, would be assigned pollution "credits." Each credit conveys the right to emit a specified quantity of SO_x. These credits can be sold to another company that is unable to meet its emission allowance. There is a financial incentive for electricity producers to achieve significant reductions of emissions, because their extra credits can be sold to companies not meeting targets. The buying of these credits by the companies that cannot meet SO_x standards will allow them to continue operation, but likely few companies would be willing to tolerate for long the continuous expense of having to buy pollution credits. Such companies would have a financial incentive to work toward reducing their own emissions. Credits have been bought and sold in private transactions and in public auctions.

NITROGEN OXIDES

NO_x has proven to be much harder to deal with than SO_x. Chemically, it is harder to remove nitrogen from a fuel than it is to remove sulfur. In the specific case of coal, much of the sulfur is present in various minerals (such as pyrite) that can be removed easily and cheaply. Nitrogen, though, is chemically bonded to the fuel molecules, and its removal would require a chemical process, rather than a relatively simple and inexpensive physical separation. Formation of thermal NO_x is a persistent problem in high-temperature combustion equipment. Our customary approach to achieving high efficiencies involves operating at high temperatures, exactly the situation that favors the formation of thermal NO_x. This situation provides a difficult dilemma for combustion engineers who strive for high system efficiencies coupled with low thermal NO_x production.

Several technologies approach the problem of NO_x control by modifying the combustion process itself. One involves the so-called low-NO_x burners, which function by regulating the available air so that pulverized coal burns in stages. In the first stage, the amount of air is limited, so that any NO_x that does form becomes converted back to nitrogen. Then, in a second stage, more air is made available to complete the burning of the coal but at temperatures low enough to minimize further NO_x formation.

Re-burning aims to reduce the formation of thermal NO_x. Coal is burned in the lower part of a boiler to provide 80%–90% of the heat needed to generate the steam. Natural gas is then added in a "re-burn" region higher in the boiler. The constituents of the natural gas, such as methane, react with nitrogen oxides produced by the coal combustion and decompose them into nitrogen. Re-burning can be combined with other techniques to reduce emissions even further. A related strategy is co-firing, in which coal and biomass are burned simultaneously. Since there is no sulfur in biomass, SO_x emissions are reduced in direct proportion to the amount that is co-fired with the coal. NO_x emissions are also reduced.

Other technologies aim to treat the flue gas to reduce or destroy NO_x before it can be emitted to the environment. One approach injects a nitrogen-containing compound, such as ammonia, into the flue gas. Ammonia reacts with NO_x to produce nitrogen and water, both being harmless to the environment. A variation of this approach uses catalysts to facilitate the reaction. Figure 27.12 provides a schematic diagram of how this might work.

Reduction of total NO_x emissions is particularly challenging because most of the sources are small, individually owned, and mobile. There are almost two billion of

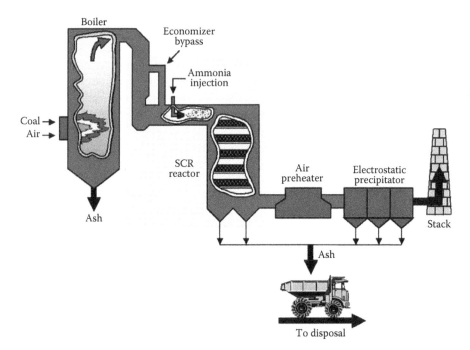

FIGURE 27.12 A schematic diagram of a coal-fired electric plant, with the various operations for controlling emissions. (Image courtesy of National Energy Technology Laboratory, U.S. Department of Energy, Pittsburgh, Pennsylvania, http://www.netl.doe.gov/technologies/coalpower/cctc/cctdp/project_briefs/scr/selcatreddemo.html.)

them in the world—cars and trucks. About half of the NO_x emissions in the United States come from the transportation sector. NO_x emitted by cars and trucks can be reduced by catalytic converters, which we discuss in Chapter 28. Catalytic converters reduce NO_x emissions by about 60% over the life of a vehicle. Since 1980, three years after the mandatory introduction of catalytic converters in the United States, NO_x emissions in North America began to level off, but the sheer number of new cars on the road every year keeps the total NO_x emissions high.

An alternative would be to eliminate the internal combustion engine altogether, or at least to reduce the use of such engines. Not too many years ago, such a suggestion might have been considered a sacrilege by many people. (It probably still is.) Now, though, there seems to be a small but steadily growing interest in electric vehicles and hybrid vehicles that combine electric and internal combustion systems. Many major international car companies market hybrid vehicles, and electric cars are increasingly coming into the marketplace. These new cars offer the possibility of major reductions in NO_x emissions (Chapter 28).

REFERENCE

1. Kennan, G.F. *Memoirs 1950–1963*. Little, Brown: Boston, MA, 1972, p. 257.

FURTHER READINGS

Gebelein, C.G. *Chemistry and Our World*. Brown: Dubuque, IA, 1997. An excellent introductory chemistry text with many real-world examples. Chapter 18, on air pollution is particularly relevant here.

McNeill, J.R. *Something New under the Sun*. Norton: New York, 2000. Chapters 3 and 4 of this book discuss what we are doing to the atmosphere, including the problem of acid rain.

Miller, B.G. *Clean Coal Engineering Technology*. Elsevier: Amsterdam, the Netherlands, 2011. A thorough treatment of coal technology, primarily for electricity generation. Chapter 9 deals particularly with various emission control strategies. This book presumes some basic chemistry knowledge.

Thorsheim, P. *Inventing Pollution*. Ohio University Press: Athens, OH, 2006. A history of how present views of pollution came about, as a result of increasing attention to the effects of emissions from coal burning during the 19th century. The topics include acid rain and the early attempts to understand it, such as those by Angus Smith.

28 Vehicle Emissions and Emissionless Vehicles

Vehicle exhaust contributes to three of the most serious pollution problems in the world today: acid rain (Chapter 27), global warming (Chapter 29), and photochemical smog, the topic of this chapter. Photochemical smog first entered the public's consciousness in the 1940s, especially in Los Angeles. Smog can be so irritating to the eyes and respiratory tract that, during the Second World War, some people thought the smog was actually a poison gas attack by the Japanese military. Within twenty years, smog was stunting tree growth eighty kilometers from the city center. By 1976, Los Angeles' air had reached officially unhealthy levels about 75% of the time, or three days out of every four.

This problem is not confined to the United States. Air pollution from cars and light trucks has reached serious proportions in at least half the cities of the world (Figure 28.1). Cities such as Athens, Budapest, and Mexico City now have to enforce emergency driving bans on hazy days to minimize health risk to their citizens. Other cities impose sizable road tolls or taxes on drivers wanting to take their vehicles into the center-city area. Car sales continue to grow in western Europe, Japan, and the United States. Twenty years ago, greatest growth took place in the growing, industrialized nations of Asia, such as South Korea and Taiwan. Rapid growth in sales of cars and light trucks is likely to occur in developing regions of Africa, Asia, and Latin America, and in the countries of the former Soviet Union and eastern Europe.

The world is closing in on two billion cars and light trucks on the roads. The United States, which has only about 4% of the world's population, accounts for half of all the cars in the world.

TAILPIPE EMISSIONS

When a hydrocarbon fuel, such as natural gas, gasoline, or diesel fuel, burns, the dominant product depends on the amount of oxygen (from air) available for the amount of fuel. In the ideal case—complete combustion—the air supply is ample to convert all of the carbon atoms in the fuel into carbon dioxide. If the air supply is not adequate to achieve complete combustion, an incomplete combustion can occur, in which the carbon atoms in the fuel are converted primarily to carbon monoxide (Table 28.1). And, if the air supply is very limited, carbon atoms in the fuel are converted to elemental carbon, normally recognized as soot. In all these cases, hydrogen atoms are converted to water.

Suppose that some of the fuel does not burn at all. It leaves the combustion system as a kind of emission called *unburned hydrocarbons*. Some fuel molecules can sweep

FIGURE 28.1   Air pollution in Mexico City, circa 2003. (From http://www.airqual.cee. vt.edu/research.htm.)

TABLE 28.1
Ratio of Oxygen to Fuel for Three Kinds of Combustion

	Oxygen/Fuel Ratio	
	Natural Gas	Gasoline
Complete combustion	2	11
Incomplete combustion	1.5	7.5
Oxygen-starved combustion	1	4

---

through an internal combustion engine without being burned. Unburned hydrocarbon emissions have other sources. Some gasoline evaporates from fuel tanks and fuel lines. Some is spilled on the ground (many of us do this occasionally at the self-service gasoline pumps of service stations) and evaporates. In all cases, volatile molecules in the fuel escape into the atmosphere.

These emissions are of concern for a number of reasons. First, carbon monoxide is lethal. Exposure to carbon monoxide can lead to death, for example, when people accidentally operate their cars in closed garages (as when tinkering with the engine at home). In some tragic cases, this is done deliberately as a form of relatively painless suicide. Carbon monoxide enters the bloodstream and disrupts delivery of oxygen to the body's tissues by combining with hemoglobin, the molecule that normally carries oxygen. Because it combines more tightly with hemoglobin than

FIGURE 28.2 Being stuck in heavy traffic can subject drivers to exposure to various air pollutants emitted from other vehicles. (From http://commons.wikimedia.org/wiki/File:Automobile exhaust gas.jpg.)

does oxygen, carbon monoxide steadily reduces the oxygen-carrying capacity of the blood. At high concentrations, carbon monoxide is deadly. At lower concentrations, it can lead to angina, blurred vision, and impaired physical and mental coordination.

Carbon monoxide is a major pollutant in all cities. Many city dwellers suffer from exposure to elevated levels of CO. Carbon monoxide from car exhausts may cause potentially fatal heart rhythm disturbances in otherwise healthy people. Joggers who exercise close to a busy highway could be affected by inhaling too much carbon monoxide. Since people sitting in nearby cars are affected, it is particularly hazardous to be in traffic jams (Figure 28.2), when about 10% of the gas emitted from the exhaust is CO.

Unburned hydrocarbon emissions represent a significant waste of fuel. These losses amount to about three thousand tonnes per *day* in the United States, just as vapor from filling station pumps or spills.

Sulfur in fuel is converted to SO_x. For gasoline-fueled vehicles, gasoline has so low a sulfur content that, for all practical purposes, we can neglect automobiles as a potential source of SO_x emissions. Further, the specifications for sulfur in diesel fuels are becoming increasingly stringent worldwide. Most industrialized countries now use diesel fuel of very low sulfur contents.

Nitrogen in fuel is converted to fuel NO_x. Regardless of whether it is fuel or thermal NO_x, it contributes to acid rain formation. Though fossil-fuel-fired electricity-generating stations also produce significant NO_x, worldwide, transportation is the largest source. Nitrogen dioxide, NO_2, is a reddish-brown gas that is toxic when concentrated; the maximum safety limit for NO_2 is 5 ppm, but even at this very low level NO_2 causes eye and lung irritation. NO_2 causes the distinctive brownish haze typical of photochemical smog. NO_2 can irritate lungs and lower resistance to respiratory infections such as influenza. NO_x emissions from automobiles concentrate mostly in the urban areas where there are the most vehicles and, of course, the most people.

HOW SMOG FORMS

A significant air pollution problem occurs when carbon monoxide, NO_x, and unburned hydrocarbons react together to form smog. The complex chemical reactions involved in smog formation are promoted by the ENERGY in sunlight, leading to the term *photochemical smog.** Smog formation is undesirable for a variety of reasons. Eye irritation occurs because some compounds produced during smog formation are chemically similar, or identical, to the components of tear gas. Respiratory tract irritation can occur even in otherwise healthy adults by exposure to smog. It is very harmful to persons with chronic respiratory problems, such as emphysema, and to infants whose respiratory tracts may not yet be fully developed. Smog is also aesthetically undesirable; thanks to the NO_x, it tends to have a dull grayish-brown color. Photochemical smog is typical in cities where sunshine is abundant and internal combustion engines exhaust large quantities of pollutants to the atmosphere.

The three *primary pollutants*—carbon monoxide, unburned hydrocarbons, and NO_x—are converted into the *secondary pollutants* found in photochemical smog. Air quality is further compromised by the formation of ozone[†] in conjunction with the breakdown of nitrogen dioxide. Ground-level ozone concentrations have increased in recent years. Since ozone is not a normal industrial pollution product, this increase almost certainly relates to photochemical pollution. The observed average atmospheric concentration of about twenty parts per billion (ppb) is below the safety limit of eighty ppb, but in photochemical smog, concentrations rise.

Air in an urban environment contains unburned hydrocarbons from car exhausts, evaporation, and spillage at filling stations. Highly reactive atomic oxygen (i.e., single atoms of oxygen, not the common diatomic oxygen molecule) reacts with hydrocarbons to form reactive, short-lived intermediate compounds that form secondary

---

* The adjective *photochemical* is used to differentiate this form of air pollution from *sulfuric acid smogs*, which are quite different in origin. Sulfuric acid smogs are produced when high-sulfur content coal is burned without emission control equipment. A droplet of water forms around a fly ash particle, and SO_x dissolves to form a solution of sulfuric acid. Sulfuric acid smogs have been implicated in major air-pollution disasters, such as the one in Donora, Pennsylvania, in 1948. Because coal is only rarely used as a fuel for home heating any more, and because coal-fired electricity-generating plants have scrubbers, sulfuric acid smogs are a thing of the past in most industrialized nations. Nowadays when the term smog is used without a modifier, it can be taken to mean photochemical smog.

† Ozone, the form of oxygen with three atoms in the molecule instead of the normal two, is the proverbial double-edged sword. In the stratosphere, ozone shields us against excessive ultraviolet radiation. Without that absorption of ultraviolet, severe sunburns would be more likely, as would the incidence of skin cancers. But, ozone is quite toxic. Its safety limit is only 0.08 ppm. At or near ground level, ozone is a serious health and environmental hazard. At concentrations common in many cities, some healthy adults and children experience coughing, painful breathing, and temporary reduction of lung function after an hour or two of exercise. The effects of long-term exposure to elevated levels of ozone are uncertain.

pollutants such as formaldehyde.* As little as 0.2 ppm of nitrogen oxides and 1 ppm of reactive hydrocarbons are sufficient to initiate these reactions.

CATALYTIC CONVERTERS

Cars and light trucks are the leading cause of photochemical smog. The rates at which the smog-forming reactions take place depend on several factors, including ambient temperature, the amount of available sunlight, and the presence of unburned hydrocarbons and relative amounts of hydrocarbons and NO_x in the air. In electricity-generating plants, scrubbers and baghouses can help remove or reduce potential emissions. From an emissions perspective, such plants represent *stationary sources*. One virtue of a stationary source is that, as the name implies, it doesn't move around. It stays put, and we can add hardware to it to help minimize its impact on the environment. A second virtue is that, usually, we need not worry about the size, weight, or complexity (disregarding, for the moment, cost) of emission control devices on stationary sources. Automobiles and other vehicles are *mobile sources*. Here the size, weight, and complexity—and of course the cost—of emission control devices are indeed a concern.

The strategy relies on destroying the emissions before they can exit the tailpipe. The potential emissions of special concern are those just discussed: carbon monoxide, unburned hydrocarbons, and thermal NO_x. We try to destroy these potential emissions in a device called a *catalytic converter* (Figure 28.3). Some of the earliest ideas about design and use of catalytic converters came from Eugene Houdry, better known as the inventor of catalytic cracking for gasoline production. The basic technology for the catalytic converter came into use in the 1970s and has improved steadily since then. The catalytic converter is designed to accelerate three chemical reactions: carbon monoxide reacting with oxygen to produce carbon dioxide and water; unburned hydrocarbons also reacting with oxygen to make the same products; and NO_x decomposing into nitrogen and oxygen. The first two of these reactions seek to *add* oxygen to other molecules, and the third seeks to *remove* oxygen. It is a very tricky business to do both (i.e., add and remove oxygen) with the same catalytic unit. How do we achieve this?

To answer this question, we first need to consider a bit of jargon. When the exact amount of oxygen needed for complete combustion is available, we call this condition *stoichiometric combustion*. If there is more air (oxygen) present than is needed for stoichiometric combustion, this condition is *fuel-lean* combustion. If there is not enough air (oxygen) available for stoichiometric combustion, this condition is *fuel-rich* combustion. At fuel-rich conditions, with not enough air (oxygen) for stoichiometric combustion, emissions of carbon monoxide and unburned hydrocarbons will be high. In moving from fuel-rich to stoichiometric to fuel-lean combustion, more and more air (oxygen) becomes available and carbon monoxide and unburned

---

* Pure formaldehyde is a gas. The compound is perhaps best known as its solution in water, usually called formalin. The water solution is used as a preservative for biological specimens and in embalming fluids. It is also used as a germicide and fungicide. The odor of formalin is probably unforgettable to those who have spent much time doing dissections in biology laboratories. Formaldehyde itself is extremely irritating to the mucous membranes of the respiratory system. It may also be a carcinogen.

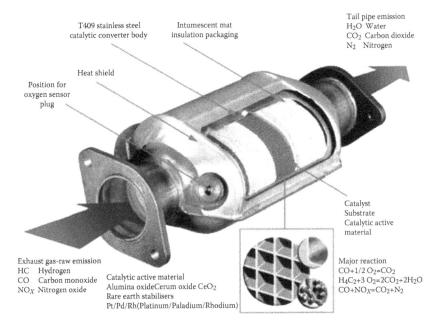

T409 stainless steel catalytic converter body

Intumescent mat insulation packaging

Tail pipe emission
H₂O Water
CO₂ Carbon dioxide
N₂ Nitrogen

Heat shield

Position for oxygen sensor plug

Catalyst
Substrate
Catalytic active material

Exhaust gas-raw emission
HC Hydrogen
CO Carbon monoxide
NO_X Nitrogen oxide

Catalytic active material
Alumina oxideCerum oxide CeO₂
Rare earth stabilisers
Pt/Pd/Rh(Platinum/Paladium/Rhodium)

Major reaction
$CO+1/2 O_2=CO_2$
$H_4C_2+3 O_2=2CO_2+2H_2O$
$CO+NO_X=CO_2+N_2$

FIGURE 28.3 A schematic diagram of a catalytic converter used in automobile exhaust systems. (From http://www.cateran.com.au/cateran/data.aspx; http://www.redbackmufflers. com.au/Cateran/images/technical.jpg.)

hydrocarbons will decrease. As we move from fuel-lean to stoichiometric to fuel-rich conditions, the amount of heat produced decreases, the temperature in the engine cylinder will be reduced, and thermal NO_x formation drops. Figure 28.4 summarizes this.

The catalytic converter works best when the concentrations of all three potential pollutants are high, around the point at which the curves in Figure 28.4 meet. The point where we get the highest simultaneous concentration of carbon monoxide, unburned hydrocarbons, and NO_x is right around stoichiometric combustion. In other words, the point of stoichiometric combustion in the engine is where the catalytic converter works best. It also happens that near-stoichiometric combustion is where the engine works best. Thus, keeping the engine tuned up, and running near the stoichiometric condition, helps to reduce emissions to the environment (because this is where the catalytic converter works best), to get the best performance from the engine, and, likely, to reduce the need for major maintenance on the engine.

There is little question that the catalytic converter is a major technological success. Fifteen years after the mandated introduction of the catalytic converter in the United States, the average car emitted 96% less carbon monoxide, 96% fewer hydrocarbons, and 76% less nitrogen oxides than did the cars before catalytic converters were mandated. Despite that, vehicle exhaust pollution remains a serious and growing problem around the world. This is *not* a result of poor design or improper functioning of the catalytic converter. The problem, rather, is that there are a lot

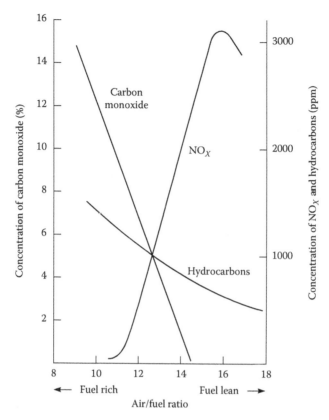

FIGURE 28.4   This graph shows how levels of tailpipe emissions vary as combustion conditions in the engine change from fuel rich to stoichiometric to fuel lean.

more people, and more people able to buy cars, and those who own them are driving more. The increasing number of cars on the roads around the world, and the greater use of these cars, has offset these emission reductions per car achieved by the use of catalytic converters.

REFORMULATED GASOLINES

Some thirty years ago, it was already apparent that, even with converters, further reductions in emissions were needed. Although there likely remains some room for improvement, the present-day catalytic converter is approaching the limits of what can be done in the tailpipe of the car. The next step is actually to change the fuel—that is, reformulate the gasoline being burned.

The approach to *reformulated gasolines* has come largely from the United States, and is based on several observations. The lower the boiling point of a substance, the easier it is to evaporate. Evaporation losses can be reduced by eliminating the lowest boiling components—in other words, by shifting the boiling range of the gasoline upward. Aromatic compounds contribute to smoke and soot formation and are suspect carcinogens. The amount of such compounds in gasoline is severely limited. *Oxygenates*, compounds containing at least one oxygen atom, such as methyl tertiary-butyl ether (MTBE)* and ethanol, are, in a sense, already partially combusted. When they burn in the engine their combustion temperature is lower, which helps reduce formation of thermal NO_x. Furthermore, oxygenates bring some oxygen in with the fuel (chemically combined with fuel molecules), in addition to what comes in with air, shifting the combustion to fuel-lean conditions. As a result, carbon monoxide and unburned hydrocarbon formation are reduced. As an added bonus, many oxygenates have very high octane numbers, and their addition boosts the octane rating of gasolines (which would otherwise be reduced by removal of aromatics). Thus the production of reformulated gasoline involves shifting the boiling range up, removing aromatics, and adding oxygenates.

Currently the favored oxygenate is ethanol, to be discussed in more detail in Chapter 31. Ethanol can be produced from biomass sources such as sugar cane and corn, which are, in principle, renewable resources. Use of biomass-derived ethanol as a gasoline additive is now also encouraged by the European Union. Given increasing concerns about global climate change (Chapter 29) and long-term availability of petroleum (Chapter 30), it may be that using ethanol as an additive to gasoline will begin a transition to ethanol as a replacement for gasoline.

LEVs, HYBRIDS, AND ELECTRIC CARS

The catalytic converter was, and is, a superb technological advance in fighting air pollution caused by motor vehicles. The use of reformulated gasolines adds another step toward a cleaner environment. There still remain, though, areas where smog presents a problem. The catalytic converter represents a change to the conventional automobile. Reformulation changes the fuel. If these two together are not adequate to improve air quality to acceptable levels, then the last major step to be taken is to change the vehicle itself. Protection of air quality by addressing emissions from motor vehicles has led to consideration of low-emission vehicles (LEVs), and their evolutionary descendants, ultra-low-emission vehicles (ULEVs), super-ultra-low-emission vehicles, and zero-emission vehicles (ZEVs). From there, the possibilities are to do away with the internal combustion altogether, or to combine a small gasoline engine with some other form of propulsion (usually electric motors).

Terms such as LEV, ULEV, and ZEV often tend to be used in a rather loose and general sense. However, in some places, they have strict legal definitions. For a

---

* In the 1980s and 1990s, the use of MTBE as a gasoline additive was considered a splendid advance in the fight against air pollution from vehicles. Its future seemed so bright that there were concerns that there might not be enough production capacity to meet the expected demand. Then it was discovered that, if oxygenated gasoline leaks from storage tanks and gets into the water supply, some of the MTBE in the gasoline will dissolve in the water. Even small quantities of MTBE in water give it an offensive odor and taste. Concerns have also been raised about the possible toxicity of MTBE ingested with drinking water. Though the human health effects of MTBE may be less severe than once feared, nevertheless the use of MTBE is being eliminated in many places. Likely, MTBE will join a huge array of other technological developments of which it could be said, "It seemed like a good idea at the time."

FIGURE 28.5 Ferdinand Porsche and the world's first hybrid car. The Lohner-Porsche Mixte Hybrid was the first hybrid vehicle, developed in 1901 by Ferdinand Porsche. It was a series hybrid using wheel hub motors mounted in each wheel, and powered by electricity delivered from both batteries and a small generator. (From http://en.wikipedia.org/wiki/File:Lohner_Porsche.jpg.)

vehicle to qualify as, for example, an ultra-low emission car, it must meet specified and quantitative values for emissions. Super-ultra-low emission vehicles are hybrids.

Ferdinand Porsche* built the first hybrid vehicle (Figure 28.5), one that used both an electric motor and an internal combustion engine, in 1899. Hybrid vehicles can be thought of as being electric cars that are equipped with a small internal combustion engine in addition to the battery-driven (or fuel-cell-driven) electric motor. This gives the immediate advantage that they are not limited in driving range by the energy in the batteries. The range of the hybrid car is comparable to that of conventional gasoline-powered vehicles. Indeed, hybrid vehicles offer virtually all of the advantages of gasoline vehicles combined with most of the advantages of electric vehicles. A hybrid vehicle has electric batteries and an electric motor connected to the drive wheels, along with a gasoline-powered engine. The gasoline engine has sufficient POWER for reasonable driving performance and confers the added advantage that there is no need to "plug in" the car to recharge the batteries. The batteries are kept charged by the generator connected to the engine. Fuel economy is claimed

---

* Ferdinand Porsche (1875–1951) played an extraordinary role in the development of automobiles. Not only did he create the first hybrid car, he is responsible for the Volkswagen Beetle, the Mercedes-Benz SS series, and many highly successful vehicles that carry his own name. His talents extended to the design of several of the most successful tanks used by the German army in the Second World War, the Junkers Ju-88 night fighter and tactical bomber, and the V-1 "buzz bomb." He served out his own military service, in the early 1900s, as the chauffer for Archduke Franz Ferdinand of Austria-Hungary. In 1914, with a different driver at the wheel, the Archduke was assassinated in Sarajevo, the act that set off the cataclysm of the First World War.

to be to 3.4 L/100 km; some anecdotal reports suggest even better results, but many indicate that closer to 6 L/100 km is more realistic. Of course, the fuel economy of any vehicle depends greatly on the conditions under which it's driven (the common comparison being heavy city traffic vs. open highways) and on the driving habits of the owner.

A hybrid car is not a ZEV, because the gasoline engine will still produce some emissions. However, because the hybrid will run on electricity some, even much, of the time, it offers very low emissions compared to conventional gasoline-fueled cars. Hybrid car technology has evolved to a point at which commercial models are available from most of the major car companies. Now that hybrids are increasingly used by the public, we will soon have a good body of experience to learn about their long-term durability and maintenance. Hybrids are comparatively complex because they have two drive systems, so may prove to be expensive to maintain.

Getting more and more LEVs into the total vehicle fleet can be required by legislation at state, regional, or national level. However, genuine acceptance is more likely to come from the public's wanting to own such cars, not being compelled to do so. A significant shift away from traditional gasoline-fueled internal combustion engine vehicles will occur when consumers decide that they really want hybrids or electric cars. What will that take? The gasoline-engine vehicle cannot be criticized on the basis of its performance; even small, inexpensive vehicles can easily reach speeds well above any legal limits. Today's vehicles also offer an extraordinary range of amenities and accessories. It's unlikely that hybrid or electric vehicles would be able to demonstrate clear superiority to gasoline-fueled vehicles either in speed or in comfort.

Consumer acceptance is a necessary criterion for the success of new vehicles, but not the only one. Modern gasoline-fueled vehicles are remarkably reliable. They usually require little major maintenance for many years and can now go for long periods between routine tune-ups. We are used to that. New kinds of vehicles will have to approach that level of technical reliability. Similarly, most of us, except perhaps in small towns and rural areas, are used to having a gasoline station within close proximity, often open round-the-clock every day of the year. A few minutes spent at the pump suffices to refuel the vehicle. We are used to that, too. Ideally, new vehicles will have to be equally easy to refuel or recharge, and there will have to be an infrastructure in place to allow that to happen. A major switch in consumer preference for a new type of vehicle won't be triggered by vastly improved comfort or speed, and there may even be some inconvenience in refueling or recharging. Likely, a wide acceptance of electric or hybrid vehicles may require a significant change in consumers' attitudes, a change perhaps brought about by concerns for the environment.

It is unrealistic to expect that there will be an instantaneous change-over from gasoline vehicles to electrics or hybrids. Rather there will likely be a long period of transition when all of these kinds of vehicles are on the road. The most likely "transition car" from the gasoline vehicle to an all-electric vehicle is the hybrid.

Early in the twentieth century, electric cars, using lead-acid batteries, were actually more popular than gasoline-powered vehicles (Figure 28.6). However, once car owners began to drive their newfangled vehicles over longer distances, it soon became apparent that gasoline offered an enormous advantage in energy density.

FIGURE 28.6 Thomas Edison shown with an electric car, around 1918. (From http://www. nps.gov/nr/twhp/wwwlps/lessons/25edison/25visual1.htm.)

The energy density is the amount of energy provided per volume, or per weight, of an energy source. Lead-acid batteries are heavy. (This can be verified by anyone who has ever changed the battery in his or her own car, or simply by feeling the weight of a battery in the auto parts section of a store.) Unfortunately, a lead-acid battery provides about 1% of the energy that can be obtained by burning an equal weight, or equal volume of gasoline. As a result, gasoline-fueled vehicles have a much greater range, and quickly established their dominance in transportation. Ten years ago, for every one electric vehicle on the road, there were about eighty thousand gasoline-fueled cars and light trucks. Today, there are over a dozen makes or models of electric cars available, at least in some countries. The leader, in terms of sales, appears to be the Nissan Leaf (Figure 28.7).

Electric cars are perceived to have significant range and speed limitations. Both of these problems result from limitations in present battery technology. In the most modern technology, lead-acid batteries have been replaced by lithium batteries. Lithium batteries for automobile use are similar in concept to the batteries used in mobile telephones and laptop computers. Their much lower weight than a comparable lead-acid battery provides a tremendous advantage. Driving ranges of about 120 km seem reasonable, though some manufacturers indicate more in their promotional literature. The average commuting distance is about twenty kilometers, and 90% of all vehicle trips for any purpose are still under a hundred kilometers in length. On the other hand, for those who must frequently drive long distances, or who must have some sort of emergency capability in a vehicle, electric cars could still serve as a second vehicle. With advanced battery technologies, now a subject of a vigorous

FIGURE 28.7   The Nissan Leaf, a successful example of one of the new generation of electric cars. (From http://en.wikipedia.org/wiki/File:Nissan_Leaf_005.JPG.)

research and development effort, electric cars will be designed and built to accelerate rapidly and to cruise easily at interstate highway speeds.

Though some electric vehicles still retain a small gasoline engine, they certainly offer tremendous advantages for the possible reduction of smog, global climate change (Chapter 29), and acid rain. They have no emissions of unburned hydrocarbons, NO_x, SO_x, carbon dioxide, carbon monoxide, or soot. Furthermore, an electric motor is quieter in operation than internal combustion engines, so electric cars would also help mitigate "noise pollution" associated with vehicle traffic. However, when considering an electric car as a ZEV, it is important to remember that the adjective zero emission refers to the *vehicle*. The electricity used to charge the batteries has to come from some place. If that electricity is generated in a fossil-fuel–fired plant, there will be emissions to the environment from the plant itself. To be sure, controlling emissions from one large, stationary electricity-generating plant is much easier than trying to control emissions from millions of internal combustion engines driving around. Furthermore, the electricity-generating plant might be in a relatively remote area, and not contributing its emissions to a congested downtown. It would be possible to have a zero-emission transportation system if the electricity used for charging the batteries came from such sources as solar (Chapter 33), wind (Chapter 32), hydroelectric, or nuclear plants. Electric cars can also be designed to make their own electricity onboard in fuel cells.

All storage batteries, regardless of specific design, operate on the same general principle. During charging, electrical ENERGY is converted into chemical ENERGY. When the battery is discharging (i.e., providing electricity to us), the reverse process occurs, and chemical ENERGY is converted back to electrical ENERGY. One potentially great advantage is that it does not really matter where that original electricity came from. That is, the electricity could be generated in fossil-fuel, nuclear, solar, hydro, or wind plants, providing enormous flexibility for an overall energy economy. In contrast, choices of fuels for internal combustion engine vehicles are limited. However, in considering this advantage of electric vehicles operated by storage batteries, we must also take into account the demand for electricity that these cars would create. Replacing a fraction of the transportation energy now provided by gasoline with electricity for storage batteries would increase the total load on electricity-generating capacity. Replacing *all* gasoline-derived energy with electricity would likely triple our electricity demand, requiring substantial investments in new generating capacity.

Additional advantages of electric cars come from the motors and drive systems. Electric motors are inherently more efficient than internal combustion engines. When an electric car is stationary, the motor does not run, so does not consume energy. In a conventional vehicle the internal combustion engine is still running, consuming fuel, and producing emissions when the engine is idling. Electric propulsion systems are much simpler—in terms of the number of components—than those for internal combustion engine vehicles, so are also lighter and smaller. That means they are also less costly to produce and, with fewer parts, need fewer repairs. Periodic tune-ups are eliminated. Electric cars can be designed to take advantage of so-called regenerative braking. In this technology, when an electric car slows down, the braking energy is recaptured and returned as electricity to the battery. In an internal combustion engine car, the energy generated (often as heat) in braking is wasted.

The charging time, for completely discharged batteries in an electric car, is estimated to be about six to eight hours. Therefore the most convenient period for the owner to recharge them would be overnight. Some commuters have to put up with very long commutes, possibly close to the vehicle's driving range. For those situations, vehicles could be recharged during the day while the driver is at work, if shopping malls and places of employment offered "plug-ins" for people to recharge their vehicles. The lengthy charging time would pose a serious limitation for using electric cars in long-distance travel. However, lithium batteries also offer a "fast charge" cycle, in which the battery could be recharged to approximately 80% capacity in about a half-hour. Doing so does not seem to be recommended as standard practice. An intriguing concept to solve this problem is that the car owner would own the vehicle itself but not the battery pack. When the batteries were getting low, the driver could pull into a "battery station" where the entire set of discharged batteries would quickly be swapped for another complete set of charged batteries—no doubt for an appropriate fee.*

For all their potential virtues, electric vehicles are not likely to eliminate gridlock, traffic accidents, urban sprawl, or any of the other impacts discussed in Chapter 20. Taking care of those problems will not be solved by a change in vehicle technology. Rather, major societal changes, and changes in individual behavior, will be needed. Such changes may not be made willingly.

It may prove that the future of electric and hybrid vehicles lies not with batteries, but with fuel cells. A fuel cell converts chemical ENERGY directly into electrical ENERGY, just like a battery does, and, also like a battery, has no potentially harmful emissions if operated using hydrogen. The fundamental idea is implicit in the name:

---

* Lest this scheme seem too fanciful, in the early days of domestic electric lighting there was a comparable idea. The lights and appliances in a house would be operated from a group of storage batteries in the basement. When a battery became depleted, the householder, or, more likely, a servant, would leave the battery by the front door, and a delivery man, coming by on some set schedule, would replace it with a freshly charged battery. This scheme was referred to as the "milk bottle plan" since it was quite similar to the once-common system for home delivery of milk.

a *fuel* is converted to electricity in a *cell*. The earliest concept of the fuel cell derives from the work of Humphry Davy in the eighteenth century. Little was done to convert Davy's original concept into a practical, commercial device until the advent of the space program. Fuel cells were quickly found to be useful for operating the electrical equipment of space vehicles. Fuel cells have been used to generate electricity for large communities (e.g., in sections of Tokyo) and to run buses. Today fuel cells, like batteries, are under intense research and development in numerous laboratories around the world. Many designs are being studied. The fundamental concept involves conducting, in the fuel cell, a reaction chemically similar to combustion, but instead of releasing the ENERGY of the reaction as heat, it is produced as electricity instead. One promising reaction is that of hydrogen. Hydrogen readily burns in air by uniting with oxygen to produce water and heat. The same reaction takes place in the fuel cell, but now the reaction occurs by the relatively slow transfer of electrons. These electrons are transferred via a current in an external circuit, which can be made to do WORK. Fuel cells can be stacked together to increase both the voltage and the current. They have no moving parts, are quiet, and environmentally clean. Fuel cells operating on hydrogen produce only water as an emission.

Fuel cells offer the advantage of the battery in an electric vehicle, but without its shortcomings. As long as fuel (e.g., hydrogen and oxygen in the example above) is supplied to the cell, it will continue to operate. In a sense, we can think of a fuel cell as a kind of battery that will never go dead. Though hydrogen is thought to be the ideal fuel for a fuel cell, there are problems to be overcome in handling and storing hydrogen for vehicle use, particularly to give the vehicle a reasonable driving range. The alternative to storing hydrogen onboard is to make it as needed. The vehicle would be fueled with a material easier to handle and store such as gasoline, methanol, or methane. Then, as the vehicle is operating, the original fuel is "re-formed" to provide hydrogen. For example, methane will react with steam to produce a mixture of carbon monoxide and hydrogen. The carbon monoxide by-product needs to be absorbed or destroyed somehow. However, a ready supply of hydrogen is available without needing to store it on board the vehicle. An advantage of relying on gasoline is that the fueling infrastructure is already in place. Furthermore, being able to "fill up" one's fuel cell vehicle at a conventional, familiar gas station may help greatly with consumer acceptance. A second advantage is that gasoline has a much greater energy density than hydrogen, thus providing a much greater driving range. Critics of the onboard re-former concept say that it converts the car into a rolling chemical factory, with attendant need for careful maintenance.

FURTHER READINGS

Fletcher, S. *Bottled Lightning*. Hill and Wang: New York, 2011. This book discusses the technology of lithium batteries, including the applications in electric cars. Just as the use of petroleum has geopolitical implications, so does an increasing use of lithium—a topic also covered in this book.

Gebelein, C.G. *Chemistry and Our World*. Brown: Dubuque, IA, 1997; Chapter 18. A well-illustrated introductory college-level chemistry text, with plenty of real-world examples. This chapter discusses air pollution as related to energy use, including photochemical smog.

McNeill, J.R. *Something New under the Sun*. Norton: New York, 2000; Chapter 3. An excellent book, subtitled 'An environmental history of the twentieth-century world.' This chapter treats the urban history of the atmosphere, including the impacts of emissions from automobiles on urban air quality.

Moore, J.W.; Stanitski, C.L.; Wood, J.W.; Kotz, J.C.; Joesten, M.D. *The Chemical World*. Saunders: Fort Worth, TX, 1998; Chapter 14. This book is an introductory textbook for college-level chemistry courses. A good discussion is provided on the topics of chemical reactions in the atmosphere and urban air pollution.

Schlesinger, H. *The Battery*. Smithsonian Books: New York, 2010. An excellent discussion of the development of batteries, from the beginnings to modern technology. Not specifically directed to electric cars, but does provide a good overview of batteries and how their development impacted various areas of technology.

Sperling, D.; Gordon, D. *Two Billion Cars*. Oxford University Press: New York, 2009. This book covers, among other things, electric cars, hybrids, ethanol, fuel cells, and the possible need to change consumption and driving habits.

29 Global Climate Change

If rain, which is so necessary to vegetable productions, and if fire, which is so necessary to animal life and the internal constitution of this earth, have such an intimate connection with the atmosphere of the globe, it is surely an object for the philosopher, to trace the chymical process of nature by which so many operations are performed.

—Hutton[1]

OUR PLANET IS A GREENHOUSE

During the daylight hours, a growing plant is warmed by ENERGY in sunlight. At night, when the heat of the sun is temporarily no longer available, the plant will cool. It does this because heat flows spontaneously from a warm object—in this case, the plant—to a cooler one—the air surrounding the plant. The plant loses heat by convection, when currents of warm air rise from the plant, and by radiation, when the plant likely radiates heat as infrared radiation. An energy balance equation applies to this plant, just as to any other SYSTEM:

$$E_{in} - E_{out} = E_{stored}$$

For this system, E_{in} is the ENERGY provided by sunlight and E_{out} is the ENERGY lost by convection and radiation of infrared light.

A plant in a greenhouse (Figure 29.1) will also be warmed by the ENERGY provided by sunlight. The sun emits ENERGY in many regions of the electromagnetic spectrum. Much of the ENERGY received on Earth is in the "visible" region of the spectrum. Ordinary glass is, obviously, transparent to visible light. (Glass would be a much less useful material to us if it weren't transparent.) ENERGY brought in by the sun's light helps to warm the contents of the greenhouse (Figure 29.2). Once again, when daylight hours are over, the warmth of the sun is no longer provided and the plant will cool. In this case, though, two factors intervene to retard the cooling process. First, the roof of the greenhouse helps to trap convection currents of warm air and keep them inside the greenhouse. Second, and more importantly, the glass helps trap infrared radiation.

Any object hotter than its surroundings will attempt to come to some equilibrium temperature by transferring heat to those surroundings. The region of the spectrum in which a hot object radiates depends upon its temperature. Some objects radiate heat in the visible region, an example being the heating coil on an electric stove set on "high." We can see this, and speak of an object like the heating coil as being "red hot." Objects only slightly warmer than their local environment do not radiate in the visible region, but instead radiate in the lower energy (longer wavelength) infrared region of the spectrum. Humans, for example, with body temperatures of

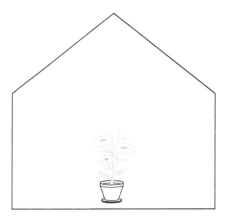

FIGURE 29.1 A plant growing in a greenhouse provides a simple model for understanding planetary-scale climate change.

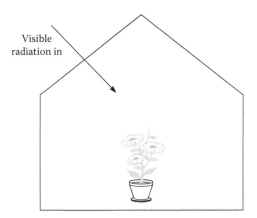

FIGURE 29.2 A significant source of ENERGY supplied to the greenhouse derives from sunlight coming from the glass walls and roof.

37°C, most of the time are only slightly warmer than our surroundings, so we radiate in the infrared. The essence of operation of a greenhouse is that, while ordinary glass is transparent in the visible, it is nearly opaque in the infrared. Radiated heat is trapped inside the greenhouse (Figure 29.3). In the case of a real greenhouse, a second mechanism helps retain heat inside: convection is limited because the roof keeps currents of warm air from rising completely out of the greenhouse. The energy balance equation also applies to this new SYSTEM (i.e., plant plus greenhouse):

$$E'_{in} - E'_{out} - E'_{stored}$$

In this equation, the "prime" signs (') have no significance other than to signal that we're writing the equation for a different SYSTEM.

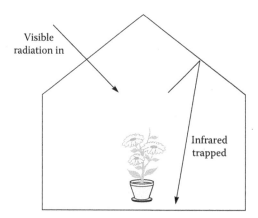

FIGURE 29.3   Some ENERGY is trapped in the greenhouse because the glass is opaque to infrared radiation. This trapping of ENERGY reduces the amount that would otherwise be lost by being radiated into the environment.

The ENERGY coming in is the same in both SYSTEMS; it's the ENERGY in the sunlight. But, E'_{out} is smaller than E_{out}, because some of the heat has been trapped in the greenhouse by its roof and by the glass being opaque to infrared radiation. If the ENERGY coming in is the same in both cases, but the ENERGY going out has decreased for the greenhouse, then inevitably E'_{stored} must be larger than E_{stored}. That is, ENERGY stored in the greenhouse is larger than the ENERGY stored in the plant not in the greenhouse. This is easy to verify by making up some simple numbers for the various terms in the two energy balance equations and doing the necessary arithmetic. For example, suppose the incoming solar radiation is 100, in some arbitrary units, and that the plant by itself loses 60 such units, whereas with the greenhouse only 25 units are lost. Then the ENERGY stored by the outside plant is 40 units, while the ENERGY stored in the greenhouse is 75 units. More ENERGY is stored in the greenhouse.

The observable effect of the larger value of E_{stored} for the greenhouse relative to the plant outside is that the greenhouse is warmer. More ENERGY is stored as heat. We sometimes do greenhouse experiments ourselves. Almost everyone has had the experience of leaving a car parked in direct sunlight on a hot summer day with all the windows up. After several hours, the interior of the car becomes hot enough to be painful to touch, or to melt candies or plastics left inside. In this case, the car's roof and windows trap much of the heat provided by the sun (again by limiting heat loss via convection and infrared radiation), and the temperature inside soars. Of course, the ability of the greenhouse to maintain an artificially warm temperature inside is the reason why we build greenhouses in the first place—to provide an environment for growing fresh flowers, fruits, or vegetables in the winter or for growing plants in regions of cool climates that normally require tropical or subtropical climates.

Each day our planet receives ENERGY from the sun (Figure 29.4). The energy reaching the surface of the planet is mostly in the visible and infrared regions of the spectrum. This radiation is absorbed by the Earth, warming the continents and oceans. The current average temperature of the planet, about 15°C, is much higher

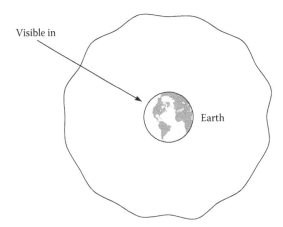

FIGURE 29.4   The primary source of ENERGY for our planet derives from sunlight coming through the atmosphere.

than the −270°C of outer space. Consequently, the Earth will try to transfer heat spontaneously to the cooler (in this case, frigid!) surroundings of outer space.

Our atmosphere is like glass, in the sense that it is transparent to visible radiation. During the night, those objects that were heated during the day will cool by radiating their heat to outer space, in the form of infrared radiation (Figure 29.5). Thus, our planet has achieved its own energy balance, in which the amount of ENERGY stored as heat maintains the average temperature that we experience. The principal components of our atmosphere, nitrogen and oxygen, are essentially transparent to infrared radiation. However, the atmosphere also contains small amounts of other gases that absorb some infrared radiation (Figure 29.6). Without these gases, the average temperature of our planet would be substantially lower than it actually is. The average temperature of the Earth results, in part, from the trapping of infrared radiation in the atmosphere by

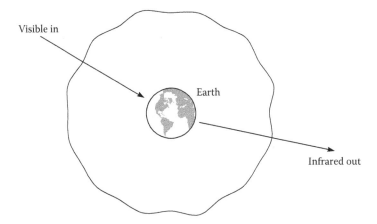

FIGURE 29.5   At night, the part of the planet that is no longer in direct sunlight will cool by radiating infrared back into space, through the atmosphere.

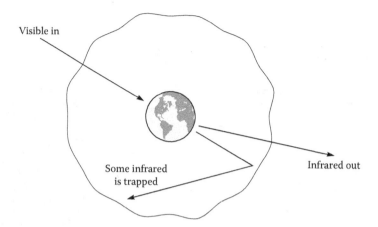

FIGURE 29.6 Some of the infrared radiation that otherwise would carry heat back into outer space is trapped by components of Earth's atmosphere.

some of the atmosphere's components. Because these components—carbon dioxide and water vapor being examples—act somewhat like the glass of a greenhouse, such infrared-absorbing gases are called *greenhouse gases*. By extension, this warming of Earth by trapping infrared is commonly called the *greenhouse effect.*

Infrared radiation does not become trapped permanently in molecules of greenhouse gases. If that were to happen, the absorption of more and more infrared radiation would add more and more ENERGY to a molecule, until enough ENERGY had been absorbed to cause the molecule to break apart. When a molecule absorbs infrared, it contains more ENERGY than it had previously, just like water can absorb WORK from a pump and move to a higher state of gravitational potential energy (a fancy way of saying that the water is pumped uphill). Eventually, the water will release that ENERGY when it flows back downhill. A molecule of a greenhouse gas gets pumped up an "energy hill" by absorbing infrared, increasing its chemical potential energy. At some point the molecule will release ENERGY by reverting back to its preferred "downhill" state. The ENERGY that is released is radiated equally in all directions. Some of that ENERGY, therefore, is radiated right back to Earth's surface and does not escape into outer space. The net effect is exactly the same: the E_{out} term becomes smaller and E_{stored} increases.

If there were no greenhouse effect, then the average temperature of Earth would be a balance between the energy provided by the sun and that radiated back into outer space. Earth would have an average temperature of −18°C, so that most of the water on the planet would be frozen year-round. It is very questionable whether life as we know it could have evolved in such an "ice world." But even if it had, certainly our lives would be much different today if we had to cope with such temperatures. The greenhouse effect is essential in keeping our planet habitable for the species that have evolved here.

The composition of the atmosphere plays the crucial role in the Earth's enjoying an average temperature about 33°C warmer than would be expected by balancing solar radiation with infrared losses. Though water vapor and carbon dioxide represent

FIGURE 29.7 John Tyndall (1820–1893) lecturing at the Royal Institution. Tyndall was one of the first to explain the trapping of infrared radiation by components of the atmosphere. (From http://en.wikipedia.org/wiki/File:TyndallLecturingAtTheRoyalInstitution,1870, LondonIllustratedNews.jpg.)

only very small fractions of the total composition of the atmosphere (some 10,000–50,000 ppm and 350 ppm, respectively), they provide significant contributions to the greenhouse effect. The notion that the atmosphere might be involved in trapping some of the sun's heat originated with the French mathematician and physicist, Joseph Fourier.* About two hundred years ago, Fourier developed the comparison of the function of the atmosphere to that of the glass in a greenhouse. However, he did not identify which components of the atmosphere were responsible, nor how they acted. Sixty years after Fourier's original idea, John Tyndall (Figure 29.7), an English physicist,[†] demonstrated that carbon dioxide and water vapor absorb heat

---

* Jean-Baptiste Joseph Fourier (1786–1830) was one of the most brilliant intellects of his era, the late eighteenth and early nineteenth centuries. He made many original contributions to mathematical physics, including areas of heat conduction, harmonics in music, and prediction of tides. In addition, Fourier was keenly interested in ancient Egypt and accompanied Napoleon on his ill-fated invasion of Egypt in 1798. Fourier also published books on Egyptology. Fourier's discovery of the greenhouse effect came from his original calculations suggesting that an object the size of Earth and with its known distance from the sun should be much colder than it really is, assuming that the only source of energy was sunlight.

[†] John Tyndall (1820–1893) was one of many outstanding experimental physicists in nineteenth-century England and was a good friend of Michael Faraday. Perhaps Tyndall's most famous contribution is his explanation of why the sky is blue: Some of the sun's rays are scattered by molecules in the atmosphere. He also did experiments on the concept of spontaneous generation—the notion that persisted for a long time in science that living organisms could arise spontaneously from nonliving matter, for example, that bacteria found on decaying food somehow originated "spontaneously" from the food. Tyndall's experiments were the final nail in the coffin for this deservedly dead idea. Tyndall's interest in studying Alpine glaciers led him to become an early enthusiast for the sport of mountaineering.

and calculated the effect that would result from these two compounds in the atmosphere. In principle, any molecule that can absorb infrared radiation is potentially a greenhouse gas. Many such substances, in addition to the carbon dioxide and water, include methane, the principal component of natural gas; nitrous oxide, one of the potential contributors to NO_x; ozone; and chlorofluorocarbons, formerly used in refrigeration and air conditioning.

The atmosphere does not provide a mechanism only for warming. Other atmospheric effects can act to cool the planet. Besides acting as a greenhouse gas, water vapor also forms clouds, which help cool Earth by reflecting sunlight. Aerosols from volcanoes also absorb and reflect radiation, and usually act to cool the climate. When a major volcanic eruption occurs, for example Mount Pinatubo in the Philippines in 1991, unusually cool summers and harsher winters can persist for several years.

The composition of the atmosphere has a major role in determining the average global climate. As we've seen, without naturally occurring carbon dioxide and water vapor in the atmosphere, Earth would be a grim place to live. Any changes to the composition of the atmosphere could have potentially drastic effects on the environment and on life. We refer generally to such impacts as *global climate change*. Current attention focuses on increases in greenhouse gas concentrations. Because there now exists an abundance of evidence that the planet is warming; the impact is also referred to as *global warming*.

The warming of the planet by some 33°C as a result of the carbon dioxide and water vapor that occur naturally in the atmosphere is the *natural greenhouse effect*. It exists and occurs without human intervention. Without naturally occurring greenhouse gases in the atmosphere,

> *... the warmth of our fields and gardens would pour itself unrequited into space, and the sun would rise upon an island held fast in the iron grip of frost.*
>
> —Tyndall[2]

We know that carbon dioxide captures and returns the infrared radiation coming from the surface of the Earth. If the concentration of CO_2 in the atmosphere were to increase, the effect of this increase would reduce the value of E_{out} in Earth's energy balance. More than a century ago, the Swedish chemist Svante Arrhenius (Figure 29.8)* calculated that doubling the concentration of CO_2 would result in an increase of 5°C–6°C in the average temperature of the planet's surface. This would be global climate change on a grand scale!

---

* Svante Arrhenius (1859–1927, Nobel Prize, 1903) made numerous contributions to the newly developing discipline of physical chemistry, most notably in the areas of the behavior of electrolyte solutions and in the theoretical treatment of rates of chemical reactions. He taught himself to read at age three, quickly mastered arithmetic, and started school in the fifth grade. Arrhenius received his doctoral degree from the University of Uppsala in Sweden. His dissertation was considered to be fourth rate by the faculty, but they later raised their evaluation to third rate. It is said to have taken the personal intervention of Wilhelm Ostwald, one of the world's greatest chemists of that era, to persuade the Uppsala faculty that Arrhenius's work on the dissociation of electrolytes in solution did indeed have merit. It certainly did; it was the basis for Arrhenius's Nobel Prize in Chemistry.

FIGURE 29.8   Svante Arrhenius (1859–1927), who was the first scientist to calculate the warming effect that could be attributed to absorption of infrared radiation by carbon dioxide and water vapor. (From http://commons.wikimedia.org/wiki/File:Svante_Arrhenius_01.jpg.)

By the influence of the increasing percentage of carbonic acid in the atmosphere we may hope to enjoy ages with more equable and better climates, especially as regards the colder regions of the earth, ages when the earth will bring forth much more abundant crops than at present for the benefit of rapidly propagating mankind.

—Arrhenius[3]

Atmospheric CO_2 concentrations have been increasing steadily for well over a century. The effect of increased atmospheric CO_2 is to trap even more infrared and thus to augment or enhance the amount of warming of the planet beyond that provided by the natural greenhouse effect. This additional heating by increased concentrations of CO_2 is then called the *enhanced greenhouse effect*. Since the increased CO_2 primarily results from human activities, it is also called the *anthropogenic greenhouse effect*.

A significant fraction of the increased CO_2 results directly from human use of energy, and especially from the combustion of fossil fuels. In the United States, which has the second-highest total CO_2 emissions (behind China), and the second-highest per capita emissions (behind Saudi Arabia), fossil fuels provide 99% of the transportation energy and about 75% of the primary energy used to generate electricity. A strong push to reduce anthropogenic CO_2 emissions—either through legislation and regulation or through collective societal decisions—will result in the enhanced greenhouse effect being the greatest of all issues to impact the use of energy in the future, and the sources from which we obtain it. Many complex issues are involved: scientific and technical issues related to the utilization of energy sources and the reduction of greenhouse gas emissions and societal issues of choices relating to energy use and energy conservation. Sadly, resolving these issues is made all the more complicated by the injection of politics. In the United States, in particular, it has now become a tenet of political orthodoxy of the party representing the extreme right that global warming is not occurring,* and that those who claim that it is occurring are part of a massive fraud.[†]

AN ENHANCED GREENHOUSE EFFECT: EVIDENCE FOR GLOBAL CHANGE

Much of the focus in discussions of global climate change is on carbon dioxide, which seems most closely linked to anthropogenic sources. Over the past 400,000 years, during which there were four major glacial cycles, atmospheric CO_2 concentrations averaged about 240 parts per million (ppm), with upward or downward swings of about 20%. That period of Earth's history also witnessed four major cycles of glaciation and subsequent retreat of the glaciers—essentially four large swings in global climate—just with this relatively small variation in CO_2. For the period from A.D. 800 to about 1800, the CO_2 concentration in the atmosphere lay in the range 270–290 ppm. A rapid increase began in the first half of the nineteenth century, so that by 1900 the concentration had hit 295 ppm. Acceleration of the CO_2 concentration increase continued, leading to values of 310–315 ppm by the middle of the twentieth century, hitting 360–370 ppm at the dawn of the twenty-first century and 397 ppm in 2012.

Atmospheric levels of CO_2 have been measured directly since 1957. The concentration records earlier than the 1950s consist of evidence from many different sources. Analyses of ice cores serve as the primary data source because they provide a continuous record of past atmospheric composition. Based on these data, it can be said confidently that, for the last 18,000 years, atmospheric CO_2 concentrations fluctuated around 280 ppm, and that, in the relatively recent past, there has been an increase to a concentration of 360–370 ppm, with a current rate of increase of

---

* As the manuscript of this book was in final editing, the legislature of North Carolina enacted a law forbidding, for four years, anyone's acknowledging a rise in sea level as a result of global warming, particularly when considering development of property on the coast. A week before this law was passed, the sea level rise on a 1000 km stretch of the U.S. east coast, from North Carolina to Massachusetts, was reported to be some three to four times faster than the global average. A millennium before the law was enacted, King Canute (king of what are now England, Denmark, Norway, and Sweden) placed his throne at the edge of the sea and commanded the tide not to rise enough to wet his robes or his feet. Oddly enough, the tide paid no attention.

† Having spent 25 years in various positions of managing groups of scientists, I can attest to the fact that there are days when it is difficult enough to get two scientists to agree to do the same thing. The idea that thousands of scientists, representing about a dozen different scientific disciplines and coming from numerous countries, are somehow all in league to perpetrate an immense fraud deserves to be treated as the hilarious nonsense that it is. Unfortunately, demagogues play the gullible public for fools by shrieking about "the fraud of global warming," or that "it's not climate science, it's political science."

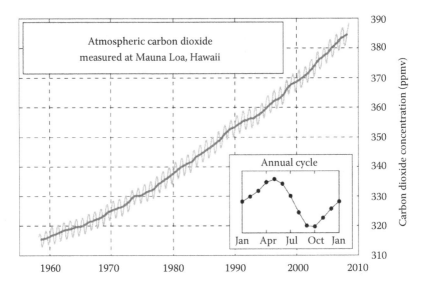

FIGURE 29.9   Changes in the concentration of atmospheric carbon dioxide over time, measured at Mauna Loa, Hawaii. Measurements at other places around the world confirm this trend. (From http://en.wikipedia.org/wiki/Carbon_dioxide_in_Earth%27s_atmosphere.)

1.5 ppm per year. Concentrations of CO_2 are now higher by at least 30% than the maximum concentrations that occurred prior to the rapid growth of CO_2 concentration in the past 150 years.

The modern rate of atmospheric CO_2 increase is accurately known from direct measurements of CO_2 in the atmosphere, as well as from analyses of air bubbles trapped in ice and firn.* The increase in atmospheric concentrations of CO_2 measured at the observatory at Mauna Loa, Hawaii, has been tracked since 1958 (Figure 29.9). On average, the rate of increase was 0.83 ppm per year during the 1960s, 1.28 ppm per year in the 1970s, and 1.53 ppm per year in the 1980s. Not only is the absolute amount of CO_2 increasing, but the rate at which the concentration is increasing is also growing. Three factors have a significant role in affecting atmospheric CO_2 concentration: growth in global population, in agricultural production, and in industrialization.

The very rapid increase in atmospheric CO_2 has occurred since the Industrial Revolution of the eighteenth century. The Industrial Revolution was fueled, both figuratively and literally, with coal. The twentieth century added its own contributions, especially in the great increase in petroleum consumption. Carbon dioxide is an inevitable product of the complete combustion of fossil fuels. Industrial production today exceeds what it was at the turn of the last century (1900) by some fifty times. In the next half-century, it will likely grow by another factor of five to ten. Much of the growth in industrial production has depended on the combustion of

---

* Firn is a form of solid water that is essentially a transition between snow and glacial ice. Snow that lasts through a summer melting season becomes firn; the firn becomes glacial ice when it freezes to an extent that liquid water can no longer trickle through it.

fossil fuels, either as a direct source of ENERGY or for electricity generation. The annual growth rate in ENERGY consumption is typically two to three percent, much of which derives from fossil-fuel sources. If present-day rates of fuel consumption continue unchanged, atmospheric concentration of CO_2 could double the level that existed in 1860 sometime between 2030 and 2050.

Earth's population tripled in the twentieth century and likely will double or triple again in the twenty-first century. Even if the additional people consumed ENERGY only at a subsistence level, they will still add to the increased use of ENERGY. However, they will, not unreasonably, aspire to something much more than a crude, impoverished subsistence-level existence. Increased ENERGY use provides the key to improving their standard of living, with readily available transportation, abundant electricity, access to education and health care, and all the other amenities of life that those of us fortunate enough to live in wealthy, industrialized nations often take for granted.

People have to eat. Conversion of forested land to agricultural use results in a redistribution of carbon from being locked up in plants or in the soil to being CO_2 in the atmosphere. Land-use conversion represents another anthropogenic factor in increasing atmospheric CO_2. Two trends drive this buildup: first, fossil-fuel combustion in agricultural machinery, fertilizer production, and transporting food to market and second, deforestation, nowadays mainly a problem in the tropics, but which a century ago occurred mostly in North America and the temperate parts of Asia. Deforestation accounts for about a quarter of the anthropogenic impact on atmospheric CO_2; fossil-fuel combustion accounts for most of the rest.

Drilling into the ice covering Greenland has provided about three kilometers worth of ice cores dating back at least 110,000 years. A second project at the Russian Vostok Station in Antarctica has yielded over 1.5 kilometers of ice formed over a period of 160,000 years. The ratio of deuterium ($_1H^2$) to ordinary hydrogen ($_1H^1$) in the ice can be used to estimate the temperature at the time the snow fell. Higher temperatures tend to increase the deuterium/hydrogen ratio in the rain or snow.* In addition, the bubbles of air trapped in the ice can be analyzed for carbon dioxide and other gases.

Around the world, average surface temperatures increased by about 0.8°C over the past hundred years (e.g., Figure 29.10), but 0.5° of that has been since 1980. When the first edition of this book was prepared in 2002, nine of the ten hottest years on record had occurred between 1987 and 1997. Now, nine of the ten hottest years on record have occurred since 2001; and all ten since 1998. For the northeastern United States, the summer of 2012 was the hottest ever. Similarly, the first edition mentioned that the 1990s were the hottest decade in about six centuries. Now the decade of the

---

* Water molecules made of atoms of $_1H^1$ have a lower molecular mass than molecules of so-called heavy water, in which the hydrogen atoms are $_1H^2$. Water made with ordinary hydrogen has a molecular weight of 18, while that of heavy water is 20. Following a general rule for molecules that are chemically similar but different in molecular weight, the lighter ones evaporate more readily. Because of this effect, there is relatively more $_1H^1$ and less $_1H^2$ in water vapor that has evaporated into the atmosphere than in liquid water in the oceans. Rain or snow that condenses from atmospheric water vapor will also reflect this enrichment of $_1H^1$. However, as average global temperatures warm, more molecules of heavy water will evaporate from the liquid state, and consequently the isotope ratio, $_1H^2/_1H^1$, in rain or snow also varies with average temperature. This is the basic concept in estimating ancient temperatures by analysis of the isotopic composition of ice.

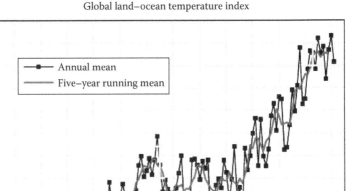

FIGURE 29.10 The global temperature anomaly since 1880–2010. It shows the same trend as the carbon dioxide concentration (Figure 29.9). (From http://en.wikipedia.org/wiki/File:Global_Temperature_Anomaly_1880-2010_%28Fig.A%29.gif#file.)

2000s can claim that distinction. Changes this big and this fast are within the natural range of variation for swings in average global temperatures. However, they rarely occur in the temperature record for the past two million years, most likely have not occurred at any time within the past ten thousand years, and definitely are unprecedented for the past six hundred years. Any anthropogenic changes in the energy balance of Earth, an increase in greenhouse gases or aerosols, will alter atmospheric and oceanic temperatures. As the atmosphere warms, changes in the amount of water vapor in the air, the amount of sea ice, and the extent of cloud cover will all work to amplify or reduce any warming that would lead to climate changes. The increase in the global mean temperature of about 0.8°C since the start of the twentieth century is associated with a stronger warming of daily minimum temperatures more than with increasing daily maximums. Consequently, there has also been a reduction in spread of temperatures noted in a 24-hour period.

For every country where data exist, the number of frost days has become fewer. This is consistent with the warming in average minimum temperature. As examples: In Australia and New Zealand, the frequency of days with temperatures below freezing decreased, with a simultaneous trend for warming in daily minimum temperatures. In northern and central Europe, the number of frost days has been decreasing since the 1930s, again associated with strong increases in winter minimum temperatures. In the past forty years, the elevation at which the temperature always remains below freezing has ascended over 150 meters up the sides of mountains in the tropics.

In the thirty years from 1940 to 1970, average global temperatures dropped by about 0.3°C. This change has been ascribed to a temporary increase in Earth's reflection of incoming sunlight back into space, caused by, among other things, an increase in volcanic activity that put more aerosols and dust into the atmosphere. However, over the long term, that temperature decrease vanished comparatively quickly, and the short-lived 0.3°C dip has now been totally overwhelmed by a steady increase in temperature.

For about the past 160,000 years, there appears to be a correlation between the concentration of carbon dioxide in the atmosphere and average global temperature. Currently, the carbon dioxide content of the atmosphere is rising. Humankind is consuming fuels—and thus producing carbon dioxide—at an unprecedented rate. The average global temperature is rising. Certainly there were periods of global warming long before humans began burning fossil fuels, long before there were humans, in fact. Further, CO_2 is a relatively minor greenhouse constituent in the atmosphere; there are other greenhouse gases besides CO_2. Generally, over the past 160,000 years, when the CO_2 concentration was high, average temperatures were high. Periods of high temperature have also been characterized by high atmospheric concentrations of methane. Such correlations do not necessarily *prove* that elevated atmospheric CO_2 and CH_4 caused the temperature increases. Presumably the converse could have taken place—high temperatures could have triggered natural phenomena that caused more CO_2 and CH_4 to enter the atmosphere. Unquestionably the fact is that the rise in CO_2 at ever-increasing rates coincides with the onset of extensive use of fossil fuels, also at ever-increasing rates. This strongly implicates carbon dioxide from human-related sources as a cause of recent global warming.

ROLES FOR OTHER GREENHOUSE GASES

Other greenhouse gases of concern include methane, nitrous oxide, ozone, and the chlorofluorocarbons (gases that formerly had been used extensively in refrigeration and air conditioning equipment). The effectiveness of these gases in contributing to global warming depends on their lifetime in the atmosphere, and their interactions with other gases and water vapor. These factors are combined into an indicator called *global warming potential* (GWP).

Methane, which has a GWP of 11, comes from a wide variety of sources. It is a major component of natural gas, so likely some methane has always leaked into the atmosphere naturally from porous or cracked rock in natural gas reservoirs. Human use of natural gas has likely led to increased emissions from leaks in gas wells, pipelines, and storage. Methane has always been released by decaying plant matter. This also represents a natural source of methane, but, as with natural gas, any human activity that increases decay of plant material also leads to increased methane release. Two such anthropogenic sources are the decaying organic matter in landfills* and

---

* Since it is unlikely the world will ever run out of garbage, and since methane is an excellent gaseous fuel, some localities have started to harness the natural methane production in landfills and use the gas as a source of ENERGY. This technology definitely works and offers a promising, albeit local source of fuel (Chapter 31), but the methane from most of the world's landfills is simply allowed to escape into the atmosphere.

decaying residue of cleared forests. Another major source of methane is cultivated rice paddies. Concentrations of methane now exceed previous high values by more than 250%. Methane has increased by 150% since the mid-eighteenth century.

The breakdown of organic matter under conditions leading to the production of methane also occurs in the digestive systems of cattle and sheep. There, bacteria decompose the cellulose of plant material and form methane. A healthy adult cow, placidly chewing its way through life, can add about five hundred liters of methane to the air every day. Some estimates suggest that methane emission from cattle and sheep are greater, by about fifty percent, than anthropogenic methane emissions. Fodder, taken in at one end of the cow, is readily converted to methane, which emerges from the other end. The tiny little digestive tracts of termites operate on a similar kind of chemistry, converting cellulose to methane. An individual termite is hardly a match for a cow as a methane source, but the termites' contribution is in their sheer numbers: there's about a half-tonne of termites for every man, woman, and child on the planet.

Over the past two decades, atmospheric concentrations of nitrous oxide have shown a slow but steady rise, about 0.25% per year. Natural sources include formation in lightning and volcanic gases—basically, natural thermal NO_x. Nitrous oxide concentration in the atmosphere is only about 320 ppb, but, molecule for molecule is over three hundred times better than CO_2 at trapping infrared. Major anthropogenic sources include artificial nitrogen fertilizers, discharge of sewage, and the burning of plant material (as when forests are cut and burned in land-use conversion). In addition to its role in the greenhouse effect, nitrous oxide at high altitudes in the atmosphere contributes to stratospheric ozone depletion. Near the surface of the Earth, the reactions of nitrogen oxides and hydrocarbons lead to the production of ozone. Ozone can also act like a greenhouse gas, but its action depends very much upon the altitude at which it occurs. It appears to have its maximum warming effect at altitudes of about 14,500 meters (i.e., in the upper troposphere). Loss of ozone has a cooling effect in the stratosphere and may also promote slight cooling at the surface of the Earth.

Chlorofluorocarbons, already implicated in the destruction of stratospheric ozone, also absorb infrared radiation, extremely strongly in fact. Because these gases have been implicated in the destruction of the ozone layer in the stratosphere, their manufacture has been stopped and their use is being eliminated. But, because an enormous number of refrigerators and air conditioners using these chemicals are still in use, atmospheric concentrations of the chlorofluorocarbons will continue to rise for some time, until those units reach the end of their working lives and are replaced by newer models. Atmospheric concentration of chlorofluorocarbons increased at about 6% per year into the 1990s but seems to have leveled off in recent years.

Atmospheric aerosols consist mainly of tiny particles of ammonium sulfate formed from sulfur dioxide released by natural or artificial sources. These particles promote global cooling by reflecting and scattering sunlight before it can reach the Earth's surface. Also, aerosols can enhance the condensation of water droplets and, consequently, enhance cloud formation. Clouds also increase reflection of incoming sunlight back into space. Aerosols counter the effects of the greenhouse gases, but have a much shorter lifetime in the atmosphere than greenhouse gases. The impact between volcanic emissions and climate was first noted by Benjamin Franklin, in an essay published in 1784.* A temporary drop in average global temperature that followed the eruption of Mount Pinatubo in 1991 was a consequence of the large volume of sulfur dioxide released by the volcano. SO_x emissions are a major contributor to acid rain. That SO_x also contributes to aerosol effects that counter the action of greenhouse gases has led to arguments—apparently made in all seriousness—that acid rain could be good for us! Suggestions have also been made for counteracting global warming by deliberately injecting aerosols at high altitudes into the atmosphere.

The role of clouds was dramatically highlighted on September 12–14, 2001. Worldwide air travel had been suspended on those three days, due to the terror attack on the United States on September 11. Airplanes, especially at high altitude, produce contrails (trails of condensed moisture from the engine exhausts); often from the ground it's easier to see the contrails than the airplane itself. With the whole world aviation fleet grounded for three days, there were no high-altitude contrails. Average ground temperature for those three days rose by about 1°C.

SIGNS OF GLOBAL CHANGE

The effect of an anthropogenic change in Earth's energy balance equation, that is, a reduction of the E_{out} term, should have as its primary impact an increase in average global temperature. However, temperature has, in turn, many impacts on weather patterns, ecosystems, glaciers, and human health. Temperature changes occurring will also be reflected in many related secondary effects.

When a crime has been committed, if the police have only one witness, there can be great uncertainty about the actual facts in the case. The witness may be totally mistaken in what he or she saw. The witness may even be lying for some reason. If there are two witnesses, the level of confidence increases somewhat, but care has to be taken to be sure that the witnesses are not somehow in collusion. When there get to be numerous witnesses, all saying essentially the same thing, and all independent of each other, then a high level of confidence can be assured. Nowadays there are numerous "witnesses" indicating the existence of global warming, and independently representing many different areas of science. There is good reason to accept with confidence that the planet is indeed warming.

Science often proceeds by creating models or hypotheses that can then be tested, either by experiments designed to do so or by collecting observations from events occurring in the world. These models or hypotheses can have the form of a statement such as "if X is true, then Y will happen or be observed." As more and more effort has been devoted to understanding global climate changes, models were able to predict changes in such phenomena as precipitation patterns, sea level rise, melting of ice, and extension of geographic ranges of certain plants or animals. (This list is by no means a complete catalog of global changes anticipated by climate models.) Nowadays many of the changes anticipated to result from global warming are indeed being observed. In some cases, the observations may not be in perfect agreement

---

* The essay is entitled "Meteorological Imaginations and Conjectures," published in the Memoirs of the Literary and Philosophical Society of Manchester. The complete essay can be found online.

with the predictions, but they are certainly "directionally correct,"[4] in that they do a good job of predicting trends. The confirmation of more and more of the "if...then" predictions of climate change modeling further increases the level of confidence that significant global warming is indeed occurring.

Changing temperature causes changes in the circulation patterns of the atmosphere and the oceans. These changed circulation patterns lead in turn to changes in weather patterns. Around the world, areas affected either by drought or by excessive wetness have increased. In the past decade, severe droughts have affected such places as Kenya, the Amazon basin, and western Australia. At the same time, other parts of the world experienced record-breaking flooding: Turkey, central Europe, and southeast Asia, as examples.

Heat waves also became increasingly prevalent, as did devastating storms, such as hurricanes. The worst hurricane year on record was 2005, the year that included Hurricane Katrina, which devastated New Orleans. Other than 2005, three of the most active years for hurricanes were 1996, 2010, and 2011.

A change in climate affects many aspects of the health of an ecosystem, such as the abundance and distribution of various species. Many biological processes undergo sudden shifts at particular "thresholds," either of temperature or of precipitation. The ability to tolerate frost or low levels of precipitation, as examples, can often determine the boundaries of the ranges of various plant and animal species. Changes in the number of days exceeding a particular temperature threshold for a certain species, or changes in the frequency of droughts or extreme precipitation, can lead to changes in some species in an ecosystem. Over a period of time, the cumulative effect of such changes could drastically affect distributions of many species.

For example, in the western parts of North America, the range of the Edith's checkerspot butterfly (Figure 29.11) shifted northward by nearly a hundred kilometers

FIGURE 29.11   Edith's checkerspot, the butterfly that is a signal of global warming. (From http://www.blm.gov/or/resources/recreation/csnm/csnm-gallery-insects.php.)

and upward on mountainsides by about 125 meters of altitude during the twentieth century. These shifts in range turn out to correlate quite well with the shifts in constant average temperature, which in the same period moved northward by about 105 kilometers and upward about 110 meters. In the Mesa Verde Cloud Forest preserve in Costa Rica, about twenty species (out of an original total of fifty) of amphibians have become extinct in the past two decades. During that time, three major periods of die-offs of frogs all occurred during unusually dry winters.

The increasing range of Edith's checkerspot results from the fact that warmer temperatures are, in effect, climbing up mountains. Unfortunately, in other parts of the world, mosquitoes, and the diseases they carry, are also moving upward. Especially in the tropics, a change in average temperature of just a few degrees can mean the difference between the likelihood of encountering freezing temperatures, or not. In the past thirty years, *Aedes aegypti* mosquitoes (Figure 29.12), which lived only at low altitudes because of the temperatures at which they can survive, have now been found above 1.5 km of elevation in northern India and 2 km in the Andes Mountains. These mosquitoes are carriers of dengue, and yellow fever may follow.* Dengue fever has already been reported at the 1.5-km mark in Taxco, Mexico.

In the 1990s, as a result of the succession of hot years, malaria outbreaks were reported in the United States as far north as New York (whereas ten years earlier the disease had been confined to California). Malaria has now also been reported in Canada, in some of the countries of the former Soviet Union, and in Korea (Figure 29.13).

FIGURE 29.12   The *Aedes aegypti* mosquito, carrier of yellow fever. An increase of its range occurs as global temperatures rise. (From http://en.wikipedia.org/wiki/File:Aedes_aegypti_CDC-Gathany.jpg.)

---

* Both dengue fever and yellow fever are very serious diseases. Dengue fever, caused by a virus carried by *Aedes*, is characterized by muscle and joint pain, eruptions on the skin, enlargement of the lymph nodes, and a decrease of white blood cells. The victim is debilitated for weeks. Yellow fever is also a viral disease and has such effects as jaundice, slowed heartbeat, and tendency to bleeding.

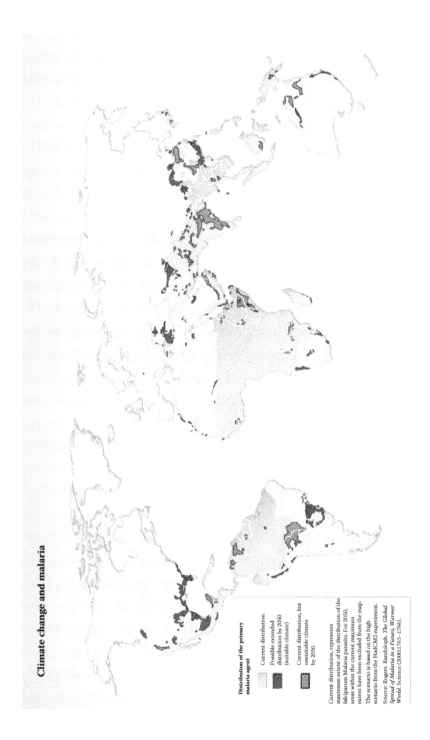

FIGURE 29.13   As the planet warms, the geographic range in which malaria can be encountered widens. (From http://www.grida.no/graphicslib/detail/climate-change-and-malaria-scenario-for-2050_bffe.)

This spread of disease is a possible by-product of the disruption of ecosystems resulting from climate change. A healthy ecosystem has a balance of species; some will surely help keep down the population of disease-carrying organisms, such as mosquitoes, by eating them. When the ecological balance is upset, as could happen as a result of climate change, then there could be a decline in species that keep pests under control. As a result, diseases borne by mosquitoes or other carriers could spread further in range.

The effect of warm temperatures ascending mountains is also manifest in the fact that, in many parts of the world, glaciers at mountain summits and at sea level are melting (Figure 29.14).

These vast piles of snow, which never melt, and seem destined to last as long as the world holds together...

—**Darwin**[5]

In 1839, not even the greatest naturalist of all time could have imagined the rapid destruction of the world's ice a century and a half into the future. Over the last forty years, the elevation at which temperatures are always below freezing has ascended over 150 meters on mountains in the tropics. The ice covering Greenland appears to be melting at a rate of about eighty cubic kilometers per hour.

Perennial snow mountains resplendent in their glory
Bow down and melt into water
The majestic oceans lose their ageless equilibrium
And inundate islands

— **Dalai Lama**[6]

POTENTIAL CONSEQUENCES OF AN ENHANCED GREENHOUSE EFFECT

Global warming could affect us in many ways. These include, but are certainly not limited to, rises in sea level, changes in agricultural productivity, and threats to human health.

INCREASING AVERAGE GLOBAL TEMPERATURES

Different climate forecasting models predict different increases in average global temperature. This does not mean that modelers are incompetent or that the models are useless. Differences in predicted temperatures arise in part from using different initial assumptions (such as how rapidly atmospheric CO_2 concentrations will increase) and different degrees of sophistication or complexity built into the model. Most predictions of future temperatures, based on an assumed doubling of atmospheric CO_2, lie in the range of a warming of 1°C–6°C; many are in the narrower range of 2°C–5°C, with "best estimates" often 2.5°C–3.5°C.

Part of the oral tradition of those involved in modeling complex physical phenomena (and the modeling of the global climate is as complex a modeling job as has

FIGURE 29.14 An example of the melting of glaciers over time, additional evidence for global warming. (From http://www.conditionsmagazine.com/archives/1567.)

ever been undertaken) says that, "All models are wrong; some models are useful." How do we know that we can put confidence in a model and that a reasonable likelihood exists of its results being accurate? Several tests of a model help build this confidence. First, the model can be tested to see if it can accurately simulate today's climate, or some features of the climate, such as the variations of seasons. Second, attention can be given to certain specific aspects of the model, such as its ability to predict the extent of cloudiness. Third, the model can be run "backward," so to speak, to determine its accuracy in calculating ancient climates of Earth. Accurate results from all of these tests provide justification of the model's credibility. All of the major climate models undergo continual testing, fine-tuning, and retesting.

The oceans provide a huge thermal reservoir—they can soak up a great deal of heat without a large increase in water temperature. The oceans will warm slowly. That is, the warming predicted on the basis of a doubling of atmospheric CO_2 might be delayed for some years, or even decades after the time at which CO_2 in the atmosphere actually has doubled. Eventually, though, the full extent of warming will occur at the time when the oceans "catch up" to the temperature increases.

If CO_2 concentrations double, New York City could expect at least fifty days a year with temperatures above 32°C (instead of fewer than twenty now typical for the city). Dallas could expect days with temperatures above 38°C for nearly one-third of the year. A government official in the administration of President G.H.W Bush (1988–1992) alleged that a few degrees' climate warming would have no more effect than the increased temperature one would experience by getting off a flight bound for Boston in Washington instead.* In other words, Boston would then have the climate that Washington does today. Among numerous issues that this glib response overlooks is this question: If a future Boston were to have the climate that Washington now does, what would Washington be like?

RISING SEA LEVEL

Doubling the CO_2 concentration in the atmosphere will raise the sea level by about 20–75 centimeters. In the period after the last ice age, some 12,000 years ago, a 5°C temperature increase accompanied sea-level rises of some ninety meters, along with significant habitat changes for many species. Sea levels have already risen 12–15 cm over the last century, mainly due to the thermal expansion of all the water in the oceans as they warm.

An increase in sea level of about five meters potentially could inundate the Netherlands, and certainly would do so for many low-lying islands. Fifty million people of Bangladesh, half the total population, live on land within five meters of sea level. Even smaller sea-level increases—in the range 0.5–1.5 m—would endanger such cities as New York, Venice, Bangkok, and Taipei. Across the world, a large percentage of people live on, or very close to, sea coasts. Eighty percent of the population of Australia lives within twenty kilometers of the coast.

The ice cap at the North Pole is melting, as first noticed during explorations in the summer of 2000. The formerly solid ice cap, which had long resisted passage by any ships other than specially equipped icebreakers, is now turning to slush (Figure 29.15). The melting of the ice sheet over the North Pole will not contribute to a rise in sea level. (This can be verified by a simple experiment of filling a glass to the brim with water and ice cubes, and watching what happens to the water level as the ice melts.) However, if the Arctic Ocean were to be free of ice, it could be warmer by some 10°C in winter. This would surely have some effects on climate in the northern hemisphere. The melting has already kicked off squabbles among the countries

---

* Former Governor George Wallace of Alabama (1919–1998) used to delight in calling such officials "pointy-headed bureaucrats."

(a)

(b)

FIGURE 29.15 Melting of the polar ice cap, showing figures for (a) September 21, 1979 (From http://www.nasa.gov/centers/goddard/news/topstory/2003/1023esuice.html#addlinfo.) and (b) August 26, 2012 (From http://svs.gsfc.nasa.gov/Gallery/ArcticSeaIceResources.html.) taken at the same time of year.

that border the Arctic Ocean, regarding which of them can lay claim to energy and mineral resources on, or below, the ocean floor, and in what locations.

In the Antarctic, the ice sheet is mostly on land. If this ice were to melt, which could potentially happen over only a few decades, the extra water added to the oceans just from melting of the Ross ice shelf would raise the sea level by nearly six meters. Such a rise would drown most of the world's major seaports. If the sea were to rise by nine meters, the United States would lose Louisiana, at least one-quarter of Florida, and parts of other states as far north as New Jersey. The impact there would be particularly severe, because much of the land that would be flooded is the heavily developed and highly populated area near New York City. Warming that was prolonged enough, and warm enough, could melt the entire Antarctic and Greenland ice sheets, bringing the sea level up another sixty meters.

Optimists believe that the total melting of the Antarctic and Greenland ice sheets would likely take centuries, and the change likely would be slow enough to allow society to respond in some rational fashion. Pessimists point out that observations made on glaciers show that they can move surprisingly rapidly for relatively short periods, that is, a few months to several years. When this happens, the bottom of the glacier melts, and the resulting layer of water allows for even easier sliding. The sliding motion generates heat, and the heat in turn causes even more melting, more sliding motion, and still more heat, in an escalating feedback process. If this were to happen to the Ross ice sheet, the predicted six meters rise in sea level would come very quickly, with disastrous results, because so little time would be available for any reasonable response.

SHIFT OF PRIME AGRICULTURAL REGIONS

The consequences for agriculture of a doubling of atmospheric CO_2 are also likely to be significant. The regions of greatest agricultural productivity would likely change. Temperate climate zones ideal for farming will shift toward the poles. In some cases, this shift will cross national borders. Combined drought and high temperatures could reduce crop yields in the midwestern United States, but the growing range might then extend further into Canada. The boundary between permafrost (or tundra) and normal soil would move north by 100–200 kilometers per degree celcius of temperature rise. Consequently, much more of Canada should be able to grow wheat while at the same time wheat production in the southern Great Plains will decrease. These changes could have enormous impacts on global geopolitics. What if Canada exported food to the United States, instead of vice versa? What if Siberia became warm enough to become a major food exporter to the rest of the world?

The species of plants grown would change as well. The corn belt in the United States would shift 160 kilometers toward the southwest or northeast for each degree celcius in average daily temperature during the growing season. Though the temperatures would permit growing corn in more northern latitudes, the soils there often are less suited for agriculture. Probably crop yields would decrease. Past experience has shown that world grain markets have responded by increasing grain prices for even a 1% change in the amount of grain coming to market. Price changes caused by drastic drops in the amount of corn, or other grains, being produced could be very severe.

Many species of trees take hundreds, even thousands, of years for substantial change of their growing range. Although individual species of trees can migrate with slowly changing climate, whole habitats usually do not, so that the species composition of an ecosystem is dramatically altered. Habitats would change unpredictably, and some species would, in effect, be stranded in ecologically unsuitable territory. Plants provide the necessary habitat that various animals depend on for their survival. Rates of species extinction could be substantially increased if climate changes occur more rapidly than ecosystems are able to adapt. As climate changes, the climate tolerances of birds could cause them to change their ranges and migration patterns. Of course, birds can migrate in periods of days or weeks, while trees or other plants may take centuries to catch up. This too will alter the nature of ecosystems.

When the last ice age ended about 12,000 years ago and a period of warming set in, major ecological changes occurred worldwide. The global warming then amounted to about 5°C. The time necessary for the ecosystems to adjust to this sustained warming was several thousand years. Now, however, the potential exists for this same level of warming to occur within a century. If it does, substantial change to plant and animal populations and their habitats are virtually inevitable. And, the more rapidly the climate changes, the more severe the ecological disruption.

THE SPREAD OF DESERTS

If atmospheric CO_2 doubles, areas of several Mediterranean countries, parts of the United States, and some of the former Soviet Union countries will probably become deserts. Much of this land is presently good agricultural land. Increasing "desertification" is already a severe problem in parts of Africa (Figure 29.16). It could also occur in the United States, in a way similar to—but far worse than—the "dustbowl" of the 1930s. An example is in central Nebraska. Today, the Sand Hills region of Nebraska is home to over seven hundred species of plants, as well as an area of wetland that supports many kinds of migratory birds. Analyses of the geological history of the Sand Hills, combined with computer modeling of climate changes, suggest that a drop of about five to ten centimeters in annual precipitation, combined with

FIGURE 29.16 An example of the spreading of deserts, in this case the encroachment of sand dunes on Nouakchott, the capital of Mauritania. (From http://pubs.usgs.gov/gip/deserts/desertification/.)

an increase in temperature rise of about 2°C–5°C, would turn the Sand Hills back to desert. That is, an area of some fifty thousand square kilometers in northern and central Nebraska would become a desert. (And this does not include changes that would certainly be occurring elsewhere on the Great Plains.)

If the Midwestern United States were to become drier as a result of climate change, the growing season would be shortened by about ten days and droughts will become more frequent. Farms that depend on irrigation will be in deep trouble, because many of the water reservoirs that would be useful for irrigation are already depleted. More Colorado River water has been allocated for irrigation purposes than actually flows in most parts of the river! According to these allocations, people can, in principle, pull more water out of the Colorado than is actually in it. A 2°C temperature rise could cause a 40% decrease in water flow relative to the present-day flow. We can be sure that this situation would ignite fierce legal battles among the farmers, corporations, municipalities, and states that depend on having that water.

Human Health Effects

An increase in average temperatures could increase the geographical range of disease-spreading insects. The result could be significant increases in such severe diseases as malaria, yellow fever, and sleeping sickness. The insect carriers of these diseases are very sensitive to weather conditions. Cold temperatures limit mosquitoes to living in seasons and in geographic regions where temperatures stay above certain minimums. Freezing kills many eggs, larvae, and adults outright. *Anopheles* mosquitoes, which transmit the malaria parasite, cause sustained outbreaks of malaria only where temperatures routinely exceed 15°C. *Aedes aegypti* mosquitoes, responsible for yellow fever and dengue fever, convey virus only in regions where temperatures seldom fall below 10°C. Higher temperatures increase the rate at which the disease-causing organisms that live inside the mosquitoes reproduce and mature. As entire geographic regions experience warming, mosquitoes could expand into territories they formerly could not tolerate because of temperature limitations. As they do so, they would bring the diseases they carry with them. Furthermore, within their survivable range of temperatures, mosquitoes breed faster and bite more often as the air becomes warmer. Warmer nighttime and winter temperatures could enable them to cause more disease for longer periods in the areas they already infest.

The late spring and early summer of 2012 saw an extensive heat wave across much of the United States, with the eventual death toll likely to be well above a hundred. In the United States, 206 people died of heat-related effects in 2011, almost seventy more than in 2010. A 2010 heat wave in Japan killed at least 130 people and sent tens of thousands to the hospital. In 2003, an extreme heat wave in Europe killed an estimated fourteen thousand people in France alone. At least two thousand more met up with the Grim Reaper in the 2006 European heat wave.

Lengthy droughts and massive rainfalls, which have become more common as the average temperature has risen, could undermine health in various ways. They could damage crops and make them vulnerable to infestation by pests and weeds, reducing food supplies and potentially contributing to malnutrition. Floods help trigger outbreaks of disease by creating breeding grounds for insects whose dried-out eggs will hatch in still water, promoting the emergence and spread of disease. The insects can make a further gain if climate change reduces the populations of predators that normally keep them in check.

THE POSSIBLE RELEASE OF MORE GREENHOUSE GASES ON A WARMER EARTH

Global warming could possibly exacerbate the release of even more CO_2 or other greenhouse gases. The ability of gases to dissolve in liquids decreases as the temperature of the liquid increases. This is why, for instance, a warm carbonated beverage is more likely to foam when it's opened than would a cold one. Increasing the temperature of the oceans will decrease the solubility of CO_2.* Substantial amounts of methane, which we've seen is a potent greenhouse gas, currently exist in environments such as mud on seafloors, bogs, and the permanently frozen soil of arctic regions (permafrost). As temperature rises, these substances will also get warmer, making it more likely that they would release methane to the atmosphere. Global warming would also cause more water to evaporate. Since water vapor is a greenhouse gas, an increasing concentration of water vapor in the atmosphere could further add to the greenhouse effect.

INCREASED PLANT GROWTH AS A BENEFIT

There seems to be a complicated interaction between the roles of CO_2 and water in plant growth. Increased atmospheric CO_2 might not be entirely bad for plants. As CO_2 concentration increases, the rate at which water is lost by evaporation from the leaves falls. CO_2 is essential for photosynthesis, but because the concentration of CO_2 in the air is relatively low, CO_2 molecules are difficult for a plant to catch. To give themselves the best opportunity for absorbing CO_2 molecules, plants keep small pores on the surfaces of their leaves open. But, at the same time these wide-open pores make it easy for water molecules inside the leaves to escape by evaporation. In other words, capturing a little CO_2 involves losing a lot of water. Plants can adapt to increasing CO_2 in one of two ways. One strategy is not to change the size of the pore openings. This allows more photosynthesis to occur (because there is more CO_2 to be captured) with the same amount of water loss. Alternatively, the second strategy is to close the pore openings somewhat. In this case, less water is lost by evaporation, and about the same amount of photosynthesis occurs. In dry conditions, increased CO_2 can essentially compensate for less water.

In greenhouses, both agricultural and wild plants show higher growth rates, between 20% and 40%, when the CO_2 concentration is doubled. The results vary

---

* There is a possibility that more CO_2 could be absorbed, rather than released. An increase in ocean temperature could promote the growth of microscopic algae (phytoplankton) and hence increase CO_2 absorption by photosynthesis. However, if the ocean is warmer, water will not circulate as well as it does now, inhibiting the growth of phytoplankton. A warmer climate would mean that the tree line would move north, bringing with it added CO_2-absorbing capacity, again due to photosynthesis. But, related reductions in rainfall elsewhere might turn areas that nowadays have abundant vegetation into deserts, reducing CO_2 absorption.

greatly from one species to another, however. Plants that have low growth rates, the case for many wild plants, don't seem to respond as well to increased CO_2 as do plants that normally have rapid growth rates. Many crop species fall into the latter category. Most of the studies so far were done for relatively short periods and in greenhouses. Some studies under field conditions show few significant increases in growth with increased CO_2. Nevertheless, artificially increased CO_2 levels in indoor growing environments are already in commercial practice to accelerate the growth of crops.* In some greenhouses, the carbon dioxide concentration is maintained at about 1400 ppm, which is more than triple the present concentration in the atmosphere.

THE ROLE OF THE GLOBAL CARBON CYCLE

The global carbon cycle (Figure 29.17) allows tracking the removal of CO_2 from the atmosphere, and the ways in which CO_2 is added to the atmosphere. Processes that add CO_2 to the atmosphere are called *sources*. They include the decay of organisms; fossil fuel combustion; decomposition of carbonate rocks, either by natural processes or in the manufacture of cement; and CO_2 coming out of solution from the ocean. Processes that remove CO_2 from the atmosphere are called *sinks*. They include photosynthesis, dissolving of CO_2 into the oceans, or the fixation of CO_2 into various carbonate forms. Changes in patterns of land use can act either as sources or as sinks. If, for example, forest land is converted to agricultural uses, such change tends to increase atmospheric CO_2. On the other hand, growth of new forest would likely serve as a sink.

Changes in the concentration of CO_2 in the atmosphere result from the relative balance between sources and sinks. When absorption into sinks occurs at about the same rate as release of CO_2 from sources, the amount of CO_2 in the atmosphere will remain steady. If the sinks absorb more CO_2 than the sources produce, then atmospheric concentration will drop. Since the concentration of CO_2 in the atmosphere

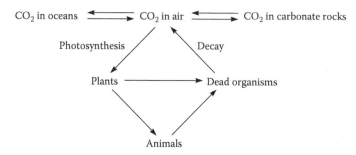

FIGURE 29.17 A simplified version of the global carbon cycle showing natural sources for atmospheric carbon dioxide, and natural sinks removing carbon dioxide from the atmosphere. Not shown are the anthropogenic sources, most critically fuel combustion.

* Some of the details of this interesting work, already being used commercially in the United States and in the Netherlands, can be found in Chapter 3 of Michael Pollan's book, *The Botany of Desire* (Random House, 2001).

has been increasing steadily for many years, this is an indication that more CO_2 is being produced from sources than can be absorbed into sinks.

Patterns of changing land use have a substantial effect on storage of carbon on land (in plants and soils), decreasing the potential store of carbon. In addition, deforestation releases one to two billion tonnes of carbon (as CO_2) to the atmosphere every year. Around the world, about 150,000 square kilometers of rainforest are cut down or burned every year. This does double harm. First, trees do a good job of removing CO_2 from the atmosphere; thus a significant carbon sink is removed. Second, regardless of whether the trees are burned to clear land quickly or are simply left to decay, the carbon locked up in the now-dead trees will be returned to the atmosphere as CO_2. Perhaps the least damaging approach would involve using the trees as lumber (so that they are not left to rot nor are burned), and to use the cleared land for agriculture, since the growing farm crops would absorb some CO_2. Even so, it may be that the ability of the land to absorb CO_2 drops by up to 80%. This reflects the fact that forests contain ten to twenty times more carbon than cropland, when compared on an equal-area basis.

Until about 1950, the expansion of agriculture and forestry in the middle latitudes of the globe dominated carbon emissions from the land. Since then, though, the conversion of temperate forests to agricultural lands has decreased. In some places, forests are even making a comeback, growing on previously logged land and on abandoned agricultural land. As a result, since the 1950s, CO_2 emissions from the tropics, which have been increasing, are now greater than from temperate regions.

The dominant anthropogenic CO_2 sources are the use of fossil fuels and production of cement. As the generation of energy and the consumption of fossil fuels increased, so did the quantity of combustion products released to the atmosphere. Since 1860, the CO_2 concentration has increased from 290 ppm nearly to 400 ppm, and the present rate of increase is about 1.5 ppm per year. Currently, fossil fuels containing five gigatonnes of carbon are burned annually.* This represents four times the rate of release in 1950 and ten times the rate in 1900. During the 1990s, about a tonne of carbon dioxide was produced each year for every human on the planet. The cumulative production of CO_2 from fossil-fuel combustion since the beginning of the Industrial Revolution (i.e., since about 1750) is about 280 gigatonnes of carbon.

The ocean contains more than fifty times as much carbon as the atmosphere. Most of the carbon in the ocean is in the form of dissolved bicarbonate and carbonate ions. The rest is organic, either as living organisms, or suspended or dissolved organic compounds. The water at the surface of the ocean is more acidic than the water in the deep ocean. As this relatively acidic water circulates, it will dissolve calcium carbonate. Doing so puts more CO_2 into the seawater. Over many years, perhaps centuries, much of this CO_2 will be released to the atmosphere.

---

* Statistics relating to the global carbon cycle use two different units. One is the amount of *carbon* involved, and the other is the amount of carbon dioxide. One unit of carbon is equivalent to 3.7 units of carbon dioxide. Burning a tonne of carbon releases 3.7 tonnes of CO_2. Conversely, one unit of carbon dioxide is equivalent to 0.27 units of carbon. One tonne of CO_2 released contains only 0.27 tonnes of carbon.

However, the sea also takes up carbon dioxide in its surface layer. Some CO_2 is used by organisms living in the ocean, such as marine plants that live by photosynthesis, and reef-building corals. The oceans act as a "biological pump," taking CO_2 from the atmosphere and into the ocean at a rate of about ten gigatonnes per year. In the ocean, the plants and animals that convert CO_2 to carbonates in their skeletons, pass through their life cycles, and sooner or later die. Their skeletons sink to the ocean depths, where part of the carbonate is deposited in sediments, and the rest dissolves. Thus, the CO_2 taken out of the atmosphere slowly cycles into carbonate sediments or into the carbon reservoir of the deeper ocean. The deep ocean is thought to hold some 35 teratonnes of CO_2. The transfer of carbon between the surface and the very deep ocean is slow. As atmospheric CO_2 levels increase, the ocean responds by dissolving more CO_2 in its surface layer, and by mixing the CO_2-enriched surface waters downward through exchange with deeper waters. These deeper waters may have an enormous capacity for carbon. If so, the ocean should have the capability to absorb large quantities of CO_2 over many decades and perhaps even millennia.

On a time scale of one to ten thousand years, the ocean could absorb about 85% of the anthropogenic CO_2 that has been produced since the beginning of the Industrial Revolution. Some estimates suggest that the oceans could potentially absorb the CO_2 produced by burning the entire known amount of all the coal, oil, and natural gas available on Earth. Of course, nobody knows how the plants and animals living in the ocean would respond to such a massive uptake of CO_2. Certainly, changes in temperature or in the circulation patterns of the ocean and changes in the marine ecosystems would affect the rate of exchange of CO_2 between the atmosphere and the ocean, and thereby affect the concentration of CO_2 in the atmosphere.

Carbon on Earth is stored primarily in rocks and sediments. The oceans contain about 50 quadrillion tonnes of sedimentary carbonates, 38 trillion tonnes of dissolved inorganic carbon, and about 20 quadrillion tonnes of organic carbon. The atmosphere contains around 700 billion tonnes of carbon. Terrestrial vegetation contains some 500–900 billion tonnes of carbon, and the soil contains about twice that much. The reservoirs or sinks (the atmosphere, oceans, soil, and terrestrial biosphere) from which or into which carbon can easily be exchanged account for only a tiny fraction of the total carbon on Earth. But, it is only this tiny fraction that is available to play a role in biological, physical, and chemical processes at the Earth's surface. Anthropogenic carbon cycles among the atmosphere, oceans, and the terrestrial biosphere.

POSSIBLE POLICY OPTIONS FOR DEALING WITH GLOBAL WARMING

The global carbon cycle points to the two strategies for slowing the buildup of carbon dioxide in the atmosphere: either to reduce the rate of carbon dioxide coming from sources, or to increase that going into the sinks. Realistically, both need to be employed.

With respect to reducing CO_2 coming from sources, by far the most effective immediate approach is to increase the efficiency of processes that consume fuel, to increase the amount of useful ENERGY obtained per unit carbon dioxide emitted. Examples include switching to cars that are more fuel efficient, increasing the efficiency of electric generating plant operation, and practicing energy conservation at home. Increased energy efficiency and conservation not only reduce CO_2 emissions and other environmental impacts, but in the long run save money. Driving the same total distance, but with much greater fuel economy, means spending less for gasoline or diesel fuel. Using energy-saving appliances or increasing the insulation in a house can reduce the monthly electric bill, but with minimal change, if any, in lifestyle.

Fuel consumption could be shifted to those fuels that produce more usable ENERGY per unit weight of carbon dioxide emitted. In most cases, this comes down to using less coal and more natural gas. Some sincere individuals have argued that we must stop using fossil fuels, *right now*. That's impossible. There are no realistic short-term alternatives, and the world has far too much investment in infrastructure that relies on fossil fuels to make any radical change. Fossil-fuel use will be with us for decades, at least. Furthermore, coal is the cheapest fuel available. Economic consequences could be severe for those developing nations that have good supplies of coal if they were to switch to a different, more expensive fuel.

Third, the use of "nonfossil" or "noncarbon" sources of energy can be increased, even though such a switch to large-scale use cannot be made quickly. Two of these sources have already been discussed: hydro and nuclear. Others, to be discussed in later chapters, include the biofuels, wind energy, and solar energy. Just like every other energy source, each of these options has its advantages and disadvantages. Likely a realistic future energy policy for almost any country would involve a mixture of energy sources and an attempt to make optimum use of the resources available in that country or region. Solar might be a great idea for some places, hydroelectricity in others, and wind energy in yet others.

The second major strategy involves enhancing the uptake of CO_2 into sinks. The first step that could be made here is to slow the rate of destruction of the sinks we already have, most notably to slow the enormous rate of destruction of the Amazon rain forest. Stopping entirely the destruction of rain forests and similar natural sinks is probably hopeless.

Then, the ability of other natural sinks to absorb CO_2 can be enhanced. As an example, phytoplankton "blooms" growing over immense areas of ocean surface could potentially absorb large amounts of CO_2. Iron is an essential nutrient for some of these marine phytoplankton. "Fertilizing" the surface waters of the ocean with iron (Figure 29.18) should trigger the growth of phytoplankton. The trials of this idea have been debated, as to whether they really are effective. Those who oppose the idea, or favor a very cautious approach, point out that the ocean represents about 75% of the planet, and that humankind's record of environmental stewardship is rather dismal, so adding large quantities of iron to the ocean with the idea of changing the growth of marine plant life over large areas might not be such a great idea.

Processes can be designed to mimic the sinks in the global carbon cycle. Carbon dioxide can be fed to these processes to be kept out of the atmosphere, and ideally

FIGURE 29.18 Scientists seeding the ocean surface with iron, to stimulate the growth of aquatic plants that would help remove carbon dioxide from the atmosphere. (From http://www.oc.nps.edu/~stanton/ironex.html.)

to be locked up centuries or even millennia. These kinds of processes are broadly known as *carbon capture and storage* (CCS), or as *carbon dioxide sequestration*. The CO_2 sinks in the global carbon cycle include its being incorporated into carbonate rocks, dissolved into the oceans, or used by living plants in the process of photosynthesis.

The CCS strategy of mineral carbonation involves reacting carbon dioxide (as might be present in the flue gases of a boiler) with a mineral using some conditions that force uptake of CO_2 in a relatively short time, rather than relying on the very long time scales of natural geological processes. The minerals used would be those rich in calcium, magnesium, or both, so that the carbonation process would form stable calcium or magnesium carbonates. These compounds have industrial uses, so they might be sold to offset some of the costs of the CCS process, but in the worst case they can be buried and locked away in Earth's crust for thousands of years. A candidate mineral for this process is serpertinite (Figure 29.19), very widely distributed around the world, and with available calcium and magnesium.

Naturally occurring brines (solutions of salt in water) occur in such places as oil or natural gas reservoirs. Carbon dioxide can be pumped into such reservoirs, where it dissolves in the brine much as it would in the ocean. One example is the work being done by Statoil, the Norwegian state oil company. Since 1996, Statoil and its partners have been producing natural gas that contains 9% CO_2, almost

FIGURE 29.19 Specimen of serpentinite, a mineral that has potential application in processes for capturing carbon dioxide and sequestering it as carbonates. (Photograph courtesy of Charles E. Miller, Jr., Retired geologist, State College, Pennsylvania.)

four times the amount allowed by exportation rules. This gas comes from the Sleipner field in the North Sea. Normally, gas companies that encounter CO_2-rich gas would separate the CO_2 and then vent it into the air. Instead, Statoil pumps the CO_2 down into a geologic formation nearly a kilometer below the sea-floor (Figure 29.20). If they had taken the standard approach, of venting the gas to the atmosphere, Statoil would single-handedly have boosted Norway's CO_2 emissions by 3%. Somewhat similar projects are running at In Salah in Algeria, and Weyburn, Saskatchewan. The Weyburn project sequesters CO_2 from a plant producing synthetic natural gas from coal, in Beulah, North Dakota. The CO_2 is sent to Weyburn via pipeline. Each of these three projects sequesters about a megatonne of CO_2 per year.

The development of artificial photosynthesis is in early days. The technology of photocatalysis is somewhat more advanced, though still with much development work to be done. In a photocatalysis system, carbon dioxide and, usually, water would react in the presence of a special catalyst to convert the carbon dioxide into useful chemicals or fuels, such as methanol. Energy for the reaction would be supplied by sunlight.

The current challenge of CCS is mainly one of the mismatch of scale between the prodigious volumes of CO_2 produced each year, and our ability to capture and store it. The Sleipner, In Sallah, and Weyburn projects are the world's biggest CCS projects, and each one can handle about one megatonne of CO_2. The largest single point source of CO_2 is the synthetic fuel plant in South Africa, which converts coal into an array of liquid fuels and chemical products. From an engineering and technical perspective, this plant is highly successful. It also generates about 73 megatonnes of CO_2 per year. If that synthetic fuel plant were to be duplicated today, but with CCS added to eliminate the CO_2 emissions, *one* such plant would require about 75 Sleipners or Weyburns.

FIGURE 29.20 The Sleipner project off the coast of Norway, one of the world's most successful carbon dioxide sequestration projects. (From http://www.netl.doe.gov/technologies/carbon_seq/FAQs/carbonstorage3.html.)

Presently we humans produce vast amounts of CO_2 per year, which shows no sign of abating. Much excellent research, development, and demonstration work is underway around the world to improve energy efficiency, to develop alternative "noncarbon" energy sources, and to improve CCS technologies. Yet at the same time a large fraction of the world's population, possibly between a third and a half, lack some combination of adequate and safe housing, clean drinking water, an adequate diet, transportation, and access to education or health care. The world's financial resources are limited. In light of these considerations, some sincere individuals have suggested that, overall, we would do more good for more people by investing in providing some of the basic needs—such as clean water or a minimal level of health care—and not investing enormous sums in trying to stem the tide of carbon dioxide emissions.

The problem is enormous, highly complicated, and both literally and figuratively global. The consequences of making the wrong decisions could be catastrophic. We—all of humankind—are in the process of doing an experiment on ourselves. We'll see how it turns out.

REFERENCES

1. Hutton, J. In Morton, Oliver. *Eating the Sun*. HarperCollins: New York, 2008, p. 317.
2. Tyndall, J. Quoted in Kolbert, Elizabeth. *Field Notes from a Catastrophe*. Bloomsbury: New York, 2006, p. 38.

3. Arrhenius, S. Quoted in Kolbert, Elizabeth. *Field Notes from a Catastrophe*. Bloomsbury: New York, 2006, p. 42.
4. I owe this term to John Kriak, President of GroupGenesis LLC, Johnstown, Pennsylvania, PA.
5. Darwin, C. *The Voyage of the Beagle*. Modern Library: New York, 2001, p. 215.
6. Dalai Lama. The Sheltering tree of interdependence. In Mehrotra, R. (ed.) *The Essential Dalai Lama*. Penguin Books: New York, 2005, p. 262.

FURTHER READINGS

We would not need to worry about CO_2 sinks if all of the trees could be restored to life that have been used to produce paper for all of the professional monographs, textbooks, government reports, white papers, books for laypersons, papers in professional journals, popular scientific articles, newspaper "op-ed" pieces, letters to the editor, polemics, screeds, diatribes, jeremiads, philippics, and hate mail about global climate change. Out of the tsunami of writings on global climate change, the sources listed below are useful, reasonably rational, and accessible to readers with little prior scientific background.

Alley, R.B. *The Two-Mile Time Machine*. Princeton University Press: Princeton, NJ, 2000. A very readable account, by one of the world's foremost glaciologists, of how data taken from ice cores reveal information about climate change.

Archer, D.; Rahmstorf, S. *The Climate Crisis*. Cambridge University Press: Cambridge, U.K., 2010. A good introduction to the subject for laypersons.

Berberova, N. In memory of Schliemann. In *The Tattered Cloak and Other Stories*. New Directions Books: New York, 1991. This short story by the Russian émigré was written well before concerns about global warming were in the public consciousness, but depicts an overheated, overpopulated world.

Coley, D. *Energy and Climate Change*. Wiley: Chichester, U.K., 2008. This is a textbook on the topic. Chapter 19 discusses issues of energy efficiency; Chapter 29 deals with carbon sequestration.

Henson, R. *The Rough Guide to Climate Change*. Penguin: London, U.K., 2008. A very good introduction to this topic.

Intergovernmental Panel on Climate Change. *Climate Change 2007*. Cambridge University Press: Cambridge, U.K., 2007. This is the authoritative report on the topic. It is available in several sections, dealing with the scientific basis; the impacts of, and possible adaptation to, climate change; mitigation of climate change; and a synthesis or summary report. An additional document provides a summary for policymakers.

King, D. *Global Warming: A Very Peculiar History*. Salaria Book Company: Brighton, CO, 2009. The author is a former chief scientific advisor to U.K. government. The subtitle is *Keeping Calm and Carrying On as Usual Is Not an Option*, one of the themes of this short book.

Kolbert, E. *Field Notes from a Catastrophe*. Bloomsbury: New York, 2006. A series of interviews with researchers, environmentalists, various people affected by climate change.

Larson, E. *An Empire of Ice*. Yale University Press: New Haven, CT, 2011. A remarkable story of the explorations of Antarctica a century ago, conducted under extremely brutal conditions. Even in early years of the last century, the explorers noted evidence for retreat of the glaciers.

Levitt, S.D.; Dubner, S.J. *Superfreakonomics*. HarperCollins: New York, 2009. Chapter 5 discusses many aspects of global climate change, and possible responses to it.

Mann, C. *1493*. Alfred A. Knopf: New York, 2011. Chapter 1 discusses the "little ice age," a period of unusual global *cooling* in the period 1550–1750. The little ice age is often cited by climate change deniers as a period in recent human history when severe cooling was experienced. Chapter 1 includes, among other things, an explanation of the origin of the little ice age in a period of extensive *re*forestation and decreasing CO_2 emissions—increased sinks and reduced sources. This shows that the global carbon cycle functions exactly as expected, and in fact provides further credence for models explaining global warming.

Mann, M. *The Hockey Stick and the Climate Wars*. Columbia University Press: New York, 2012. A discussion of the political assault on climate science, primarily in the U.S. The author has direct experience in vitriolic attacks by politicians of the extreme right and by the assortment of "climate change deniers" in the public.

Mann, M.; Kump, L. *Dire Predictions*. DK Publishing: New York, 2008. This very extensively illustrated book is a guide to the key findings of the Intergovernment Panel on Climate Change report, intended for laypersons.

Maslin, M. *Global Warming: A Very Short Introduction*. Oxford University Press: Oxford, U.K., 2009. As the title promises, this is a relatively short introduction, but comprehensive and easily readable.

McNeill, J.R. *Something New under the Sun*. Norton: New York, 2000; Chapter 4. An important book, providing a history of the 20th century with focus on environmental issues. This chapter presents a history of the atmosphere from regional and global perspectives, including global climate change.

Oreskes, N.; Conway, E. *Merchants of Doubt*. Bloomsbury Press: New York, 2010. It is difficult to read this book without being both troubled and angry. It chronicles a history, over about forty years, of how a few scientists and scientific advisors misled the public on such critical issues as connections between smoking and lung cancer, between chlorofluorocarbons and the ozone hole, between SO_x, NO_x and acid rain, and now global climate change. Remarkably, it is often the very same people involved, moving from issue to issue.

Schwartz, A.T; Bunce, D.M.; Silberman, R.G.; Stanitski, C.L.; Stratton, W.J.; Zipp, A.P. *Chemistry in Context*. WCB/McGraw-Hill: New York, 1997; Chapter 3. This chapter discusses the chemistry of global warming, with abundant illustrations and reasonably recent data.

30 Fossil Energy
Reserves, Resources, and Geopolitics

ENERGY RESERVES AND RESOURCES

Some ENERGY sources offer the prospect of being inexhaustible. Water and wind are two such sources. Some more sources include biomass, solar energy, and nuclear fusion (all to be discussed in later chapters). However, the common sources of ENERGY that people in many countries rely on are electricity and liquid fuels for transportation. These usually come from fossil fuel sources. Most of us take for granted an unlimited supply of electricity; if there is any limitation at all, it is in our ingenuity for plugging appliance upon appliance into the available outlets. Similarly, we presume the ability to purchase unlimited amounts of gasoline or diesel fuel, anywhere, anytime. But fossil fuels are not inexhaustible. And, unlike biomass, fossil fuels cannot be regenerated on human time scales. Someday, in some sense of the term, we will run out.

The concepts of reserves and resources to be discussed in this section apply to *any* ENERGY source (and indeed to other natural substances as well, such as ores). The *reserve* is the amount of a material that can be recovered economically with known technology. Reserves usually are established by detailed exploration. A *resource* is the entire amount of the material known *or estimated* to exist, regardless of the cost or technological developments needed to extract it. In other words, the reserve is what we're sure of, but the resource includes what we think is there. By these definitions, the amount of reserves is always less than the amount of the resource.

One's personal finances can serve as an illustration of the concepts of reserves and resources, and the various categories of each. A financial reserve would consist, first of all, of the amount of cash in hand, the amount of money in checking and savings accounts at the bank, and whatever cash advance could be gotten from credit cards. All these amounts can be—in principle—determined quite accurately, and they represent money that one can count on having. This money represents *proved reserves.*

In addition to proved reserves, money could be raised by selling personal possessions, such as a car; or by liquidating investments, such as stocks or bonds. This is a bit trickier than emptying a purse or wallet and bank account. First, stock and bond markets fluctuate daily, so that while certainly some money can be realized by liquidating investments, it's not certain in advance how much could be realized on a particular day. Second, while used cars, antiques, and collectibles do have assessed values, if forced to sell a car in a hurry, one can't necessarily count on holding out to receive the "book value." These situations represent cases where there is reasonable certainty that some money can be obtained, but not exactly how much. This money represents *probable resources.*

It might be possible to obtain loans from banks, family members, or friends. Likely, various sums of money would be available, but with very little knowledge in advance of exactly how much could be obtained. This money counts as a *possible resource.* The amount of the possible resource is not very certain. Similarly, the ability to "extract" the resource (i.e., money) from these sources is not certain, either. How many of your "friends" would really provide an unsecured loan?

Hypothetical resources include things like winning a lottery. Some pot of money is available to be won. However, the jackpot amount fluctuates daily, as do the (exceptionally slim) odds of winning it. Finally, there are *speculative resources.* For example, a long-lost relative may suddenly die and leave all of his or her money to you. This does, on rare occasions, happen to people. But in an effort to raise money, a person does not necessarily know that this source exists, nor have any idea of how much loot the old codger has, anyway. These concepts are summarized in Figure 30.1.

There is a pattern among these classes of reserves and resources, which relate the assurance of receiving the money, the certainty as to the exact amount of each type that might be realized, and the grand total amount available (Figure 30.2). The proved reserve might amount to, say, $1000, but that amount is known to the exact penny, with absolute certainty that it is indeed yours. Counting up all the potential money, right through the speculative resources, nothing says you can't claim to be worth $50 billion. After all, long-lost Aunt Hildegarde might croak this very day, leaving you $49,999,999,000! (Then again, she might not, or she might be somebody else's aunt.)

The same situation exists with ENERGY sources, but with one major difference. They require increased attention to the question of "extraction," a second dimension that incorporates the economic feasibility of extracting or recovering the material

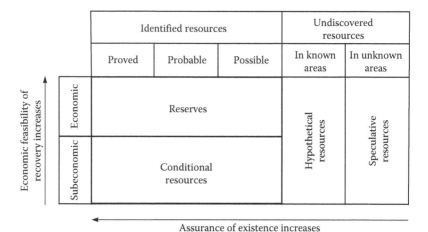

FIGURE 30.1   Reserves and resources are categorized according both to the certainty of their existence and to the economics of their recovery.

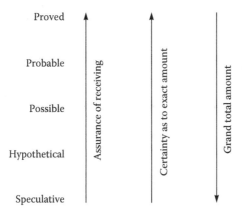

Proved

Probable

Possible

Hypothetical

Speculative

Assurance of receiving

Certainty as to exact amount

Grand total amount

FIGURE 30.2 Relationships exist among the various categories of reserves and resources in terms of the total amount, the certainty of the exact amount, and the assurance of recovery.

from the Earth. This dimension had little importance in example of trying to raise money, because presumably the "technology" for "extracting" money would involve relatively simple phone calls, personal visits, or electronic banking.

The amount of a particular energy source that is proven, probable, or possible, *and* economically recoverable constitutes the reserves. The amount proved, probable, or possible but "subeconomic" (meaning that, with today's technology, it would be so expensive to extract them we couldn't afford to do it) represents conditional resources. The remaining amounts then represent the hypothetical resources and the speculative resources.

The amount of an energy source that is counted in any given category—as reserves, for example—changes over a period of time. In fact, sometimes the reserve of an energy source, even one in use continuously, will appear to increase. Several factors contribute to the fluctuation in the numerical value of the reserve.

First, the "fineness" of the estimate is constantly being revised. For example, one method of exploring for coal involves drilling test holes into the Earth. In an early stage of exploration, drilling is done at very wide intervals, which sometimes may be several kilometers apart (Figure 30.3). In this example, we see that two test holes would have encountered a coal seam of some particular thickness. Perhaps the test holes are one kilometer apart. To estimate the amount of coal that has been discovered in this way, we might assume that the seam thickness is constant between the two drill holes. However, with further exploration and increased drilling at smaller and smaller spatial intervals, we might get much different pictures (Figure 30.4). The accuracy of the estimates of reserves is often related to the complexity of the subsurface geology. The reserves in a large deposit lying in a region of relatively simple geology often tend to be underestimated. In contrast, deposits in regions of complex geology tend to be overestimated.

A second reason for the fluctuation in reserves is that continued exploration for any energy source could discover entirely new fields. Particularly at times when energy is scarce and prices and profitability are high, an energy company might find that it pays off to mount a new campaign of exploration.

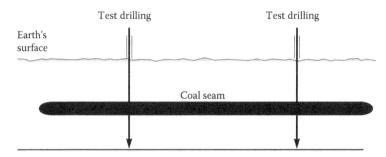

FIGURE 30.3 When exploring for a source of energy, such as coal, an initial reconnaissance might be made at intervals of a kilometer or more. The actual amount to be found in between the test drillings has to be estimated.

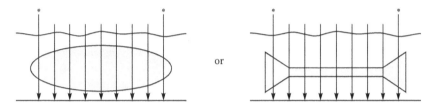

FIGURE 30.4 When more detailed test drillings are made in an exploration program, drilling at smaller intervals than shown in Figure 30.3, it may prove that the amount actually found is much greater than, or much less than, originally estimated. (*Indicates the original test drillings.)

Third, improvements in the technology of extracting an energy supply from the Earth could shift subeconomic conditional resources into the reserves category. As improved technologies are brought to bear, resources once considered impractical to extract can become economic. For example, steady improvements are being made in drilling for oil and gas, even drilling horizontal holes under the Earth's surface. Improved drilling technology may make feasible the recovery of oil or gas that once was inaccessible. Similarly, new mining technologies, perhaps involving robot miners, could extract coal from much thinner seams than are accessible with today's mining technology, or could extract coal seams that are under water in flooded mines.

When comparing the availability of energy sources in one country with those of another, it is important to recognize that the definition of *reserves* varies widely from region to region. A particularly egregious example was the former Soviet Union, which routinely published wildly optimistic "guesstimates" of its oil reserves. This practice seems to be continuing with agencies in the independent countries that formed from the breakup of the Soviet Union. Unfortunately, some analysts and economists have misinterpreted (i.e., believed in) these estimates of "proved" reserves.

WE DEPEND ON OIL

Most of this chapter focuses on petroleum, for three reasons. First, a petroleum product—gasoline or diesel fuel—is the one form of energy, other than electricity, that most people use in daily life. Second, petroleum is an energy source for which many industrialized nations have a significant dependence on imports. Third, relationships between many petroleum-exporting countries and the importing countries have been prickly or even hostile. Some petroleum-exporting countries could, or already do, have internal political instabilities. Petroleum has become a tool of geopolitics—the "oil weapon."

> *Many commodities which are most useful to us are esteemed and desired but little. We cannot live without water, and yet in ordinary circumstances we set no value on it. Why is this? Simply because we usually have so much of it that its final degree of utility is reduced nearly to zero. We enjoy every day the almost infinite utility of water, but then we do not need to consume more than we have. Let the supply run short by drought, and we begin to feel the higher degrees of utility, of which we think but little at other times.*

—Jevons[1]

Let's begin by reviewing the massive dependence on petroleum products. Automobiles and light trucks use gasoline. Heavy trucks, many buses, and many kinds of agricultural machinery (e.g., tractors) use diesel engines, and hence diesel fuel. Most small ships also use diesel engines. Large ships rely on steam turbines, with steam generated by burning fuel oil. Airplanes, depending on the type of engine, use jet fuel or gasoline. Some electricity is generated by burning fuel oil, rather than coal, in boilers. Some homes are heated using fuel oil, LPG, or kerosene-fired space heaters. Most plastics and synthetic fabrics, now virtually ubiquitous in daily life, come from by-products of the petroleum industry (indeed, the chemicals used to make these synthetics are known as *petro*chemicals). Modern industrial society is "hooked" on petroleum.

> *Comrade Stalin correctly said that whoever has oil predominates.*

—Kirov[2]

OIL ECONOMICS

Petroleum prices deserve close attention, because of the dependence of so many aspects of an industrialized society on petroleum. As an example, when the price of petroleum rises, the cost of food goes up, because most food is brought to market in trucks or by railways. The cost of an airline ticket goes up, because airlines use jet fuel. Plastic items get a little more expensive, because they are made from petroleum derivatives.

These facts have led to one group of forecasters, sometimes called "Simonists" in honor of the economist Julian Simon,* who believe that, because technological advances and increasing wealth will always improve abundance, prices of

---

* Julian Simon (1932–1998) developed economic arguments that long-term declines in the prices of various raw materials will mean that the future scarcity of resources will not come to pass, and there will always be abundant supplies of the things we depend on for modern industrial society—oil, steel, coal, and aluminum as a few examples. His ideas are published in several books, beginning with *The Ultimate Resource* (Princeton University Press, 1981). Simon was skeptical of human impacts on global climate change and other environmental issues, such as the ozone hole. Unfortunately, Simon was a victim, for most of his life, of mental depression that was so severe that he was able to work only a few hours per day.

FIGURE 30.5   Thomas Malthus (1766–1834), who argued that population will grow faster than the resources available to sustain it. (From http://en.wikipedia.org/wiki/File:Thomas_Malthus.jpg.)

commodities will continue to fall forever, and that there will always be abundant supplies of, for example, cheap oil, coal, and steel. A second group of forecasters, who are sometimes called "Neo-Malthusians" for the early economist Thomas Malthus (Figure 30.5),* anticipate that sooner or later we will begin to run out of these materials, and prices will rise drastically.

From the end of the Second World War until the early 1970s, the price of oil, in inflation-adjusted dollars, enjoyed a gradual, but perceptible drop (Figure 30.6). Prices increased drastically in the 1970s, beginning with the oil embargo resulting from the 1973 Arab-Israeli war. Up until the "oil price shock" of 1979 (occasioned in part by the Islamic revolution in Iran), oil prices were dictated by the ministers of the Organization of Petroleum Exporting Countries (OPEC) under long-term contracts

---

* Thomas Malthus (1766–1834) was a British economist of the late eighteenth and early nineteenth centuries whose best-known contribution was the idea that, sooner or later, a population will grow faster than its food supply, and therefore poverty (as well as attendant social ills and possible outright starvation) is inevitable. If unchecked, a population will grow geometrically (e.g., 2, 4, 8, 16...), whereas arable land brought into cultivation for food production will grow only arithmetically (e.g., 2, 4, 6, 8...). Since the amount of arable land is limited, at some point there will be far more people than there is land and crop production to feed them. Malthus's original argument has been extended beyond food to any finite resource; at some point the demand for that resource exceeds its availability, resulting in economic and social problems. His book, *An Essay on the Principle of Population*, first published in 1798, is an enduring classic. Supposedly his students at the East India Company College called him "Population" Malthus, or sometimes just "Pop."

FIGURE 30.6 A history of petroleum (crude oil) prices for the past 150 years. Crude oil prices from 1861 to 2011. (From http://commons.wikimedia.org/wiki/File:Crude_oil_prices_since_1861_%28log%29.png.)

with the major oil companies. That arrangement broke down during 1979–1980, because some of the oil-producing nations found that they could sell their oil to traders at prices far above that set by OPEC. In the collapse of oil prices that soon followed, oil buyers became unwilling to purchase cargoes on which the price had been set when the supertankers were halfway around the world. The buyers feared that the value of the oil would drop during its lengthy ocean voyage. One result of this situation was that American companies instead bought oil from relatively nearby Venezuela and Mexico. And, thanks to that situation, Saudi oil prices dropped by about two-thirds between 1981 and 1985. By the mid-1980s, prices were very nearly back to their pre-1973 level.

In the mid-1980s, oil prices were relatively high. One effect of those prices was to reduce demand for oil. As soon as the price of oil dropped, demand increased again. One reason for this rise in demand in the 1980s was that, as oil fell below $20 per barrel, it was used to replace natural gas in some applications, such as a fuel in power plants. In the late 1990s, crude oil inventories were very nearly at maximum capacity. As a result, the demand for new oil production fell, and prices plummeted as well.

When oil prices fell in 1999, to about half their previous level, there was not a corresponding upward swing in demand. This was ascribed to relatively mild winters in 1996/1997 and 1997/1998 (reducing demand for heating oil), and to the economic recession, facetiously called the "Asian flu," which hit many of the nations of Asia in the late 1990s. In addition, a decade's worth of improvements in using natural-gas-fired turbines to generate electricity made it undesirable to switch back to oil, even when the price of oil fell. In 1999, the price of oil was about $10 per barrel (or about $15 per barrel in constant 2012 dollars). Some optimists expected that it would soon be cut in half again, hitting the $5 mark, and that it would be 2010 before oil climbed back to $22 per barrel. In fact, oil had already topped this price in 2000. In 2010, oil was actually around $80 per barrel and has occasionally spiked at over $100 per barrel. Figure 30.5 provides a graph of oil prices in constant 2010 dollars over the past 45 years.

But man cannot live by trust alone. He has to have oil at reasonable prices as well.

—The Times of Zambia[3]

The economics of oil also has other, hidden costs, over and above the current official trading price. The burning of fossil fuels may be having long-term effects on the environment, most notably through global climate change. This has led some people to question whether the cost of cleaning up oil's environmental, political, and social costs should also be taken into account in setting the "true" price of oil. Costs such as the continued US military presence in the Persian Gulf and the cost of correcting environmental problems are sometimes known as external costs, or "externalities." When such externalities are then rolled into the cost of the oil itself, the process is referred to in "bureaucrat-speak" as "internalizing the externalities." It's been estimated that every dollar's worth of oil imported into the United States from the Persian Gulf region costs an additional two dollars in the defense budget to assure that the sea lanes remain open to the supertankers carrying the oil.

From the low point in 1999, oil prices have marched upward steadily, spiking at $133 per barrel in 2008 (adjusted to 2012 dollars). If Julian Simon is right, the current high oil prices will at some point come back down to the $20–$30 range, just as they had done by the mid-1980s after the oil price shocks of the 1970s. If Thomas Malthus is right, the steady upward swing we have seen since the turn of the century is just the beginning.

LIFETIME ESTIMATES

With quantitative information about an energy source in hand, its lifetime—how long it will last—can be estimated. The calculation can be done simply: the total amount available is divided by the amount used per year. As an example, if a country has an economically recoverable coal reserve of 100 gigatonnes and consumes 0.5 gigatonnes per year, the lifetime would be 200 years (i.e., 100/0.5 = 200). This calculation should seem quite straightforward, and it is, if only the arithmetic is considered. A few keystrokes with a calculator and it's done.

These lifetime calculations must be interpreted with extreme caution. Why? First, it is important to know what number was used as the "total amount available." The example above used the economically recoverable reserve. Suppose that instead someone made the calculation based on the estimated total resource. To continue the example for this fictitious country, suppose that the estimated total coal resource is 3000 gigatonnes. With a consumption of 0.5 gigatonnes per year, as above, a very different number results: 6000 years (i.e., 3000/0.5). In other words, this calculated lifetime of the very same material, coal, is thirty times longer than in the first calculation!

Second, using the rather conservative value of economically recoverable proved reserves assumes that there will be no fluctuation in the amount of reserves. Historical data clearly show this assumption to be untrue. The quantitative estimate of reserves changes with time, for reasons that were discussed above.

Third, how does one pick, or estimate, the amount to be used each year, especially if the lifetime is very long? The examples just discussed assumed that the annual consumption rate of coal would not change for at least two hundred years. For virtually all ENERGY sources in all countries, this assumption is historically untrue. Using coal in the United States as an example, over a period of about sixty years, from 1913 to 1973, the annual rate of coal consumption hovered around 450 megatonnes per year. It dropped somewhat during the Great Depression and increased during the Second World War, but for a six-decade period, the 450-megatonne number was reasonably accurate. After 1973, coal consumption doubled, to about 900 megatonnes per year, over a period of about fifteen years. Now, it appears to have leveled off again at the 950-megatonne figure. So, what number should be picked for annual rate of consumption? And, how should it be adjusted for possible growth or shrinkage of consumption?

Several questions should be asked regarding any estimate of the lifetime of some energy source. First, what is meant by "total amount"? Is it the economic, proved reserves; is it the total resource; or is it something in between? Second, what approach was used to estimate the "amount used per year"? Is it assumed to be constant, or is it adjusted for growth? If it's adjusted for growth, how was it adjusted? To return to the example of a lifetime of 200 years for our coal supply, it should really be expressed as follows: The lifetime of country X's *present economically recoverable coal reserve* is 200 years *at present annual rates of consumption*. A statement of this sort provides the information needed to interpret the number.

Using some 2012 data for oil in the lifetime equation, proved reserves are 1478 trillion barrels and annual consumption is about 33 billion barrels per year, worldwide. This suggests a lifetime of the world's proved oil reserves, at the present rate of consumption, to be 45 years. This simplistic calculation provides us the rather dire prediction that we will run out of oil within 45 years, shortly after midcentury. Some forecasters have suggested that the lifetime may be as small as ten years; however, such forecasts have been made with regularity at least since 1920, and we have yet to run out. (And in fact, in the nineteenth century there were similar forecasts that we would run out of coal within ten years, also made on a fairly regular basis throughout that century.) Yet, consider this prediction from 1950:

> *Despite those gloomy predictions that our oil reserves will soon be exhausted, Dr. Gustav Egloff of the Universal Oil Products Company believes that in the year 2000 we will be refining 16,000,000 barrels of petroleum daily. He predicts that most of our clothing in the next half century will be made basically from petroleum.*

> **—Benford**[4]

This prediction was made for the United States. The prediction for refinery throughput was remarkably close, though more than half of those barrels of petroleum were imported. And, materials derived from petroleum do indeed, as predicted, have a significant role in synthetic fiber production.

Needless to say, we have yet to run out of oil (or coal, or natural gas). So what is the problem? Why have these estimates been so consistently wrong? Are the people who have made them scoundrels or dolts? The lifetime, estimated according to the above equation, is a ratio of two numbers, the reserve and the consumption rate. A change in *either* of these numbers will cause a change in the estimated lifetime. With increased geological exploration, and more detailed knowledge, hypothetical resources become conditional resources, conditional resources become reserves, and possible or probable reserves become proved reserves. Improvements in extraction technology can be responsible for conditional resources becoming economical to recover, and hence adding to the reserve base. More extensive exploration, improved extraction, or both will increase the reserves, and, as a mathematical consequence, increase the lifetime, assuming that there has been no change in consumption. But, changes in consumption rate also impact lifetime estimates. If consumption rate drops, as might be caused, for example, by price increases or supply shortages, then the estimated lifetime will increase. If consumption rate goes up, lifetime drops. Both the reserve base and the rate of consumption are likely to be changing at the same time, and for valid reasons. Thus the estimates of lifetimes are also changing continually. Both of these changes—in reserves and in consumption rates—are addressed in more detail later in this chapter.

Is the calculation of a lifetime of 45 years for the world's oil reserve true? There are, of course, two possible answers. The optimistic answer is *No*, because predictions that the world is running out of oil, made at least since the 1920s, haven't come true yet. For all we know, there may be enormous quantities of oil yet to be found. Furthermore, we will not literally run out of oil, because as the supply declines, the price will rise, and thus reduce the rate of consumption. The pessimistic answer is *Yes*, mainly because since the 1970s the rate of discovery of new oil reserves is less than the rate of production and consumption of oil. We cannot realistically presume that current oil production could be constant for 45 years and then stop overnight. Such production could be maintained only if it were continually matched by new discovery, which, as we will see, is now far from the case. In all oil fields, production declines in the latter half of a field's life, a point that will be revisited in the discussion of Hubbert's peak.

> *To say anything about the long-term prospects of crude oil availability is made invidious by the fact that some thirty or fifty years ago somebody may have predicted that oil supplies would give out quite soon, and, look at it, they didn't. A surprising number of people seem to imagine that by pointing to erroneous predictions made by somebody or other a long time ago they have somehow established that oil will never give out no matter how fast is the growth of the annual take. With regard to future oil supplies, as with regard to atomic energy, many people manage to assume a position of limitless optimism, quite impervious to reason.*
>
> —Schumacher[5]

INCREASING THE PETROLEUM RESERVE

Ideally it would be good to have the value of lifetime for any energy source to be as large as possible. Mathematically, there are two ways to do this. One is to increase the total amount available. The other is to decrease the amount used per year.

One factor that will increase the total oil reserve is to find more oil. Some scientists and oil industry analysts argue that we have not yet totally explored the world, and that improved satellite imaging, computer-based data analysis, and other techniques will lead to new finds of oil.

The cheerful doctrine that 'Something always turns up' is admissible only in a condition of profound ignorance, and we are now far past that condition—perhaps unhappily for ourselves....

It is, of course, true that not every acre of the earth's surface has been explored, but in a looser sense we do know what it contains, and that looser sense is enough. First, take an example near home. There are rich coal fields under us in parts of England, and it is not impossible that there are some more not yet discovered, but nobody would be likely to maintain that there is any chance whatever that there is 10 times as much undiscovered but mineable coal under us as that already known. In this loose sense we do know the amount of English coal.

—Darwin[6]

In addition, vast quantities of so-called nonconventional oil are available. First, it is already well established that there are huge deposits of heavy oils and bitumen* in Canada, Venezuela, and Siberia. Canada, in particular, has steadily expanded production from the oil sands in Alberta, to the extent that Canada is now the leading source of oil imports to the United States. Second, there still exist prospects for enhancing production from oil reservoirs already discovered. One such approach, which relates directly to issues of carbon dioxide sequestration discussed in the previous chapter, is to inject CO_2 into oil wells to stimulate production. Third, some possible contributions from very small fields, or oil reservoirs in very deep water, might also add to the total. A small number of scientists believe that there may be vast pools of oil buried deep inside the Earth, formed not by the generally accepted kerogen maturation processes we discussed in Chapter 18, but rather from other geochemical pathways that may not even involve starting from living organisms.[†]

Certainly, improved technology is steadily being developed for recovering the oil we already know exists. In the 1960s, oil companies assumed that about 30% of the oil in a field was typically recoverable; now, however, they expect to recover 40%–50%.

---

* The word *bitumen* is used loosely to refer to a variety of extremely viscous, very high boiling hydrocarbons that have the approximate look and consistency of road tar. Potentially the most important bitumens are the oil sands (also called tar sands), which occur in many parts of the world and, collectively, contain immense quantities of hydrocarbons. In Canada alone, the tar sand resource is estimated to be equivalent to about two trillion barrels of oil. The oil sands are deposits of bitumen with sand or sandstone.

[†] Over the years, a few scientists have suggested that much, or all, of the world's petroleum does not come from the geochemical alteration of biologically derived organic matter in the relatively recent past, but has some other, nonbiological origin. One of the first to suggest such a concept was the Russian chemist Dimitri Mendeleev, best known as one of the discoverers of the periodic classification of the elements. In recent years, the most visible proponent of such ideas was Thomas Gold, who hypothesized that the majority of the world's petroleum represents "preplanetary" material that was entrapped in place when the Earth first coalesced many billions of years ago. The most recent argument in favor of this view is Gold's book *The Deep Hot Biosphere* (Springer-Verlag, 1999). A very substantial majority of geochemists reject the idea of a nonbiological origin of petroleum.

Finally, as the price of oil goes up, in response to its appearing to become more scarce, it becomes more profitable to explore for oil, and to extract some of the poorer quality oils that otherwise might have been left in the ground.

Anything can be done in the oil field. If you think there's oil somewhere, and have the dollars and desire to find out, it will be done. There is no such thing as 'I can't.'

—Bass[7]

LESSENING DEPENDENCE ON PETROLEUM

What factors will decrease the amount of oil used per year? The best approach by far is through energy conservation. As the price of petroleum and its products rises, more and more people, and industries, take steps to curtail use. This includes, as examples, switching to more fuel-efficient cars, or using more insulation or turning thermostats back in the home to burn less heating oil. Reducing oil consumption can be achieved with more efficient use of oil or by substituting alternative energy sources for oil. Together, these strategies offer the quickest and cheapest way of reducing dependence on petroleum, and should at the same time maximize competition and innovation.

The experience in the United States provides a useful example. The Corporate Automobile Fuel Efficiency (CAFE) standards mandated in 1975 pushed cars from 13 mpg (18 L/100 km) to 27.5 mpg (8.5 L/100 km) in 1986, saving five million barrels of oil a day. In President Carter's term (1977–1981) and the five years following it, oil imports from the Persian Gulf region fell by 87%. In the period from 1977 to 1985, the gross domestic product (Chapter 1) of the United States rose by 27%, while at the same time total oil imports dropped by 42%. That saving single-handedly took away one-eighth of OPEC's market. Worldwide, the oil market shrank by about a tenth, with OPEC's share drastically reduced and OPEC's output being cut about in half. More fuel-efficient cars were the most important cause. In 1985, President Reagan rolled back the CAFE standards. This lessening of car and light-truck efficiency standards doubled American oil imports from the Persian Gulf. The additional oil consumed as a result is equivalent to all the oil thought to be in the Arctic National Wildlife Refuge. Unfortunately, federal policy in the 1980s discouraged energy efficiency. To be sure, concern for energy efficiency appeared needless after the 1986 oil price crash, which brought ten years of cheap oil. At the same time, budget cuts in federally funded energy research and development slowed down technological innovation in energy productivity. Even as late as the Persian Gulf War in 1991, if the President G.H.W. Bush had required that the average car get 32 mpg (7.3 L/100 km), that measure alone would have displaced all Persian Gulf oil imports to the United States.

Because of experiences like those described above, it's argued that the biggest untapped oil fields in the world are not underground; they are hovering about 50 cm off the ground—in the fuel tanks of our vehicles. Chapter 29 argued that the first and most important practical step that can be taken to reduce CO_2 emissions is energy conservation. Further, the first and most important practical step that can be taken to reduce petroleum consumption is also energy conservation.

Unfortunately, some of the petroleum that we do have simply becomes wasted. About 2.5 million cubic meters of oil end up in the ocean every year, from a variety of sources, all of which represent sloppy habits. Some comes from used engine oil being poured down the drain. Some is from gasoline or lubricating oil that accumulates in streets or parking lots as a result of leaks and is then carried away in storm runoff. Some is lost in leaks from offshore oil drilling. About 140,000 cubic meters spill from oil tankers and 60,000 more from oil drilling rigs.

In addition, there are likely to be changes in patterns of ENERGY use. Electric cars and hybrid vehicles are now on sale in many parts of the world. Some fuels not derived from petroleum are being investigated, such as biodiesel, derived from plant oils, and ethanol from corn or sugar cane (Chapter 31). These changes, combined with increased energy efficiency, suggest a long-term lessening of dependence on petroleum. Further, the share of oil in the total consumption of primary energy has been falling steadily, in favor of increases in natural gas and in nuclear energy. This trend is particularly evident in Japan and Europe. In Japan, the share of oil in total energy consumption fell by more than 21 percentage points between 1973 and 1996, whereas the share of natural gas went up by more than ten percentage points during the same period. Now, with increasing discovery and exploitation of vast amounts of unconventional gas, especially shale gas, there may be even shift away from petroleum to other energy sources.

OIL PRODUCTION

Three points need to be considered to project future oil production. The first is the quantity of oil that has been extracted to date, that is, the cumulative production. Second, an estimate of reserves tells us the amount of oil that can be pumped out of the known oil fields before they have to be abandoned. Third, a good estimate is needed of the amount of oil that still remains to be discovered and extracted from the Earth. The sum of these three numbers is the *ultimate recovery*, which is the total number of barrels of oil that will have been extracted by the time oil production comes to an end many decades or centuries into the future.

The rate at which any well—or any oil field, or any country—can produce oil always rises to a maximum and then, when about half the oil is gone, begins falling gradually back to zero. The oil geologist M. King Hubbert* discovered that, in any large region, extraction of a finite resource rises along a bell-shaped curve that peaks when about half the resource is gone (Figure 30.7). This assumes that no legislative or regulatory restraints have been placed on extracting the oil, so that its extraction is essentially unhindered. (Indeed, the graph of extraction of *any* finite resource starts at zero, rises to a peak, and ends at zero.) The maximum point in the graph has now come to be known as *Hubbert's peak*, also known as *peak oil*.

---

* M. King Hubbert (1903–1989) spent most of his career working for Shell Oil, at their research facility near Houston, and after retiring from Shell, followed by a second career with the US Geological Survey. Hubbert thought that the long-term energy future lay in solar and nuclear energy. Over the course of his career, Hubbert received numerous prestigious scientific awards. One of Hubbert's best pieces of advice is his statement that, "Our ignorance is not so vast as our failure to use what we know."

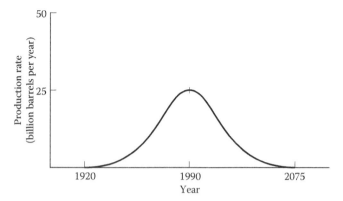

FIGURE 30.7 The Hubbert curve is a model for the depletion of any finite resource. The curve shown here relates to the depletion of oil, which is forecast to occur sometime around mid-century.

Hubbert projected that oil production in the lower 48 states of the United States would reach a maximum around 1969, with a year or two leeway either way. In fact, oil production peaked in 1970, well within the prediction. The 1973 oil embargo, resulting from the Arab-Israeli war of that year, was an epochal event in the United States. Gasoline prices skyrocketed, supplies were short, and sales were sometimes limited to a fixed (small) quantity per customer. Long lines at gasoline stations were punctuated by the occasional fist fight, stabbing, or shooting among aggrieved individuals waiting in line. Enterprising drivers carried a length of rubber tubing, to be able to siphon gasoline from other people's cars, usually foregoing the small detail of obtaining permission to do so. The 1967 oil embargo, resulting from another Arab-Israeli war, is almost totally forgotten. The difference is that in 1967 the United States single-handedly broke the embargo, by ramping up domestic production. In 1973, downhill from Hubbert's peak, that couldn't be done.

In all oil fields, peak production occurs at approximately the same time as mid-point of the total yield (Figure 30.8). (Again, this relationship does not include those situations in which artificial legislative, regulatory, or other arrangements deliberately restrict production.) Remarkably, that prediction does not shift very much even if estimates are off somewhat, even by a few hundred billion barrels. Therefore, what we must pay particular attention to is not when all the oil is gone, but rather the point at which *half* is gone. That point will represent peak oil production.

The total world oil production, in the entire history of the oil industry, was 800 billion barrels by the end of 1997. It's estimated that the industry will be able to recover only about another trillion barrels of conventional oil. Thus the ultimate recovery is about 1.8 trillion barrels. Currently, the demand for oil worldwide is rising at about 2.5% per year. This increase is driven largely by the expanding economies of countries such as China and India. The world's midpoint of depletion will come when 900 billion barrels have been produced (i.e., half the ultimate recovery), which is likely to be sometime in this decade. Once the approximately 900 billion barrels have been consumed, oil production is likely to begin to fall soon thereafter. This point

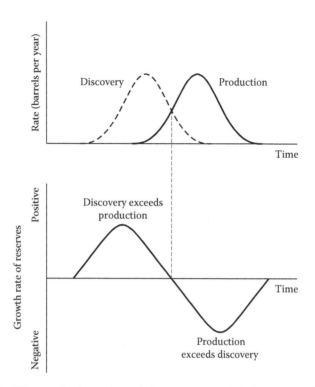

FIGURE 30.8 When production and completion are compared, the important point is reached when half of the total amount of the energy source (oil, in this example) has been produced.

might, though, be pushed out by some years as a result of the major worldwide economic recession that began in 2008 and is still affecting much of the world.

But most of the oil that has been produced so far, as well as that most likely to be produced over the next twenty years, is conventional oil. Much of it, especially from the Middle East, flows at high rates from giant fields. This oil is easy, cheap, and fast to produce. We should not lose sight of the fact that there is also a great deal of nonconventional oil. As we've seen, this material includes heavy oils, bitumens, oil shales,* and that amount of oil that could be obtained by enhanced recovery techniques.† The Orinoco oil belt in Venezuela contains a trillion barrels of heavy oil. Oil sands and

---

* Oil shale is an organic-rich rock that contains no free oil. In fact, the rock isn't shale, either. However, oil can be produced by heating, a process usually called "retorting," to drive oil out. Worldwide there are probably five to six times more recoverable reserves of oil locked up in oil shale than there are in conventional liquid petroleum. Unfortunately, producing oil from oil shale has formidable processing, economic, and environmental problems and is not economically competitive with conventional petroleum. The previous great flurry of interest in oil shale occurred in the late 1970s in response to the "oil crisis" of the time.

† The petroleum industry has an array of advanced recovery or enhanced recovery methods aimed at producing additional oil from wells or stimulating production of oil from very small accumulations. These methods include, as examples, injecting steam into wells, flooding wells with chemicals to reduce the interactions of oil with the surrounding rocks, "acidizing" to decompose carbonate rocks, or fracturing rocks surrounding the well. Injecting carbon dioxide, already mentioned in the main text, both enhances oil production and sequesters carbon dioxide at the same time.

shale deposits in Canada and the former Soviet Union may contain the equivalent of more than 300 billion barrels of oil. Certainly, these nonconventional oils will help as production of conventional oil passes its midpoint. Unfortunately, all of this nonconventional oil is more difficult and more expensive to produce than conventional oil.

World oil reserves have increased steadily over the past thirty years. Extrapolating that trend into the future would allow the conclusion that oil production will continue to rise unhindered for decades to come, still increasing well past 2020.* However, almost 80% of the oil produced today flows from fields discovered prior to 1973, and the great majority of those fields are already in declining production. During the 1990s, oil companies discovered an average of seven billion barrels a year of new oil; but typical production worldwide is about twenty billion barrels a year—three times the rate of discovery. Now, the ratio is even worse—we are using about four times as much oil as we discover. Of course, advanced oil extraction technologies will buy a bit more time before production begins to fall. There is some relief to be had in the prospect of extracting more oil from existing fields by advanced technology and in the development of nonconventional oil. However, it is very difficult to quantify the amount and timing of such contributions. So, from an economic perspective, the timing of when the world will run completely out of oil is not really important; what truly matters is the timing of when production will begin to decline. Beyond that point, oil prices are almost sure to rise unless the world's demand for oil declines at the same time. Consequently, the question seems not to be whether, but when, worldwide oil production will start to decline, ushering in what might be called the "permanent oil shock era."

The techniques that Hubbert used to predict peak oil in the United States with remarkable accuracy could also be applied to the world's oil. Doing so would allow us to predict peak oil for the entire world. The arithmetic of the calculation is not difficult, but selecting the proper numbers to use for making the calculation is chancy. Not all countries have governmental agencies, such as geological survey organizations, which can provide accurate statistics. Some countries have been accused of fiddling the data. Depending on whose set of numbers one uses for making the calculation, world peak oil may hit sometime between about 2005 and 2040. Some estimates suggest that we have already passed peak oil worldwide. (Just like hiking in mountains, it's not certain that the peak has been passed until you're well on the downhill side.) Others indicate that peak oil may be about 25 years into the future. The exact date is not so important as the potential of peak oil causing major shifts, relatively soon, in patterns of energy consumption.

THE RACE BETWEEN DISCOVERY AND PRODUCTION

Oil must be found before it can be produced.

—Campbell[8]

---

* Mark Twain's essay on the length of the Mississippi River is the finest discussion of the perils of making long-range extrapolations from limited data that has ever been written. It should be required reading for anyone who uses or manipulates numerical measurements to determine trends. The essay appears in Chapter XVII of his *Life on the Mississippi*, available in many editions. In it, Twain "proves," on the basis of twenty years' worth of data, that, about a million years ago, the Mississippi River must have been 1,300,000 miles (two million kilometers) long.

In the mid-1970s, there was a fundamental change in the pattern of petroleum use. The rate of finding new oil each year (i.e., the quantity of oil discovered per year) became, for the first time ever, smaller than the annual rate of using oil. The rate of finding oil is analogous to depositing money into a bank account. Similarly, the rate of using oil is analogous to withdrawing money or writing checks. In essence, we are taking money out of the bank (petroleum out of the ground) faster than we are putting it into the bank (i.e., discovering more oil). This discussion is not unlike our previous discussion on energy balances (e.g., Chapter 4), and particularly the balance in daily nutrition. Withdrawing more than is being put in can only "work" if we already have some money in the bank, or some ENERGY available in our body, to start with—or some oil in the ground. Regardless of the original size of those reserves, if we persistently take out more than we put in, sooner or later we'll go broke. In *exactly the same way*, it doesn't matter how much oil was originally in the oil reserve—if we consume it faster than we find new oil, sooner or later we'll consume it all.

Several factors may have a role in the decline in rate of discovery. First, there may indeed be less oil, and the oil that does remain is more difficult to extract. Most areas of the world have already been explored, at least to some extent. Some believe, therefore, that the last ever major oil discovery, like Alaska's North Slope, has already been made, and that it is highly unlikely that ever again there will be another big oil discovery. Further developments in technology are needed to exploit the hard-to-extract reserves of oil. Finally, another factor may be policies of oil companies regarding exploration vs. acquisition. For most oil companies, it is easier to "find" oil simply by buying another company, thereby obtaining that company's oil, than it is to mount a major campaign of exploration.

Increased oil prices, which would lead to higher sales prices for petroleum products and, ideally, higher profits for the corporation and a greater return to its stockholders would make it worth gambling on increased oil exploration. On the other hand, low oil prices depress exploration and drilling activity, especially in marginally producing areas such as the United States. By nationalizing their oil industries, and then by doing their best through OPEC to keep prices high in the 1970s and 1980s, the Middle Eastern nations encouraged oil development elsewhere. With oil so profitable in those days, prospectors were motivated to search even remote, inhospitable parts of the world. Now, though, virtually the whole world has been mapped for oil. The extent of oil exploration has now been such that only locations in extremely deep water and the polar regions remain to be fully explored. The partial melting of the ice in the Arctic Ocean, as a result of global warming, might lead to new discoveries there.

WHAT DOES "RUNNING OUT" REALLY MEAN?

We are all going to run out of oil very soon. Our country sooner than others, to be sure, but all of us down to nothing, Venezuela and Arabia and China too, and it will be very interesting to see how we handle it.

—Bass[9]

What does it mean that we are going to run out of oil? Will we ever pump what is literally the last drop of oil out of the Earth? The answer to the second question is, of course not. A day will not come when assorted members of the Saudi royal family are clustered around the end of a pipeline, waiting for *the* very last drop of oil to dribble out.

We will never truly run "out of" anything. As any energy source becomes increasingly scarce, its price will rise. This price increase will have several effects. It will become increasingly profitable to explore for additional sources. And, it will become increasingly profitable to invest in the research and development needed to develop new extraction technologies. Both of these activities add new reserves. At some point it becomes economically feasible to switch to a totally different source of energy. This could happen when the price of petroleum rises to a level that potential investors accept as being more-or-less permanent, allowing them to make long-term decisions about investments and profits.

The world is not running out of oil. What we are likely to run out of, and soon if the Neo-Malthusians and Hubbert are correct, is the end of the era of abundant, cheap oil. Oil prices have risen, in constant dollars, since about 1999. From one economic perspective, resources such as oil are not truly finite in size, but merely finite at a certain price. In other words, if we want more oil, all we need to do is drill more wells or develop some better extraction technology, or come up with some package of economic incentives to make it more profitable to produce oil. However, the reservoirs of conventional oil, especially the prolific ones in the Middle East, were formed only at certain periods in the geological history of the Earth, and formed under a limited set of geographical and geological circumstances. We can successfully find and extract oil now only if all the necessary geological conditions have already been met. Unfortunately, there is nothing we can do in the way of technology, or in providing economic incentives, which can have any effect whatsoever on the geological events that happened millions of years ago. Oil is indeed a finite resource.

> *...when United States production peaked, that didn't mean the end of the oil age, since the U.S. could still import oil. But when the global production peaks...it means the beginning of the end of the economy as we know it.*
>
> **—Jensen**[10]

It's sometimes pointed out that the Stone Age didn't end because we ran out of stones. The horrific problems of reliance on horses for transportation were discussed earlier (Chapter 17). A major change occurred around the turn of the twentieth century, but not because we ran out of horses.

> *And then the problem vanished. It was neither government fiat nor divine intervention that did the trick. City dwellers did not rise up in some mass movement of altruism or self-restraint, surrendering all the benefits of horse power. The problem was solved by technological innovation. ...the electric streetcar and the automobile, both of which were extravagantly cleaner and far more efficient. The automobile, cheaper to own and operate than a horse-drawn vehicle, was proclaimed an environmental savior.*
>
> **—Levitt and Dubner**[11]

A shift away from dependence on petroleum will be helped by the advent of some better technology for transportation.

However, we must also face a second major concern, besides the issue of how much oil remains. That second concern is where the remaining oil is located—who's got it, who needs it, and what some of the implications are regarding the global distribution of oil. We now change our attention to aspects of the geopolitics of oil.

GEOPOLITICS

The world's major reserves of conventional oil are in the Middle East. In terms of oil reserves, the top five countries in the world are Saudi Arabia, Iraq, United Arab Emirates, Kuwait, and Iran (Figure 30.9). Western relationships with Iran have been prickly since 1979. Now, with Iran pursuing what many observers believe to be nuclear weapons capability, and long-range missile capability, relations are positively poisonous. Some observers of foreign affairs believe that Saudi Arabia may be ripe for a fundamentalist Islamic revolution; if that happens, its oil exports may be curtailed drastically.

FIGURE 30.9 A map of the important oil-producing regions of the Middle East. (From http://wikitravel.org/en/File:Map_of_Middle_East.png.)

To illustrate the rich endowment of oil in the Middle East, consider the Ghawar field in Saudi Arabia. It was discovered in 1948. The field is about 240 km long and, at its widest, is about 40 km wide. The oil is contained in a layer of porous limestone about thirty meters thick. This single oil field in Saudi Arabia is estimated to contain eighty billion barrels of oil, an amount that represents more oil than is known to exist in any other single location anywhere, and represents three times the remaining oil in the entire United States. Though the Ghawar field contains an extraordinary quantity of oil, it's just one of many oil fields in Saudi Arabia. There are nearly sixty oil fields in Saudi Arabia, but oil is being produced from only seven. That production, from essentially one-eighth of all Saudi oil fields, is enough to satisfy all of the country's customers. All of the Saudi oil fields together contain about one-fourth of the world's oil reserves. Not only are these oil fields in Saudi Arabia huge, they are also highly productive and inexpensive to operate. The oil in the rocks under the desert is held there at high pressure. Because of that, once the reservoir rocks are penetrated by the drilling bits, the oil easily gushes to the surface without having to be pumped. The oil travels by pipeline, still under pressure, to separator plants where the natural gas that is dissolved in the oil is released and separated. As the pressure is reduced, natural gas bubbles out of the oil, in the same way that carbon dioxide fizzes from a carbonated beverage when a can is opened. Any hydrogen sulfide in the oil is also removed. Then the oil is piped to ports at Yanbu, Ju'aymah, or Ras Tanura.

There are some significant economic and geopolitical consequences of the fact that about half the world's remaining oil is in five countries in the Middle East. These five countries collectively are the only ones that can vary their oil output according to prevailing market conditions. And, when their market share is high, they can trigger a price crisis. This is exactly what happened in 1973. At that time, these five Middle Eastern members of OPEC were able to hike prices not because oil was scarce (in fact, only about one-eighth of the world's reserves of conventional oil had been consumed to that point) but because they had managed to corner nearly 40% of the market.

Then came discoveries on the North Slope and in the North Sea. Those new oil fields, coupled with a drop in demand for oil (occasioned at least in part by high prices), seriously reduced the "clout" of these countries in the oil market. Prices collapsed. OPEC's share of the market was cut in half (to about 18%). From a peak of $283 billion in oil revenues obtained in 1980, OPEC's revenues fell to $77 billion in 1986, then rising again to $132 billion in 1995. If the 1995 revenues are adjusted for inflation and expressed in terms of 1980 dollars, then they would be less than $100 billion, about a third of what was amassed in 1980. Furthermore, non-OPEC oil production (not counting the countries of the former Soviet Union and not counting the United States) tripled between 1976 and 1995. This increase in supply of non-OPEC oil came at the expense of OPEC's market share. Because there are no new oil fields comparable to a North Sea or a North Slope, the market share enjoyed by the Middle Eastern countries slowly increased. However, by 2015 or even sooner, they will be close to the midpoint of their depletion. Physical shortages of conventional oil would develop around 2020.

By 2015–2020, the Caspian Sea region (Figure 30.10) could become the world's second largest source of oil, after the Middle East. Its three principal producers are Azerbaijan, Turkmenistan, and Kazakhstan (sometimes somewhat facetiously called

FIGURE 30.10 A map of the Caspian Sea region, another important oil-producing region, and another potential geopolitical trouble spot. (From http://commons.wikimedia.org/wiki/File:Caspianseamap.png.)

"the Stans"). These countries provide oil and natural gas to Europe via a pipeline system that runs across Russia. If the Caspian region becomes a major player in world energy supply, the way the Middle East is today, then surely the United States and other energy-importing counties would have a strong interest in the stability and security of the region. It seems rather unlikely that western nations increasingly dependent on Caspian oil would blithely allow countries like Russia, with grave economic problems and the potential to revert to an authoritarian, nationalistic government, or staunchly anti-American Iran, to have a dominant role in the operation and security of the pipelines.

One possibility would be to find different routes to get the pipelines to consumers outside the region. Little would be served, from the perspective of the United States or Europe, by sending the Caspian oil northward into Russia. The only alternatives to the westward route that has the problems just mentioned would be to head to the east, or to the south. A route from Turkmenistan to Pakistan would have to pass through

Afghanistan, a nation whose political future has been uncertain for a long time, and still is. A pipeline project heading through Iran would likely be impossible to finance, because of Western economic sanctions resulting from Iran's nuclear program.

There remains another knotty problem to be resolved: What is the Caspian, and who owns its resources? A decision on what exactly the Caspian is has a remarkable bearing on the answer to the second half of the question. For Russia and Iran, the Caspian is a lake. For Azerbaijan and the "Stans," it is an inland sea. This dispute may seem to be an overly pedantic exercise in geographic nomenclature, but its outcome has important consequences. If the Caspian is indeed a lake, then its resources would have to be shared equally among all the surrounding countries, regardless of what they might be able to claim as their territorial waters. On the other hand, if the Caspian is a sea, then its resources (including, of course, oil) would be divided according to the extent of each country's territorial waters. These are determined by projecting the length of each nation's shoreline out into the Caspian. Since the lengths of the shorelines of the countries surrounding the Caspian are not equal, then neither would be the distribution of the shares of oil resources.

It will be interesting to see how China will use its oil in world politics. There have been indications that China plans to rely on its enormous coal reserves for its domestic energy needs and use its oil for export. Strong Asian-Pacific economies, as in Japan and South Korea, and some of the rapidly developing economies, as in Singapore and Malaysia, are mostly in countries with little or no indigenous energy sources. Therefore, China has the opportunity to use its oil to affect the course of events in the Pacific Rim. On the other hand, China is voraciously acquiring rights to oil, and many other natural resources, elsewhere, especially in Africa.

Europe and Japan have virtually no oil resources. Some oil from the North Sea is available in Europe, but virtually all of Japan's oil is imported. Despite that, both Europe and Japan have modern, stable industrial economies. What are the lessons of this? Should an oil importer like the United States worry about the amount of oil imported, since Japan and many of the European countries have strong economies but yet import over 90% of their oil? A major difference between these countries and the United States is that, in Japan and virtually all European countries, gasoline is taxed very heavily. In some places, it costs the equivalent of about $2.50/L. This high price encourages the use of small, highly fuel-efficient cars and a heavy reliance on public transportation. Indeed, in some countries, the gasoline tax helps pay for the excellent public transportation systems. Many of the railway systems use electric locomotives, rather than diesel-engine locomotives. Some countries, particularly France, have made a significant investment in nuclear generation of electricity.

NATURAL GAS

Natural gas is a premium fuel, for several reasons. It has a high calorific value and instant on/off capability and is clean burning and relatively easy to handle. It is an excellent fuel for domestic cooking and heating and has become the preferred fuel for "peak-load" power plants. These small plants, usually producing 50 MW of power or less, use natural-gas-fired combustion turbines to operate the generators. Natural gas produces the lowest carbon dioxide emissions per unit of energy than any of the fossil fuels. These coupled with the fact that natural gas is a clean fuel means there is a push to use natural gas in a variety of applications, including domestic heating, peak-load power plants, natural gas vehicles (such as city buses), and hybrid vehicles, some of which have both electric motors and natural gas engines.

The major areas of the world possessing large reserves and resources of natural gas are the countries of the former Soviet Union (which collectively have about 37% of the total), Iran (with about 25%), the United States, Algeria, Saudi Arabia, Canada, and Mexico. The total in all the Middle Eastern countries amounts to about as much as is in the countries of the former Soviet Union.

If we examine the historical trends for lifetime estimates of conventional natural gas, a story not unlike that for petroleum emerges. Table 30.1 summarizes some lifetime estimates made for natural gas at various times during the past half-century. And of course, the explanation is the same, the outcome of the race between discovery and production.

Europe is not a major producer of natural gas. Only a few countries in Europe that produce oil from the North Sea also produce their own gas. But, because natural gas is a premium fuel, there is significant demand in Europe. Europe basically has two options.

The first involves depending on Russia and the other countries of the former Soviet Union. Nowadays the former Soviet Union exports natural gas to Europe via pipelines. This situation has some potentially positive outcomes. First, it provides a flow of hard currency into the former Soviet Union countries, which may help them to obtain consumer goods and the various services needed to improve living standards. This could in turn lead to maintaining some political stability in these countries, which is of great interest and advantage to the more modernized countries of western Europe. However, there remains the specter of someone like Vladimir Putin seizing absolute power in Russia, and stopping the exports of natural gas to western Europe for political reasons. The second option relies on Algeria. This country exports gas to southern Europe (e.g., to Italy) via pipelines on the seafloor of the Mediterranean. At present, Algeria has serious internal problems with Islamic

TABLE 30.1
**Lifetime Estimates for Natural
Gas—Invariably Proven Wrong**

Year Estimate Was Made	Estimated Lifetime (Years)	Predicted Year to Run Out
1950	30	1980
1955	24	1979
1960	21	1981
1965	18	1983
1970	14	1984
1975	12	1987
1980	11	1991

fundamentalists seeking to disrupt and overthrow the government. What might happen to natural gas exports if the Islamic fundamentalists come to power in Algeria is highly questionable.

A consideration affecting the world trade in natural gas is that the easiest way to transport it is via pipelines. Countries possessing abundant natural gas, but which are very distant from major gas markets, may not have a major role in international trade in natural gas. This generally seems true of the countries in the Middle East, with little domestic demand for natural gas (because of the sparse population) and long distances to major markets elsewhere.

However, natural gas does not need to be shipped as *gas*. Recent years have seen increased interest in so-called gas-to-liquids technology (GTL). The overall strategy is to convert natural gas into a liquid product, either near the gas wells, or at a site in the country where the gas is being produced. Then, the liquid product can be shipped in tankers. The liquid products resemble the distillation fractions of petroleum.

The only way that large quantities of natural gas can be shipped (as cargo, distinct from transportation via a pipeline network) is to chill the gas to temperatures low enough to cause it to condense into a liquid (called liquefied natural gas, LNG). The temperatures required to produce and store LNG are about $-170°C$. This practice raises some safety concerns. Suppose a leak allowed some of the LNG to vaporize. Mixtures of 5%–15% of natural gas in air are explosive. What if an LNG tanker blew up in the harbor of a major city? The loss of life and property damage would be enormous. For that reason, some places now advocate the building of transfer points well offshore, where the LNG can be vaporized back to the gaseous state, and then transported the remaining distance to an on-shore terminal via pipeline.

The foregoing discussion has focused on conventional natural gas. In the past decade, much interest and excitement has been generated by unconventional gas, most especially shale gas. As discussed previously, this is gas trapped within non-porous rock, usually shale. Immense quantities are believed to exist in China, South Africa, the United States, and elsewhere. Optimistic projections suggest that shale gas could provide at least half the natural gas consumption of the United States by 2020, and world production will make this kind of gas a major factor in the world energy economy. Pessimists point out that gas production requires hydraulic fracturing of the shale, which might release various chemicals into the local water supply. Methane is a potent greenhouse gas, so some worry that the various operations in producing shale gas will add to greenhouse gas emissions.

If the reserve estimates for shale gas are reasonably accurate, and if this gas can be produced with reasonable care for the environment, then there is little doubt that shale gas will indeed have a very large impact on the ways in which energy is used in those countries having reserves or able to export it to neighbors. Shale gas likely is the most important energy development in decades.

COAL

Three areas of the world account for 90% of the world's total coal reserves: the countries of the former Soviet Union, the United States, and China. However, the total amount of coal in the world is so huge that even nations with a tiny percentage of the total world reserve are major coal-exporting nations. This includes Australia (the world's leading coal-exporting nation), South Africa, and Poland. Each of these three countries has, at best, about 1% of the total world coal reserve. The United States is both an importer and an exporter of coal simultaneously, exporting to the Pacific Rim and to Europe, and importing from Colombia.

Coal has significant environmental burdens. It produces the highest CO_2 emissions per unit of energy of any of the fossil fuels. It has the highest sulfur content. It is the only fossil fuel to produce significant quantities of ash when burned. Despite the occasional spectacular oil spill, coal extraction leads to more environmental disruption than the extraction of other fuels, and with more cost in human life. If mined land reclamation laws do not exist, or are not enforced, environmental devastation of surface mining could last decades or possibly even centuries.

So why bother? Coal is the least expensive energy source in terms of dollars per megajoule of useful energy produced. Modern industrial societies don't just require abundant energy; they require abundant *cheap* energy. There exists, worldwide, an immense investment in infrastructure for coal utilization, most notably in electric generating plants. The world has neither the financial wherewithal nor the manufacturing capacity to retrofit all of these plants to handle other fuels, let alone replace the complete plant with a new one, in a short span of years. Some countries may have abundant coal reserves but lack indigenous petroleum or natural gas. For them, it might make good economic sense to rely on domestic coal rather than on imported gas or oil.

Substantial strides are being made in the so-called clean coal technology, aimed at the large-scale utilization of coal with very low emissions from the plant. Scrubbers, low-NO_x burners, units for capturing ash, and carbon dioxide capture would make it possible to create electric generating plants, or synthetic fuel plants, with very low emissions compared to today's technology. Of course none of this technology is free. A concerned citizenry needs to understand that building coal plants to have minimal impact on the environment will add both to the capital investment and to the continuing operating costs. What is needed is the collective will to pay those costs, in the form of an increased electric bill, and the insistence that elected or appointed officials work to make sure that environmental issues are addressed with the best available technology.

Technology exists to convert coal to substitute natural gas and to synthetic crude oil. In principle, coal-derived gaseous and liquid fuels could replace natural gas and petroleum. Likely this would involve a much higher cost for those forms of energy. There is ample evidence that substitute or synthetic fuels derived from coal are technologically feasible. During the Second World War, Germany supplied much of its enormous liquid fuel needs from coal. Of course, under the extremely harsh Nazi dictatorship, and the exigencies of fighting a two-front war, economic considerations of the cost of these coal-derived liquids were not an issue. For much of the second half of the twentieth century, South Africa supplied much of its liquid fuel needs, along with most of its chemical products, from coal. Until 1994, world revulsion at the apartheid regime made it difficult for South Africa to purchase oil on open world markets. South Africa has negligible oil resources, but is abundantly endowed with coal. Again, international issues, in this case the embargo on oil sales to the apartheid regime, made the economic considerations of synthetic oil production from coal a secondary issue. Since 1994, the South African synthetic fuel industry has continued production with what seems to be great success. In the United States, a plant for producing substitute natural gas from coal has been running since the mid-1980s. China has just recently started up a plant for making diesel fuel from coal.

Europe has a centuries-long tradition of coal use. However, many of the European coal mines are by now very deep and, in many cases, in poor-quality coal seams. Many mines have been closed, or might expect closure in the not-distant future. The energy options in Europe include the use of natural gas and oil from the North Sea, importing natural gas from Russia, and nuclear energy.

REFERENCES

1. Jevons, W.S. Quoted in Heilbroner, R. *Teachings from the Worldly Philosophy*. Norton: New York, 1996, p. 214.
2. Kirov, S. Quoted in Knight, A. *Who Killed Kirov?* Hill and Wang: New York, 1999, p. 109.
3. From an editorial in *The Times of Zambia*, February 26, 1974; quoted by Arnold, G. *Africa: A Modern History*. Atlantic Books: London, U.K., 2005, p. 428.
4. Benford, G. *The Wonderful Future That Never Was*. Hearst Books: New York, 2010 p. 54.
5. Schumacher, E.F. *Small Is Beautiful*. Harper and Row: New York, 1973, p. 126.
6. Darwin, C. Quoted in Gregory, K. *The First Cuckoo*. Akadine Press: Pleasantville, NY, 1997, pp. 236–237. (*This* Charles Darwin is the grandson of the great naturalist who was one of the founders of the theory of evolution.)
7. Bass, R. *Oil Notes*. Houghton Mifflin: Boston, MA, 1989, p. 62.
8. Campbell, C. How secure is our oil supply? *Science Spectra* 1998, 12, 18–24.
9. Bass, R. *Oil Notes*. Houghton Mifflin: Boston, MA, 1989, p. 59.
10. Jensen, D. *Endgame*. Seven Stories Press: New York, Vol. I, 2006, p. 109.
11. Levitt, S.D.; Dubner, S.J. *Superfreakonomics*. HarperCollins: New York, 2009, p. 10.

FURTHER READINGS

Arnold, G. *Africa: A Modern History*. Atlantic Books: London, U.K., 2005. This splendid history of the past half-century in Africa covers far more than just petroleum. However, there are many discussions of the effects, primarily by Western countries or oil companies, at destabilizing emerging nations in Africa in pursuit of cheap oil.

Baer, R. *Sleeping with the Devil*. Crown Publishers: New York, 2003. The effect of "petrodollars" on Saudi Arabia, an extremely conservative country with a very poor human rights record and support for fundamentalists. Many Americans forget that the terrorists who attacked the United States in 2001 were Saudi citizens. Nonetheless, this book documents the continuing efforts to align the interests of the U.S. and Saudi Arabia.

Brown, L. *Plan B 3.0* W.W. Norton and Company: New York, 2008. This book's message is that "business as usual" is not an option, because we have reached tipping points in both natural systems, and in political systems. Chapter 2 discusses peak oil; Chapter 11, the need for energy efficiency.

Coll, S. *Private Empire*. Penguin Press: New York, 2012. This book discusses the ExxonMobil corporation, which is big enough to have its own foreign policy. If ExxonMobil were a country, so that its annual revenue counted as gross domestic product, it would be among the top thirty countries of the world.

Conaway, C. *The Petroleum Industry: A Nontechnical Guide*. PennWell: Tulsa, OK, 1999. Well described by its title. Chapter 2, on petroleum origins and accumulation, and Chapter 3, on petroleum exploration, are particularly relevant.

Crane, H.; Kinderman, E.; Malhotra, R. *A Cubic Mile of Oil.* Oxford University Press: New York, 2010. The world uses a volume of oil equivalent to one cubic mile each year. There is a need for improvements in energy efficiency, and in energy conservation.

Deffeyes, K. *Beyond Oil.* Hill and Wang: New York, 2005; and *When Oil Peaked.* Hill and Wang: New York, 2010. These two books discuss Hubbert's peak, along with other relevant issues such as how oil formed, prospects for natural gas, oil sands, oil shale, and the need for transportation efficiency.

Kandiyoti, R. *Pipelines.* I.B. Taurus: London, U.K., 2008. A very interesting discussion of how international oil pipelines help to shape political realities. The focus is mainly on central Asia, the Middle East, Europe, Russia, and China.

Klare, M. *Rising Powers, Shrinking Planet.* Henry Holt and Company: New York, 2008. A book discussing the geopolitics of energy, including oil. Russia, which "lost" the Cold War, is once again becoming nationalistic and arrogant. China and the United States are likely to compete all the more for available energy supplies. Klare argues on behalf of international collaboration as the best way forward.

Kunstler, J.H. *The Long Emergency.* Atlantic Monthly Press: New York, 2005. A dismal view of the future, in which two catastrophes converge: global warming and oil depletion. In Kunstler's view, there is no way that various alternative forms of energy will allow society to rely on energy at the scale we do now.

McPhee, J. *Annals of the Former World.* Farrar, Straus and Giroux: New York, 1998. The best book on geology ever written for the nonscientist or nonspecialist reader. Much information on oil geology, including resource issues. Highly recommended.

Phillips, K. *American Theocracy.* Viking: New York, 2006. The first three chapters discuss the role of oil in world affairs, and the impacts of America's dependence on oil, particularly imported oil.

Simmons, M. *Twilight in the Desert.* Wiley: Hoboken, NJ, 2005. Based on a careful analysis of numerous papers published in the professional literature of petroleum engineering, Simmons argues that Saudi Arabia is already "overproducing" oil and can no longer be depended upon to ramp up production to stabilize world markets.

Sperling, D.; Gordon, D. *Two Billion Cars.* Oxford University Press: New York, 2009. This book discusses, among other topics, possible improvements in energy efficiency in cars, along with the need for transportation planning.

Tertzakian, P. *A Thousand Barrels a Second.* McGraw-Hill: New York, 2007. In 2006 the world oil consumption exceed the rate of one thousand barrels per second. Sooner or later there will be a breakpoint, which will create new challenges for a world with a different energy economy.

Yergin, D. *The Quest.* Penguin: New York, 2011. A discussion of the geopolitics of oil, likely sources and supplies of oil, and possible alternatives as we move to the future.

Zubrin, R. *Energy Victory.* Prometheus Books: Amherst, NY, 2007. This book points out how "petrodollars" help to fund terrorists, or get funneled to nuclear weapons programs in rogue nations. Breaking the dependence on petroleum would be an excellent way to defeat terrorists.

31 Renewable Energy from Biomass

Most of the industrialized world depends heavily on fossil fuels for both electricity generation and transportation. Two major factors will put tremendous pressure on fossil fuel use in the twenty-first century. The first is continued concern about global climate change, or the greenhouse effect (Chapter 29). Although CO_2 is not the only greenhouse gas, human activities are not the only source of CO_2 emissions, and fossil fuel combustion is not the only human activity that generates CO_2, concerns about global climate change have already led to increased interest in energy sources and technologies that do not result in a net increase in atmospheric CO_2 concentrations. The second factor is the concern about reserves, resources, and geopolitics (Chapter 30). The potentially short lifetime of remaining reserves, especially of oil, makes it prudent to consider other sources of ENERGY that are not likely to be depleted. ENERGY sources that are not subject to depletion are called *renewables*. Examples include wind, water, solar, and biomass. The term *renewables* implies that this source of ENERGY is continuously being renewed, either through natural events, as in wind energy or solar energy, or through human intervention, such as the energy crops discussed later in this chapter. Such ENERGY sources that are harvested but then regrown are also known as *sustainable energy sources*. Renewables do not rely on localized concentrations in seams or reservoirs, as do coal, oil, or natural gas, but rather can be utilized almost anywhere. Therefore renewables are not nearly so likely to be impacted by concerns of geopolitics and possible embargoes as are the fossil fuels.

The caution in considering renewables lies in the issue of time scale. We might argue that an annual plant—soybeans for example—harvested this year could be replaced with another fully grown plant next year. This is not the case with trees, which would require several years, even for the fastest-growing species. Even fossil fuels are, in a sense, renewable, provided that we have enough patience (millions of years' worth).

> *Wood is a renewable resource, but for making lumber or paper, not energy. The same adjective can even be applied to fossil fuels like oil and coal, if your time scale for renewability is measured in tens of millions of years. As a good rule of thumb, any resource that captures solar energy in the form of chemical bonds—coal, natural gas, oil, even biofuels—can always be consumed faster than it can be "renewed."*
>
> **—Rosen**[1]

If there were such a thing as an "ideal" ENERGY source, it might be something that is essentially limitless, so that we need not worry about lifetime estimates; produces little or no pollution and no radioactive waste, eliminating, or at least reducing, many environmental concerns; produces no *net* increase in the atmospheric CO_2; and is a domestic resource, eliminating any nation's concern about dependence on foreign sources. Various ENERGY sources meet, or are thought to meet, these criteria. The renewables come closest.

INTRODUCTION TO BIOMASS

Most materials that contain carbon, or carbon and hydrogen, will burn, and could, at least in principle, be used as fuel. *Biomass* refers to any type of animal or plant material—that is, materials produced by life processes—that can be converted into ENERGY or used directly as ENERGY sources. Broadly, biomass falls into three categories: wastes (e.g., straw, bagasse, garbage), standing forests (e.g., firewood), and energy crops (e.g., corn grown to produce ethanol). In principle, biomass should be inexhaustible, renewable in the literal sense of the word, provided that we grow a new plant to replace each one harvested for energy. In principle (though not necessarily in practice, as we will see later), biomass should have no net effect on atmospheric CO_2, because removal of CO_2 by photosynthesis in the growing "replacement" plant should exactly balance CO_2 production when the biomass is consumed. When biomass material is used for ENERGY production it is sometimes called a biofuel.

With renewed interest in biomass in the early twenty-first century, it may be easy to lose sight of the fact that, for all but a small fraction of human history, biomass has been the predominant source of fuel, well into the mid-nineteenth century. The most important biomass source was wood. Dried plants, oils extracted from plant parts, dried animal dung, and even animal fat also contributed to domestic heating, lighting, and cooking needs until the latter part of the nineteenth century. The rise of coal as an important fuel for stationary steam engines and steam locomotives through the nineteenth century and then the increasing importance of petroleum products as transportation fuels in the twentieth century were responsible for displacing biomass as an important energy source.

Nowadays, biomass supplies about 15% of the world's primary energy. At least two billion people depend almost exclusively on biomass for their energy supplies. Most of these people live in developing nations where the per capita energy consumption is low; altogether, their total energy demand is still relatively small. Biomass supplies over 35% of the energy in developing nations but has the potential of supplying even more, by improving biomass production and doing a better job of utilizing agricultural wastes. Some nations far surpass the average of 35% for developing countries; India, for example, obtains more than half of its energy from biomass, and Tanzania, virtually all.

Biomass fuels utilize the chemical ENERGY fixed by photosynthesis and stored in the molecules within plants. Consider the carbon cycle in nature (Figure 31.1). The key process is photosynthesis—the combining of carbon dioxide with water to produce sugars and oxygen, facilitated by light.

The ENERGY of sunlight facilitates the transformation illustrated in Figure 31.2. Since the sugars are in a STATE of relatively high chemical potential energy, at

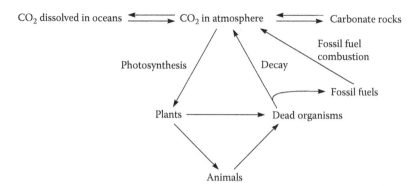

FIGURE 31.1 Biomass production begins with the important step of photosynthesis in the global carbon cycle.

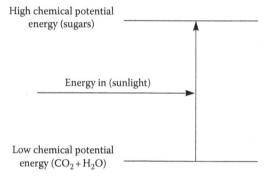

FIGURE 31.2 ENERGY from sunlight is the input needed for the reverse energy diagram converting carbon dioxide and water into simple sugars.

some time later the reverse transformation can occur (Figure 31.3). The chemical potential energy created by the plant during photosynthesis can be released to create heat for domestic cooking and heating, to produce industrial process heat, or can be converted to other ENERGY forms: electricity, or gaseous or liquid fuels. Therefore,

Biomass is a way of storing solar energy for later use.

To create the chemical potential energy stored in biomass requires sunlight and water, nutrients for the growing plants, and land on which to grow the plants. The maximum possible photosynthetic conversion of sunlight to chemical energy via photosynthesis is only about 7%. Just as real heat engines never actually achieve the ideal Carnot efficiency, neither do plants achieve this level of maximum conversion. Sugar cane is about 2% efficient in converting available solar energy to biomass, and corn about 1% efficient. More likely, though, only about 0.3% of

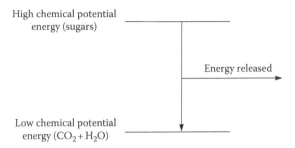

FIGURE 31.3 When biomass is used as an ENERGY source, the heat liberated comes from the conversion of plant material, such as sugars, back into carbon dioxide and water. Combined with Figure 31.2, biomass can be considered as a way of storing solar energy for use later.

FIGURE 31.4 Plants living in water, such as the algae shown here, tend to do a better job of efficient conversion of sunlight into biomass than do land-based plants. (Courtesy of UTEX—The Culture Collection of Algae at the University of Texas at Austin.)

the solar energy falling on land is stored as sugars and other compounds in land plants. Plants growing in water, such as algae (Figure 31.4), come much closer to the ideal maximum.

Biomass as an ENERGY source has a number of attractive features. It is indigenous in most parts of the world. A country that can grow its own ENERGY need not be fearful of, or held hostage to, countries owning large supplies of a scarce resource such as oil. Biomass is renewable. Agricultural fields produce every year, and trees can be regrown over longer periods, say, ten to a hundred years. Biomass provides a way of

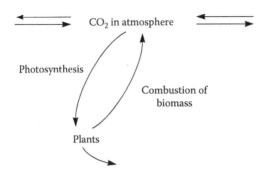

FIGURE 31.5 Biomass represents a "short circuit" in the global carbon cycle, reducing the cycle simply to photosynthesis and combustion. In principle, though not likely in practice, as much carbon dioxide would be removed from the atmosphere as was produced by burning the biomass.

conveniently storing energy, which is not true of other renewables. Wood, for example, can be cut and stacked. But we can't store wind. One of the major challenges in solar energy is the fact that the sun doesn't shine at night; we can't store sunshine. Biomass captures CO_2 during growth. Biomass utilization represents a way of short-circuiting the global carbon cycle (Figure 31.5) so that in principle (though perhaps not in practice) if we grow exactly as many new plants as are harvested for energy, then we will remove CO_2 from the atmosphere by photosynthesis as rapidly as we produce it by combustion. The net effect should be no *net* change in atmospheric CO_2 concentration. Production of biomass is one of the few methods of obtaining energy sources, which does not arouse much public opposition, at least so far.

IS BIOMASS CO_2 NEUTRAL?

Biomass utilization involves burning directly plants or products derived from plants, short-circuiting the global carbon cycle. The basis of that "short circuit" is that, in principle, CO_2 released during combustion would be cycled back into plant materials by photosynthesis. The crucial assumption is that growth of a new crop following harvesting of biomass will offset all of the carbon released from the harvesting and use of the biomass.

It is questionable whether this ideal can be achieved in practice. There would have to be no loss of carbon compounds from the soil (the loss of which would also produce CO_2). The regrown biomass, of whatever kind, would have to contain the same amount of carbon as the harvested biomass. If the new plants—those intended to replace the ones just harvested—are of a different kind, this ideal may not be met. To be sure, it is also possible to gain, rather than lose, in carbon uptake. Furthermore, it takes ENERGY to cultivate, harvest, and process the biomass, as well as to transport it. Likely the farm machinery and transportation vehicles rely on devices burning gasoline or diesel fuel. If electricity is used in the processing, it may have come from generating stations that burn fossil fuels. The CO_2 emissions from these sources must be taken into account in considering whether the biomass energy system is truly CO_2 neutral.

ENERGY CROPS

Today, much of the biomass that is used as an energy source is actually a by-product of other crops, or other operations. Examples include bagasse, the waste material left from sugar cane (Figure 31.6) processing, and paper mill wastes, which satisfy more than half the energy needs of the pulp and paper industry. To go beyond this—to make biomass a significant contributor to our energy economy—will require production of biomass on a much larger scale. The need for larger-scale production leads to the concept of "energy crops." Two categories of energy crops provide examples: so-called short-rotation woody crops and herbaceous crops. Short-rotation woody crops include poplar (Figure 31.7), willow, and, in warm climates, eucalyptus. Switchgrass and sorghum (Figure 31.8) are examples of herbaceous energy crops. The best choice of an energy crop depends on the specific local conditions—prevailing climate and quality of soil, as examples—and the expected use of the crop.

Plants selected as energy crops should be efficient in producing an energy-rich product, and producing it quickly. Broad-leaf plants have short-growing seasons and are efficient at photosynthesis. Such kinds of plants would be preferred. Rapidly growing annual or perennial grasses that can be harvested with existing farm equipment are also being considered. Examples of such plants include sorghum,

FIGURE 31.6 Bagasse is the residue from the processing of sugar cane to make sugar. It is a low-cost, or even no-cost, biomass fuel. Workers in this sugar cane processing plant are moving bagasse from the crushers on the upper level into the chute "in the center of the lower level. (From http://en.wikipedia.org/wiki/File:Engenho_da_Calheta_437.jpg.)

FIGURE 31.7 Poplar trees, an example of a short-rotation wood crop. The James River Corporation is already using a dedicated feedstock supply system to meet part of its energy needs in Oregon. (From http://www.nrel.gov/data/pix/searchpix.php.)

FIGURE 31.8 Sweet sorghum. (Photograph courtesy of Mandy Gross, Oklahoma State University Robert M. Kerr Food & Agricultural Products Center, Stillwater, Oklahoma.)

switchgrass, and Napier grass.* Ideally, energy crops would be grown much like ordinary food crops. If evaluated in terms of potential profit, sorghum may prove to be the best energy crop. However, large-scale cultivation could increase soil erosion. Perennial plants offer the advantages that their roots help hold soil in place and, once the perennials have established themselves, they are highly resistant to weeds. Plants chosen as energy crops should ideally require little or no fertilizer, because considerable energy is used in producing fertilizers. The plants should be ones with good resistance to insect attack and plant diseases. Like fertilizers, pesticides and herbicides require large amounts of energy to produce. The plants should need little water, because in many farming areas of the world, water is often in short supply.

In addition to the direct advantage of providing biomass as a source of energy, these crops offer other advantages as well. Lands not suited for food crops, or lands that have deteriorated to a point of no longer being suitable for food crops, could possibly be returned to productivity. Worldwide, hundreds of millions of hectares might be made productive again. Energy crops could reduce soil erosion and water run-off from land. Rural economic development could be promoted by planting energy crops. Cultivating energy crops could be a valuable source of income for farmers. The energy crops could provide a steady supplemental income, or allow some unused land to be farmed without requiring much investment in new equipment.

Of course there are also perceived disadvantages. Some energy crops may require large quantities of water. Plants for energy crops that are both fast growing and hardy could become nuisances if they spread outside the farm. They could displace native plant species, upsetting the local ecosystem and having negative impacts on the animal life as well. Growth of energy crops could affect biodiversity via the destruction of habitats. There is also an issue of energy crops competing with, or displacing, food crops, referred to as the food vs. fuel debate. In many developed countries, there is already competition for the use of land, water, and nutrients among various forms of food production, including animal farming. If a significant new market for energy crops were to develop, farmers could choose to grow energy crops on their prime land to achieve the highest possible yields and profits. Their doing so could result in some displacement of conventional food production. This becomes of even greater concern in those countries where food is already in short supply.

Farmers will need to consider their business in a different light. Ever since the time in prehistory when the first farmer produced more food than was needed just for himself and his family, and managed to sell or barter the excess, farmers have been in the business of selling to people or corporations who were purchasers of food

---

* Sorghum is the name for various kinds of cornlike grasses that include, as examples, milo and kafir. Varieties of sorghum have been used in Africa and Asia since prehistory, where to this day they provide food for millions of people. Sorghums generally have good drought resistance. Switchgrass is a perennial grass that can grow up to two meters high. It is currently grown in the American and Canadian midwest for livestock fodder and has been considered as a potential feedstock for ethanol production. One of the more ambitious schemes to have been floated suggested planting 15 million hectares of switchgrass on the Canadian prairies, using it to produce ethanol to fulfill Canada's gasoline needs. Napier grass, sometimes also called elephant grass or Uganda grass, is a tropical forage grass of very high yield, so high in fact that in some countries it is considered a weed. Some clumps can grow to over six meters tall. It is a native of tropical Africa, but now occurs also in South America, the Philippines, Hawaii, and in some of the southern regions of the continental United States.

products. If energy crops succeed, farmers will now be selling to customers such as electric utilities, fuel-processing companies, or even chemical companies, a change that could be a challenge. A related issue is whether a specific crop has more value as food or as fuel. As a crop is growing, the farmer would need to decide whether it would be better to sell that crop as a food supply, or to sell it as a fuel. Should corn, for example, be grown for food or as a source of fuel (ethanol)? In addition, what about crop surpluses? They can be stored against future shortages, or they could be burned as a source of low-cost biomass. However, if the latter course were chosen, we might then ask why land should be devoted to growing surplus crops in the first place. Perhaps that same "extra" land—that is, the land producing the surplus— could be put to better use in growing a better-quality energy crop.

In an energy economy heavily reliant on biomass, energy production will ultimately be limited by the amount of available land on which biomass can be grown. Estimates of land requirements for biomass to make a significant contribution to a country's energy economy vary a good bit, depending on the initial assumptions. Regardless, large amounts of land would be needed, though some energy crops may not need to be grown on prime cropland. Concerns about the availability of land as a constraint, and about displacement of food production for energy production, may be lessened by considering aquatic plants as energy crops.

Algae, seaweed, and other aquatic plants can be intensively grown in certain areas of the sea or in inland lakes or ponds. Doing so would eliminate the competition of energy crops for land with conventional agriculture. Marine algae, such as kelp, have been considered as potential renewable resources. The water hyacinth (Figure 31.9), nowadays considered to be a weed, has been studied as a feedstock for producing gaseous fuels. Aquatic weeds are a hazard in some waterways, especially in tropical or subtropical regions, and often have to be harvested anyway, just to keep the waterway open. The energy value is then a useful by-product.

POSSIBLE DISADVANTAGES OF BIOMASS

As with *any* energy source, biomass has both advantages and disadvantages, sometimes more perceived than real. Nevertheless, several concerns have been raised about biomass energy, particularly if it were to be used on a scale much larger than at present.

One issue is the efficiency of use of solar energy. As an approximation, biomass can be considered to convert 1% of the incoming solar energy into plant material through photosynthesis. In comparison, the efficiencies may be 10%–20% for conversion of solar energy to electricity via photovoltaic cells and 30% for a generating plant that uses solar energy to generate steam. (Both of these technologies are discussed in Chapter 33.) Both biomass and solar can require large amounts of land. Solar, with its greater efficiency of conversion, could provide more ENERGY per area of land or require less land to produce the same amount of ENERGY.

Biomass is not emission free. The issue of whether biomass is really "CO_2 neutral" has already been mentioned. Burning of biomass will produce substances that could lead to air pollution if emitted to the environment: carbon monoxide, nitrogen oxides, and especially particulates such as soot and ash. The amounts produced vary

FIGURE 31.9   Water hyacinth. (From http://wric.ucdavis.edu/information/aquatic/wh.htm.)

with the specific combustion technology used, and with the specific biomass source. Actual emissions to the environment will also depend on what, if any, emission-control equipment is used. Emissions from biomass-fueled electricity-generating plants are likely to be similar to emissions from coal-fired power plants, with the notable exception that biomass units would produce very little SO_x. Particulate emissions would likely be the most serious problem.

Barring extensive adoption of aquatic energy crops, production of biomass inevitably requires land. The amount of available land has to be considered in assessing the ultimate potential of biomass fuels on a global scale. Also needing consideration is the trade-off between use of land for food crops vs. its use for energy crops. Some estimates of land requirements are very large indeed. For example, a small electricity station using solid biomass as the fuel to raise steam might require six hundred hectares of land (i.e., six square kilometers) to grow enough biomass to generate one megawatt of power. There may not be enough arable land in Europe—even if it were *all* used for energy crops and not food—to supply Europe's energy needs entirely from biomass. Taking the wildly optimistic assumption that we could grow all energy crops worldwide as fast as sugarcane grows in Brazil, meeting the world's energy needs with biomass would require that two-thirds of the land now used for food crops be diverted to energy crops.

A final issue is the energy balance for biomass conversion, that is, the amount of ENERGY liberated when biomass is used, or converted into a different fuel form, relative to the amount of ENERGY required to produce the biomass in the first place. A particularly fierce debate has centered on the energy balance for production of ethanol, especially using corn starch as the feedstock for the process. When the energy liberated in using biomass is expressed as a ratio to the ENERGY used in growing and converting it, the value of that ratio needs to be greater than 1. If not, then more ENERGY was used up in growing the biomass (and, in the case of ethanol, converting the harvested biomass to ethanol) than is "paid back" by using the biomass. Possibly barring very peculiar and short-lived circumstances, it makes no sense at all to rely on an ENERGY source that gives us less energy back than was invested in growing it.* In such cases, we would likely be better off not using ENERGY to grow, harvest, transport, and process biomass, but rather to use that same energy directly.

WOOD

Wood was probably the first fuel to be used on a large scale. Until the nineteenth century, it was the predominant fuel in the world. As recently as 1800, about 95% of the world's energy came from wood. Wood was finally replaced by coal as the major energy source in 1880. Even in relatively recent times, wood has supplied a large amount of energy for certain countries. In the Soviet Union in the 1920s, the bitter and brutal civil war had paralyzed much of the coal and oil industry. As a result, about 85% of the energy used by the young Soviet Union came from wood, until coal mines were reopened. During the last year of the Second World War, when most of Japan's industry had been destroyed and energy imports were cut off, about half of the energy used came from wood. Cooking with wood accounts for roughly 60% of all energy consumed in the African Sahel (the semiarid region south of the Sahara). We've seen (Chapter 7) that wood is not a very desirable energy source with respect to natural gas or oil, but for most impoverished people in developing countries, it may be the only choice available.

The heating value of wood, on an equal mass basis, is not as good as competing fuels. The major components of wood are cellulose† and lignin.‡ The molecules

---

* This situation is reminiscent of the jokes about used-car salesmen who make the argument that, "We lose money on every car we sell, but we make up for it in volume [of sales]." It can't be done. Mathematics is inexorable; it wins every time. The use of an energy source that requires a greater investment to produce than is repaid when it is utilized will sooner or later lead to major economic problems.

† Cellulose is one of the major structural materials in plants; as examples, various kinds of wood contain 30%–40% cellulose, and cotton, about 90%. It is also one of the most abundant organic compounds on Earth. At least a billion tonnes a year of cellulose (not counting the many other components of plants) are produced each year via photosynthesis. Its molecular structure consists of several thousand glucose units linked together (glucose is the primary product of photosynthesis). Complete hydrolysis of cellulose, if we could find a way to do it easily, would produce the simple sugar glucose.

‡ Lignin is also a major constituent of most plants, but much different in composition from cellulose. It is composed of three chemical "building blocks," not one as in the case of cellulose, and, rather than being simple sugars, the building blocks of lignin are aromatic compounds that can be considered derivatives of benzene. The properties of a particular lignin depend on the relative proportions of its three basic constituents. Lignin occurs in all woody plants and accounts for a quarter to a third of the (dry) mass of such plants. Lignin provides much of the structural rigidity of such plants. It is much more resistant to chemical degradation than is cellulose; through a series of geochemical transformations it eventually transforms into a major component of coal.

of these materials contain abundant numbers of oxygen atoms. When plant matter containing cellulose and lignin gets converted into coal, many of the oxygen atoms are removed by the "cooking" inside the Earth. As a result, coal contains much less chemically combined oxygen than does wood. When either fuel is burned, the carbon atoms are ultimately converted to carbon dioxide. But, when carbon atoms are bonded to oxygen atoms already present in the fuel, those carbon atoms are, in a sense, already partially oxidized. As a result, less energy is liberated from the fuel containing the greater amount of oxygen. Though there is variation among different kinds of coals and different kinds of wood, as a rule-of-thumb, wood has half the heating value of coal, and one-fourth the heating value of natural gas. With a poor-quality wood, this fraction would drop to an eighth (looked at from a different angle, replacing natural gas by wood would require burning four to eight times as much wood as gas). The actual heating value of wood depends on the kind of tree from which it came, and whether it was allowed to dry before burning. A good grade of dry wood will provide about 14 MJ/kg.

Uses of Wood as an Energy Source

Certainly the most straightforward way of using wood as an ENERGY source is to burn it. Unfortunately, the small-scale burning of wood in domestic applications is very inefficient. At best, only 10%, possibly closer to 5%, of the heat liberated from the wood is used effectively in a traditional open fire. Much good could be done in designing stoves that would be simple enough and cheap enough so that people could make them on their own, with simple tools and materials (Figure 31.10). An enclosed fire in a stove that regulates the air intake and directs the hot combustion gases into a chimney could likely increase efficiency by at least a factor of four.

FIGURE 31.10 A woman cooking over a traditional stove using biomass fuel. This photograph is from Cameroon, but similar scenes can be found throughout much of the developing world. (From http://de.wikipedia.org/wiki/Datei:Cameroon_2005_-_cooking_woman.jpg.)

The same heating and cooking could be achieved with one-fourth to one-fifth the wood consumption of an open fire. This saving would be important from an environmental perspective (e.g., as it impacts deforestation, discussed below) and an economic one, because in some developing countries 40% of household expenditures are for wood.

On a commercial, rather than domestic, scale, direct combustion of wood is also the cheapest and most straightforward option. Wood combustion could be used simply as a source of heat, or to generate steam. In areas where lumbering is a major industry, electricity-generating plants burn a wood residue called "hog fuel" to produce ENERGY. The pulp and paper industry is a leader in so-called cogeneration processes, which provide the simultaneous production of heat and electricity for industrial use. If it's burned in well-designed and well-run combustion equipment, wood will produce less NO_x emissions than coal or fuel oil. Wood has a low sulfur content, so will also produce less SO_x emissions.

A second approach to wood utilization involves pyrolysis, the breaking down of the wood by heat. If wood is heated to about 250°C in the absence of air (so that it won't catch on fire), the moisture and volatile components are driven off, leaving a solid product that has a high carbon content—charcoal. Charcoal is a good-quality solid fuel, which can be transported and handled more economically than an equivalent amount of wood. In some countries, charcoal is the most widely used household fuel in urban areas. The gaseous and liquid products that are driven off during the pyrolysis also have value. At one time, for example, methanol (discussed in the next section) was recovered from wood in this way. If the gaseous and liquid products are used as well as the charcoal, the ENERGY obtained is equivalent to an overall efficiency up to 80%.

A third approach to wood utilization is gasification, the conversion of wood into a gaseous fuel. Wood, reacted with steam and oxygen, produces a gaseous fuel that is mainly carbon monoxide and hydrogen. This fuel is burned in a gas turbine, similar to a turbojet engine. The turbine is connected to an electricity generator. The hot combustion gases leaving the turbine are used to produce hot water that is used for heating nearby homes. A pioneering plant in Värnano, Sweden, produced electricity plus hot water with 80% efficiency, and no SO_x emissions. During the years of the Great Depression and then the Second World War, small gasifiers were used on motor vehicles to convert wood or charcoal to gaseous fuel for the engine (Figure 31.11). These units required considerable maintenance, but nonetheless did operate reasonably well.*

---

* During the 1930s and again during the Second World War, there were efforts in a number of countries to convert automobiles to run on a gaseous fuel produced onboard from wood or charcoal (or coal). Many of these vehicle conversions were done by private individuals with home workshop equipment. The gas producers and accessory equipment take up considerable room in the vehicle and require careful maintenance. And, in some cases the converted vehicle looks pretty goofy. Nevertheless, they do function, and it is entirely possible to use wood as an automobile fuel, via its conversion to a gas. Details are provided in the book *Producer Gas for Motor Vehicles* (John Cash and Martin Cash, Lindsay: Bradley, IL, 1997). Many excellent photographs can be seen in *Wartime Woodburners* (John Ryan, Schiffer Military History: Atglen, PA, 2009).

FIGURE 31.11   An automobile equipped with a unit to gasify wood, to produce a fuel gas for the engine. This photograph was taken in Berlin during the Second World War. (From http://fr.wikipedia.org/wiki/Fichier:Bundesarchiv_Bild_183-V00670A,_Berlin,_Auto_mit_Holzgasantrieb.jpg.)

PROBLEMS WITH THE USE OF WOOD

Rapid population increase in developing nations has caused extensive cutting of forests for fuel. As a forest is cut back from the population center, more time and effort must go into foraging for fuel. In some cases, this may involve five to ten hours of work per day, every day. Usually it is the women who have this job. Their labor is lost to other, potentially more productive, activities such as child rearing, farm work, or work in the home. As more effort goes into finding fuel, less goes into raising crops, tending animals, or light manufacturing, establishing a vicious cycle of increasing poverty and lowered standard of living.

Large-scale wood use leads to deforestation. A forest is not just trees or wood—it is an entire ecosystem. As forests are cut, animal habitat is destroyed along with the trees. One effect may be to reduce the habitat of birds that are useful in keeping insects in check. In addition, trees are very helpful for holding soil in place—absorbing moisture plus holding soil. Loss of trees leads to increasing water runoff and soil erosion. The soil erosion leads in turn both to the loss of valuable topsoil and to water pollution from suspended soil and mud washed into rivers or lakes. Mudslides can occur. The desperate need for energy combined with exploding populations is causing people to destroy their own environment. Some thirty years ago wood was being cut in Africa at a rate of about two million hectares per year. About 90% of this wood was burned as fuel. Deforestation and desertification is widespread and increasing in Africa. In those past thirty years, the southern edge of the Sahara desert has moved at least 120 kilometers further south.

Where wood is heavily used as an energy source, destruction of forests often leads to their replacement by open fields or, where "slash and burn" agriculture is practiced, by scorched earth tactics. Transformation from forest land to open fields or scorched earth causes a change in the fraction of incoming sunlight that is reflected back from the surface (the *albedo*). A change in the albedo plays a role in climate change. The destruction of the Amazon rainforest, which is probably the most awful global ecological catastrophe now in progress, is due largely to slash and burn agriculture. If only deforestation in the tropics is considered, in the last two million years the albedo is estimated to have changed by 0.001 units (0.1 W/m^2). In the last quarter of the twentieth century, the change was 0.00035 units. Over a period of 25 years, humans have caused a change of about a third of the value that took two million years in nature; at this rate, some seven centuries from now the albedo change would be 0.01 unit, enough to change the climate by 2°C.

Wood requires both time and, especially, energy to be harvested. When highly mechanized wood-harvesting equipment is used, the energy expended in cutting the wood is likely greater than the energy available from burning it. That is, if we consider the energy consumed in gasoline for chainsaws, diesel fuel for operating the logging equipment, gasoline or diesel fuel used to transport the wood to the point of use, and energy consumed by the workers themselves, this sum is likely greater than the energy liberated when the wood is burned. In other words, the total energy available to society has actually decreased!

When we burn wood, low combustion temperatures lead to low NO$_x$ emissions. SO$_x$ emissions are negligible, because wood normally has a very low sulfur content. Wood fires can produce unburned particulates, both of ash and of soot. In addition, the slow heating of wood in the fire can result in "creosote" tars accumulating in the chimney (Chapter 6). When the weather problem known as a thermal inversion* occurs, large smoke clouds from wood fires can build up in the atmosphere and cause severe local air pollution problems.

Is Wood "Renewable"?

Although wood is classified as a "renewable," in fact it may take twenty to more than a hundred years to grow a new tree to replace a mature tree cut down, depending on the species of tree and its growth rate. Without very strict forest management in place for ten years or more we can't regrow trees as fast as we cut them. Wood is a renewable resource if we assume that a new tree is planted for each one cut down and that we measure in units of decades. It is obviously not true that wood is a renewable resource in the short term, because a seedling one or two years old is not equivalent to a mature tree.

---

* A thermal inversion is an atmospheric phenomenon in which a layer of warm air moves on top of a layer of cooler air that is close to the ground. This event is especially bad in regions with significant air pollution, because the upper, warmer air layer acts as a "cap" that prevents the ground-level air, laden with pollutants, from rising and dispersing. This causes the air pollutants to collect near ground level (where, of course, we live). If the thermal inversion persists for several days, ground-level pollutant levels can become high enough to cause serious problems. Eventually, high winds or rain will break up the layer of warm air and allow the pollutants to disperse.

FIGURE 31.12   Coppicing, a practice for fast harvesting of wood. This example is from Old Wood, near Bunny, England. Coppicing is a traditional method of woodland management, in which trees are repeatedly cut down to near ground level. The resultant shoots can be used for fuel for charcoal, and many more uses. (From http://commons.wikimedia.org/wiki/File:Coppicing_in_Old_Wood,_near_Bunny_-_geograph.org.uk_-_1750835.jpg.)

The least desirable way to use wood is to harvest trees that have been growing a long time. Fortunately, there are practices that allow trees to be used and renewed on relatively short time frames. Some tree species, such as poplars, eucalyptus, silver maple, and black locust, can be used as energy crops by the practice known as coppicing. In this approach, year-old trees—willow being a good example—are cut down near the ground. This causes the cut stumps to sprout about a half-dozen new shoots, each of which grows rapidly over the next three to five years (Figure 31.12). Depending on the species used and how rapidly trees grow in the chosen location, the tree shoots would be harvested every two to eight years. The harvested coppice wood can be dried, chipped, and used as a fuel. Since new shoots grow from the cut stumps, replanting is required only once for every three or four times that the coppice wood is harvested.

Competition from the forest-products industry could limit the amount of wood harvested as an energy crop from existing commercial forests. Environmental concerns about destruction of old-growth forest ecosystems and the apparent devastation caused by clear-cutting may limit, or even eliminate, opening of forests not presently being commercially exploited for the harvesting of energy crops.

Agroforestry, a combination of food and wood production on the same land, offers a way to increase yields of food, firewood, and fodder for cattle. The experience of African nations such as Kenya and Nigeria has shown that mixing

FIGURE 31.13 A *Leucaena leucocephala* tree, known around the world by many common names, such as wild tamarind. This tree is growing near Kaupo, on the island of Maui, Hawaii. (From http://commons.wikimedia.org/wiki/File:Starr_040514-0216_Leucaena_leucoceph ala.jpg.)

corn and leucaena trees* can increase corn production anywhere from 40% to 80% relative to that from a field growing only corn (Figure 31.13). At the same time, the yield of wood is four to five tonnes per hectare.

METHANOL

ALCOHOLS AS LIQUID FUELS FOR VEHICLES

Since the dawn of practical internal combustion engines in the late nineteenth century, liquid fuels have been the overwhelming choice as the ENERGY source for these engines and the vehicles that use them. Among the liquid fuels, petroleum derivatives, gasoline and diesel fuel, have dominated, though there was strong competition from ethanol in the early years.

---

* Leucaena trees are natives of tropical America that have now been introduced to most of the tropics. They typically grow to heights of about ten meters, and sometimes twice that in selected locations. Currently these trees are used as shade plants for coffee, cacao, and rubber crops.

Why liquid fuels? Simply because neither of the alternative choices is as suitable. Solid fuels, for example, finely pulverized coal, are not well suited for use in Otto or diesel engines. Such fuels invariably produce ash. Ash can build up in the cylinders; erode cylinder walls, pistons, and exhaust valves; or add to air pollution unless some sort of ash-collection device could be added to the vehicle. Gases can be superb fuels for internal combustion engines. However, unless the gas can be liquefied or compressed to high pressure, it is difficult to carry enough gaseous fuel on board a small vehicle to obtain much practical driving range. On large vehicles that travel short distances, and that have access to specialized fueling facilities (an excellent example being city buses), compressed natural gas is proving to be useful fuel choice. Liquid fuels provide clean-burning characteristics similar to gases, with a density close to that of solid fuels, to allow an ample fuel supply to be carried on the vehicle.

Reasons for a possible switch away from petroleum derivatives as transportation fuels have been discussed previously: concern for long-term supplies of petroleum or dependence on imports (Chapter 30), concern about urban air quality (Chapter 28), and concern about anthropogenic CO_2 emissions (Chapter 29). Likely replacements for petroleum are two alcohol fuels: methanol and ethanol. Both can be made from biomass.

Why alcohol fuels? They can be burned in existing spark-ignition engines with minimal modification. Cars and light trucks already on the road could be adapted to these fuels. Car-makers could continue to use existing production facilities and methods to build engines, eliminating the enormous capital investment needed to tool up to make an entirely new kind of engine. Though not so good as petroleum, the alcohol fuels have sufficient energy density (i.e., megajoules produced per kilogram or liter of fuel burned) that fill-ups are not needed so often as with uncompressed gaseous fuels. Their use would not add undue weight to the vehicle, as heavy tanks for compressed gas might. An enormous infrastructure already exists for the handling, storage, and transporting of liquid fuels (i.e., gasoline). Conversion from petroleum to alcohol fuels could be effected with little change in the transportation-fuel infrastructure. Finally, we already have the know-how to make both methanol and ethanol in large quantities at reasonable cost.

Producing methanol and ethanol could become an important future role for biomass. Improvements in the conversion of biomass to these liquids, and the possible introduction of less expensive energy crops, would be the two key developments that could open the very transportation market for these fuels. Since the alcohols can be burned in existing combustion systems with only slight modification, they could be an energy source providing a relatively smooth transition between our current energy economy that is dominated by fossil fuels and an energy future that relies on nearly pollution-free energy sources such as solar energy and wind-generated electricity.

PRODUCTION OF METHANOL

In the past, methanol was produced from wood. This gives methanol its common name, "wood alcohol." Hardwoods—beech, hickory, maple, or birch—were the favored sources. Heating wood in the absence of air vaporized a whole array of substances that produced a gas useful for illumination, a tar containing dozens of useful chemicals, and a liquid called pyroligneous acid. The solid portion of the wood remained behind as charcoal. Methanol can be recovered from the pyroligneous acid. While this method unquestionably works, because it has been the main source of methanol in the past, the yield of methanol is dismal—perhaps 2%. That is because so much of the wood winds up in other products.

Today most methanol is made from natural gas. Natural gas is reacted with steam to produce a mixture of carbon monoxide and hydrogen, called synthesis gas. This gas mixture reacts in the presence of a catalyst to produce methanol. The importance of this process is that its long-term use is not dependent on the availability of cheap natural gas. If for any reason natural gas became in short supply, or prices became unacceptably high, synthesis gas can be made from other sources. Virtually any carbonaceous material can be converted to synthesis gas.* Biomass is a candidate for future production of methanol. (So is coal.†)

ADVANTAGES OF METHANOL

Generally a standard spark-ignition automobile engine can be modified to run on methanol. (Some of the modifications are required because methanol can cause deterioration of rubber and plastic parts.) Indeed, methanol is already used successfully as a fuel in racing cars. Methanol has a very high octane number, about 110, so that it can be used in engines of very high compression ratio. Cars designed specifically to run on methanol would gain a 10%–20% benefit in thermal efficiency relative to similar cars designed for gasoline, because of the higher compression ratio. A high compression ratio also means that a methanol-fueled car could have higher acceleration than comparable gasoline-fueled cars (which is why methanol is favored in racing cars). Methanol also works well in blends with gasoline. A blend of 85% methanol and 15% gasoline (called M85) can be used in current car engines with very minor modifications. It is also possible to devise a "flexible fuel vehicle" that could run on gasoline or M85.

---

* The exact ratio of hydrogen to carbon monoxide in the synthesis gas depends heavily on the relative amounts of hydrogen and carbon in the material that reacted with steam. If there is proportionately too much hydrogen and not enough carbon monoxide, the composition is "shifted" by reaction with carbon dioxide: $H_2 + CO_2 \rightarrow H_2O + CO$. On the other hand, if there is proportionately too little hydrogen and too much carbon monoxide, the composition can be "shifted" in the other direction by reacting with steam: $CO + H_2O \rightarrow CO_2 + H_2$. The versatility and the beauty of synthesis gas is twofold: synthesis gas can, in principle, be made from *any* carbonaceous material and the composition of the synthesis gas can be shifted to *any* desired proportion of hydrogen and carbon monoxide, depending on whatever we intend to synthesize from it.

† Coal can be converted to synthetic liquid fuels in several ways. The two main strategies are direct liquefaction, in which the coal is reacted with hydrogen, usually in the presence of a catalyst and a solvent; and indirect liquefaction, in which the coal is first converted to synthesis gas and then the synthesis gas in turn is converted to liquid products. During the Second World War, much of the liquid fuel supply used for the German war effort was produced from coal by these two processes, which led to a campaign of bombing synthetic fuel plants by the American air force. In South Africa today, a significant amount of liquid fuels and chemicals is produced by the indirect liquefaction of coal.

Use of methanol would potentially reduce air pollution associated with motor vehicles. A methanol–air mixture has a lower flame temperature than a gasoline–air mixture, leading to much less thermal NO_x production. Methanol contains no sulfur, so there would be no SO_x emissions. Methanol burns well at fuel-lean conditions, helping to reduce CO emissions, and yet does not produce soot at fuel-rich conditions. Methanol evaporates more slowly than gasoline, helping to reduce evaporative emissions of unburned hydrocarbons. At the same time, though, this makes starting a methanol-fueled engine more difficult in cold weather.

We already know how to make methanol in large-tonnage quantities; many large-scale plants exist all around the world. Most use natural gas as the feedstock. This existing methanol capacity may be very useful when coupled with the methanol-to-gasoline (MTG) process. This process converts methanol directly to high-octane gasoline in excellent yields. The MTG process does not involve substituting methanol for gasoline, nor is the product some kind of ersatz gasoline—it *is* gasoline. An MTG plant, with methanol made from natural gas, ran successfully in New Zealand for about a decade. Although it would be sensible to have the methanol synthesis and the MTG integrated in one plant, it is not necessary to do so. In the event of major shortages of petroleum, the output of the world's existing methanol plants could be diverted to MTG plants, even if the latter were in a different location. While this would create major disruptions in existing methanol markets, it could lead to a relatively rapid ramping up of gasoline production that does not require petroleum supplies.

Methanol can also be used in fuel cells, making it a possible fuel for hybrid vehicles (Chapter 28) or in electric vehicles that use fuel cells as the source of electricity. Whether methanol would see use in these applications depends on continued development of effective methanol fuel cells, and on continued public acceptance of hybrid or fuel-cell electric vehicles.

DISADVANTAGES OF METHANOL

The heating value of methanol, on a volumetric basis, is about 21 MJ/L; gasoline is about 36 MJ/L. Methanol provides only about 58% of the ENERGY as a comparable volume of gasoline, or 58% of the fuel economy. For example, a car that uses 9 L/100 km with gasoline might require 16 L/100 km on methanol. This has two consequences. First, either the driver would have to put up with stopping at a "methanol station" somewhat more frequently than at a gasoline station, or else the fuel tank on the methanol vehicle would have to be larger. Second, a person driving the same distance would have to purchase substantially more (almost twice as much) methanol as gasoline. To be truly cost competitive, methanol would have to sell at about half the cost per gallon as gasoline.

A second major concern with methanol is safety. Accidental ingestion of methanol can lead to blindness or death. Drinking about forty millimeters of methanol can cause death; half that amount can lead to permanent blindness. Effects at lower levels of exposure include severe headaches, visual impairment, and convulsions. Though few might ever drink methanol, the same problems can be caused by inhalation of the vapors or, insidiously, by absorption through the skin.* Being splashed with methanol as a result of an accidental spillage could result in this happening. In addition to the direct health effects, methanol burns with a nearly colorless flame. It might be hard to see an accidental methanol flame, unless something has been deliberately added to the methanol to make the flame luminous. (One such additive could be a small amount of gasoline.)

A vehicle set up to burn gasoline would need some adjustment of the fuel injection system because of the different air:fuel ratio needed to burn methanol. Comparing methanol with octane, one methanol molecule requires 1.5 molecules of oxygen to burn completely; but one octane molecule requires 12.5 oxygen molecules for complete combustion.

Unlike gasoline, methanol is soluble in water. If water is accidentally mixed with gasoline (e.g., in a storage tank), a layer of gasoline will float on a layer of water. It would be possible to separate the gasoline from the water. Methanol would form a solution with the water. Thus it could be possible to have water blending into a methanol fuel (much more severely than with gasoline) causing problems for running the engine. Methanol contaminating engine oil can severely degrade its lubricating characteristics of the oil. In extreme cases, this could result in serious maintenance problems for the engine.

Finally, although methanol can be produced cheaply today, in large amounts, we must remember that today's methanol production uses cheap natural gas as the starting material. It may prove difficult for biomass to compete with natural gas as a feedstock for methanol synthesis without substantial tax incentives. Using biomass would require a more complex, and therefore more expensive, device to make synthesis gas, relative to when natural gas is the feedstock. Especially now with more and more shale gas becoming available, biomass might not be competitive.

ETHANOL

Ethanol is also called ethyl alcohol or grain alcohol. A very large family of alcohols exists, but when the word *alcohol* is used by itself (e.g., as in the term "alcoholic beverages"), almost always it is ethanol that is meant. A major source of ethanol is the fermentation of starches or sugars in plants, such as corn and other grains—hence its common name of grain alcohol.

Since ethanol burns readily, it deserves consideration as a liquid fuel for motor vehicles. In the early years of the twentieth century, ethanol competed with gasoline as an automotive fuel. Henry Ford built a small number of "fuel-flexible" Model Ts that could run on ethanol or gasoline. During the First World War, Ford was a strong proponent of increased alcohol fuel production, and attempted to convince President Wilson that American cars should be converted to run on ethanol instead of gasoline.

---

* Sometimes people in advanced stages of alcoholism, especially those who are really down-and-out, will drink methanol by inadvertence. These folks, in their desperation for a drink, may find some wood alcohol and assume that "alcohol is alcohol." Doing so can literally be a fatal mistake. A small amount of wood alcohol in the system can cause incurable blindness or death. Sad to say, such events still occur, judging by occasional reports in the news media.

During the war we completely blockaded Madagascar for two whole years because of their pro-Vichy loyalties, yet when our troops arrived there in 1942 they found life going on quite serenely.

The cars were running more or less happily on rum, of which Madagascar is a large producer.

—Anderson[2]

And consider this prediction from 1950:

Instead of driving to California in their car—tear-drop in shape and driven from the rear by a high-compression engine that burns cheap denatured alcohol—the family of A.D. 2000 use the family helicopter, which is kept on the roof.

—Benford[3]

PRODUCTION OF ETHANOL

Interest in ethanol comes from the fact that it can be renewable, plus the fact that it can also be produced from waste materials such as sugar beet or sugar cane pulp and potato peelings, as well as from surplus crops. *Fermentation* is the process that converts plant materials to ethanol. The process of fermentation goes back at least ten thousand years. It is likely the second-oldest chemical process to be exploited by humans (the oldest being combustion). And, like combustion, fermentation was probably discovered independently by many early prehistoric societies. Beer was prescribed as a medicine in Sumer at least four thousand years ago. The oldest set of laws still preserved—those written by Hammurabi in Babylon sometime around 1770 BCE—included regulations for drinking establishments.

In the fermentation process, starches or simple sugars from biomass are decomposed by yeasts or other microorganisms (more specifically, by enzymes produced by these microorganisms) to produce ethanol. About 85% of the chemical potential energy in the molecule of sugar is retained in the two molecules of ethanol. Sources of starch are mainly grain crops but also include root plants, such as potatoes. Sugar sources include cane and beet sugar, and sorghum. The ethanol business in the United States nowadays is based on corn, whereas other grains and sugarcane are used elsewhere. The high starch content of corn makes fermentation relatively easy. Second, in most years so much corn is grown that millions of tons are left over as surplus, so can be purchased by the ethanol industry at a low price. Duckweed (Figure 31.14) represents a potential aquatic plant source for ethanol.* Duckweed grows rapidly. It can grow in brackish (i.e., slightly salty) or polluted water. Duckweed plants are

---

* Duckweeds are floating aquatic plants of various kinds that range from the so-called giant duckweed that grows into thick, dense masses to another variety believed to be the smallest flowering plant on Earth. One of the duckweeds has the intriguing name of mud midget. Generally the duckweeds are considered to be invasive plants in aquatic ecosystems. Harvesting and using the duckweeds would likely be beneficial, not only for the use of the biomass, but also for reducing their impact on the local ecosystem. Because even the giant duckweed is, individually, a relatively small plant, duckweeds would not have to be chopped up before being fed to a biomass conversion or combustion system.

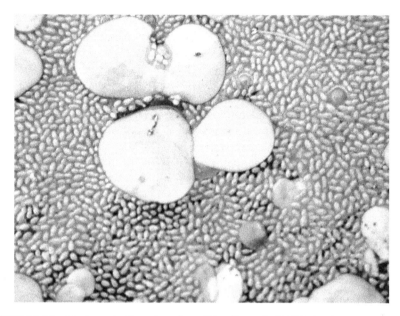

FIGURE 31.14   Duckweed. (From http://en.wikipedia.org/wiki/File:Duckweeds.jpg.)

small enough that they can be pumped through pipes and do not need to be chopped up prior to the fermentation step.

ADVANTAGES OF ETHANOL

A potential long-term advantage of ethanol lies in the versatility of the fermentation process. Virtually any biomass material that provides starches or sugars could, in principle, be fermented to ethanol. Fuel ethanol production would not be reliant upon, nor hostage to, a single type of energy crop. Not only could energy crops such as grain grown specifically for ethanol production be used, but also grain wastes (that might, e.g., have otherwise rotted in storage), or some food wastes, such as potato peelings.

It's been known for a century that ethanol can be a suitable fuel for automobile engines. Henry Ford had many character defects, but idiocy was not one of them. Like its chemical sibling methanol, ethanol can be made in high tonnage by known technology. It has a high octane number (about 110), which makes it well suited to high-compression-ratio spark-ignition engines (but not diesels). Ethanol burns cleanly, with reduced thermal NO_x emissions and no fuel NO_x or SO_x. A 90:10 blend of gasoline and ethanol, sometimes called *gasohol* (discussed further below) and now usually known as E10, can be run in standard automobile engines with no modifications.

DISADVANTAGES OF ETHANOL

As a fuel, ethanol has a heating value of 28 MJ/L. Thus it has about three-fourths of the energy of gasoline on a volumetric basis. This is not so bad as methanol.

It means, though, that a car using 10 L/100 km with gasoline would get 13 with ethanol. Ethanol shares with methanol the issues of water solubility and a need to modify the fuel system to accommodate a different fuel-to-air ratio.

Raising corn for ethanol production is very energy intensive—plowing, fertilizing, harvesting, and transporting the corn all require energy. The growing of corn contributes to erosion of agricultural land; topsoil washed away eventually contributes to water pollution. The distillation step required to separate ethanol from the water in which it was produced also requires large amounts of energy (as heat), generally supplied from petroleum or natural gas. Fertilizers used in corn production, often in massive amounts, are made from petroleum-derived chemicals. A breakthrough in chemical engineering that would replace the distillation step with some other operation for recovering the ethanol from the fermentation mixture would be helpful, as would further progress in alternative biomass-based processes that don't even require distillation.

We cannot claim that ethanol is truly a CO_2-neutral energy source. The farm machinery and transportation equipment all emit CO_2, as docs the generation of the heat needed for distillation. Substituting biomass or solar energy for fossil energy wherever possible in the ethanol production system would help to bring ethanol closer to being truly CO_2 neutral.

As is the case with virtually any biomass energy source, we must address the questions of whether there is enough biomass and enough land. Growing crops for energy takes land or crops away from food production. With a large percentage of the world's population malnourished and in some cases even starving, does it make sense to divert lots of agricultural land to the production of "energy crops"? Though it's important to remember that many other biomass sources can be converted to ethanol, corn is the dominant feedstock in the United States. If much corn were diverted to ethanol production, the price of corn as a food would likely rise. Furthermore, since much corn is used as animal feed, it is also likely that the price of meat would rise.

Plants make glucose as the primary product of photosynthesis. Then, glucose is stored in two ways inside the plant: as starch, for long-term energy storage, and as cellulose, to provide mechanical strength to the plant. Humans have the capacity to digest starch, but not cellulose. In other words, we can eat potatoes and corn, but not wood. If cellulose could be used as the starting material for ethanol production (i.e., converting cellulose back to glucose and then fermenting the glucose), this would have a major benefit for the food vs. fuel argument. Ethanol produced using cellulose is sometimes called *cellulosic ethanol*. It is no different chemically from ethanol made from any other source. The adjective *cellulosic* simply signals that the original biomass was a cellulose source, and not starch. Cellulose reacts with water—to revert to glucose—much more slowly than does starch. (That's a good thing, since cotton is 100% cellulose, and we would prefer not to have cotton garments disintegrate while being washed.) A challenge remains to find easy, quick chemical treatments to convert cellulose back to glucose. A breakthrough in that area, permitting large-scale production of cellulosic ethanol instead of starch-based ethanol, would be a very significant boost for ethanol.

Ethanol in the United States and Brazil

Beginning in the 1970s, in response to the OPEC oil embargoes, the United States developed gasohol, a blend of 10% ethanol and 90% gasoline. This was by no means a new idea, since gasoline–alcohol blends were widely available in the Midwestern United States during the 1930s (Figure 31.15). The ethanol came from agricultural products or agricultural wastes, such as the residue from processing sugar beets. The idea was simply to use ethanol to extend supplies of gasoline. No engine modifications are needed to run on gasohol. Anecdotal reports on gasohol experience were highly controversial. Some drivers swore by it; others swore at it, and claimed it caused harm to the engine and fuel system. In the 1970s and early 1980s, some car companies voided warranties if alcohol fuels were used. The negative reports probably stemmed from two issues: Alcohol can soften or dissolve some rubber or plastic parts in the engine or fuel system, but which are impervious to gasoline. Ethanol absorbs water, which can lead to increased corrosion of fuel system components and poor combustion performance. Now, gasoline sold in the United States contains 5%–10% ethanol as an oxygenate and octane booster (Chapter 19).

At about the same time that the United States began making gasohol, Brazil—one of the world's largest sugar producers—found a world situation where sugar was very cheap, and oil, which Brazil had to import, was very expensive. Brazil undertook a major national program, one far larger than in the United States, on alcohol fuels.

FIGURE 31.15 Gasoline station selling ethanol, in the midwest United States. (Photograph of Earl Coryell filling station courtesy of Mary-Jo Miller, Nebraska State Historical Society, Lincoln, Nebraska.)

FIGURE 31.16   Sugar cane. (From http://commons.wikimedia.org/wiki/File:Sugar_Cane_rows.jpg.)

Two blends were developed: 22% ethanol:78% gasoline (E22) and 98% ethanol:2% gasoline. (The reason for using the odd 98:2 ethanol:gasoline blend, and not pure ethanol, was to keep people from drinking it.) Eventually about 60% of Brazil's liquid fuel needs came from ethanol.

By 1990, about four million vehicles were running on the 98:2 blend, and another six to eight million on the 22:78 blend. The program was a great success while sugar was cheap and oil expensive. When oil prices dropped and sugar escalated, the program was a financial disaster, but has since rebounded. By the late 1990s, Brazil was making nearly 15 billion liters of ethanol per year from sugar cane. By 2011, that figure was up to 21 billion liters. This is easily the most successful program for alternative liquid fuels from biomass anywhere in the world. The Brazilian car industry has developed flex-fuel vehicles, capable of operating on gasoline, E22, and pure ethanol (E100) without modification. Numerous makes and models of these vehicles are available now.

The great success of the Brazilian ethanol program comes from numerous factors. The feedstock is sugar cane (Figure 31.16), in which the juice already contains simple sugars—no conversion of starch or cellulose is needed. Sugar cane grows readily, in some places two crops per year, with relatively "low-tech" agriculture (especially in comparison to corn). Bagasse, the residue from squeezing the juice out of cane, is burned on-site as the fuel source for the process heat needed in fermentation and distillation. In a sense, this energy source is free.

PLANT OILS AND BIODIESEL

A trip down the cooking-oil aisle of any well-stocked supermarket will show numerous kinds of oils derived from seeds, nuts, or other plant parts. Olive oil, peanut oil, corn oil, palm oil, and sunflower oil provide examples. Many kinds of

FIGURE 31.17   Olives from these trees can be processed to olive oil, which can be used in food or can be converted to biodiesel fuel. (Reprinted from first edition, with permission from Whole-in-One Catalog.)

plants produce seeds, nuts, or other parts that can be sources of oils (Figure 31.17). The technology for extracting these oils is well known and very old, going back at least to the time of Vitruvius. Olives crushed in waterwheel-operated mills provided a source of olive oil to the ancient Romans. Over the centuries, millions of home cooks have made empirical observations that attest to the fact that these oils will indeed catch on fire and burn.

While fats and oils are usually considered as plant products, animal fats also are good fuels. Lard, for example, makes an excellent boiler fuel, about comparable with a heavy grade of petroleum-derived fuel oil, but with near-zero sulfur content. Meat-processing plants could use lard on-site as a fuel to make heat or steam. During the depression of 1837 in the United States, prices of farm products became so low that lard was used as fuel on steamboats on the Mississippi River.

Since it is easily demonstrable that these oils will burn, from time to time there has been interest in their use as fuels. As early as 1900, a demonstration diesel engine was operated in Paris on peanut oil. In 1912, Rudolf Diesel predicted that plant oils would be as important as petroleum. The Japanese navy used soybean oil as a fuel in the battleship *Yamoto*, the largest and most powerful battleship ever built. The oil "crises" of the 1970s renewed interest in alternative sources of liquid fuels for motor vehicles. Generally, the plant oils are better suited to operation of diesel engines than spark-ignition engines. Nowadays, though, most plant oils used as fuel are transformed into a product, biodiesel, which is better suited, in many ways, than the direct use of plant oils. Biodiesel fuel made from oils from rapeseed (Figure 31.18) and soybeans (Figure 31.19) provides some of the diesel fuel used in Europe and the United States, respectively.

Like other forms of biomass, plant oils are, in principle, renewable. We need only grow a new crop of plants to replace the ones harvested for the oil to assure

FIGURE 31.18 Rapeseed (canola) oil is another oil that can be used in foods or for biodiesel production. (Reprinted from first edition, permission from Photo Researchers, Inc., New York, New York.)

FIGURE 31.19 Fields of soybeans in South Carolina. (From http://agriculture.sc.gov/fieldcrops.)

a continuous supply. And, like other forms of biomass, plant oils would, in principle, be CO_2 neutral. The by-product of plant oil production is a high-protein "meal" that can be used as cattle feed.* Biodiesel from plant oils produces no SO_x or fuel NO_x emissions. Thermal NO_x emissions are relatively low. Production of plant oils requires much less energy than production of ethanol from corn. In particular, no distillation step is needed. If the seeds or nuts were crushed in a water-operated mill, the CO_2 emission associated with production could be quite low.

Compared to conventional petroleum-derived diesel fuel, plant oils themselves have a higher viscosity and a lower cetane number (about 33). However, an even more serious problem with the direct use of plant oils is the very short operating life of a diesel engine before a major overhaul is needed. It's not unreasonable to expect 10,000 hours of operation from an engine running conventional diesel fuel before a major engine overhaul or rebuilding is needed. In contrast, some reports suggest that the comparable figure is 600 hours for a diesel engine running on sunflower oil and 100 hours on linseed oil.[†]

Problems with the direct use of plant oils in diesel engines have generally been solved by converting the raw oils to biodiesel, by reacting the oils with methanol. This is a genuine chemical reaction, not a physical mixing. The chemical process is called *transesterification*. If the methanol were to be derived from biomass as well, instead of natural gas as used nowadays, then this would be an "all-biomass" process.

The molecules in biodiesel are roughly the size of the molecules in diesel fuel. Biodiesel has viscosity comparable to conventional petroleum-derived diesel, and cetane numbers of about 50 (compared to about 33 for the raw oil). It appears to provide a much longer engine operating life than raw plant oils. Like regular diesel fuel, it is a safer product to handle and store than is gasoline (or ethanol). Biodiesel has to get very hot before its vapors will sustain an open flame. Biodiesel does mix with lubricating oils, so may cause some lubricity and engine wear problems.

A resource likely to be available forever is used cooking oil, such as used for preparing French-fried potatoes at fast-food restaurants. In many communities, waste cooking oil is currently sent to landfills. Instead, it would be advantageous to be able to use it (or biodiesel made from it) as a fuel for trucks. Kits of parts are available, at least in the United States, for converting vehicles from operating on conventional diesel oil to this so-called waste vegetable oil. Sending the oil to a landfill represents a cost to the restaurant; an enterprising individual might be able to set up an agreement with a restaurant to haul it away free, to be used as fuel later. An advantage of this material is that, in the event of a spill, it soon biodegrades and does not contaminate the environment.

---

* In the early years of the Soviet Union, the "cake" left from pressing oil out of cottonseeds was used as a solid fuel in ship's boilers. During the appalling conditions and mass starvation of the siege of Leningrad during the Second World War, "cake" from pressing flax seeds was used to produce an ersatz macaroni.

[†] Linseed oil, made by pressing oil from the flax, has been used for generations as an additive to varnish, paint, and putty to help these substances dry into a hard, protective material. In addition, the oil itself is often rubbed into fine woodwork (such as furniture or the wooden stocks of sporting firearms) to provide a waterproof, hard finish to the wood. There is little wonder that such a material would quickly turn to "gunk" in an engine.

Biodiesel is relatively easy to produce, even by amateurs working in a garage. This makes it possible to find a local source of, say, waste vegetable oil, and to make one's own biodiesel at home. Some of the chemicals used in the process are dangerous (e.g., methanol); so very careful attention must be given to safe handling procedures. The other area demanding strict attention is controlling the quality of the product, to ensure that it meets acceptable specifications. Aside from its relative ease of production, an attractive feature of biodiesel is that, in many jurisdictions, it is relatively free from bureaucratic oversight—unlike ethanol—so long as one does not get into the business of selling it to others.

MUNICIPAL SOLID WASTE

We will never run short of waste products. Many materials now routinely thrown away are combustible and could be used as fuels: paper, cardboard, plastics, discarded textiles, and rubber tires are examples. This aggregate of waste material has been given various names, such as municipal solid waste, eco-fuel, or refuse-derived fuel.

Using municipal solid waste as an energy source accomplishes several useful things. It provides useful ENERGY that otherwise might literally have been thrown away. It extends the useful life of landfills by reducing significantly the volume of material to be put into the landfill. Burning municipal solid waste instead of putting it in a landfill can also provide economic benefits. It might actually cost less to burn waste than to put it in a landfill. In addition, it might be possible to make money, or receive some form of credit, for ENERGY produced from burning the waste, as opposed to the certainty of incurring a cost for adding it to a landfill.

Municipal solid waste can be burned in combustion equipment designed to run on this kind of fuel (Figure 31.20). Incinerators reduce the total volume of waste to about 10%–12% of the original volume burned, helping reduce the volume of material to be put in landfills. While much of what we throw away is combustible, some of it isn't. Glass and metals won't burn. Removing noncombustible material beforehand improves the use of municipal solid waste as a fuel. This could be done where the waste is generated—in the home—if the majority of householders cooperate with such a program. The alternative would be to do the separation after the collected waste has been taken to the incinerator location. Making the separation there could provide a possible source of employment and income for unskilled workers.

Burning municipal solid waste leads to problems in pollution control not shared by other forms of biomass or by the fossil fuels. Some plastics produce potentially harmful emissions when they burn. Poly(vinyl chloride) is almost ubiquitous. Examples of its uses include PVC pipe, garden hose, plastic films (such as food wrap), artificial leather, and plastic toys. Burning poly(vinyl chloride) can form hydrogen chloride, which dissolves in water to become hydrochloric acid. It could also form other compounds containing chlorine, many of which are health hazards. Urethane foams and acrylic fibers can produce hydrogen cyanide, which is toxic. Heavy metals that might be contained in noncombustibles not completely separated from the waste can also be health hazards if released to the environment. Mercury and lead are examples. These chlorides, cyanides, and heavy metals need to be contained in the incineration process or captured before they escape to the environment.

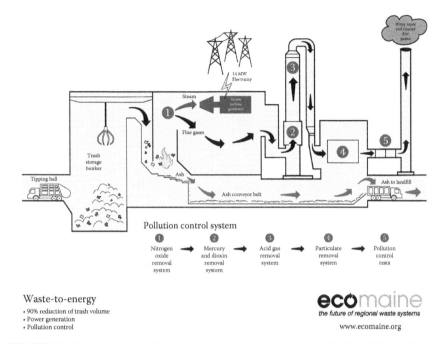

FIGURE 31.20 A schematic diagram of a waste-to-energy plant in Portland, Maine. (Image courtesy of Shelly Dunn, www.ecomaine.org.)

METHANE FROM BIOLOGICAL SOURCES

Many biological materials, for example, animal manure, produce methane. This process occurs naturally, as with waste organic matter accumulates in a landfill, or by deliberate intent in equipment designed for the purpose. The reaction is facilitated by anaerobic bacteria (i.e., bacteria that do not use oxygen for their life processes). The gaseous fuel produced by the anaerobic digestion of wastes is sometimes called *biogas*. The important component is methane, an excellent fuel that is also the main component of natural gas.

Many kinds of biomass could be (and are) converted to biogas. These include animal waste (e.g., manure from farms or feedlots); the organic portion of municipal solid wastes (garbage, specifically the organic portion); agricultural and food processing residues, such as bagasse, sawdust, and scrap wood; sewage; and various kinds of industrial wastes. The very simple technology uses a large tank (the digester) from which air can be excluded (Figure 31.21). Usually a mixer slowly stirs the digesting material. The necessary anaerobic bacteria might already be in the original waste being digested, such as animal manure, or they may be intentionally introduced. The digester operates at a working temperature in the range 35°C–60°C. Gaseous products—the biogas—bubble to the surface and can be collected for use.

In addition to producing the desired biogas as an ENERGY source, other advantages derive from the digestion of wastes. The process can be used as a waste treatment operation to reduce pollution hazards and odors. Many animal feedlots, put

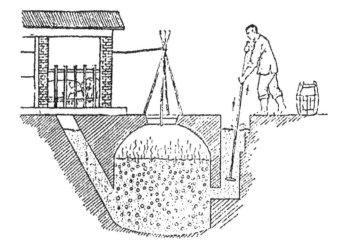

FIGURE 31.21 A simple digester for production of biogas from human or animal waste, in China. (Image courtesy of Steven Harris, KnowledgePublications, Turtle Creek, Pennsylvania, http://www.KnowledgePublications.com and http://www.USH2.com.)

bluntly, stink. They are concentrated sources not only of what is known by the somewhat more polite euphemism "olfactory pollution," and of water pollution from storm water runoff carrying wastes into the water system. Biogas production can be of some benefit to the public health. The digestion process can kill bacteria and parasites present in the waste and that could be threats to health. The solid by-product that remains after digestion is a fertilizer rich in nitrogen that can be returned to the land.

Anaerobic digestion normally yields one molecule of CO_2 with every molecule of CH_4. Because CO_2 is noncombustible, biogas of this composition has a much lower heating value than natural gas unless the CO_2 is removed. Doing so might ultimately release this CO_2 into the atmosphere unless special provisions were made to capture and sequester it. Continued research on biogas production has yielded ways of modifying the process so that an excess of methane, relative to CO_2, can be produced. The methane concentration in the gas could be over 70%. More than 85% of the chemical potential energy of the original waste material being digested is retained in the biogas.

Organic waste sent to a landfill will eventually produce methane when it becomes buried deeply enough to allow anaerobic processes to occur. Methane leaking out of landfills has several potential problems. First, methane is a substantially more potent greenhouse gas than carbon dioxide (on a molecule-for-molecule basis, methane is about eleven times better at absorbing infrared). Second, methane is flammable and, in some concentrations, methane/air mixtures are explosive. Digesting wastes rather than consigning them to a landfill, or, at the least, deliberately tapping into the landfill to capture the so-called "landfill gas" can therefore provide several advantages.

Biogas production from digestion of wastes is becoming an attractive technology for some developing nations where there are very high population densities (which translate directly into a large, unending supply of human waste and a large energy demand). In China, over 35 million digestion systems have been installed, many of which are family-sized units, to help meet the cooking, and heating of small rural communities. Over a hundred million people now cook with biogas. The United States now has over 120 biogas plants, the majority of which are used to collect the gases produced in landfills. Disney World, a site ideal for collection of high concentrations of human waste, has an experimental digestion unit to produce a gas about 95% CH_4. Germany now has three thousand sites in operation.

Biogas from anaerobic digestion of biomass has been used successfully in two applications: energy self-sufficiency on farms, providing energy for farm vehicles, electricity, and heat; and energy sufficiency for running a municipal waste system—vehicles and heating. When the biogas has been available to operate collection vehicles, its use essentially eliminates the energy cost of collecting garbage. Biogas could be used directly for heating or in internal combustion engines. Biogas production offers an economical route to on-site electricity production wherever sufficient wastes are available and there is an identified demand for the electricity.

HYDROGEN

Hydrogen can be produced from several sources. Methane (natural gas) offers an attractive choice because it contains the highest proportion of hydrogen by weight, 25%, of any chemical compound. Reacting methane with steam makes synthesis gas; "shifting" the synthesis gas by reacting with more steam converts the carbon monoxide to carbon dioxide. Removing or sequestering the CO_2 leaves essentially pure hydrogen. This is presently the most popular route to hydrogen.

Gasifying fuel oil, biomass or coal as the starting material will get us to the same point. Because these materials contain proportionately more carbon and less hydrogen than does methane, more CO_2 will be produced. Some bacteria can produce hydrogen directly from biomass under anaerobic conditions (i.e., without the intervening production of methane and its subsequent conversion to hydrogen). Presently, the yields of hydrogen are low, but as research progresses it may prove possible to make hydrogen directly from biomass just as methane is made from wastes nowadays.

The ultimate hydrogen resource is water. Splitting the water molecule produces hydrogen (and oxygen) directly. Electricity is usually the energy source for this process; the splitting apart of a molecule by electricity is called *electrolysis*. About three-fourths of the Earth's surface is covered by water. Immense quantities are available. This fact alone suggests that we need not be concerned about running out of hydrogen. If the planet ever runs out of water, we will have a lot more pressing worries than how to drive to the local discount emporium.

On a volumetric basis, liquid hydrogen has a heat of combustion of 10 MJ/L. This does not compare well with other fuels, even with methanol. This is due to the very low density of hydrogen. When compared on a weight basis, however, hydrogen looks exceptionally good, providing 120 MJ/kg. This is more than double the heating value per kilogram of methane, quadruple that of coal, and ten times better than wood.

Hydrogen combustion produces only water as a product and can be considered to be nonpolluting. In high-temperature combustion systems, the possibility of making thermal NO_x still must be taken into account, but almost all applications for hydrogen envision its use in fuel cells, not combustion systems. Use of hydrogen, in fuel cells or combustion, produces no CO_2. If the hydrogen were made by electrolysis of water, and the electricity were made from hydro-, nuclear, or solar plants, then there would be no CO_2 emission at all. If water is the source of the hydrogen, then it is perpetually renewable, since water is produced again during the combustion.

Of course, some problems exist. The first is how to make hydrogen cheaply enough to be cost competitive with present petroleum-derived fuels. Electrolysis of water would require very inexpensive electricity. Reaction of fossil fuels with steam is known technology, indeed in commercial practice, but adds both cost and energy, raising the question of why we wouldn't use the fossil fuel directly in the first place. Water will break apart at extremely high temperatures, several thousands of kelvins. It might be possible to obtain such temperatures by focusing solar energy onto a tiny volume.

Breaking a water molecule, regardless of how it's done, requires a certain amount of ENERGY. When the hydrogen burns with oxygen, that same amount of ENERGY will be released again, in principle. If that could be achieved, we would "break even" in terms of ENERGY, theoretically recovering the same ENERGY from burning hydrogen that was used to make it. In that sense, we could regard hydrogen as an "ENERGY carrier." That is, we would invest a certain amount of ENERGY at, say, a hydroelectric plant to produce a quantity of hydrogen. Later, at some different location, our investment would be repaid when that ENERGY is liberated by the burning of the hydrogen. We must also consider that some ENERGY, regardless of how little, is going to be required simply to move the hydrogen to its point of use. And a far more formidable problem is that no process of any kind will work with 100% efficiency. Even if we merely electrolyzed water and burned the hydrogen right on the spot (disregarding what strange motivation might lead us to do this), we would never recover all of the ENERGY expended to make the hydrogen.

A third issue is handling and storing hydrogen. Under ordinary conditions, hydrogen is a gas of very low density. To store a sufficient quantity of hydrogen to provide any reasonable amount of energy (e.g., in a car), hydrogen would have to be liquefied or stored as a compressed gas at very high pressures. To liquefy hydrogen at atmospheric pressure requires that we cool it to $-253°C$ (20 K). "Slush hydrogen" is a mixture of solid and liquid hydrogen—a hydrogen frappe—held at about $-260°C$. This material can be handled and stored by an organization having the financial wherewithal and technical resources of NASA, but is hardly suitable for the average homeowner commuting to work or driving to the supermarket. Storing compressed gaseous hydrogen requires very strong high-pressure containers, likely to be very heavy and adding substantially to the weight of the vehicle. Since H_2 is the smallest known stable molecule, it has a knack of worming its way through very tiny pores or holes, and can easily leak. The best approach to handling and storing hydrogen seems to be finding a material that will adsorb and hold large quantities of hydrogen, but from which the hydrogen can easily be removed (desorbed) as fuel is needed.

Very likely the best material for this application will turn out to be one of the forms of carbon, an element of which most of the forms are themselves of low density.

Hydrogen is already being transported by rail, truck, and pipeline and stored in liquid or gaseous forms with a superb record of safety. Hydrogen-fueled city buses have been demonstrated successfully in Chicago and Vancouver. These buses used compressed hydrogen to operate fuel cells. Since buses are bigger and heavier than cars to begin with, they can more easily accommodate the weight of the storage tanks and of the fuel cells.

Fuel cells hold the key to hydrogen's future as a large-scale energy resource. Continuing engineering development of fuel cells is bringing them closer to the market for both stationary uses, in the on-site generation of electricity, and transportation uses, in hybrid or electric vehicles. As improvements in fuel-cell technology move ahead together with improved systems for purifying and storing hydrogen made from fossil fuels or biomass, interest in hydrogen as an energy source will continue to evolve.

REFERENCES

1. Rosen, W. *The Most Powerful Idea in the World*. Random House: New York, 2010, p. 147.
2. Anderson, F.C. Quoted in Gregory, K. *The Second Cuckoo*. Akadine Press: Pleasantville, NY, 1997, p. 199.
3. Benford, G. *The Wonderful Future That Never Was*. Hearst Books: New York, p. 150.

FURTHER READINGS

Boyle, G. *Renewable Energy*. Oxford University Press: Oxford, U.K., 2004. An edited collection of chapters, each prepared by various experts in their respective fields. Chapter 4 discusses biomass energy.

Brown, L. *Plan B 3.0*. W.W. Norton and Company: New York, 2008. This book presents the message that "business as usual" in energy is no longer an option. Chapter 12 covers aspects of renewable energy.

Bryce, R. *Gusher of Lies*. Public Affairs: New York, 2008. Among the topics presented in this book is a negative view of the possibilities of corn-based ethanol as a significant contributor to the energy economy of the United States.

Cash, J.D.; Cash, M.G. *Producer Gas for Motor Vehicles*. Lindsay: Bradley, IL, 1997. This is a reprint of a book originally published in 1942, now available in an inexpensive paperback edition. It describes ways of converting wood into a gaseous fuel, and even how to run an automobile on such a fuel.

Coley, D. *Energy and Climate Change*. Wiley: Chichester, U.K., 2008. Chapter 25 of this textbook is a detailed discussion of biomass energy.

Fine, D. *Farewell, My Subaru*. Villard: New York, 2009. A chronicle of the author's adventures after trading his Subaru on a diesel-engine truck, and converting it to run on waste vegetable oil.

Kemp, W. *Biodiesel: Basics and Beyond*. Aztext Press: Tamworth, ON, 2006. For anyone wishing to make biodiesel at home, this is probably the best available book, with detailed instructions and precautions. An excellent "how-to" guide.

Mousdale, D.M. *Biofuels*. CRC Press: Boca Raton, FL, 2008. Chapter 6 provides a solid overview of the chemistry and production of biodiesel.

Schobert, H. *Chemistry of Fossil Fuels and Biofuels*. Cambridge University Press: Cambridge, U.K., 2013. Several chapters in this textbook deal with photosynthesis, ethanol, biodiesel, and wood. A prior knowledge of chemistry is assumed.

Smil, V. *Energy at the Crossroads*. MIT Press: Cambridge, MA, 2003. Subtitled "global perspectives and uncertainties." Chapter 5 discusses, among other topics, biomass energy.

Starbuck, J.; Harper, G.D.J. *Run Your Diesel Vehicle on Biofuels*. McGraw-Hill: New York, 2009. Many books have been published in recent years for laypersons who are reasonably handy with tools and who are interested in producing biodiesel fuel at home and/or converting a vehicle to run on biodiesel. This book has excellent illustrations and checklists of tools needed, and steps to be followed.

32 Electricity from Wind

WHERE WIND COMES FROM

Winds are generated by the uneven heating of the Earth and its atmosphere by the sun. The Earth is generally warmest at the equator. As the air there is heated, it expands. (Remember Charles and Gay-Lussac—increasing the temperature of a gas increases its volume.) The same mass of air now occupies a larger volume, another way of saying that the density of that portion of air has decreased. Because the density has decreased, the air will rise. This expansion by heating and reduction of density is the origin of the bit of folk wisdom that "hot air rises."

Two effects follow from the rising of heated air. First, cooler air from elsewhere will flow into the region, to replace the heated air that has ascended into the atmosphere. This flow of air is felt by us as a wind. The second effect is that the heated air that has risen into the atmosphere will travel at high altitudes until it cools. As it cools, its volume decreases (Charles and Gay-Lussac again) and it becomes more dense. In doing so, it sinks to lower elevations.

Essentially, a gigantic "conveyor belt" of air movement has been established. Air masses that benefit by being warmed by the sun rise and are replaced by cooler air flowing into the region. At the same time, the warmed air travels away from the equator, cools, and sinks, replacing some of the cooler air that had moved toward the equator. These large circulation patterns, the "conveyor belt" model, owe their origin to the differential heating of Earth by the sun.

Wind energy is an indirect form of solar energy.

This model focuses on a global scale and explains how atmospheric circulation can distribute energy, originally obtained from the sun, toward the poles. The other global-scale phenomenon that contributes to the pattern of winds is the rotation of the Earth. The rotation adds the effect of stirring or swirling into the pattern of air transport established by the "conveyor belt."

On a local scale, this simplistic model becomes much more complex. Each component of the local landscape—soil, rocks, plants, and water—has different abilities to absorb solar energy or to reflect or reradiate it back into the atmosphere. Each, then, experiences a different degree of heating. Further, different landscapes will also have different features of topography: Some places may be flat, others hilly or mountainous. Some may incorporate large bodies of water. Urban landscapes may be marked by numerous tall buildings. As a result, the frequency, direction, and speed of winds can vary greatly from one location to another.

WIND ENERGY IN THE RECENT PAST

We've seen in Chapter 9 how wind was an important energy source in the Middle Ages and Renaissance. Although the rise of the steam engine and the Industrial Revolution that the steam engine helped foster diminished the role of wind as a significant source of energy, wind never left the picture entirely.

> *One of the great curiosities in Zealand, the flourishing Holland colony in Ottawa County, Michigan, is the great, awkward and unmanageable concern called the Windmill. This is a monstrous wooden pile in the form of an octagon tower. The mill is moved by the force of the wind striking against four winding slats, covered with canvas. They were sawing, or attempting to saw, while I was there. Occasionally, with a fair wind, the saws would strike a few minutes quite lively, then draw a few slower strokes and then entirely stop, perhaps for half an hour. An enterprising individual is now putting up a steam sawmill, which will do a better business.*
>
> *—Scientific American*[1]

Windmills continued to be important well into the twentieth century, especially as sources of ENERGY in rural regions throughout the world. There does not seem to be an accurate census of how many windmills were in use, but estimates range well into the millions. These windmills provided ENERGY for crucial chores such as pumping water for livestock or for irrigation. Even now it's possible to see the occasional windmill when driving through rural areas where farming or ranching is important (Figure 32.1).

FIGURE 32.1 Windmills, an important source of mechanical WORK in the Middle Ages, never really disappeared. They are still useful in rural areas, for example, pumping water for livestock. (Reprinted from first edition, permission from Phototake, New York, New York.)

FIGURE 32.2 This windmill in Malawi was built by a young boy, William Kamkwamba, using parts from a junkyard. It is the first of what has come to be known as "William's windmills," a useful source of ENERGY for impoverished rural areas. (From http://commons. wikimedia.org/wiki/File:William_Kamkwambas_old_windmill.jpg.)

In the late nineteenth century, the technology of generating electricity had been worked out by Gramme and other pioneering electrical engineers (Chapter 13). Soon thereafter, windmills were operating small generators to produce electricity either for recharging batteries or for operating small electrical devices on the farm (Figure 32.2). For example, batteries recharged by the wind made it possible for farmers and ranchers living far from electricity distribution lines to get news and entertainment by radio.

Use of windmills to produce electricity dates from the late 1880s. The first installation of a windmill to recharge batteries for electric lighting was at a home in Marykirk, Scotland. Shortly thereafter, a mansion in Cleveland may have been the first major dwelling in the United States to be lighted by electricity; the generator was operated by a windmill mounted on a nearby tower that was about twenty meters (five to six stories) tall and weighed about 36 tonnes (Figure 32.3). It generated twelve kilowatts. A half-century later, there were hundreds of thousands of so-called "home light plants" operating in the United States, mostly in the windy Great Plains.

These applications of wind energy for local or on-site electricity generation were killed off by the steady spread of electrical grids that brought electricity from central generating plants to all parts of the country. The process was hastened by the rural electrification programs of the 1930s, especially in the United States. That supply of electricity, coming via distribution lines from plants tens or hundreds of miles away, was ample for all the many applications of electricity around the farm or ranch. Small wind-operated generators fell into disuse.

FIGURE 32.3 Charles Brush's windmill in Cleveland, in 1888. It was used to generate electricity for the house seen in the background. (From http://en.wikipedia.org/wiki/File:Wind_turbine_1888_Charles_Brush.jpg.)

In Europe, Denmark has been the leader in applications of wind technology. At the same time—roughly around 1890—that wind was being used for small-scale electricity generation in the rural United States, the Danish engineer Poul la Cour (Figure 32.4)* was developing similar devices. Indeed, la Cour is the founding father of the wind-to-electricity technology that we will discuss later in this chapter. By the time of the First World War, several hundred wind-operated generators were in use in Denmark, mostly along the coast.

The easy availability of cheap oil after the Second World War was a major factor in eliminating European interest in the use of wind to produce electricity. However, that period of cheap oil lasted only about a quarter-century, until the oil price shocks of the 1970s. The Danish response to the rapid run-up in oil prices in the early 1970s

---

* Poul la Cour (1846–1908) is almost unknown today, but is essentially the founder of modern wind energy technology. He was a prolific inventor, with many ideas relating to telegraphy (the nineteenth-century internet) including ways of sending multiple messages over the same line, and an optical telegraph that used different frequencies of light to transmit messages. He also developed a hydrogen storage system. His work on using windmills to generate electricity is his most enduring achievement.

FIGURE 32.4   Poul la Cour (1846–1908), the Danish engineer who was a pioneer in wind turbine development. (From http://www.folkecenter.net/gb/rd/wind-energy/48007/poullacour/.)

provided the impetus for the revitalization of the wind industry. The Danish government took several steps to encourage a shift back to the use of wind. These steps included a 30% subsidy on the purchase of wind machines and a requirement that electricity utilities purchase the electricity that these devices produced.

WIND TURBINES

Gramme and his contemporaries were responsible for the engineering development of Faraday's scientific discovery—that a moving magnetic field generates an electric current—into the practical generators that are still used to produce virtually all of the electricity that we use today. Only the small fraction that we derive from batteries and the tiny fraction from solar energy do not come from modern variants of the "Gramme machine." Since Gramme's generator involves rotating an electric conductor in a magnetic field, the essence of electricity generation is that of finding ways to turn the generator as cheaply and reliably as we can. We need to find a reliable source of rotary mechanical WORK to operate the generator. This WORK is provided to the generator by a turbine, a device designed to capture some of the kinetic energy of a moving fluid and convert that ENERGY into rotating mechanical WORK.

FIGURE 32.5 The future of wind energy lies in the wind turbine, which captures kinetic energy from the wind to operate a generator. (Reprinted from first edition, permission from Photo Researchers, Inc., New York, New York.)

As we've seen, the fluid that passes through the turbine is the so-called working fluid. We have already met several working fluids: water in hydroelectric plants, steam in fossil fuel or nuclear plants, and combustion gases in jet engines. Since wind represents a fluid (air) in motion, it could, in principle, be used as the working fluid in a turbine. We use the term *wind turbine* to refer to machines that use the ENERGY of wind specifically to supply WORK to an electricity generator (Figure 32.5). Nowadays the term *windmill* is usually restricted to mean devices that use the wind for other applications of mechanical WORK, such as pumping water or making flour.* The focus of interest for applications of wind ENERGY in the twenty-first century is entirely on the wind turbine. There seems to be no interest in reviving windmills.

Installations using windmills and waterwheels could be configured such that the axis of the wheel was either vertical or horizontal. The same is true with wind

---

* Actually neither the wind turbine nor the windmill represents the most powerful of wind-operated devices. That honor goes to the large sailing ships of the eighteenth and nineteenth centuries, the best of which could extract as much as 7500 kilowatts from the wind.

FIGURE 32.6 Horizontal-axis wind turbines on a wind farm near Buffalo Ridge, Minnesota. The black blades on these Zond wind turbines help to shed ice and frost in the winter. (From http://www.nrel.gov/data/pix/searchpix.php?getrec=06330&display_type=ver bosc&search_reverse=1.)

turbines. In a horizontal-axis machine (Figure 32.6), the turbine blades look some-what like the propeller on an airplane. The axle on which the turbine blades are mounted is connected, usually via a gearbox, to the axle of the generator. The generator, gearbox, and any necessary components of the electrical system or control systems are all mounted in a nacelle, which sits on top of the turbine tower, directly behind the turbine blades. The electricity that is generated is conducted down to ground level by a cable.

The vertical-axis wind turbine (Figure 32.7) represents the alternative design. They look like enormously large eggbeaters, having two aluminum blades attached to a central, vertical mast. The vertical-axis wind turbines have several design advantages relative to the horizontal-axis type: The gearbox and generator are on the ground, making maintenance much easier; wind coming from any direction can be used without having to turn the unit into the wind; and the blades are much lighter in weight than those on horizontal-axis machines.

Other considerations of design of wind turbines are the generating capacity of an individual unit, and whether they would be constant- or variable-speed devices. The industry seems presently to have adopted wind turbines in two general size ranges: small machines that would be used by individual farms or households that are not connected to the electricity distribution grid, and larger machines generating

FIGURE 32.7 A vertical-axis wind turbine, on the Magdalen Islands, Quebec, in the Gulf of St. Lawrence. (From http://en.wikipedia.org/wiki/File:Darrieus-windmill.jpg.)

electricity for distribution by an electric utility. Smaller units have generating capacities typically in the range 0.5 to about 50 kW, whereas larger machines, intended to feed electricity into a grid, would be in the range 50 to about 600 kW.

The occurrence of gusts, during which the wind speed could double and then perhaps drop to even lower than original value in a matter of seconds, has to be taken into account in designing wind turbines. Constant-speed wind turbines must be designed and built to endure higher mechanical loads on their drive trains and gearboxes as the wind speed increases. A variable-speed machine can operate at a range of wind speeds while still delivering constant output. Fluctuating mechanical loads imposed by the winds are more easily withstood by blades and drive train. The drive train can be less rugged and thus made lighter and less expensively. A further advantage of the variable-speed wind turbines is their ability to deliver electricity over a wider range of wind speeds. A disadvantage is an increased cost for the necessary electrical control system and so-called "power conditioning" equipment needed to provide electricity output of constant quality even while the speed of the turbine is changing. In areas that have fairly constant wind speeds, there would be ample reason to stay with the simpler constant-speed devices.

Current commercial designs of wind turbines can operate over a wide range of wind speeds, from a little under 15 kilometers per hour to about 100 km/h. Rapid strides have been made in improving reliability. Some pioneering designs of the 1980s were out of service more than half the time. Nowadays the "down time" for a wind turbine is in the range of 2%. The theoretical maximum efficiency of converting wind ENERGY to WORK, the *Betz limit*, is 16/27, essentially 59%. Just as the Carnot efficiency tells us the theoretical maximum efficiency of a heat engine, and the real efficiencies of actual operating engines are lower than the ideal, in the same way the efficiency of actual wind turbines is some 50%–70% of the Betz limit. The conversion of mechanical WORK to electrical ENERGY in the generator is quite high (as typical of most generators), above 90%.

Though the future of wind as an energy source seems to be in the development of wind farms (discussed below), each with hundreds of wind turbines having individual capacities in the several hundred kilowatt range, some remarkable small designs have been developed that are tributes to human ingenuity. The Savonius rotor (Figure 32.8) is a low-tech device that seems to take its design inspiration from the Islamic vertical mill (Chapter 9). At first glance, a sketch of the Savonius rotor might give the impression that the device is nothing more than an oil drum sawn in half lengthwise and opened outward to make two "cups" capable of catching the wind. In this case the first impression is exactly correct, for that is in fact the easiest way to make a Savonius rotor. A virtue of this device is that it can be constructed for very little money by do-it-yourselfers with common tools. Though the Savonius rotor is not nearly so efficient as the more sophisticated, large commercial devices, it can be built and installed in impoverished villages and used, for example, to operate water pumps. Another remarkable design is a collapsible, portable wind turbine used

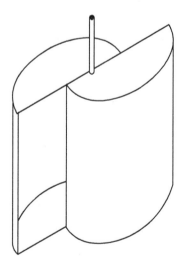

FIGURE 32.8 A conceptual diagram of a Savonius rotor. A simple one can be made by cutting an oil drum in half vertically, and putting the two halves together as shown. (From http://en.wikipedia.org/wiki/File:Savonius_turbine.gif.)

by Mongolian nomads. The spread of blades is only about 2 m. The turbine can be taken down, packed up, and moved to the next camp. These small portable units have generating capacities of about 50–250 watts.

WIND AS A MODERN ENERGY SOURCE

If we start with the wind energy theoretically available, ignoring for the moment engineering issues of building practical wind turbine devices to capture all that energy, and ignoring societal issues of where wind turbine installations might actually be located, then the potential is extraordinary. In the United States, the total amount of kinetic energy dissipated by winds across the United States is some forty times larger than the annual energy consumption. In other words, it could—in theory—be possible to satisfy all of the energy needs of the United States by extracting just one–fortieth of the energy from the wind. Limitations come from assuming reliance on today's wind turbine machines. Their efficiencies are limited to some 50%–70% of the Betz limit. Also, wind turbine towers are at most about a hundred meters tall; winds at higher elevations from the ground are not utilized.

It is not likely that every possible site for wind-to-electricity energy conversion will be developed. Some wind sites may be so remote that it would be impractical, or not cost effective, to connect them to a distribution grid. Other sites may be in environmentally sensitive areas where society might decide to limit development. Indeed, land-use issues may most likely be among the most contentious issues for wind energy development. Still others might be in highly populated urban areas where the costs of acquiring the land, razing existing buildings, and developing the site might be prohibitive.

Worldwide, the total wind flow is ample to supply several hundred times the world's total annual energy consumption. The issues discussed above would apply to estimating how much of this potential might actually be tapped, and are even more difficult to take into account. For example, some countries have very strict limitations on land use, while others have essentially none. It seems reasonable to expect that over the next several decades about 10% of total worldwide electricity production could be supplied by wind.

> *Take any given space of the earth's surface…and all the power exerted by all the men, and beasts, and running water, and steam, over and upon it, shall not equal the one hundredth part of what is exerted by the blowing of the wind over and upon the same space.*

> —Lincoln[2]

PRIME LOCATIONS FOR WIND ENERGY DEVELOPMENT

Some of the best locations are over oceans, but generally near to the shore. This raises the possibility of offshore generation of electricity from ocean winds. Such offshore generation might eliminate, or at least reduce, some potentially contentious issues of transmission rights-of-way and alternative uses for the "land" (or, in this case, ocean) surface. Another area of highest potential for wind energy development would be large expanses of flat or gently rolling plains. In some places, ridge-and-valley topography offers good wind resources, because the ridges act as channels or conduits for the wind flow. In some countries, such winds have been exploited for millennia by migrating birds, and probably for decades by glider and sailplane enthusiasts.

In choosing a site for a wind turbine installation, having wind that blows reasonably constantly throughout the day and throughout the year is usually much more important than finding a place that has high peak wind speeds but has them only occasionally. Detailed data on wind speeds, directions, and duration over a number of years are needed to help make a proper selection of a site. More than just the nature of the wind is important in selecting a site. Obstacles such as hills, various kinds of vegetation, and buildings or other structures can greatly alter the wind profiles, and even create regions of strong turbulence, such that a simple average wind-velocity distribution will not give an adequate description of what might be encountered at the site.

The feasibility of developing a particular site depends not only on the average wind speed and current land use but also on more complicated factors such as daily and seasonal variations in wind speed. Wind never blows at a steady speed for 24 hours a day throughout the year. Wind is almost always variable. The POWER generated depends critically on the speed; electric POWER depends on the cube of wind velocity—doubling the wind speed increases POWER by a factor of eight. As a result, an accurate assessment of the pattern of winds at a prospective site is vital to success. Questions such as whether there is a windy season, and, if so, when; the strength of the strongest gusts; and how long "calms" might last are all important parts of this assessment.

Wind speed and direction often do generally follow daily and seasonal patterns that are remarkably predictable despite some occasional wide fluctuations. Had this not been true, the ability of sailors to travel and explore the oceans and to use them for trade routes would have been seriously hampered. Often a database of several years' worth of wind speed measurements will provide a fairly accurate assessment of the amount of energy that a wind turbine site can be expected to deliver at any given time of day or year. Assuming that such data are available to allow making reasonable predictions of wind energy, then a more important, and sometimes more complicated, issue emerges: how well the wind electricity generation correlates with the timing of needs for electricity by the utilities. If a wind installation can provide electricity at times when demand is highest, so much the better; if wind generation is highest when demand is lowest, the system is "out of synch" and may not be so worthwhile an investment.

Once sites with appropriate wind potential have been identified and assessed, but before the first kilowatt-hour of electricity can be generated and sold, much work needs to be done. Land leases have to be negotiated and land-use permits obtained. Agreements have to be set up for the transmission of the electricity and its purchase. Turbines must be designed, manufactured, and installed, along with an array of electricity collection equipment, substations, and interconnection facilities. All of this hardware has to be installed on site, tested, and maintained.

ELECTRICITY FROM WIND: THE CURRENT SCENE

Commercial generation of electricity from wind nowadays relies on the wind farm concept. A *wind farm* is a location containing large numbers—from dozens to hundreds—of wind turbines, each with a generating capacity of 100–500 kW (Figure 32.9).

Wind farms are now in operation in many countries around the world. There is still a market for the smaller machines that are used individually to supply electricity for a single user, such as a ranch, rather than for a utility.

Worldwide, the wind industry is steadily growing. Currently, China and the United States lead the world in terms of installed generating capacity. The Gansu wind farm in China produces 5000 MW. The largest single facility in the United States is the

FIGURE 32.9 For large-scale electricity generation, many turbines would be clustered together in large wind farms. (Reprinted from first edition, permission from Photo Researchers Inc., New York, New York.)

Alta Wind Energy Center in California, producing about 1000 MW. The Chinese and American installations are all onshore facilities. In terms of offshore generation, collectively the countries of Europe are in the forefront. Britain has the largest single offshore wind stations, one at 370 MW and another about 300 MW. Total European capacity is about 2400 MW. The European Wind Energy Association has ambitious plans to increase this nearly fifty-fold, with 100,000 MW (100 GW) to be produced by wind energy by 2030. Many other countries are involved in wind energy development in onshore or offshore sites, or both. There remains a continuing, albeit modest, demand for small units to be installed in remote or rural locations.

In Europe, the current leader in wind technology is Denmark, appropriate as the home country of Poul la Cour. Denmark deserves the credit for taking the lead, worldwide, in revitalizing wind technology in the 1970s. In response to the two oil price shocks of that decade, Denmark made a conscious, national decision to switch to renewables, most notably biomass and wind. About half the wind turbines in use worldwide are of Danish design or manufacture. Currently, Denmark's wind turbines supply close to 30% (4000 MW) of Danish electricity demand. The national energy plan calls for reaching 50% by 2020.

ADVANTAGES OF WIND FOR ELECTRICITY GENERATION

Wind is a very environmentally friendly energy resource. It is essentially pollution free. There is no air or water pollution, no acid precipitation nor smog. There is no pollution of groundwater and drinking water supplies. There are no SO_x emissions to attack and degrade plants. There is no radioactive waste, fly ash, or scrubber sludge. There are no carbon dioxide emissions, nor emissions of any other greenhouse gases, so there is no impact on global climate change. About the only imaginable effect on the environment is a very slight reduction in the strength of the wind as it passes through a wind farm. A wind farm poses no significant threat to the safety of the public. Overall, it is hard to imagine an energy source more benign to the environment than wind technology.

Each individual wind turbine requires a relatively short time to build. In fact, they can be built on an assembly line. If that were done, it would be possible to produce enough wind turbines to generate a thousand megawatts in a year. In comparison, the construction time to build a thousand-megawatt fossil-fuel power plant would be several years and for a nuclear plant about ten years. Fossil or nuclear plants must be erected on the site where they will be operating, with components brought in by truck or rail. Furthermore, wind turbines are modular. Once a site for a wind farm has been selected, new turbines can be added quickly and relatively inexpensively as electricity demand increases. In comparison, fossil or nuclear plants would usually be built in increments of several hundred megawatts at a time, at a cost well into the hundreds of millions of dollars.*

---

* There is one significant, and increasingly important, exception to this. Natural gas can be used to fire a *gas* turbine connected to a generator in a design similar to that of an aviation gas turbine (jet) engine. Small, natural-gas-fired gas turbine plants can be built relatively quickly and at much lower cost than the several-hundred-megawatt behemoths. This approach to electricity generation might become increasingly popular if the apparently vast resources of shale gas can be exploited. Nevertheless, even these plants probably do not offer the potential flexibility of expanding a wind farm a few turbines at a time.

FIGURE 32.10   An advantage of wind farms is that all of the land, except for the bases of the towers, can be put to other uses, such as farming. (From http://commons.wikimedia.org/wiki/File:Llyn_Alaw_Wind_Farm_-_geograph.org.uk_-_102122.jpg.)

Although the total area occupied by a wind farm may be large, the only actual use of the ground surface is the "footprint" of each wind turbine. Over the area of an entire wind farm, the turbines themselves use no more than about 5% of the total land. The remaining land can be put to other uses, including farming (sheep may safely graze), ranching, forestry, or recreation (Figure 32.10). Because the turbines themselves occupy only a small fraction of the land area, wind energy development is ideally suited to farming areas. In Europe, farmers plant crops right up to the base of the turbine towers. In California, cows graze close to the turbine towers. The leasing of land for wind turbines provides benefits to landowners through extra income and increased land values.

Wind energy also has a profound advantage relative to solar energy, which is perceived to be one of the other environmentally benign renewable energy technologies (Chapter 33). Solar suffers the undeniable disadvantage that the sun doesn't shine at night. On the other hand, wind can blow day or night, and on sunny or cloudy days. Winds are often at their strongest and most reliable during the dark and stormy nights of winter, precisely when most of us have our greatest demand for electricity.

POTENTIAL DISADVANTAGES OF WIND AS AN ENERGY SOURCE

The very first issue, perhaps so obvious that it would be easy to overlook, is that wind turbines only work in areas that are windy. Some places of the world ideal for wind energy development, such as the northern Great Plains in the United States, are far from major population and industrial centers. There could possibly be significant investment required for transmission lines to get the electricity from where it is generated to where it is needed. But even when a suitably windy site is located, we must still recognize that, in most places, the wind does not blow all the time, nor does it usually blow steadily. This intermittent or episodic nature of the wind causes a variety of problems. Nuclear, fossil-fuel, and hydro plants can operate 24 hours per day and can generate electricity on demand. Wind farms, necessarily dependent on wind, might only produce an average output of one-fifth of the rated capacity. In this specific respect, these other energy sources have a significant advantage relative to wind.

On a very short time scale, minutes or even seconds, gusts bring sudden, sharp changes in wind speed. On a time scale of hours, there are lulls and freshenings; an example that we all experience is the way the wind picks up, dies down, and picks up again as a sequence of rain storms pass through the area. On the time scale of a day, there are variations such as a drop in wind speed at sunset. Still more variations occur on scales ranging from several days to months to years. A wind gust can put serious mechanical strains on an individual turbine, but the effects on a wind farm are usually mitigated by having numerous turbines spread across the area of the farm. Local weather variations can be compensated for by having several wind farms, located some distance (e.g., a hundred kilometers) apart.

To take into account long periods of calm, a wind energy system needs backup generating capacity. There would be a need for backup that could come online in an hour or so. To compensate for the entire wind farm being becalmed for a long period of time, backup capacity that essentially duplicates the whole system would be needed. In addition, an energy storage system might be needed in those areas where there is a mismatch between the times of peak generation from the wind turbines and times of peak demand by the customers. These issues might be addressed, in favorable cases, by having multiple wind farms located some distance apart. If numerous, scattered wind farms were linked into a single grid, a drop in wind speed in one area might be compensated by an increase in wind somewhere else. The costs of backup and energy storage systems have to be factored into the total cost of the wind installation.

Because of these concerns about the intermittent nature of wind, there once was a general rule of thumb that no more than about 20% of a nation's electricity supply should come from wind. If the percentage is much higher than that, there is a perceived risk of occasional electricity supply shortages. However, some places, notably Denmark, have successfully surpassed that rule-of-thumb limit. It is unlikely that any country or region will ever become 100% reliant on wind, but the prospects seem much brighter than they did ten to twenty years ago.

The issue of land use is seen, by different camps of course, to be either an advantage of wind or a significant disadvantage. Electricity production using wind turbines requires a great deal of space. Because of the turbulence created by the rotating blades, the turbines have to be placed between 100 and 300 meters apart. A reasonable-sized wind farm will have at least a hundred wind turbines. The turbines occupy less than 5% of the area of the wind farm, so the rest of the space can be used for such activities as farming or ranching. Concerns arise when the total area set aside for the wind farm must be weighed against other competing uses of the same land. Perhaps the most contentious problem is setting aside for a wind farm land that might otherwise have been used for housing developments. Concerns would also surely arise if roads for access to the wind turbines had to be cut through forests, and more so if the forest itself were to be cleared to make way for the wind farm.

Noise is another contentious issue relating to wind farms. There's no getting around the fact that wind turbines make noise. The "silent turbine" has yet to be designed. As a result, it is likely that regulations might be established to limit the noise production from a wind farm, and to determine how close turbines may be sited to occupied buildings. Again, Denmark seems to be leading the way; the Danes have a regulation that the noise level at the house nearest to any turbine must be less than 45 decibels,* which is about the level of a conversation.

Birds can be killed by flying into the spinning blades of a turbine. At Altamont Pass, birds as large as golden eagles have been killed. No utility would want to face the public relations nightmare of being accused of wiping out endangered birds. However, bird kills are certainly not unique to wind energy systems. Birds are killed or injured by flying into transmission and distribution lines coming from other kinds of electricity-generating plants too. But by far the most effective killer of birds is glass—windowpanes. Not all wind farms have experienced bird-kill problems. Ones particularly likely to experience bird kills are those that happen to lie on a bird migration route, and for which the land around the wind turbines has a large, active population of the small mammals that are favored prey of the raptors.

A final argument raised against wind farms is "visual pollution." Though these installations produce no tangible pollution in the form of emissions to the air or water, many people simply object to the sight of tens or hundreds of wind turbines, and their associated transmission lines, on the landscape.† A wind farm of modest output might cover, say, eight square kilometers of land and incorporate fifty to sixty wind turbines, each well over thirty meters tall. That wind farm would produce perhaps 25 MW of power, whereas a new coal or nuclear plant would be at least 1000 MW. To help assure that the turbines capture as much wind as possible, it is good to locate the wind farm on a hilltop. Thus a situation develops in which people for many miles around wind up looking at dozens and dozens of turbines. Unlike many other kinds of utility or industrial installations—even junkyards—a screen of tall, thick trees cannot be used to hide a wind farm from view, because it would defeat the whole purpose, by screening the wind, too. The fact that some offshore winds can be especially powerful suggests the potential of locating wind farms on the ocean surface. Offshore wind farms (Figure 32.11) could potentially negate many of the concerns relating to land use and noise, yet they too are accused of being a form of visual pollution.

---

* The decibel is a unit of sound intensity. Like the pH scale we've met in Chapter 30, it is a logarithmic scale. The faintest detectable (by humans) sound is arbitrarily assigned a value of one. The number of decibels between any two sound intensity levels is taken to be ten times the logarithm of the ratio of the two intensities. A faint whisper or very light rustling of leaves is about 10 decibels; ordinary conversation about 40; light to moderate traffic, 60; and an aircraft engine, about 120. The aircraft engine is 10^{12} times more intense than the faintest detectable sound (that is, 1,000,000,000,000 times!).

† There is much truth in the cliché that beauty lies in the eye of the beholder. Though cries of "visual pollution!" have been raised against proposed wind farms because people find them unsightly or downright ugly, it is fair to say that others find some of the turbine designs visually attractive. This seems especially true for the vertical-axis "eggbeater" design that is thought to be attractive in a futuristic sense, perhaps an emblem of a future of clean, renewable energy.

FIGURE 32.11   An offshore wind farm, this one in the Øresund, near Copenhagen. (From http://en.wikipedia.org/wiki/File:DanishWindTurbines.jpg.)

COMPARISON OF WIND ENERGY WITH OTHER SOURCES OF ELECTRICITY

Cost is always of concern for any energy system. At least two measures of cost must be considered. The first is the cost to build and install the generating system. The second is the actual cost of the electricity that is eventually delivered to the consumers.

Cost estimates must always be treated with caution, because they are usually made on the basis of numerous assumptions, and unless those are clearly known, it may not be obvious how a final estimated cost was developed. Further, it might be that different kinds of assumptions were made for, say, a wind farm and a nuclear plant, leading to the proverbial "apples vs. oranges" comparison.

By 2010, the cost for the capital investment to build and install an onshore wind farm had reached a point of being second lowest among the likely sources of large-scale electricity generation. Currently the lowest capital cost would be for a natural-gas-fired plant. Offshore wind installations are much more expensive, being more than double the cost, in dollars per kilowatt of capacity, than an onshore installation. The estimated operating and maintenance cost for an onshore wind farm is lower than any alternatives except solar photovoltaics, hydro, and natural gas.

The major costs in the wind power industry are in capital equipment, since the turbines do not need "fuel." The wind-generated electricity will be considerably more expensive if the site chosen for the wind farm is located far from existing transmission lines, so that they too would have to be included in the cost of the facility. The maintenance costs on modern wind turbines are extremely low. If some way could be devised to account for the environmental and social damages caused by fossil-fuel or nuclear plants, these additional costs would enhance the economic advantages of wind.

The *capacity factor* is the amount of electricity actually generated (e.g., over the course of a year) divided by the maximum generation at the peak capacity of the plant. As an example, if a plant had the maximum capacity of 200 million kWh per year and actually did generate 100 million kWh, its capacity factor would be 50%. The capacity factors for wind farms range from about 25% to 50%. In comparison, nuclear plants have capacity factors of 70% and coal, 75%–80%. However, depending on the specific wind farm site, the electricity actually generated might be no more than about 25% of the maximum possible. A typical fossil fuel plant with a capacity factor of 75%, in comparison, can produce two or three times as much electricity per unit of generating capacity as the wind plant. So, to replace a 500 MW coal-fired plant a utility would need to install over 1000 MW of wind farm capacity. One might argue that this issue is counterbalanced by the fact that the "fuel"—the wind—for the wind farm is free, certainly not true of the coal plant.

Availability represents the percentage of time the unit is actually operating (i.e., not including scheduled maintenance shutdowns and unexpected outages). For example, if a plant operates for 7000 hours in a year its availability would be—given 8760 hours in a year—80%. Currently the availability is about 90% for wind, which is equivalent to fossil fuel or nuclear plants. Optimistic projections are eventual availabilities of 95%–96%. Achieving these availabilities will require more rugged designs, better materials of construction, and more careful attention to maintenance.

The same cautions about comparing estimated capital costs also apply to comparing the cost of electricity from different sources. A further complication is that, by the time the consumer actually receives a charge of so much money per kilowatt-hour of electricity, the cost of generation might have been changed by virtue of various taxes, tariffs, subsidies, surcharges, or other costs. These additional costs vary from one country to another and from one state or region within a country. Generally, in terms of cents per kilowatt-hour, electricity from wind is about the cheapest, though hydro and nuclear are close behind. Electricity from wind is less than one-third the cost of electricity from solar.

The characteristics of the local wind resource are an important factor in determining the value of the electricity produced. The value of electricity may be reduced if winds are at their peak when daily electricity demand is lowest, or may increase if wind generation is high at the same times that demand is high.

On balance, wind is an essentially pollution-free method of generating electricity that appears now to be very cost competitive with conventional sources. It seems reasonable, therefore, that

Wind is the renewable energy resource with fewest technical, economic, or environmental hurdles to making a significant contribution to our electricity supply.

REFERENCES

1. From the August, 1849 issue of *Scientific American Magazine*, 281(2), 10, Reprinted in the August, 1999 issue.
2. Lincoln, A., quoted in Rosen, W. *The Most Powerful Idea in the World*. Random House: New York, 2010, p. 323.

FURTHER READINGS

Armstrong, F.; Blundell, K. *Energy Beyond Oil*. Oxford University Press: New York, 2007. This book is an edited collection of chapters written by experts in various fields of energy. Chapter 5 deals with wind energy.

Boyle, G. *Renewable Energy*. Oxford University Press: Oxford, U.K., 2004. Another collection of chapters by various experts in their respective fields. Wind energy is discussed in Chapter 7.

Bryce, R. *Power Hungry*. Public Affairs: New York, 2010. An argument that the likelihood of developing an energy economy based on "green" energy, particularly wind and solar, is based on myths. This author argues for an energy future involving natural gas and nuclear.

Coley, D. *Energy and Climate Change*. Wiley: Chichester, U.K., 2008. Chapter 22 of this textbook discusses wind energy.

DeBlieu, J. *Wind*. Houghton Mifflin: Boston, MA, 1998. This is a broad-ranging book about wind itself, not specifically focused on energy. Chapter 9 does discuss wind farms, including personal visits and interviews by the author.

Smil, V. *Energy at the Crossroads*. MIT Press: Cambridge, U.K., 2003. Chapter 5 discusses the possibilities for wind energy, along with biomass, nuclear, and solar photovoltaics.

Walker, J. F.; Jenkins, N. *Wind Energy Technology*. Wiley: Chichester, U.K., 1997. This handy paperback book is probably the best reasonably short introduction to the field.

33 Energy from the Sun

A list of criteria for the ideal energy source would likely include these: It should be environmentally friendly, causing little or no harm to the environment; it should be inexhaustible, or least renewable, eliminating worry about issues of reserves, resources, and lifetimes; and it should be domestic, eliminating worry about geopolitics and possible import restrictions.

Hydroelectricity has many advantages (such as no "fuel" costs and the fact that it is essentially pollution free). While the actual electricity generation itself—water flowing through a turbine connected to a generator—is very clean, the impact of dams on ecosystems is now of great concern. Wind energy shares many of the advantages of hydro and is becoming an increasingly cost-effective and reliable source of energy. Yet another energy source meets the criteria set out above: solar energy. Solar energy shares many of the advantages of hydropower: there is no cost for the "fuel," it is certainly inexhaustible, and there is virtually no pollution. In addition, solar energy can be used in any area of abundant sunshine, whereas hydropower is restricted to sites where there are water sources. It can be used by any country or any region, eliminating many of the geopolitical issues concerned with unevenly distributed resources such as petroleum.

The sun's energy derives from the conversion of mass to energy according to Einstein's equation, $E = mc^2$ (Chapter 23). We will defer a discussion of why this equation applies until the next chapter. The mass loss associated with the nuclear processes in the sun amounts to about four million tonnes per second. Since c^2 is itself a huge number, the energy production per second in the sun is so large that we have no way of translating it into our normal experience. The sun radiates this energy into all directions in space (Figure 33.1). Since Earth is a tiny "dot" 150 million kilometers from the sun, virtually all of the energy produced by the sun misses the Earth. In fact, the fraction of the sun's energy hitting the Earth is only one two-billionth (1/2,000,000,000th!) of the sun's total output.

This extremely tiny fraction would not seem worth bothering with. And yet, almost all the ENERGY that we use derives in some fashion from the sun. Growing plants require sunlight to provide the ENERGY for photosynthesis. Our use of food, and of firewood or other biomass products for ENERGY, depends directly on the sunlight-driven photosynthesis process. Fossil fuels derive from the remains of once-living organisms preserved in the Earth's crust; those organisms once required sunlight and photosynthesis. Water-driving waterwheels or hydroelectric turbines are supplied by Earth's hydrological cycle, driven by the ENERGY in sunlight. Wind results from differences in the amount of heating of Earth's atmosphere by the sun. Only the ENERGY provided by nuclear fission reactors, and the comparatively tiny amount of ENERGY derived from chemical processes in batteries do not derive, directly or indirectly, from the ENERGY of the sun. Furthermore, if we could collect completely this

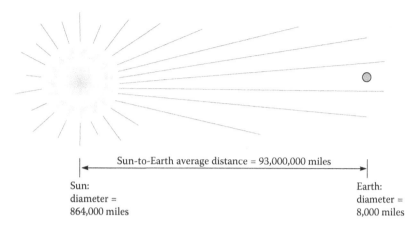

FIGURE 33.1 From the perspective of the sun, Earth is a tiny dot about 150 million kilometers away. Only an extremely small fraction of the sun's ENERGY arrives on Earth.

tiny fraction of the sun's total energy that falls on Earth, and convert it into useful ENERGY or WORK with 100% efficiency, we could provide the current annual ENERGY consumption of the whole world in only forty minutes. Of course, we haven't quite learned how to do that yet.

ENERGY FROM THE SUN TO EARTH

The energy radiated by the sun into all directions of space in an hour is equivalent to the amount of heat that would be released by burning 400,000,000,000,000,000,000,000 tonnes of coal. About 1/2,000,000,000th of this amount comes into the upper atmosphere; not all of this energy actually makes its way to Earth's surface. One-third is reflected back directly into space, and the atmosphere absorbs another one-fourth. Clouds, the ice caps, and the oceans cause the reflection of solar radiation back to space. Energy absorbed in the atmosphere helps drive the hydrological cycle and weather patterns. The remaining energy that actually falls on Earth's surface provides warmth, evaporates water into the hydrological cycle, drives the winds, and provides the energy for photosynthesis.

Not all parts of Earth's surface receive the same amounts of solar radiation, nor does solar radiation at one particular spot remain constant through the year. As we see every day, solar radiation is at a maximum at some time during the daylight hours but is zero at night. Lack of solar radiation at nighttime represents a significant complication in the use of solar energy. Solar radiation is less in winter than in summer. This derives from two factors: In winter the sun appears lower in the sky than in the summer so that the POWER provided by the sun is less; and in winter the available daylight hours are fewer. A third complication comes from weather patterns: Even in a given season at a given location on Earth, available solar radiation can vary from day to day because of changes in the prevailing cloud cover.

The effects of latitude and local weather patterns result in considerable differences in available solar radiation in a particular country. In a large country such as the United States, for example, the average amount of sunlight received in a year varies by about a factor of two from New England and the Northeast to the desert Southwest states. In Great Britain the mean annual solar power received changes significantly through the course of the year, with a distribution month-by-month that peaks in June. Further south, the nations in northern Africa, such as Egypt and Sudan, receive about three times as much solar energy as Britain, with a month-by-month distribution that shows much less variation.

The promise and challenge of solar energy derive from two issues. First, it is a clean, environmentally friendly form of energy that will last forever (at least in terms of the longevity of the human species) and that is inexhaustible. Second, even after the tiny fraction of the sun's energy is received by Earth, and further filtered by reflection and absorption, the amount that does hit Earth's surface potentially could provide enormous quantities of energy throughout the world. Sunshine delivers energy as heat and light to Earth at a rate of some 15,000 times the entire ENERGY consumption of the world. In the course of three weeks, Earth receives as much energy from sunlight as is stored in our entire reserves of coal, oil, and natural gas. The total amount of solar energy falling on the Earth is more than enough for present human needs and amply sufficient to provide any anticipated future demand. For developing nations, where per capita energy use (and standard of living) is still low, development of solar energy could provide an opportunity to improve the standard of living to that of the wealthier nations with relatively little environmental disruption.

Though solar energy offers considerable promise, there remain significant problems on the road to its large-scale commercial use. First, solar energy is not a concentrated form of ENERGY, in contrast to, say, the ENERGY in a flame or in current of electricity. Considerable space and ingenuity are required to capture large amounts of solar energy. We must inevitably cope with the fact that the sun does not shine at night. Neither solar energy nor its two direct products, heat and electricity, can be stored easily. Because of the effects of latitude, local weather conditions, and terrain, different parts of the world, and of individual countries, receive different amounts of sunlight. In some parts of the world, such as the northern portion of the United States, Canada, and much of Britain and Europe, solar radiation is strongest in the summer, when the energy demand is relatively low, and weakest in winter, when there are significant energy demands for heating and lighting.

By 2025, the worldwide demand for fuel is likely to increase by 30% or more, and that for electricity will more than triple. Even with more efficient production and use of our present energy sources, and with vigorous measures for energy conservation, it seems inevitable that new sources of energy will be required to meet the world's demand. Solar energy, if developed for large-scale use, could provide a significant portion, perhaps more than half, of the electricity demand and could displace some of the requirement for fuel.

Solar energy will have two major categories of application: providing space heat and hot water for buildings and generating electricity. Each of these major categories has within it two broad strategies. Approaches to building heat and hot water generation are divided into passive and active methods. So-called indirect and direct strategies are proposed for electricity generation. Each is discussed in the sections that follow.

PASSIVE SOLAR HEATING OF BUILDINGS

Two types of solar energy systems are used to heat and cool houses: passive and active. A *passive solar energy system* uses the entire building—walls, floors, windows, and roof—to collect, channel, and circulate heated or cooled air. In its simplest form, a passive solar system uses no mechanical devices or other moving parts (such as pumps or fans) to distribute heat. Passive solar takes advantage of the easily observed fact that rooms with a window facing the sun are heated when sunlight shines in. In the United States, Canada, and northern Europe, this form of solar energy can contribute some 10%–20% of the annual heating requirements of the house. A passive solar structure can be built and landscaped so that it becomes, in effect, one large solar collector. The passive solar approach operates by providing for direct admission of solar radiation, retaining heat by thermal insulation and heat storage in building components, and distributing heat input by natural heat-transfer processes of conduction, convection, and radiation. A well-constructed passive solar building offers advantages of simplicity, reliability, durability, and economy.

In the classical Greek drama *Prometheus Bound*, Prometheus, the god of fire, tells how he first found the people of Earth, lacking "the knowledge of houses turned to face toward the sun." The playwright, Aeschylus, was not referring to his own people, the ancient Greeks, who had an appreciation for solar energy. The Greeks were apparently the first to make extensive use of passive solar design. Their houses were constructed of adobe walls some 45 centimeters thick on the northern side to keep out cold north winds (Figure 33.2). The southern side, in contrast, faced a portico supported by wooden pillars that led to an open courtyard. This design permitted sunlight to enter the main living areas of the home through the portico. The solar radiation was trapped in the earthen floor and thick adobe walls. Both the floor and walls provided a useful form of heat storage and allowed for a gentle release of solar warmth throughout the night. These buildings first appeared in Greece over 2000 years ago. Their design spread to the nations of classical Islam, where they became common throughout the past millennium.

In addition to his writings on waterwheels, the Roman architect Vitruvius (Chapter 8) also provided guidelines for orienting houses, public buildings, temples, and even whole cities to the sun. Roman remains in various parts of Europe and the Mediterranean countries demonstrate that the Romans used solar water heating. Many Roman baths obtained water from man-made channels open to the sun and lined with grooved black slate so that water absorbed heat from the sun-warmed slate as it ran through the grooves.

To provide the simplicity and reliability that constitute the potential advantages of passive solar, a building requires both careful design and the right materials and components. The most important building concept is the orientation of the house to the south to take advantage of the sun during the winter. (Throughout this discussion we assume that the building is in the Northern Hemisphere; the directions would be reversed for a home in the Southern Hemisphere.) Once the building is properly oriented, two things are necessary for a passive solar system: windows and thermal mass. Windows let the solar energy in and can help trap heat by preventing it from

FIGURE 33.2 The remains of this early Greek house show the thickness of walls, which helped in passive solar heating. (From http://www.flickr.com/photos/salyangoz/2897343608/.)

being reflected back to the atmosphere. So-called thermal mass serves to collect the heat, store it, and pass it on to the air space in the building. Both of these components of the passive solar system are part of the structure of the building, justifying the notion that a well-designed passive solar building is, in effect, one large solar collector. Other relatively simple measures of building design and construction, such as providing for natural ventilation and sensible landscaping, with overhead shading for summer cooling, can reduce a typical building's annual heating and cooling costs by about 15%–25% while at the same time adding little or nothing to its construction cost. (However, the cost of retrofitting an existing building for passive solar features, especially if it is not properly oriented toward the south, is usually prohibitive.)

On sunny days a south-facing window gains more energy from sunlight than it loses through infrared radiation from the surfaces in the room, and the larger the window, the warmer the room. Most of the windows in a passive solar building should face south, or south and east, and should be large to maximize the amount of solar radiation admitted to the building. The north wall should have only few, small windows, or none at all, to minimize any energy losses through the windows. Double- or even triple-paned windows can be used to help hold heat during the winter.

While large windows are desirable to maximize the amount of solar radiation admitted, when the sun is not shining, or at night, the windows also allow heat to escape and can actually cool the room. One solution to this problem uses windows that reflect radiation back into the room. Glass itself traps infrared better than it does visible light, but special coatings on the glass can enhance this effect. Thick, well-insulated drapes or blinds can also be closed to help reduce heat loss at night or on cloudy days.

Sunlight passing through the windows should strike surfaces that absorb the energy and can later radiate it back into the rooms as heat, causing a greenhouse effect. Thick walls and floors made of tile, stone, concrete, adobe, or brick serve to store heat. They can be thirty centimeters or more in thickness. Stone or brick fireplaces also provide this effect. After dark, the heat absorbed in these thick floors and walls is radiated back into the building space. To help further with heat retention, wall studs are 5×15, 5×20, or even 5×25 centimeters to provide space for more insulation than could be obtained with the customary 5×10-cm studding.

An alternative to relying simply on direct solar heating is to use a specially designed interior wall for thermal storage. This type of structure was developed by Felix Trombe in the 1950s and is often called a *Trombe wall* (Figure 33.3). One design involves large concrete wall with blackened outer surface located immediately inside the south-facing double-glazed window. Incoming radiation heats the air in the narrow cavity between the wall and the window but is mostly absorbed by the wall. Heat absorbed by the wall causes the interior wall temperature to rise and to transfer heat energy into the house indirectly by conduction, convection, and radiation. The air between the glazing and the wall rises as it is warmed and is distributed throughout the building. If the wall is vented at top and bottom, the heated air can be circulated around the house either by natural convection currents or, if need be, by a fan.

FIGURE 33.3 A Trombe wall, visible behind the windows of this building, which is the visitor center at Zion National Park, Utah. (From http://www.nrel.gov/data/pix/searchpix. php?getrec=10020&display_type=verbose&search_reverse=1.)

At night, heat losses by radiation out through the window can be greatly reduced by use of screens or blinds across the window. Trombe walls have disadvantages of high construction costs and reducing the available space within the room.

Without doubt, one of the biggest problems with any solar building design is heat storage. When the sun is shining, it provides more heat than a building can comfortably use. On the other hand, some method has to be used to store heat for use later, on cloudy days and at night. Put in other terms, the problem for solar space heating is that when you need the heat, you often haven't got the sun; when you don't need much heat, there's plenty of sun. Unfortunately, technology for long-term storage of heat—in support of any kind of energy system—does not yet exist. Many solar homes use a second, backup energy system in combination with the solar design.

Heating is not the only possibility with a passive solar system. Natural air conditioning can also be obtained. With appropriate building design, the heat of the sun induces convection currents that draw cool air into the building and reduce the inside temperature. Islamic architecture has used this principle for centuries. Many Islamic buildings have a "chimney" that draws up hot air and brings air into the building past the north-facing surfaces that remain cool throughout the day. A modern version of this concept uses a Trombe wall to create the air movement. The hot air is vented to the outside, while the incoming air is cooled by north-facing, heavy masonry surfaces.

ACTIVE COLLECTION OF SOLAR ENERGY FOR HEAT AND HOT WATER

An *active solar energy* system does not rely simply on the design of the building itself, but rather employs solar collectors, usually mounted on the roof, which capture the energy of sunlight to heat either air or a liquid that is pumped to a heat storage unit. An active solar heating system, by its design, achieves three things: the collection or trapping of solar energy; transferring heat from the collector to a fluid that distributes heat where it can be used; and, usually, some means of storing the collected heat for use later when the sun is not shining. Most active solar systems require pumps or blowers and so cannot operate independently of other energy sources (usually electricity).

Scientists and inventors have always been intrigued with the possibility of finding ways to capture and concentrate the energy of sunlight. The first recorded application of solar energy is the story of Archimedes, who in 213 BCE allegedly used an array of mirrors to focus sunlight on the wooden ships of a Roman invasion fleet (Figure 33.4), setting them afire and forestalling, for a time, the Roman conquest of Syracuse. Likely the story is apocryphal, though it is certain that Archimedes did write a book called *On Burning Mirrors*. Many authors of modern times question whether the technology for constructing and focusing a large array of mirrors was available to Archimedes. However, some two thousand years later the French naturalist George Buffon used an array of 168 mirrors, each 15 centimeters on a side, to ignite a woodpile nearly sixty meters away, suggesting that it just might have been possible for Archimedes to set fire to a wooden ship if indeed he had the ability to make and focus large arrays of mirrors.

FIGURE 33.4   An artist's idea of Archimedes and his burning mirrors, setting fire to Roman ships. A painting by Giulio Parigi, in 1600. (From http://fr.wikipedia.org/wiki/Fichier:Archimedes-Mirror_by_Giulio_Parigi.jpg.)

Supposedly the use of highly focused rays of the sun provided a means of welding metals during the Middle Ages and the Renaissance. Items too big to be heated and welded on the blacksmith's forge posed a real problem, because there was nothing like the modern electric or gas welding torches in those days. Special mirrors were used to "burn" a weld in metals such as copper. Leonardo referred to these devices as "fire mirrors."

Early chemists used specially shaped mirrors as a source of heat in the laboratory. A sixteenth-century process for making perfume involved placing flowers in a flask of water and using concentrated solar heat to extract the fragrant essences from the flowers into the water. At roughly the same time, Leonardo da Vinci proposed a large mirror (a *really* large mirror—some six kilometers across!) that would generate boiling water for a dyeing factory. Like many of Leonardo's other inventions, this too was many years ahead of its time, and it is unlikely that technology was available at the time to build such a device. Toward the end of the eighteenth century, Antoine Lavoisier experimented with solar furnaces in a search for a powerful source of heat. With a 130-centimeters diameter lens, he was able to melt platinum, which means that he achieved temperatures of about 1780°C.

The first solar water heaters were simple metal tanks painted black and tilted toward the sun. In 1891, Clarence Kemp patented the "Climax": a combination of metal heating tanks to collect heat from sunlight with enclosing insulating boxes that helped retain the heat. The Climax became the first commercial solar water heater, sold in models from 120 to 2700-liter sizes. The competing "Day and Night," invented by William Bailey in 1909, offered as a selling point its ability, as its name implied, to supply hot water during both the day and night. This unit consisted of two parts, a solar heater and a water storage tank that was placed next to the kitchen stove to keep the water warm at night. Nowadays the most common means of collecting the sun's rays for heating or producing hot water is with a *flat-plate solar collector*. The concept dates to the pioneering work done during the middle of the eighteenth century by the Swiss scientist Nicolas de

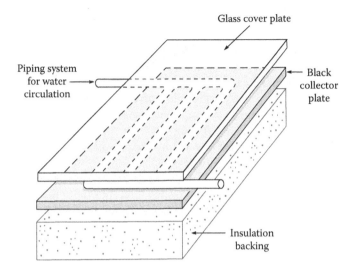

FIGURE 33.5   The flat-plate solar collector is a simple, easy-to-build device for heating water.

Saussure.* His design, which still forms the fundamental basis of the modern flat-plate collector, consisted of glass plates spaced above a blackened surface enclosed in an insulated box. This early design achieved temperatures as high as 150°C.

There are many different designs of flat-plate collectors, but all are similar in fundamentals (Figure 33.5). Any modern flat-plate collector has four main parts: an absorbing surface that faces outward to receive the sunlight, a system of pipes that carry the heated fluid, a sheet of glass on the front, and insulation on the sides and back.

The absorbing surface is a flat metal plate, painted black to absorb heat. The collector plate absorbs solar energy, becoming warm in the process. The absorbed heat in turn warms a circulating fluid. Good thermal contact between plate and fluid is important. Pipes run into the box and are attached to the black plate so that heat travels by conduction through the metal and into the heat-transfer fluid. In a conventional flat-plate collector used in homes, water is by far the best choice as a heat-transfer fluid because of its high heat capacity, low cost, and relative ease of use. In active solar systems used in climates where freezing temperatures can be encountered, antifreeze can be added to the water. The glass cover functions like the glass in a greenhouse. It allows visible light from the sun to enter the collector, but traps the infrared heat emissions from the metal plate, raising the efficiency of the collector. The glass also reduces loss of heat by convection currents from the collector plate. Heat losses also occur by conduction through any material in contact with the collector. Because of this, good insulation around the sides and back is important. Normally, several flat-plate collectors will be connected in a series. The heated water is pumped either to a storage tank or to a traditional hot water heating system using radiators or baseboard units.

---

* Nicholas de Saussure (1767–1845), like so many of the scientists of that era, was active in many different fields. He studied the chemistry of fermentation, and was one of the first to determine the chemical composition of ethanol. He also discovered that plants need more than water for their growth—they also need carbon dioxide. This was an early indication of the nature of photosynthesis.

A flat-plate collector's efficiency depends on a number of factors, foremost the amount of solar energy hitting it, which in turn is affected by the collector's orientation and tilt. Collectors should be mounted on a roof with a southern exposure and, if possible, at an optimum angle determined by the latitude of the building. A collector facing directly into the sun will receive the most insolation (i.e., the amount of solar energy received on a specified area in a given amount of time). Collectors can be motorized, so that during the course of the day a control system can position the collector so that it is always facing into the sun.

Incoming solar energy is also affected by the prevailing weather—whether the day is clear and sunny, hazy, or with heavy cloud cover—and by the season of the year, which determines the height of the sun above the horizon. The sun's height affects the strength of the solar radiation when it reaches the collector surface. The design and materials of construction determine the fraction of incoming solar energy that will be transmitted by the glass, and the fraction of that transmitted energy that will be absorbed by the collector plate.

The solar collector must be plumbed into the building's water supply or heating system. In cold climates, where the water in the collector contains antifreeze, it must be kept separate from the domestic water supply, because antifreeze is poisonous. As the morning sun begins to warm the collector, the fluid will eventually reach a point at which it is hotter than the water in a "preheat" water tank. Then the circulating pump is switched on. The hot fluid from the collector is sent through a heat exchanger to heat the water in the preheat tank. This heated water can be used directly as a source of heat, or stored in a hot water tank to provide a source of hot water. A schematic diagram of a simple installation is given in Figure 33.6.

In many applications, heat from the solar collector is transferred to an insulated storage medium such as a water tank and distributed through the building as needed.

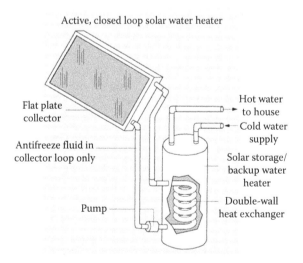

FIGURE 33.6   A schematic diagram for a closed-loop solar water heater. (From http://www. energysavers.gov/your_home/water_heating/index.cfm/mytopic=12850.)

A well-insulated six cubic meter tank can retain its heat for four to six days. If a flat-plate collector system is used to heat a large building, problems may arise because the collector may not generate a high enough temperature throughout the entire building, due to heat losses from the hot water as it circulates throughout the building. If higher temperatures are required than can be achieved by flat-plate collectors, the energy in the sunlight could be concentrated by mirrors or lenses.

In most homes, the hot water heater is one of the most costly appliances, in terms of the electricity or fuel (usually natural gas) needed to keep it operating. The energy (and hence cost) savings that might be achieved with a solar collector can vary considerably, depending on, for example, the local climate and how careful or wasteful the household is in its energy use. Rough estimates for a home in the temperate regions of the Northern Hemisphere occupied by a family that is careful in its use of energy are that a solar collector might achieve a 40% reduction in energy costs needed for hot water. In this situation, a typical domestic requirement might be one square meter of collector area, plus about forty liters of water storage for each person. A supplementary storage tank would need to be of at least one to two hundred liters capacity.

Active solar energy is growing in use around the world. Israel probably leads the world in the practical application of solar energy. Over a million solar hot water heaters have been sold and installed in the past fifty years, providing the energy for generating more than 80% of the hot water used in Israel. At least two-thirds of homes obtain hot water from solar collectors. Israeli law now requires all new residential buildings smaller than ten stories to use solar water heating. Australia has also made a serious commitment to solar energy. In some parts of Australia, solar hot water heaters are required by law, as are solar-heated distillation units for producing drinkable water from salt water. Japan has over three million small domestic solar water heating installations, with roof-mounted storage tanks and inclined flat-plate collectors. One-sixth of all houses in Greece are estimated to use solar systems. Further south, in Cyprus, more than 90% of homes now feature solar water heaters. Even in Great Britain, a nation stereotyped for its miserable weather, there are more than 40,000 solar collectors.

Active solar energy systems share some of the same problems or limitations already discussed for passive systems. Because of the variable and intermittent nature of solar energy, it is generally prudent to have a backup energy source, as well as provisions for storing some of the accumulated heat. Successful long-term storage of heat is a scientific and engineering problem that has not yet been solved. Fortunately, for many domestic hot water systems or building heat systems, it is usually only necessary for the heat energy to be stored for a few hours.

Unfortunately, commercial solar water heating systems are expensive to install in many countries, particularly when they are added to an existing building as a retrofit, and not incorporated into the design of a new building under construction. Costs vary, depending on complexity of design (i.e., whether it is to be used only to provide hot water, or for hot water and heat), materials, and local labor costs. An installation for a typical home would likely cost several thousand dollars at least. Thus payback time—the time needed to save an amount of money using present sources of energy equal to the installation costs plus the operating costs—becomes of concern. Many factors contribute to the overall payback time, including the local climate, and the local cost of the energy source (e.g., natural gas or electricity) being displaced by solar energy. The entire installation cost of a new active solar energy system would have to be paid "up front." For most of us, doing so requires borrowing money at commercial interest rates. This cost is also a factor in determining payback time. As a rule of thumb, a payback time of less than ten years is desirable; anything longer indicates that the installation of the solar energy system is prohibitively expensive. Obviously regions with sunny climates, such as many of the Mediterranean countries, are more favorable for installation of such systems.

INDIRECT CONVERSION OF SOLAR ENERGY TO ELECTRICITY

The other major application of solar energy is its use as the primary energy source for generating electricity. Solar energy provides heat, which could be used to raise a head of steam for a steam turbine/generator set. In this approach, the sun's energy is converted, via steam, into electricity. Since the ENERGY in sunlight is not converted directly into electricity, this approach is sometimes referred to as *indirect conversion*, also sometimes called solar thermal power stations. All of the principles are the same as those for a fossil-fuel or nuclear plant. A source of heat produces steam that drives a turbine. The only significant difference is the source of heat: the combustion of a fuel, the fissioning of uranium, or capturing the heat in sunlight.

To provide steam to a turbine requires significantly higher temperatures in the fluid than can be achieved using the simple flat-plate solar collector. To realize these high temperatures—certainly above 100°C and preferably well above that—from the heat in solar energy requires the use of devices that not only collect the solar energy but also concentrate it. These devices are more complex and sophisticated than those used only for building heat or hot water generation. Further, the maximum amount of energy can be collected if some provision exists to allow the collectors to track the position of the sun in the sky. A tracking system also adds to the complexity of the design.

A principal strategy for indirect solar uses a linear focusing collector. This kind of collector is more commonly known as the parabolic trough (Figure 33.7). It relies

FIGURE 33.7 The parabolic trough solar collector provides much higher temperatures than the flat-plate collector, and can be used, for example, to provide steam to operate a turbine. (Reprinted from first edition, permission from Photo Researchers Inc., New York, New York.)

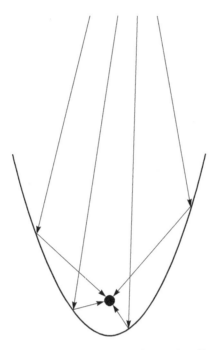

FIGURE 33.8 The principle on which the parabolic trough collector works is that rays of sunlight striking *anywhere* inside the parabola will all be reflected to the same spot.

on a fundamental principle of the parabola: Regardless of what point on the surface of the parabola is struck by a ray of the sun, the ray is *always* reflected to one particular spot, the "focus" of the parabola (Figure 33.8). As the name implies, the parabolic trough is constructed from a reflecting material having a parabolic shape. A pipe, usually painted black to help absorb heat, runs the length of the trough at the exact focus of the parabolic shape, thereby collecting all of the sun's rays that strike anywhere in the interior of the trough. The pipe is filled with a heat-transfer fluid, usually water or an organic material. The heated fluid, which could be at temperatures up to 500°C, is pumped along the pipe. The more sophisticated designs feature sensors that determine the position of the sun and relay this information to motors that keep the collectors steadily tracking the sun.

The collection area of a single parabolic trough unit may be no more than a few square meters. To generate sufficient high-temperature steam to operate an electricity-generating plant requires hundreds or even thousands of the individual trough modules. Because each module has its own heat-collection pipe, this system is sometimes referred to as a distributed collector system.

A system of distributed parabolic trough collectors lends itself easily to a wide variety of applications that require relatively low amounts of ENERGY or POWER. This system can be used to pump water for irrigation, to produce steam for industrial purposes, to generate electricity in small- and medium-sized installations, and to supply heat for residential use. The large number of individual modules that might be needed limits the distributed collector approach mainly to smaller installations. On the other

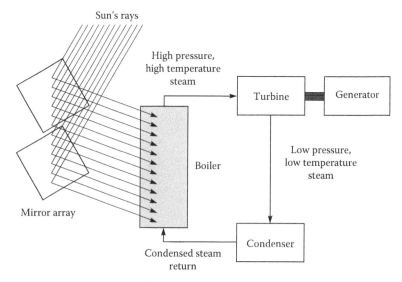

FIGURE 33.9 Mirrors (heliostats) can be used to reflect the sun's rays onto a tank of water, heating it enough to generate steam for a turbine.

hand, parabolic trough modules are well suited to "low-tech" assembly-line construction. They can be produced in a factory and shipped to the installation site, where they could be connected together, as many as needed to meet the energy requirement. This approach may offer much simpler construction, and possibly lower construction costs, than a custom-designed electricity-generating station erected on site.

The alternative to the distributed collector system uses a system of mirrors to reflect solar energy onto a single, central collector that provides enough heat to raise steam. The steam drives a steam turbine just like any other electric power plant. A conceptual sketch is given in Figure 33.9.

This system, sometimes known as the solar power tower, uses a heat-exchange fluid, usually water, held in a receiver composed of a series of blackened pipes, or in a tank, located on the top of a tall tower. The fluid is heated by solar energy reflected from thousands of individual mirrors, called heliostats (Figure 33.10). The heat-exchange fluid in the central receiver absorbs the heat from the focused solar radiation. A 100 MW solar electric facility would require about 24,000 heliostats spread out over an area of nearly five square kilometers if it were located in sunny region such as a desert. This central receiver approach is considered better suited for large-scale applications than is the distributed collector strategy.

Because the collector receives the reflected radiation from a large array of mirrors, each unit of surface area on the collector receives many times the POWER it would receive if it simply faced the sun directly. This provides a substantial concentration of ENERGY and POWER relative to that obtained by spreading out a larger number of individual parabolic trough collector modules. This concentration of the total ENERGY at the collector means that it operates at much higher temperatures, some 400°C–500°C. The result is a solar-to-electric efficiency of nearly 40%. This efficiency exceeds all but the most modern of fossil-fuel plants.

FIGURE 33.10 An entire field of heliostats is required to capture enough solar energy to operate a small electricity-generating plant. (Reprinted from first edition, permission from Photo Researchers Inc., New York, New York.)

The first large central receiver system was built in Odeillo, in southern France. Another was added at nearby Targasonne. They were 1 and 2 MW plants, respectively. These plants are very small compared with fossil or nuclear installations, where the generating capacity is in the hundreds or even thousands of megawatts. The plant at Odeillo used 63 tracking heliostats covering an area of about three thousand square meters. In the United States, the Solar One plant was built at Barstow, in California the Mojave Desert. This location averages three hundred cloudless days per year. Solar One used 1818 heliostats to reflect the solar radiation onto the boiler at the top of the "power tower." The combined reflecting area of these mirrors was about seven hectares. In the tower, a heat-transfer fluid was heated to 500°C and carried the heat to the steam generators of the power plant. The peak output of Solar One was 10 MW, again very small in comparison with standard fossil or nuclear technology. Solar One cost $142 million to build; the heliostat field accounted for almost half of the total cost. Solar One was closed in 1988 after operating successfully for six years. A disadvantage, aside from economics, was that steam pressure dropped quickly when clouds blocked the sun. The loss of steam pressure would force the plant to shut down, and it took some time to restart once sunlight returned.

Solar power towers can only be feasible in regions of abundant sunny days, such as deserts. Although the concept has been shown to be a technical success, the cost of electricity produced from such plants seems much too high to be competitive with more conventional sources and has to fall before the solar power tower concept can be economical. A proposed solution to these problems is to place a solar power tower next to an existing fossil fuel plant. The conventional plants would use solar heat where possible to operate the normal steam generators, but on days when too little solar power was available, fossil fuel could be used as the backup. An alternative would be to give up on the idea of using large, centrally located electricity-generating plants. Instead one might try to use the solar energy in small, widely distributed, plants that might have greater chance of being economical immediately.

Despite these potential problems, solar thermal power plants using steam turbines offer several advantages compared to conventional fossil or nuclear plants. The solar energy itself is free. The only potential pollution problem comes from disposal of the waste heat, though solar energy is no different from other sources of energy in this respect. The most troublesome feature of solar electricity plants is the fluctuation in available solar energy resulting from clouds and nighttime. As with any sort of solar system, an adequate means of energy storage is needed.

PHOTOVOLTAICS: DIRECT CONVERSION OF SOLAR ENERGY TO ELECTRICITY

Every method of generating electricity on a large scale discussed so far uses the kinetic energy of a fluid—water in hydro plants and steam in all others—to drive a turbine. The approach derives from Michael Faraday's original experiment showing that a moving magnetic field generates an electric current, followed by the successful engineering development by Zenobé Gramme and his contemporaries. Fossil, nuclear, and indirect solar plants all use large and technologically complicated devices basically to boil water. But why not just throw away all that hardware—boiler, turbine, generator, condenser, cooling tower—and convert the ENERGY of sunlight directly into electricity? This approach is called *direct solar electricity generation*, or *photovoltaics*.

The photovoltaic effect occurs when light strikes two dissimilar materials and produces electrical potential energy. As we will see below, certain combinations of materials respond to the ENERGY in visible sunlight. In combination, they make a photovoltaic cell, or solar cell. Such cells can be thought of as being "solar batteries," but unlike primary or secondary cells that rely on chemical reactions to produce the electrical potential, a photovoltaic cell will not "go dead," so long as sunlight continues to strike it.

The conversion of light directly to electricity, by the photovoltaic effect, was discovered by the Henri Becquerel's father, Edmond, in 1839.* An explanation of the physical basis of the photoelectric effect—how the wavelengths of light determine their ability to knock electrons out of the atoms—was developed by Albert Einstein. This research, not the theory of relativity, garnered Einstein his Nobel Prize in physics. Development of the first practical photovoltaic cell occurred in 1954, at Bell Laboratories in the United States.

The key material in most photovoltaic cells is silicon. Silicon has four electrons in its outermost "shell" of electrons. In solid silicon, the atoms are arranged in a regular, three-dimensional lattice, in which each silicon atom is surrounded by four

---

* Edmond Becquerel (1820–1891) was the second of three generations of brilliant scientists in the Becquerel family: Antoine the father, Edmond the son, and Henri the grandson. Edmond Becquerel's productive career as a physicist involved studies in, among other things, magnetism, electrolysis, luminescence, and optics. He was a very early pioneer of photography, working in that field as early as 1840. Edmond discovered the photoelectric effect at age 19, while working in his father's laboratory.

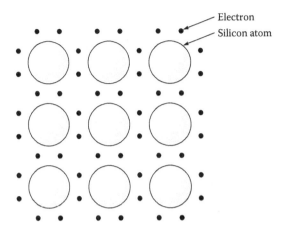

FIGURE 33.11 In pure silicon, electrons are strongly held in chemical bonds between pairs of silicon atoms.

neighboring silicon atoms. The bond between any pair of silicon atoms is formed by each of the atoms contributing one of its electrons to form a pair (Figure 33.11). The electrons are tightly held in these electron-pair bonds. For any substance to conduct an electric current, some of its electrons must be loosely held, to allow them to move when the material is subjected to an electrical potential. Since electrons are tightly held in silicon, under normal circumstances silicon is not a good conductor of electricity.

For any material that acts as an insulator (or, from the other perspective, a non-conductor), electrons are tightly bound to the constituent atoms of the material and therefore cannot move. In contrast, conductors have large numbers of electrons that are free to move under the influence of relatively small potential differences. Stephen Gray developed the broad classifications of materials as insulators or conductors more than two hundred years ago (Chapter 12). We now recognize a third category, materials with relatively few electrons capable of moving under the influence of an electrical potential, and which, therefore, sustain only small currents for a given potential difference. They conduct electricity better than does an insulator, but yet not so good as a true conductor, in a sense halfway between a conductor and an insulator. Such materials are called *semiconductors*.

Some of the atoms in a piece of silicon can be replaced with, for example, arsenic. Adding a small quantity—equivalent to about one atom in a million—of a second element to pure silicon is called *doping*. Arsenic has five electrons in its outer shell. If relatively few of the silicon atoms are replaced, the material retains the crystalline nature of silicon, but now has some "extra" electrons, thanks to the inclusion of some atoms of arsenic (Figure 33.12). If sufficient energy is provided, these "extra" electrons are able to move under the influence of a potential difference. In other words, we have created a semiconductor by doping silicon with arsenic. Because this specific doping process has provided the material with "extra" electrons, and electrons have a negative charge, we call this material an *n-type semiconductor*, where the "*n*" reminds us that the semiconducting properties are due to these extra electrons.

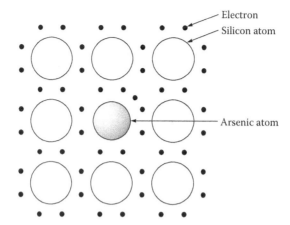

FIGURE 33.12 When silicon is doped with an element such as arsenic, "extra" electrons are contributed to the structure and are not held tightly between pairs of atoms.

In another piece of silicon, some of the atoms could be replaced with, for example, gallium (which has only three electrons in its outer shell). This material retains the crystalline nature of silicon, but which, in this case, has some "missing" electrons, because each atom of gallium contributes three, rather than four, electrons. An electron being "missing" is equivalent to there being a "hole" among the electrons in the material (Figure 33.13).

Since an electron has a negative charge, a "hole" conceptually would have a positive charge. This material, gallium-doped silicon, is also a semiconductor. In this case, the shortage of electrons means that it is the "holes," not the "extra" electrons, which migrate when exposed to an electrical potential. Since a "hole" can be

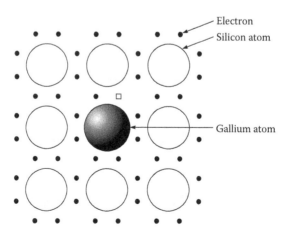

FIGURE 33.13 When silicon is doped with an element such as gallium, there are not enough electrons to provide two-electron bonds between pairs of atoms, creating electron "holes" in the structure.

considered equivalent to a positive charge we call a material such as gallium-doped silicon a *p*-type semiconductor ("*p*" for positive, since if an electron is negative, a "hole" would be positive).

Silicon is the second most common element in the Earth's crust. We need never worry about running out of silicon. The relatively high cost of solar cells is not due to scarcity of the necessary raw material, and reserves tend to be listed as unlimited. Rather, the cost comes from the requirement for purity, and in the difficulty of fabrication. The first challenge is to produce *extremely* pure silicon. Fewer than one impurity atom in a million can be allowed. This requirement demands rigorous purification and vigilant steps to ensure that the silicon stays pure. The second factor comes from the difficulties in melting and subsequently re-solidifying the extremely pure silicon very slowly and carefully to produce an almost perfect crystalline lattice. This requirement for very high-quality single silicon crystals poses great difficulties in the manufacturing cost and drives up the price.

The two different types of semiconductors can be put together, forming an *n–p junction.** When light strikes the photovoltaic material, the photons (i.e., the elementary particles of light) provide some extra energy to some of the electrons in the material. This makes it easier for those "excited" electrons to move within the material. An electron excited by light and moving away from the region it originally occupied creates a hole that is left behind. Some of the "extra" electrons in the *n*-type semiconductor could cross the junction, because they would be attracted to the positive holes. Similarly, some of the holes might migrate across the junction to the *n*-type material. We might expect that if we connected an *n*-type and a *p*-type semiconductor, the "extra" electrons ought to flow from the *n*-type into the "holes" in the *p*-type. The separation of the charges creates a difference in electrical potential energy (voltage) across the junction. However, electrons do not flow easily in a semiconductor (as implied by its name) as compared to a regular electrical conductor like copper. The energy needed to excite electrons and give them mobility turns out to be about the amount of energy available in sunlight (Figure 33.14). Thus by joining an *n*-type and a *p*-type semiconductor, connecting it into an electrical circuit, and allowing sunlight to fall on it, we can achieve direct conversion of solar energy (sunlight) to electricity. Electrons liberated or excited by the energy in sunlight will

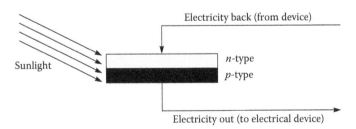

FIGURE 33.14 An *n–p* junction semiconductor provides the key for direct production of electricity from the ENERGY in sunlight—the solar cell.

* Putting two pieces together this way should work fine, but most commercial photovoltaic cells are made by taking a piece of *p*-type material and making an *n*-type dopant diffuse into it.

move under the influence of the potential difference to create a current. This current can be brought outside the cell by connecting the n-type semiconductor to the p-type with a wire (i.e., a good conductor) and be used to do some WORK. As long as the cell is exposed to light, the current will continue to flow, with the only energy source being the solar radiation.

Single photovoltaic cells made of crystalline silicon with an n–p junction have a theoretical maximum efficiency of 33.7%. Steady improvements in photovoltaic technology have brought practical cells close to this theoretical limit. The POWER output of a solar cell varies with the intensity of light. If the light intensity is cut in half, then the current will also drop by half. The voltage will decrease by only a few percent. In bright sunlight, a cell 10 cm² will give an output of about one-half volt and three amps, that is, about 1.5 watts of POWER. Voltage also depends on temperature and increases slightly (by about 0.5% per degree) above 25°C. With current technology, it would take an area of fifteen square kilometers covered with photovoltaic cells to replace a 2000 MW power station. The first significant use of solar cells was to provide electricity for NASA spacecraft. There, cost was irrelevant. The intensity of solar radiation in outer space is so high that the low efficiency of the cells was not a major limitation of their use.

Modules of photovoltaic cells are connected to form panels that in turn are assembled into groups called *arrays*. The current and voltage produced by the array depends on how the modules and panels are wired together. The array is connected to a secondary-cell battery that has two functions. First, the battery stores electricity, so that electricity will be available when the sun is not shining. Second, the battery and its control system are set up to send electricity into the external circuit at an even rate. This is vital, because the current from the array could fluctuate from one moment to the next as clouds pass overhead. With present technology, an array with an area less than a square meter could provide some tens of watts of power, enough for several small electric lights, or for operating portable radios. With about twice that area of cells, sufficient electrical ENERGY is produced to drive a small electric motor, as, for example, in a water pump.

A typical photovoltaic installation for a home would require three main components: the array of solar cell modules; a set of batteries to provide both a steady output of electricity and a means of energy storage; and, since most common household devices use alternating, rather than direct current, a device to transform the direct current output of the batteries into alternating current. Assuming a POWER output for a single cell of about five kilowatts per square meter, the array of modules would need to cover 60–80 square meters to generate enough electricity for a single family home. Assuming a roof-mounted array, an area some 6×9 meters to 6×14 meters would be needed. Normally the array would be mounted to allow for air circulation underneath. Such an installation also has to take into consideration the amount of sunlight available in the specific geographic location, the availability of electricity from a utility (as, e.g., for a backup), and a realistic estimate of the amount of electricity required by the family.

The alternative to building-by-building installations of solar panels is to set up centralized photovoltaic generating stations. Many countries now have this technology in place. Installations exist in India, China, Germany, Italy, Ukraine, and the

United States. A few of these are in the 100–200 MW range of capacity; many in the 50–100 MW range. More are under construction or in the planning stage.

Though steady progress continues in both single-building applications, as well as small-scale central generating plants, the major developments in photovoltaics have mainly focused on smaller applications in remarkably large markets. The first major consumer application was in photography, as exposure meters in cameras. When first introduced, they were sometimes called "electric eye" cameras. Solar calculators, wristwatches, and clocks are now familiar to most people. These solar cells may be small in size individually, but collectively they account for more than a quarter of the worldwide annual output of photovoltaics. Well over a hundred million of these devices have already been sold worldwide, with sales increasing steadily.

The largest market for photovoltaic cells is currently in providing on-site generation of electricity to operate equipment or appliances in communities far from utility grids. These applications include remote communications systems, such as telephone or microwave relay stations; water pumps for drinking water, livestock, or irrigation; navigational aids such as warning buoys or remote lighthouses; and roadside call boxes. Before photovoltaics were available, these devices needed to use either secondary batteries, which needed to be recharged, or small generators operated by diesel engines (the engines needing occasional refueling).

In many parts of the world, solar cell arrays provide energy for street lighting. During daylight hours, the photovoltaic electricity charges a battery. After dark the battery operates the lighting until dawn. For example, highway traffic lights in rural Alaska, far from power lines, are supplied in this fashion.

Providing electrical energy for villages in less-developed countries represents another very beneficial application of photovoltaics (Figure 33.15). More than two million villages worldwide lack electricity for their water supply, refrigeration, lighting, or other basic needs of daily life. In many cases, the cost of extending utility grids to these remote locations would be prohibitive. In the most impoverished of regions, the utility grids may not even exist in the first place. Such on-site photovoltaic systems could meet much of the need for electricity for basic human activities at lower cost than trying to expand utility grids or build new, centralized generating facilities. This application represents a market with very large future potential and chances for growth. Further, the availability of photovoltaic refrigerators to keep vital medicines or vaccines intact could literally save lives in remote, impoverished regions.

Photovoltaics have also been tried as energy sources for transportation. In 1980, the *Solar Challenger*, a specially built aircraft weighing 85 kilograms, carried a pilot safely using only solar energy. The longest flight was 28 kilometers. Solar cars continue to be developed, with sufficient interest to hold long-distance solar car races in several countries.

Many developing countries have problems that arise from the very restricted spread of electricity distribution systems, the absence of rapid communications, and the acute shortage of skilled maintenance engineers. In most villages and many towns, electricity generated by central power stations is unavailable and, at the same time, poor distribution and lack of maintenance makes local generation using diesel-engine-operated generators both expensive and unreliable. Such countries, so long as they are provided with plentiful sunshine, could avoid building (and having to pay

FIGURE 33.15   A NASA-Glenn photovoltaic system installed in Tongaye, Upper Volta. An excellent way to provide electricity when there is no electric grid infrastructure. (From http://www.nasa.gov/centers/glenn/about/history/70s_energy.html; https://newbusiness.grc.nasa.gov/advanced-energy/glenn-photovoltaic-power-system/.)

for!) large central generating stations and transmission networks by installing solar arrays to supply electricity where and when it is needed for a town, a village, a factory, or household. Photovoltaic generators could not only produce electricity more cheaply when compared to the total investment needed in new generating stations and transmission grids, but are also more reliable in terms of maintenance and upkeep.

Photovoltaic systems do not use water either for cooling or for steam generation; so they are well suited to remote or arid regions. Photovoltaics can operate on any scale, from small hand-held consumer devices to full-scale electricity plants generating at least 7 MW (Figure 33.16). Indirect solar is not applicable for small devices, but it too has been tested at the 10 MW scale. Neglecting the issues of manufacturing inexpensive photovoltaic cells, there is really only one fundamental problem with solar energy—the sun doesn't shine at night.

One answer to the storage question involves use of rechargeable batteries. Another approach involves a hybrid solar energy system, in which a low-cost fossil fuel system is added to a photovoltaic system to compensate for variations in sunlight. A combined solar–natural gas hybrid would be one of the most environmentally benign ways of using fossil fuels to generate electricity. The method called *pumped storage* uses some of the photovoltaic electricity generated during the daytime to pump water to large ponds on the top of a mountain. At night, the water is then allowed to flow back downhill and generate hydroelectricity.

We perceive that the sun does not shine at night because the part of the planet we happen to occupy rotates out of the path of the sunlight. Of course the sun does

FIGURE 33.16   A photovoltaic array of the NASA Kennedy Space Center in Florida. (From http://commons.wikimedia.org/wiki/File:Solar_power_system_at_Kennedy_Space_Center.jpg.)

not really "set" or "go out." Since sunlight is always available in outer space, we can get around the problem of the sun not shining at night by moving the photovoltaic generating station into space. The conceptual Satellite Solar Power System would consist of an orbiting satellite with huge thin wings, some twelve kilometers across, covered with photovoltaic cells. Such a satellite would generate 5000 MW of power, which then needs to be transmitted to Earth. This could be done handily if the satellite were parked in a synchronous orbit so that it could focus a beam of microwaves nearly continuously onto a receiver at a fixed location on Earth. Since one of the principal applications of microwaves is to cook things, concern has been expressed about the possibility of the beam wandering off target and microwaving the local population, plants, and animals—finally realizing science fiction's "death ray from outer space." Such a satellite-mounted photovoltaic station could cost about fifty billion dollars—roughly ten times greater than a coal-fired plant of similar capacity.

The cost of solar electricity depends on three principal factors: The first is efficiency—the fraction of ENERGY in the sunlight striking the cells that is converted to electrical ENERGY. The higher the efficiency, the fewer the cells required to produce a given amount of electricity, or the more electricity can be generated with a given number of cells. The second factor is the cost of producing the cells. The third is the overall expenditure required to install, operate, and maintain a solar energy facility. The situation is complicated by the fact that the more efficient cells are also the more expensive. Thus it may prove better, for overall cost considerations, to use a large number of cheap, but low-efficiency cells.

The cost of producing electricity dropped from three dollars per kilowatt-hour in 1974 to thirty cents per kilowatt-hour in 1990 and has dropped to about sixteen cents currently. Clearly, this represents a tremendous improvement. But, to be competitive with electricity generated in conventional fossil-fuel plants, the cost of electricity delivered to consumers must now be around six to eight cents per kilowatt-hour. Some forecasts suggest that photovoltaic electricity may drop to ten cents per kilowatt-hour within the next decade, bringing it close to being competitive with conventional sources.

Not only is the cost of producing electricity dropping, but so is the price of the cells. This price is generally expressed as dollars per peak watt, where a *peak watt* is defined as one watt of power generated when the solar radiation power is 1000 W/m^2. Forty years ago the cost of a solar cell was $1000 per peak watt. The major impetus to develop photovoltaic cells came with the growth of the American space program in the 1960s, when a lightweight energy source was needed for satellites. Prices of solar cells plummeted in the 1970s, from $300 per peak watt in 1973 down to $10 in 1983. By 1986, the cost was down to $5 per peak watt, in the $2–$3 range by the early 1990s, and hit $1 in 2009. Unfortunately, the cost of photovoltaic devices to replace the kerosene lamps and stoves still lies beyond the resources of the very poor people in developing countries around the world. Private foundations or public programs, such as via the United Nations, could possibly assist with funds for installing these photovoltaic devices.

ADVANTAGES AND DISADVANTAGES OF SOLAR AS AN ENERGY SOURCE

Solar energy is available anywhere on Earth and can be collected easily with portable devices. This advantage minimizes the problems of transporting fuel or transmitting electricity from point of production to point of use. Solar energy is nonpolluting and clean. It produces no air or water pollution. There is no problem of disposing of hazardous wastes. No mining or drilling is needed (except for the mining of a suitable source of silicon in the first place). The most important advantage of solar energy is that it is virtually inexhaustible. The energy source—the sun—cannot be owned and monopolized, as fossil fuels have become because of the unequal geographic distribution of known reserves.

Of course, solar energy also has its drawbacks. Sunlight, being a dilute energy source, requires large amounts of land for certain applications, such as central station power plants. This may pose a problem in some areas, such as where wildlife protection is a concern, or in urban areas with very high costs of acquiring land. A second disadvantage is the inconsistency of direct sunlight as an energy source. Not only night, but clouds and seasonal changes affect the rate at which direct sunlight can be collected. Erratic sunlight creates the need for methods of storing energy until it is needed. Backup systems using other energy sources would be needed to supply solar homes or offices during prolonged periods of bad weather when stored energy might be depleted. A possible problem with the large-scale use of indirect solar plants is the generation of thermal pollution, just as is true with any technology using steam turbines. Indirect solar plants also require cooling water, which may be costly or scarce in desert areas.

Ultimately, cost is the key. When the overall cost has been brought down to six to eight cents per kilowatt-hour, solar energy has the potential to come into widespread use in the generation of electricity. Another potential barrier resides in the legal system. Is access to the sun a legal right? Many places have not adequately dealt with the issue of the legal right to access to sunlight. Without such legal protection, the potential buyer of a solar collector system must think twice about committing money if a cranky neighbor's whims, the growth of trees, or the construction of a tall building nearby blocks access to sunlight. To reach solar collectors, sunlight often has to pass through air space not owned or controlled by the solar collector owner. It would seem rather prudent on the part of someone intending to install solar equipment to ensure that the necessary intervening airspace would not be subsequently blocked by the actions of other people.

FURTHER READINGS

Armstrong, F.; Blundell, K. *Energy Beyond Oil*. Oxford University Press: New York, 2007. An edited collection of chapters contributed by various authors. Chapter 8 deals with solar energy.

Berger, J. J. *Charging Ahead*. University of California: Berkeley, CA, 1997. A major section of this book, some dozen chapters, is devoted to various aspects of solar energy. Well-written and intended for readers with little previous scientific background; includes interviews with, and anecdotes of, some of the key people in the development of a solar industry, up to the end of the previous century.

Boyle, G. *Renewable Energy*. Oxford University Press: Oxford, U.K., 2004. This book is also an edited collection of chapters contributed by different authors. Chapter 2 discusses thermal applications of solar energy, and Chapter 3, photovoltaics.

Bryce, R. *Power Hungry*. Public Affairs: New York, 2010. The author contends that the energy future lies in expanded applications of natural gas and nuclear energy. He also argues that wind and solar energy will not be adequate.

Coley, D. *Energy and Climate Change*. Wiley: Chichester, U.K., 2008. A textbook that includes good discussions of solar thermal energy and photovoltaics, Chapters 20 and 21 respectively.

Madrigal, A. *Powering the Dream*. Da Capo Press: Cambridge, MA, 2011. An interesting book that points out how some "green" or renewable energy technologies are surprisingly old, such as the use of solar hot water heaters a century ago, before the First World War. The key discussion is how some ideas manage to survive into the marketplace, and others don't make it.

Schwartz, A.T.; Bunce, D.M.; Silberman, R.G.; Stanitski, C.L.; Stratton, W.J.; Zipp, A.P. *Chemistry in Context*. WCB/McGraw-Hill: New York, 1997, Chapter 9. A well-illustrated and useful introductory text on chemistry. This chapter has a good section on photovoltaics.

34 Nuclear Fusion
Bringing the Sun to Earth

Light and warmth provided by the sun are obvious aspects of our daily existence. Early humans, and likely our prehuman ancestors, must have speculated about the nature of the brilliant, glowing ball that appears to signal the start of day, traverses the sky, and then disappears again. Many early societies must also have speculated about the changing of the seasons—why the sun appears lower in the sky, and for a shorter period of time, each day until a turning point is reached and the sun begins gradually to regain its former vigor, but only until the next turning point, when the cycle starts over again. The apparent movement of the sun across the sky is an illusion created by the rotation of the Earth; that part of the planet we happen to be standing on gradually rotates out of the path of the sun's rays. Progression of the seasons derives from the revolution of the Earth around the sun, and the Earth's axis of rotation being tilted somewhat relative to the plane of Earth's orbit. During the course of the year, our part of the planet receives greater or lesser amounts of sunlight—so is warmed to greater or lesser degrees—and the tilt of Earth's axis relative to the plane of its orbit creates the appearance of the sun rising higher or lower in the sky. These explanations of the sun's behavior have been known for centuries and derive from astronomical observations and the geometry of Earth's orbit. It took far longer to develop the understanding of the source of the sun's energy.

WHERE THE SUN GETS ITS ENERGY

It must have seemed to early humans that something in the sky is on fire. On Earth, fire provides heat and light; it was not illogical to presume that this "thing" in the sky that gives heat and light to our whole world must itself be some kind of fire. For millennia the nature of the sun, the source of its "fire," and its curious course across the sky provided the material for colorful myths and religious beliefs. With the dawn of modern science in the seventeenth century, scientists too began to speculate on what this energy source must be. They realized that the sun produces a prodigious quantity of energy, warming and lighting a comparative speck of dirt orbiting some 150 million kilometers away, decade after decade, century upon century, millennium to millennium. Nothing in the scientific experience of phenomena on Earth seemed capable of accounting for this unimaginably vast outpouring of energy. The nineteenth century was fueled by coal. It should not be surprising that one of the more inspired guesses from that era was that the sun must be some gigantic lump of coal, blazing away up in the sky. (Lest we be too quick or too harsh in criticizing those early scientists with the hindsight of our much greater knowledge in the twenty-first century, it's good to remember that our knowledge of nuclear processes, dating from the discovery of Becquerel, is little over a hundred years old, and even into the early years of the twentieth century some eminent scientists adamantly refused to accept that matter is made of atoms.)

The first step toward understanding came from a determination of the sun's composition. In the 1850s, Robert Bunsen* and Gustav Kirchhoff[†] invented a device, the spectroscope (Figure 34.1), which could determine the identities of the elements based on the kinds of light they emit when heated to incandescence. Bunsen has become much better known for his invention of the handy gas-fired heater—the Bunsen burner—used by generations of chemists and chemistry students. Kirchhoff

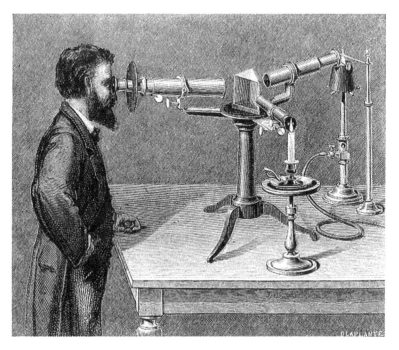

FIGURE 34.1 Gustav Kirchhoff (1824–1887) working with his spectroscope. (From http://commons.wikimedia.org/wiki/File:Kirchhoffs_improved_spectroscope.jpg.)

---

* Robert Bunsen (1811–1899) is best known today for the Bunsen burner, a small, gas-fired heat source that was once ubiquitous in chemistry laboratories the world over. As a matter of personal principle, Bunsen never took out patents on any of his inventions. Bunsen's scientific contributions were much more extensive and profound than the eponymous burner. He discovered two elements—rubidium and cesium—and developed a battery that used a relatively inexpensive carbon electrode in place of the platinum electrodes that were extensively used in his era. Bunsen nearly killed himself experimenting with arsenic compounds and lost the sight of one eye when one of these compounds blew up. Bunsen was a gifted teacher, devoted to the advancement of his students.

[†] Gustav Kirchhoff (1824–1887) is perhaps best known now for his laws pertaining to electric circuits, which play a central role in electrical engineering. Kirchhoff developed these laws during his time as a graduate student at the University of Köningsberg, first as a paper for a seminar and then expanded into his dissertation. His fruitful collaboration with Bunsen led to major advances in spectroscopy and optics.

provided the fundamental mathematical analysis of the characteristic light emission from glowing objects. The Bunsen–Kirchhoff spectroscope was a major advance in analytical chemistry, and its use quickly led to the discovery of several previously unknown chemical elements. But since the spectroscope required only the light from a glowing object, and not the object itself, astronomers could use it to study the light from stars. There was no need to obtain a sample of the star's material and bring it back to Earth for analysis—the star's light would suffice. During the solar eclipse of 1868, the French astronomer Pierre Janssen (Figure 34.2) noticed some unusual features in the spectrum of the sun. In England, the team of Norman Lockyer, an astronomer, and Edward Frankland, a chemist,* realized that Janssen had observed the spectral "fingerprint" of a chemical element unknown (at the time) on Earth. Frankland proposed the name helium, from the Greek word for sun, *helios*, for this new element.

FIGURE 34.2 Pierre Janssen (1824–1907), the French astronomer who was one of the discoverers of helium. (From http://en.wikipedia.org/wiki/File:Pierre_Janssen.jpg.)

---

* Pierre Janessen (1824–1907) is mainly remembered as a solar astronomer, measuring various phenomena associated with the behavior of the sun. He also studied the transit of Venus, an event that did not recur until 2012. Norman Lockyer (1836–1920) did much of his research at home, using a small 16 cm telescope. In addition to his personal scientific contributions, he founded the journal *Nature*, now one of the most prestigious scientific journals in the world. Edward Frankland (1825–1899) began his career as a druggist's apprentice. He contributed greatly to the chemical and bacteriological analysis of water, and for many years produced monthly reports on the quality of the water supply to London.

Spectroscopic chemical analysis of the sun revealed that the dominant component, by far, of the sun is hydrogen. Helium was second in abundance. The Earth's solid crust consists of compounds of silicon, aluminum, iron, and calcium in combination with oxygen. Hydrogen and helium are gases at all but extremely low temperatures. In the 1920s, the English astronomer Arthur Eddington* recognized that, if the sun were gaseous throughout, as we might infer from its composition, it could be stable only if its interior temperature was extremely high, some millions of kelvins. At the extreme temperatures and pressures in the interior of the sun, atomic nuclei could actually be forced together. At the temperatures and pressures we ordinarily encounter in everyday life, no way exists for atomic nuclei to be forced together, because of the natural electrical repulsion between two nuclei of the same charge. However, under the extreme conditions of temperature and pressure in the interior of the sun, nuclear reactions that could not take place at ordinary conditions would become spontaneous. Any nuclear process that is made to proceed at extreme temperatures is called a *thermonuclear reaction*.

In 1938, Hans Bethe,[†] a German-born physicist and one of the many brilliant scientists who emigrated to the United States in the 1930s to escape the Nazis, recognized that the process must be one of nuclear fusion—the joining together of small nuclei, such as hydrogen, to produce larger ones. Bethe worked out a sequence of processes, the net result of which would be the conversion of $_1H^1$ to $_2He^4$. Bethe proposed an entire series of reactions, in which, as examples, hydrogen reacted with carbon to build up nitrogen and then oxygen (the third-most abundant element in the sun). The oxygen nucleus broke apart to form helium and carbon, helium being a net product of the process and the carbon being available to begin a new cycle. Thus the net effect of the reactions converts hydrogen to helium. Subsequently other scientists have refined Bethe's model, and added other possible reactions, but it was Bethe's work, building on Eddington's insights, which provided the first real understanding of the source of the sun's energy.

We can consider a very simple hypothetical fusion reaction represented by

$$_1H^2 + _1H^2 \rightarrow _2He^4$$

This equation can be seen to conform to the rules for balancing nuclear equations (Chapter 23), but, in contrast to radioactive decay and fission processes, here the

---

* Arthur Eddington (1882–1944) was a pioneer in developing an understanding of the processes really at work in the interiors of stars. His observations of the 1919 solar eclipse helped to confirm Einstein's theory of relativity. Eddington also wrote numerous books, usually very popular, explaining astronomy and relativity for the general public.

† Hans Bethe (1906–2005, Nobel Prize, 1967) was a stellar theoretical physicist who made contributions to multiple areas of physics and continued to publish original scientific research in his 1990s. He served as the head of the theoretical division in the Manhattan Project, leading to the atomic bomb, and later helped with the development of the hydrogen bomb. However, Bethe eventually campaigned vigorously against the nuclear weapons race between the United States and the Soviet Union, and against testing nuclear weapons in the atmosphere. Bethe is also remembered for participating in the authorship of a paper on the synthesis of elements during the Big Bang creation of the universe. The other authors were a graduate student, Ralph Alpher, and Alpher's mentor, the Russian-American physicist George Gamow. The paper has become immortal as the "Alpha-Beta-Gamma" paper.

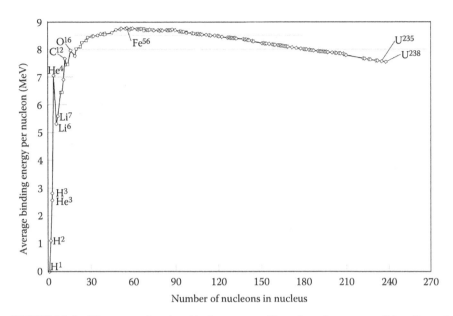

FIGURE 34.3 The curve of nuclear binding energy. (From http://commons.wikimedia.org/wiki/File:Binding_energy_curve_-_common_isotopes.jpg.)

product nucleus is *larger than* the reacting nuclei. As with radioactivity or fission processes, two ways can be used to consider the energy release from this fusion reaction.

First, the curve of binding energy, Figure 34.3, is so much steeper on the left-hand side, that there will be a substantial release of energy in a fusion process, much more than in fission. Although the change in the number of nucleons—the horizontal step across the curve—is smaller than most radioactive processes and very much smaller than in fission, the very steep slope of the curve means that the energy release—the vertical leap upward—is extremely large compared to radioactivity or fission.

Alternatively, Einstein's equation for the equivalence of mass and energy applies. To make a very scrupulous accounting of the equation shown above, the mass of the two $_1H^2$ nuclei is 4.02820. The mass number of the $_2He^4$ produced in the reaction is 4.00280. The loss of mass is 0.0254, which amounts to 0.63%. In contrast, in the fission of $_{92}U^{235}$ the mass loss is about 0.056%. The mass loss in fusion is proportionately much greater than in fission, by about a factor of ten. In other words, on a weight-for-weight basis, over ten times as much ENERGY is available in nuclear fusion as in nuclear fission via Einstein's equation, $E = mc^2$.

The energy released by fusion of hydrogen to helium in the sun is so enormously large that, while scientists can calculate the quantities of energy produced and hydrogen consumed, they are almost impossible to contemplate. We know from Einstein's equation that the energy of the sun must be obtained at the expense of its mass. The sun converts about 590 million tonnes of $_1H^1$ to $_2He^4$ every *second*. The corresponding mass loss, m, is 4,200,000 tonnes per *second*! Yet despite this seemingly prodigious conversion of mass to energy, the sun's mass is so large that it has already been radiating energy for some five billion years, and astronomers believe that it will continue to do so for about five billion years more.

We have seen (Chapter 33) that only two primary sources of ENERGY are available to us: solar and nuclear. Although our direct use of the sun's ENERGY is a quite small proportion of total energy consumption, water, wind, food, firewood, biomass, and the fossil fuels can all be traced ultimately to solar ENERGY. Now we see that solar ENERGY derives from thermonuclear processes—nuclear fusion—in the sun. So, in the last analysis, all of the ENERGY that we use in any form derives from processes involving atomic nuclei.

HARNESSING THE ENERGY OF NUCLEAR FUSION

Although nuclear fusion reactions require enormous temperatures, on the order of tens of millions of kelvins, to be initiated, they are spontaneous under such conditions. Exactly like every other spontaneous process, we can, in principle, insert some device into this SYSTEM that will allow us to capture some of the ENERGY released from fusion as useful WORK. As we will see, the main problem lies in creating the temperatures required to ignite the spontaneous thermonuclear reactions.

A half-century ago, only one device could generate, on Earth, the necessary temperature to start a thermonuclear reaction—an atomic bomb. Recognizing the potential for ten times as much energy to be released from fusion as from fission, and having available a "match" to light the nuclear fire (the atomic bomb), scientists of the late 1940s and early 1950s envisioned a bomb that would make use of the energy released from fusion. At first it was called the "super" (as in super-bomb), but has become known as a thermonuclear bomb, or, more familiarly, the hydrogen bomb or H-bomb. Even before the first demonstrated explosion of the hydrogen bomb by the United States in 1952, nuclear scientists recognized that such a weapon would be of unprecedented destructive power. Many hesitated to work on the "super" or even actively protested its development.

When we turn from methods of blowing each other up to finding ways of harnessing the ENERGY release from fusion, several issues arise: What will we use as the fuel? How will we achieve the temperatures needed to ignite it? How can we find a container to hold the equivalent of a miniature sun, or a controlled hydrogen bomb? How can we capture useful energy from the process?

FUEL

To achieve nuclear fusion, two nuclei must be smashed together. To bring this about requires significant energy, because normally two objects with similar electric charges (in this case, first the negative charges due to the electrons and then the positive charges due to the protons in the nucleus) will normally repel each other. The nuclei must have sufficient kinetic energy to overcome the electrical repulsion. The greater the positive charge on the nucleus, that is, the more protons in the nucleus, the greater will be the electrical repulsion to be overcome. For this reason, it would be best for the fuel to consist of nuclei having the smallest possible number of protons—namely, one. The element of atomic number 1 is hydrogen, so hydrogen is the fuel of choice for nuclear fusion.

The simplest approach would be to use $_1H^1$ as the hydrogen fusion, since that is the most common isotope of hydrogen. $_1H^1$ is also the principal fuel for generating the sun's energy. Unfortunately, though, the temperatures required to ignite a $_1H^1$ fusion reaction are much too high to be achieved in any sort of fusion reactor that is likely to operate on Earth.

Deuterium, $_1H^2$, will undergo fusion at a temperature lower than that needed for $_1H^1$ fusion. The even heavier isotope, tritium ($_1H^3$), will undergo fusion at still lower temperature. However, $_1H^3$ is unstable and would be difficult to collect in reasonable quantities. That leaves $_1H^2$ as the best possible fuel.

The overall fusion process can be represented as

$$5_1H^2 \rightarrow {}_2He^3 + {}_2He^4 + {}_1H^1 + 2\,_0n^1$$

Much of the energy liberated in the fusion is actually carried off by the neutrons.

The total energy that would be produced by the complete fusion of one gram of $_1H^2$ is 238 thousand gigajoules (238 terajoules). However, deuterium is not the common isotope of hydrogen. Only one of every 6500 hydrogen atoms occurring in nature is the deuterium isotope. The deuterium must be extracted from the water and separated from the ordinary isotope ($_1H^1$). The technology for doing this is now well known. While it may seem at first sight that only one atom out of 6500 is not very much of the desired isotope, there is still enough deuterium available so that the complete fusion of the deuterium that could be obtained by processing one liter of water would still yield about 10 GJ of energy.

Any tritium formed in the process will also undergo fusion by reacting with another deuterium nucleus. It is also possible to begin the fusion reaction using a mixture of deuterium and tritium as the fuel, rather than only deuterium. This mixture will ignite and begin to fuse at a lower temperature than if pure deuterium were used. A disadvantage of tritium is that it is radioactive and does not occur in nature. Fusion reactors that rely on tritium as part of the fuel must use a radioactive material. Tritium is a β-emitter with a half-life of twelve years. It can be made in various nuclear processes.

IGNITION

Most experimental fusion reactors are based on the fusion of deuterium with tritium nuclei. This requires very severe conditions. At the normal temperatures of chemical reactions or in chemical engineering practice, which are seldom above 1000°C, deuterium and tritium atoms exist with their nuclei and electrons just as any other atom. When these atoms approach each other, they reach a point at which their approach is so close that the electron clouds around each atom repel each other. (Like charges repel!)

Fusion is much harder to initiate than fission, because it requires that two nuclei be brought extremely close together. To overcome the electron repulsion between atoms, their electrons must be stripped away. This can be done by heating the deuterium–tritium mixture to very high temperatures, generating a *plasma*. Temperatures of about forty million kelvins are required to effect this. A plasma is a cloud of ionized gas. In the plasma, the nuclei can approach each other without the interference of the electron clouds. But again, because each nucleus has a positive charge, the nuclei will repel each other as they approach. The challenge is to give each nucleus enough kinetic energy to overcome the repulsion energy and let the nuclei slam into each other. The speeds needed to achieve sufficient kinetic energies require in turn extremely high temperatures, which heretofore have been found only in the center of the sun or other stars. For deuterium fusion, generating the kinetic energy required to overcome the repulsive energy of the nuclei is done by heating the plasma to temperatures on the order of 100 million kelvins. In the case of a deuterium–tritium mixture, the necessary temperatures are still above 50 million kelvins.

This leads straightaway to the first engineering problem: How do we heat the plasma to 100 million kelvins? There are at least three approaches. One uses very high electric currents, analogous to the resistance heating of the filaments in light bulbs or in the heating elements on electric stoves. High-temperature plasmas conduct electricity; currents can be millions of amperes. To achieve even more intense heating, radiofrequency systems similar to those in microwave ovens can be used. A second approach uses a battery of extremely powerful lasers (Figure 34.4). This so-called inertial system uses a mixture of deuterium and tritium frozen into small pellets less than one millimeter in diameter. One of these is placed in the middle of a vacuum chamber and hit from all sides by intense laser beams. The laser bombardment takes place in a small fraction of a second, heating the fuel rapidly without expanding it. The outer layer of the pellet vaporizes almost instantly and expands outward with great force. This very rapid vaporization with its outward force also creates a force pushing against the remaining interior of the pellet, causing it to implode. The great density and heat of the imploding pellet ignites the fusion reaction. A third method is to inject a beam of $_1H^2$ or $_1H^3$ nuclei into the plasma. (In a hydrogen bomb, the high temperatures necessary for fusion are achieved by first detonating an atomic bomb, which is an integral part of the hydrogen bomb.)

It may not seem so at first sight, but thermonuclear fusion has much in common with the more familiar processes of chemical combustion. In both cases, a high

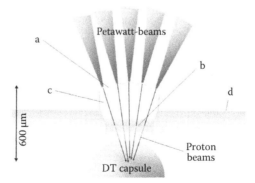

FIGURE 34.4 An assemblage of lasers used to initiate a fusion reaction. Schematic for a proton fast ignition scenario with a peripheral cavity consisting of production target (a), main heat shield, 50 mm (b), and secondary heat shield, 10 mm (c) next to the main fusion cavity (d). (From Geissel, M. et al., *Nucl. Instrum. Method Phys. Res. A*, 544, 55, 2005.)

temperature is necessary to start the reaction, and the ENERGY released by the reaction maintains a high enough temperature to spread the fire. In an ordinary chemical reaction, some of the ENERGY stored in chemical bonds is released when the atoms of the reacting molecules recombine into new, more tightly bound product molecules. In thermonuclear fusion reactions, some of the ENERGY stored as nuclear binding energy is released to us as the nuclei recombine into new, more tightly bound nuclei. The significant differences between these chemical and nuclear processes are the temperatures at which they occur and the amounts of ENERGY released for a given mass of reactants.

REACTORS

We might say that operating a fusion reactor is an approach to controlling and exploiting the energy of a hydrogen bomb. This analogy isn't exact, because there's no way of having a fusion explosion in a reactor. Certainly the engineering problems to achieve controlled nuclear fusion are much more formidable than to build and operate a fission power reactor.

The second major problem of achieving controlled nuclear fusion, once ignition has been achieved, is how to contain the tremendous temperatures of the plasma. (This is analogous to an ancient joke about two chemists. The first claims to be trying to invent an acid so powerful it will dissolve any known substance. The second immediately asks, "Oh yeah? So what are you going to keep it in?") A hundred million kelvins is not only way above the melting point of any known material, it's also way above the boiling point of any material. In fact, 5000 K is about enough to melt or vaporize any known substance. We might expect that a plasma reaction, once started, could burn a hole through anything.

Since plasmas conduct electricity, they can be confined and controlled by electric and magnetic forces. The plasma in a fusion reactor can be confined by a properly shaped magnetic field. This approach to containing a plasma undergoing fusion relies on the observation first made by Ørsted: A moving electric current generates a magnetic field. Since the plasma has an electric charge, when it is set in motion it will generate a magnetic field (Figure 34.5). If we enclose the magnetic field in an even stronger magnetic field of the same polarity (like poles repel!) then this strong, external magnetic field will tend to compress the plasma. The compression serves two purposes: First, it pushes the plasma away from the walls of the reactor vessel. Thus the materials the reactor is made of do not have to do the impossible job of sustaining a temperature of millions of degrees. This process is called *magnetic confinement*. Second, by forcing the plasma into a smaller volume, the density of nuclei (i.e., the number of nuclei per unit volume of the plasma) is substantially increased. This in turn increases the likelihood that two nuclei will collide and react. Magnetic confinement is one of three major strategies being pursued to achieve practical, controlled fusion reactions.

In practice, most experimental fusion reactors are, in effect, a hollow doughnut made of copper (Figure 34.6). Very strong electromagnets are mounted on the outside for magnetic containment. This type of reactor is called a *tokamak* (from the Russian: *tok* means current, and *mak* or *mag* is short for magnetic). The fundamental

Magnetic field generated by moving plasma

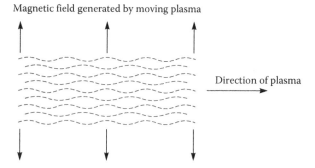

Direction of plasma

FIGURE 34.5 A plasma is a collection of electrically charged particles in motion and generates its own magnetic field just as Ørsted once demonstrated with a compass and Voltaic pile.

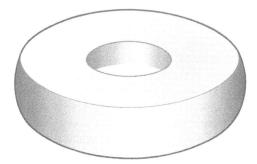

FIGURE 34.6 The tokamak reactor is a doughnut-shaped device in which the plasma is confined and fusion reactions can occur.

concepts of the tokamak reactor were developed in the early 1950s by Igor Tamm and Andrei Sakharov (Figure 34.7),* then at Moscow University.

The main magnetic field of the tokamak forms a "magnetic doughnut." The magnetic field is formed inside a doughnut-shaped chamber by electromagnets around the outside of the chamber. Plasma nuclei inside the tokamak spiral around the magnetic field lines and don't touch the chamber walls.

Using a magnetic field gets away from the possible problem of finding a material to confine the extreme temperatures of the plasma. A magnetic field is not made of

---

* Igor Tamm (1895–1971, Nobel Prize 1958) made numerous contributions to twentieth-century physics. With Sakharov, he developed the principles of the tokamak in 1951. The first such reactor in the Soviet Union was built in the late 1960s. Andrei Sakharov (1921–1989, Nobel Prize 1975) was not only a brilliant physicist but today is also remembered for his resistance to the Soviet state. Sakharov is regarded as the "father" of the Soviet H-bomb. Like Hans Bethe, he eventually advocated an end to the atmospheric testing of nuclear weapons and the nuclear arms race. He was one of the founders of the Committee on Human Rights in the USSR. In the time-honored tradition of nitwitted totalitarian thugs everywhere, the Soviet government first denied him permission to travel to receive his Nobel Prize. When Sakharov protested the Soviet invasion of Afghanistan, he was arrested, shipped off to internal exile, imprisoned in a hospital, and, for a time, force-fed. Sahkarov's *Memoirs* (Random House: New York, 1995) is well worth reading.

FIGURE 34.7 Andrei Sakharov (1921–1989), the Soviet physicist and crusader for human rights, who helped develop the concept of the tokamak reactor. (From http://www.wikiberal. org/wiki/Fichier:120216a.jpg.)

any material substance, can exist at any temperature, and can exert forces on moving electrical charges. If so-called magnetic walls could be designed with appropriate strength, the magnetic field alone could contain the plasma and produce the necessary fusion temperatures. The plasma will not melt or burn a hole through the walls of the reactor in part because it is at an extremely low density. The density of nuclei in the plasma is only about 1/10,000th of the normal density of the atmosphere. The total amount of heat in such an exceptionally low-density material is very small. Furthermore, even if the plasma did contact the walls of the reactor, such contact would only slow the ions and, consequently, cool the plasma. Cooling due to contact with the container wall would stop the reactions completely. This also adds an inherent level of safety to the fusion reactor, because if the slightest upset goes wrong with magnetic confinement, the plasma will contact the walls, cool, and shut down the fusion reactions.

A system somewhat related to the tokamak is the magnetic mirror system, which uses open-ended reactors rather than the "magnetic doughnut." This approach is based on the fact that a plasma cannot pass through a magnetic field. The reactor is set up such that the external magnetic field is strongest at the open ends. This strong magnetic field at the ends of the reactor is called the "magnetic mirror" (Figure 34.8). It turns, or reflects, the plasma back toward the center of the reactor where the magnetic field is weaker. The plasma will reflect back and forth between the magnetic mirrors at the reactor ends.

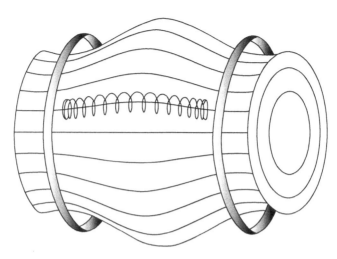

FIGURE 34.8 A schematic diagram of the magnetic mirror concept for confining a plasma in a fusion reactor. (From http://farside.ph.utexas.edu/teaching/plasma/lectures/node21.html.)

A second major approach to fusion reactor design relies on inertial confinement. As discussed above, intense pulses of laser light heat and compress a tiny sphere of deuterium and tritium. Though the laser pulses last only a few billionths of a second, they are nevertheless sufficient to vaporize the surface of the fuel particle. As the surface expands outward with explosive force, the core simultaneously experiences enormous inward forces that cause it to implode. The implosion compresses the fuel to a density more than twenty times that of lead (the fuel density would be about 220 g/cm^3). The intense compression caused by this implosion causes the temperature to rise to that needed to ignite fusion. Successive ignition of a series of such fuel pellets dropped steadily into the laser beams in a fusion reactor might produce a steady stream of electrical power. The success of this technique clearly requires precise timing of the dropping of the pellets, along with the development of more efficient lasers.

Energy Capture

Finally, if the fusion reaction can be ignited, and then confined, successfully, a way still must be found for withdrawing or capturing a portion of the energy release in the form of useful ENERGY. At least two methods of capturing some of the ENERGY in a useful form are under consideration. One is the direct conversion of the current of charged particles from the reaction into electricity. A second is capturing the neutrons liberated from the fusion reaction as heat in a liquid lithium blanket surrounding the reactor. Heat is then converted to electricity by raising steam to operate a conventional steam turbine and generator.

Withdrawing ENERGY from a fusion reactor relies on Faraday's principle that a conductor flowing or moving in a magnetic field generates an electric current. In this instance the plasma is the conductor; allowing it to flow through a

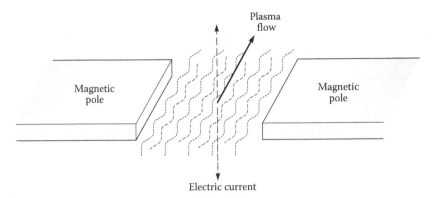

FIGURE 34.9 The flowing, reacting plasma in a fusion reactor can be used to generate an electric current.

magnetic field generates an electric current (Figure 34.9). This approach is called *magnetohydrodynamics.*

Most of the energy of nuclear fusion is in the kinetic energy of the product neutrons. If these neutrons could be stopped and captured, their kinetic energy would be converted into heat. One scheme for doing this would surround the reactor with a blanket of lithium metal to capture the neutrons. Doing so serves a double purpose. First, neutrons release their kinetic energy as heat, which can be used to raise steam. Second, as the lithium captures neutrons, it "breeds" tritium via the reaction

$$_0n^1 + {}_3Li^6 \rightarrow {}_2He^4 + {}_1H^3$$

Unfortunately, the tritium and the high-energy neutrons from the fusion reactions make the reactor structure radioactive. However, the quantity of radioactive material in a fusion reactor would be about a hundred times less than in a fission reactor. The tritium would be recycled into the reactor as fuel. Tritium has a half-life of 12.6 years, compared with the very long half-lives of some highly radioactive fission products (e.g., 24,400 years for the very dangerous plutonium).

So far, the energy output of most experimental fusion reactions is much smaller than the energy needed to heat and confine the plasma. Even at 100 million kelvins, more energy must be put into the plasma than is given off by fusion. At about 350 million kelvins, the fusion reactions finally reach a point at which they can produce as much energy as is put into generating the plasma. This point is called *break-even*. So far, break-even in experimental reactors has been achieved for no more than a fraction of a second. In 2011, an experimental device using 192 lasers fired, for an "instant," 500 TW of power (which is more than the total electricity consumption of the United States) into a 1–mm deuterium–tritium pellet without achieving break-even. No one has ever built, let alone operated, a practical fusion power reactor that has achieved break-even for sustained periods of time. But, there remain profound reasons to hope that a breakthrough will be achieved and the large-scale development of fusion energy will become practical.

COMPARING FUSION AND FISSION AS
SOURCES OF NUCLEAR ENERGY

Generating electricity from nuclear fission processes brings with it numerous concerns, as discussed in Chapter 25. The prospect exists to have a serious accident, releasing radioactive material to the environment, as at Fukushima, Three Mile Island, Windscale, or Chernobyl. Even with fuel reprocessing, fission reactors still produce waste material that has to be stored safely, someplace. The total supply of uranium is limited. The possible benefit of nuclear fission energy is that the amount of ENERGY per mass of uranium undergoing fission is enormous, compared to fossil fuels—larger by at least a factor of a million. It could be of great interest and advantage to find an alternative way to release ENERGY from some sort of nuclear process, while at the same time overcoming the concerns we now have with fission processes.

One advantage of fusion is that deuterium is extremely abundant (in water) but uranium or thorium used in fission reactors are relatively rare metals. Together, uranium and thorium account for about 10–12 parts per million of the Earth's crust. To be sure, the energy that could be liberated from the fission of these elements represents about ten times as much energy as could be obtained from the Earth's total supply of coal, oil, and gas. However, only a small part of Earth's supply of these fissionable fuels can be extracted from the crust with reasonable ease. Furthermore, just like the fossil fuels, the reserves of uranium and thorium are not evenly distributed around the world. Fusion, on the other hand, relies primarily on deuterium, which is available in water anywhere.

In contrast to fission processes, there is no net production of radioactive material. Tritium is radioactive, but would be captured and recycled into the reactor, so it would be retained in the system. Indeed, it's an advantage that tritium is produced in the reaction, because its half-life is so short that it does not occur in nature. If it were to be used as a fusion fuel, and were not actually made during fusion, it would have to be produced anyway, via some other nuclear process. High-energy neutrons produced in fusion would be dangerous if emitted to the environment, but in an operating fusion reactor they would likely be captured in the blanket of lithium. The neutrons may make some of the internal parts of the reactor radioactive. Nevertheless, a fusion reactor likely will produce relatively little radioactive waste, substantially less than a fission reactor. The only fusion products are helium and ordinary hydrogen (i.e., $_1H^1$), which are stable and safe. There is no problem with production of radioactive waste, or of fissionable materials that terrorists or rogue nations could use to produce atomic weapons.

Overall, then, fusion offers numerous advantages relative to fission as a source of nuclear energy. Extracting deuterium from water is not as disruptive to the environment, nor nearly so dangerous, as mining uranium, enriching it, and processing it into fuel for fission reactors. Fusion reactors do not contain a large inventory of radioactive material, in contrast to the numerous fuel assemblies in a fission reactor. The danger of a reactor accident is virtually eliminated. If any part of the system fails, the plasma will cool very rapidly and the fusion reaction will stop almost instantly. Fusion reactors cannot become supercritical and get out of control because, unlike fission, there is no critical mass for fusion. There is no need to rely on human intervention (such as scramming a fission reactor) or computer-based control systems taking over. Any component failure means the reaction stops. There is no air pollution, nor any CO_2 emissions, because there is no combustion. The only product is chemically inert helium. Storage of radioactive waste is much less of a problem with a fusion reactor. Much or all of the tritium will be recycled, and the only other likely radioactive material might be some internal parts of the reactor. The half-life of any tritium that might be regarded as waste is a good bit shorter than many fission by-products. There are no fissionable by-products (such as plutonium) that could be used to manufacture weapons, so nuclear weapons proliferation is not an issue with fusion. The problem of thermal pollution, characteristic of conventional steam turbine plants, could be avoided if the option of direct generation of electricity by magnetohydrodynamics is chosen.

PROMISE AND THE FRUSTRATION OF FUSION ENERGY

If fusion reactors could ever achieve break-even and operate in that way for some time, so that they would be reliable, long-term suppliers of electricity, the promise is truly fabulous. Development of safe, reliable, and cost-effective nuclear fusion energy would be a technological breakthrough for humankind on the same level as the invention of printing and development of the worldwide web.

Estimates of the amount of energy that could be obtained from fusion, and comparisons with other energy sources, necessarily depend on the assumptions made in doing the calculations, such as the reserves of various fuels and possible future rates of energy consumption. Because of that, projections of the benefits that could be obtained from fusion energy vary somewhat. Regardless of whether estimates agree in exact detail, if nuclear fusion could be made to succeed as a commercially viable energy source, the energy requirements of humankind would be solved for the foreseeable future.

Let's start with a liter of water. By fusion of the $_1H^2$ contained in that water, we would obtain as much energy as we would get by burning 285 liters of gasoline. Twenty liters of water contain one gram of deuterium, which when fused releases as much energy as 9500 liters of gasoline. Complete fusion of the deuterium in a cubic meter of water would yield energy equivalent to 2000 barrels of oil. Each cubic kilometer of ocean contains enough deuterium to be roughly equivalent to all of Earth's oil reserves. Each tonne of deuterium produces sixty million times more energy than a tonne of coal.

If we took the very parochial view that we would use deuterium fusion to supply humanity's energy needs, it would take about 150 billion years, at present rates of energy consumption, to use up the deuterium in the oceans. Since the universe is only some 15 billion years old, there's nothing to worry about in terms of there being enough deuterium to meet our energy needs. Obviously, this notion is pretty fanciful. Suppose we add two constraints: first, that we would be able to use only 1% of the deuterium in the oceans and, second, that everyone in the whole world would be able to use energy at the present rate at which it's consumed in the United

States. Then the best we can hope for is ENERGY to last the planet for the next three hundred million years. Depending upon the values used for total reserve or resource and annual consumption rates, fossil fuels have apparent lifetimes of, at best, some few centuries. Long before we come close to running out of deuterium, the sun will have lived through its natural life cycle and have extinguished all life on Earth—perhaps even the Earth itself.

But when considering this fantastic promise of fusion energy, we must also keep in mind the challenges that have to be met. Basically, there are three. The first is to create and heat the plasma at temperatures over a hundred million kelvins. The second is to keep enough plasma away from the container walls for long enough to permit abundant reactions to occur. The most serious problem facing fusion is that of achieving a net production of energy. A fusion reactor needs energy for generating the plasma and for operating the electromagnets used for magnetic confinement. A commercially viable fusion reactor should produce enough electricity to satisfy these needs and have some left over to sell to the public. So far, experimental fusion reactors are only able to satisfy their own energy needs for about a second at best. The third challenge is to design a practical, safe, and economic fusion reactor. Magnetic confinement fusion reactors are now close to break-even. Capital costs for commercial fusion reactors are difficult to estimate, but may be within a factor of two of the costs of other power plants. The cost of fuel would be cheap. Lasers and other technologies needed for inertial confinement fusion are very complex and troublesome, so it remains very questionable whether inertial confinement can compete with magnetic confinement, and in fact whether inertial confinement would ever be commercially viable as a source of energy. Optimists and proponents of fusion believe that the technology may be commercially available within the next 25–50 years. Others are not so sure, and sometimes paraphrase an old joke to argue that "fusion is the energy source of the future, and it always will be."

All over the world, scientists are making big strides in their efforts to harness the H-bomb for peaceful ends.

Scientists estimate that sometime within the next 10 to 20 years [this was written in 1959!] *the first full-scale, power-producing fusion reactor will go into operation. Even this first crude reactor will have a power output comparable to the huge hydroelectric plant at Hoover Dam.*

That moment, if it comes, will be a pivot point of history. Nations need never fear that their power sources will run dry. For a time, obviously petroleum will remain as a source of mobile power, and coal will continue to provide industry with heat. You can't change a way of life overnight. But eventually, according to the experts, these will give way to stored electrical power derived from the fusion reaction.

At that supreme moment, man can look ahead through the halls of time for literally billions of years and still see a plentiful supply of fuel. Billions of years! When new kilowatts are needed, they simply will be plucked from sea water. Inexhaustible and inexpensive power on earth will give man the means of navigating the stars.

—Benford[1]

REFERENCE

1. Benford, G. *The Wonderful Future That Never Was*. Hearst Books: New York, pp. 186–188.

FURTHER READINGS

Armstrong, F.; Blundell, K. *Energy Beyond Oil*. Oxford University Press: New York, 2007. A collection of chapters by various experts in the respective fields. Chapter 7 provides a discussion of nuclear fusion.

Bloomfield, L.A. *How Things Work*. Wiley: New York, 1997. This is a fine introductory textbook on physics, centered around examples of how real-world devices function. Chapter 18 provides information relevant to nuclear fusion.

Coley, D. *Energy and Climate Change*. Wiley: Chichester, U.K., 2008. Chapter 27 of this textbook discusses fusion.

Hewitt, P.G. *Conceptual Physics*. Addison-Wesley: Reading, MA, 1998. This is an introductory text on physics, well illustrated with a remarkable number of conceptual diagrams and cartoons. Chapter 33 provides information on fusion.

Ngô, C.; Natowitz, J. *Our Energy Future*. Wiley: Hoboken, NJ, 2009. Chapter 10 includes a useful discussion of fusion, including its difficulties, its long-term possibilities, and safety issues.

Seife, C. *Sun in a Bottle*. Viking Penguin: New York, 2008. This is an interesting history of the development of nuclear fusion, from the days of the hydrogen bomb to early in this century. Its subtitle is "The strange history of fusion and the science of wishful thinking."

Glossary of Symbols, Abbreviations, and Acronyms

AC   alternating current, a pattern of flow of electricity, in which the direction of flow changes periodically, usually fifty or sixty times per second.

AD   Anno Domini, used in reference to dates since the birth of Christ.

amu   atomic mass units, a scale for ranking and comparing the relative masses of atoms and atomic particles.

BCE   before the common era, used in reference to dates prior to the year AD 1 (also see AD). Numerically equivalent to BC.

Btu   British thermal unit, the amount of ENERGY required to raise the temperature of one pound of water by one degree Fahrenheit.

Btu/lb British thermal units per pound, the amount of ENERGY liberated per unit weight of a material, usually in reference to heat produced by a burning a fuel.

BWR   boiling water reactor, a type of nuclear fission reactor.

C   the nutritional calorie, or "big calorie," equal to one thousand calories (which see).

°C   degrees Celsius, a measure of temperature.

c   the speed of light.

CAFE   the corporate average fuel economy, a U.S standard that measures the average efficiency of a manufacturer's vehicles for a given model year, weighted by the amount of sales of each model.

cal   calorie, the amount of ENERGY required to raise the temperature of one gram of water one degree Celsius.

cc   cubic centimeters (also written as cm^3), commonly used as a measure of volume of liquids, or the cylinder volume in small engines.

CCS   carbon capture and storage, a strategy for dealing with potential emissions of carbon dioxide.

cm   centimeters, a measure of length or distance.

DC   direct current, a flow of electricity in which the direction of flow does not change; compare AC (above).

DIY   do it yourself. Contrast HAP (below).

E   ENERGY

E10   a numerical designation for gasohol, or any other mixture of 10% ethanol with 90% gasoline, intended for use as a liquid fuel.

E22   similarly to E10, a liquid vehicle fuel consisting of 22% ethanol and 78% gasoline.

E100   pure ethanol, i.e., 100% ethanol.
ECCS   emergency core cooling system, a safety mechanism in nuclear fission reactors.
ESP   electrostatic precipitator, a device used in electricity-generating plants to control the emissions of fine particles of ash.
esu   electrostatic units, a system of units for measuring electrical charge, current, and potential.

°F   degrees Fahrenheit, a measure of temperature.
FGD   flue gas desulfurization, a process for capturing sulfur oxides produced in combustion processes before they are emitted to the environment.
ft   foot.

g   gram, a measure of mass.
g/cm^3 grams per cubic centimeter, a measure of mass per unit volume, i.e., density.
$g \cdot cm/s^2$ gram-centimeter per second squared, a unit of force also called the dyne.
GDP   gross domestic product, the measure of the value of all the goods and services produced within a country over some period of time, usually a year.
GJ   gigajoules, one billion joules, a measure of ENERGY.
GTL   gas-to-liquids, a fuel processing strategy that converts gaseous fuels, such as natural gas, into synthetic liquid fuels.
GW   gigawatts (a billion watts, or a thousand megawatts), a measure of POWER, usually in connection with electricity.
GWP   global warming potential, a measure of the effect that a particular gas in the atmosphere has in causing global climate change.

h   hours
ha   hectare, a measure of land area.
HAP   hire a pro, the alternative to DIY (above).
HDI   human development indicator, a measure of the quality of life within a country that includes economic factors such as the GDP (which see) but also other factors such as access to health care and education.
hp   horsepower, a unit of POWER used most often in connection with engines.
Hz   hertz, a measure of frequency, such as electromagnetic waves (e.g., radio, microwaves) or of electricity in AC systems.

in   inch.

K   kelvins, the unit of temperature on the absolute scale.
kg   kilogram, a measure of mass.
kg/cm^2 kilograms per square centimeter, a measure of mass per unit area, or pressure, as in water pressure.
kg/kW   kilograms per kilowatt, a measure of how massive an engine is relative to its POWER output.

kg/kWh   kilograms per kilowatt-hour, a measure of how massive a device, such as an engine, is relative to its ENERGY output.

kJ   kilojoule, a measure of ENERGY.

km   kilometer, a measure of length or distance.

km^2 square kilometers, a measure of area.

km/°C   kilometers per degree Celsius, a measure of changing temperature over some distance, as an indication of global climate change.

km/h   kilometers per hour, a measure of speed.

kPa   kilopascals, a measure of pressure. Normal atmospheric pressure is about one hundred kilopascals.

kV   kilovolts, a measure of electrical potential energy.

kW   a measure of POWER, commonly with regard to electricity or internal combustion engines.

kWh   a measure of ENERGY, commonly of electrical energy.

kW/ha   kilowatts per hectare, or POWER per unit area, the so-called indicator value for assessing the environmental impact of hydroelectric projects.

kW/m^2 kilowatts per square meter, a measure of POWER produced per unit area.

L   liter, a measure of volume.

L/km   liters per kilometer (often expressed as L/100 km), a measure of vehicle fuel efficiency, the number of liters of fuel required to drive 100 km. The higher the number, the less efficient the vehicle.

L/min   liters per minute, a measure of the flow rate of fluids, usually liquids.

L/s   liters per second, a measure of the flow rate of liquids.

lb   pound, a unit of weight.

LEV   low-emission vehicle.

LFTR   liquid fluoride thorium reactor, an emerging design of nuclear fission reactor that relies on thorium rather than uranium.

LNG   liquefied natural gas, a material produced by cooling natural gas to about −170°C, usually done for long-distance shipping of natural gas.

LPG   liquefied petroleum gases, usually propane and/or butane, produced as a byproduct of petroleum refining and sold as a fuel in regions not connected to a natural gas supply.

m   meters, a measure of length or distance.

m   the mass loss in a nuclear process.

m^2 square meters, a measure of area.

m^3 cubic meters, a measure of volume.

m^3/ha   cubic meters per hectare, a measure of volume per unit land area, usually used in connection with the amount of wood that can be grown or harvested per hectare of land.

m/s   meters per second, a unit of speed.

M85   a mixture of 85% methanol with 15% gasoline, for use as a liquid fuel.

MJ   megajoules, a measure of ENERGY.

MJ/kg   megajoules per kilogram, a measure of the amount of ENERGY produced per unit mass of a fuel, usually in reference to heat liberated by burning a fuel.

MJ/L   megajoules per liter, a measure of the amount of ENERGY produced per unit volume of a liquid fuel, usually in reference to heat liberated by burning a it.

mm   millimeters, a measure of size or length.

MPa   megapascal, a unit of pressure. An approximate conversion is that one megapascal is ten times the value of atmospheric pressure.

mpg   miles per gallon, another indication of vehicle fuel efficiency (see L/km). The larger the mpg figure, the more efficient is the vehicle.

mrem   millirems, a unit of radiation dosage equal to one-thousandth of a rem (which see).

MTBE   methyl tertiary-butyl ether, a compound once thought to be a useful additive to gasoline to enhance octane number and reduce tailpipe emissions.

MTG   methanol-to-gasoline, a fuel conversion process that chemically converts methanol into high-octane gasoline.

MW   megawatts (one million watts), a measure of POWER, usually in connection with electricity or very large engines.

MWh   megawatt-hours, a measure of ENERGY, commonly in connection with electricy.

n a type of semiconductor in which conduction is due to electrons in the material.

NASA   National Aeronautics and Space Administration, a U.S. government agency whose mission includes not only space technology, but also studies of Earth, particularly the atmosphere.

NIMBY   not in my backyard! A slogan of opposition to energy projects or other large construction projects.

NMR   nuclear magnetic resonance, a technique of chemical analysis used, e.g., to help determine the structures of molecules or to identify substances based on their molecular structures.

OPEC   the Organization of Petroleum-Exporting Countries, a cartel intended to coordinate production and pricing of petroleum. Membership currently consists of about a dozen countries with collective oil production exceeding thirty million barrels per day.

p a type of semiconductor, in which conduction is due to "holes" in the electronic structure.

PBR   pebble-bed reactor, an emerging design of nuclear fission reactor.

pH   literally, the power of hydrogen, a measure of the acidity or basicity of substances, usually solutions in water.

ppb   parts per billion, a measure of the concentration of a substance, often in solution or in air.

ppm   parts per million, a measure of the concentration of a substance, often in solution or in air. One part per million is one thousand parts per billion (ppb, which see) or 0.0001%.

PVC   poly(vinyl chloride), a ubiquitous plastic used in an enormous variety of consumer products, including food wrap, raincoats, and pipes.

PWR   pressurized water reactor, a type of nuclear fission reactor.

R   röntgen, a measure of radiation dosage, especially of X-rays or γ-rays. A röntgen is almost equal to a rad, 1 R = 0.96 rad.

rad   a measure of the ENERGY absorbed in a body from radiation, equal to 0.01 joules per kilogram.

RBMK   literally, "reactor bolshoy moshchnosti kanalniy," a type of nuclear fission reactor used at Chernobyl.

rem   "röntgen equivalent man," a measure of radiation dosage in humans.

rpm   revolutions per minute, a measure of speed of rotation, usually of an engine or piece of machinery.

s   seconds

T temperature

TJ   terajoules, one trillion joules, a measure of ENERGY.

TMI   Three Mile Island, a site of nuclear reactor accident near Harrisburg, Pennsylvania.

TW   terawatts, one trillion watts, a measure of POWER.

ULEV   ultra-low emission vehicle.

V   volts, a unit of electrical potential energy.

W   watts, a measure of POWER.

W/m^2 watts per square meter, a measure of power production per unit area, as in photovoltaic cells; also a measure of albedo, the reflecting power of an area on Earth's surface, useful in calculating available solar energy.

ZEV   zero emission vehicle

GREEK SYMBOLS

α   a form of radiation consisting of the nuclei of helium atoms, i.e., two protons and two neutrons.

β   a form of radiation consisting of high-energy electrons.

γ   high-energy electromagnetic radiation such as X-rays.

η   efficiency.

μm   micron, or micrometer, a unit of size or length. One micron is one millionth of a meter.

Ω   ohms, a measure of electrical resistance.

$Ω/cm^2$ ohms per square centimeter, a measure of electrical resistivity (the distinction between resistance and resistivity is explained in Chapter 12).

Index